Elements of Distribution Theory

This detailed introduction to distribution theory uses no measure theory, making it suitable for students in statistics and econometrics as well as for researchers who use statistical methods. Good backgrounds in calculus and linear algebra are important and a course in elementary mathematical analysis is useful, but not required. An appendix gives a detailed summary of the mathematical definitions and results that are used in the book.

Topics covered range from the basic distribution and density functions, expectation, conditioning, characteristic functions, cumulants, convergence in distribution, and the central limit theorem to more advanced concepts such as exchangeability, models with a group structure, asymptotic approximations to integrals, orthogonal polynomials, and saddlepoint approximations. The emphasis is on topics useful in understanding statistical methodology; thus, parametric statistical models and the distribution theory associated with the normal distribution are covered comprehensively.

Thomas A. Severini received his Ph.D. in Statistics from the University of Chicago. He is now a Professor of Statistics at Northwestern University. He has also written *Likelihood Methods in Statistics*. He has published extensively in statistical journals such as *Biometrika, Journal of the American Statistical Association*, and *Journal of the Royal Statistical Society*. He is a member of the Institute of Mathematical Statistics and the American Statistical Association.

CAMBRIDGE SERIES IN STATISTICAL AND PROBABILISTIC MATHEMATICS

Editorial Board:

R. Gill, *Department of Mathematics, Utrecht University*
B.D. Ripley, *Department of Statistics, University of Oxford*
S. Ross, *Epstein Department of Industrial & Systems Engineering, University of Southern California*
B.W. Silverman, *St. Peter's College, Oxford*
M. Stein, *Department of Statistics, University of Chicago*

This series of high-quality upper-division textbooks and expository monographs covers all aspects of stochastic applicable mathematics. The topics range from pure and applied statistics to probability theory, operations research, optimization, and mathematical programming. The books contain clear presentations of new developments in the field and also of the state of the art in classical methods. While emphasizing rigorous treatment of theoretical methods, and books also contain applications and discussions of new techniques made possible by advances in computational practice.

Elements of Distribution Theory

THOMAS A. SEVERINI

CAMBRIDGE
UNIVERSITY PRESS

CAMBRIDGE UNIVERSITY PRESS
Cambridge, New York, Melbourne, Madrid, Cape Town, Singapore, São Paulo

Cambridge University Press
40 West 20th Street, New York, NY 10011-4211, USA

www.cambridge.org
Information on this title: www.cambridge.org/9780521844727

First published 2005

Printed in the United States of America

A catalog record for this publication is available from the British Library

Library of Congress Cataloging in Publication Data

Severini, Thomas A. (Thomas Alan), 1959–
Elements of distribution theory / Thomas A. Severini.
    p.   cm. – (Cambridge series in statistical and probabilistic mathematics ; 17)
Includes bibliographical references and index.
ISBN-13: 978-0-521-84472-7
ISBN-10: 0-521-84472-X
1. Distribution (Probability theory)   I. Title.   II. Series: Cambridge series on
statistical and probabilistic mathematics ; 17.
QA273.6.S48   2005
515′.782–dc22   2005016479

ISBN-13   978-0-521-84472-7 hardback
ISBN-10   0-521-84472-X hardback

To My Parents

Contents

Preface

Distribution theory lies at the interface of probability and statistics. It is closely related to probability theory; however, it differs in its focus on the calculation and approximation of probability distributions and associated quantities such as moments and cumulants. Although distribution theory plays a central role in the development of statistical methodology, distribution theory itself does not deal with issues of statistical inference.

Many standard texts on mathematical statistics and statistical inference contain either a few chapters or an appendix on basic distribution theory. I have found that such treatments are generally too brief, often ignoring such important concepts as characteristic functions or cumulants. On the other hand, the discussion in books on probability theory is often too abstract for readers whose primary interest is in statistical methodology.

The purpose of this book is to provide a detailed introduction to the central results of distribution theory, in particular, those results needed to understand statistical methodology, without requiring an extensive background in mathematics. Chapters 1 to 4 cover basic topics such as random variables, distribution and density functions, expectation, conditioning, characteristic functions, moments, and cumulants. Chapter 5 covers parametric families of distributions, including exponential families, hierarchical models, and models with a group structure. Chapter 6 contains an introduction to stochastic processes.

Chapter 7 covers distribution theory for functions of random variables and Chapter 8 covers distribution theory associated with the normal distribution. Chapters 9 and 10 are more specialized, covering asymptotic approximations to integrals and orthogonal polynomials, respectively. Although these are classical topics in mathematics, they are often overlooked in statistics texts, despite the fact that the results are often used in statistics. For instance, Watson's lemma and Laplace's method are general, useful tools for approximating the integrals that arise in statistics, and orthogonal polynomials are used in areas ranging from nonparametric function estimation to experimental design.

Chapters 11 to 14 cover large-sample approximations to probability distributions. Chapter 11 covers the basic ideas of convergence in distribution and Chapter 12 contains several versions of the central limit theorem. Chapter 13 considers the problem of approximating the distribution of statistics that are more general than sample means, such as nonlinear functions of sample means and U-statistics. Higher-order asymptotic approximations such as Edgeworth series approximations and saddlepoint approximations are presented in Chapter 14.

I have attempted to keep each chapter as self-contained as possible, but some dependencies are inevitable. Chapter 1 and Sections 2.1–2.4, 3.1–3.2, and 4.1-4.4 contain core topics that are used throughout the book; the material covered in these sections will most likely be familiar to readers who have taken a course in basic probability theory. Chapter 12 requires Chapter 11 and Chapters 13 and 14 require Chapter 12; in addition, Sections 13.3 and 13.5 use material from Sections 7.5 and 7.6.

The mathematical prerequisites for this book are modest. Good backgrounds in calculus and linear algebra are important and a course in elementary mathematical analysis at the level of Rudin (1976) is useful, but not required. Appendix 3 gives a detailed summary of the mathematical definitions and results that are used in the book.

Although many results from elementary probability theory are presented in Chapters 1 to 4, it is assumed that readers have had some previous exposure to basic probability theory. Measure theory, however, is not needed and is not used in the book. Thus, although measurability is briefly discussed in Chapter 1, throughout the book all subsets of a given sample space are implictly assumed to be measurable. The main drawback of this is that it is not possible to rigorously define an integral with respect to a distribution function and to establish commonly used properties of this integral. Although, ideally, readers will have had previous exposure to integration theory, it is possible to use these results without fully understanding their proofs; to help in this regard, Appendix 1 contains a brief summary of the integration theory needed, along with important properties of the integral.

Proofs are given for nearly every result stated. The main exceptions are results requiring measure theory, although there are surprisingly few results of this type. In these cases, I have tried to outline the basic ideas of the proof and to give an indication of why more sophisticated mathematical results are needed. The other exceptions are a few cases in which a proof is given for the case of real-valued random variables and the extension to random vectors is omitted and a number of cases in which the proof is left as an exercise. I have not attempted to state results under the weakest possible conditions; on the contrary, I have often imposed relatively strong conditions if that allows a simpler and more transparent proof.

Evanston, IL, January, 2005
Thomas A. Severini
severini@northwestern.edu

1

Properties of Probability Distributions

1.1 Introduction

Distribution theory is concerned with probability distributions of random variables, with the emphasis on the types of random variables frequently used in the theory and application of statistical methods. For instance, in a statistical estimation problem we may need to determine the probability distribution of a proposed estimator or to calculate probabilities in order to construct a confidence interval.

Clearly, there is a close relationship between distribution theory and probability theory; in some sense, distribution theory consists of those aspects of probability theory that are often used in the development of statistical theory and methodology. In particular, the problem of deriving properties of probability distributions of statistics, such as the sample mean or sample standard deviation, based on assumptions on the distributions of the underlying random variables, receives much emphasis in distribution theory.

In this chapter, we consider the basic properties of probability distributions. Although these concepts most likely are familiar to anyone who has studied elementary probability theory, they play such a central role in the subsequent chapters that they are presented here for completeness.

1.2 Basic Framework

The starting point for probability theory and, hence, distribution theory is the concept of an *experiment*. The term experiment may actually refer to a physical experiment in the usual sense, but more generally we will refer to something as an experiment when it has the following properties: there is a well-defined set of possible outcomes of the experiment, each time the experiment is performed exactly one of the possible outcomes occurs, and the outcome that occurs is governed by some chance mechanism.

Let Ω denote the *sample space* of the experiment, the set of possible outcomes of the experiment; a subset A of Ω is called an *event*. Associated with each event A is a probability $P(A)$. Hence, P is a function defined on subsets of Ω and taking values in the interval $[0, 1]$. The function P is required to have certain properties:

(P1) $P(\Omega) = 1$
(P2) If A and B are disjoint subsets of Ω, then $P(A \cup B) = P(A) + P(B)$.

(P3) If $A_1, A_2, \ldots,$ are disjoint subsets of Ω, then

$$P\left(\bigcup_{n=1}^{\infty} A_n\right) = \sum_{n=1}^{\infty} P(A_n).$$

Note that (P3) implies (P2); however, (P3), which is concerned with an infinite sequence of events, is of a different nature than (P2) and it is useful to consider them separately. There are a number of straightforward consequences of (P1)–(P3). For instance, $P(\emptyset) = 0$, if A^c denotes the complement of A, then $P(A^c) = 1 - P(A)$, and, for A_1, A_2 not necessarily disjoint,

$$P(A_1 \cup A_2) = P(A_1) + P(A_2) - P(A_1 \cap A_2).$$

Example* 1.1 *(Sampling from a finite population). Suppose that Ω is a finite set and that, for each $\omega \in \Omega$,

$$P(\{\omega\}) = c$$

for some constant c. Clearly, $c = 1/|\Omega|$ where $|\Omega|$ denotes the cardinality of Ω.

Let A denote a subset of Ω. Then

$$P(A) = \frac{|A|}{|\Omega|}.$$

Thus, the problem of determining $P(A)$ is essentially the problem of counting the number of elements in A and Ω. □

Example* 1.2 *(Bernoulli trials). Let

$$\Omega = \{x \in \mathbf{R}^n : x = (x_1, \ldots, x_n), x_j = 0 \text{ or } 1, \quad j = 1, \ldots, n\}$$

so that an element of Ω is a vector of ones and zeros. For $\omega = (x_1, \ldots, x_n) \in \Omega$, take

$$P(\omega) = \prod_{j=1}^{n} \theta^{x_j} (1 - \theta)^{1-x_j}$$

where $0 < \theta < 1$ is a given constant. □

Example* 1.3 *(Uniform distribution). Suppose that $\Omega = (0, 1)$ and suppose that the probability of any interval in Ω is the length of the interval. More generally, we may take the probability of a subset A of Ω to be

$$P(A) = \int_A dx. \qquad\qquad □$$

Ideally, P is defined on the set of all subsets of Ω. Unfortunately, it is not generally possible to do so and still have properties (P1)–(P3) be satisfied. Instead P is defined only on a set $\mathcal{F}$ of subsets of Ω; if $A \subset \Omega$ is not in $\mathcal{F}$, then $P(A)$ is not defined. The sets in $\mathcal{F}$ are said to be *measurable*. The triple $(\Omega, \mathcal{F}, P)$ is called a *probability space*; for example, we might refer to a random variable X defined on some probability space.

Clearly for such an approach to probability theory to be useful for applications, the set $\mathcal{F}$ must contain all subsets of Ω of practical interest. For instance, when Ω is a countable set, $\mathcal{F}$ may be taken to be the set of all subsets of Ω. When Ω may be taken to be a

Euclidean space $\mathbf{R}^d$, $\mathcal{F}$ may be taken to be the set of all subsets of $\mathbf{R}^d$ formed by starting with a countable set of rectangles in $\mathbf{R}^d$ and then performing a countable number of set operations such as intersections and unions. The same approach works when Ω is a subset of a Euclidean space.

The study of theses issues forms the branch of mathematics known as measure theory. In this book, we avoid such issues and implicitly assume that any event of interest is measurable.

Note that condition (P3), which deals with an infinite number of events, is of a different nature than conditions (P1) and (P2). This condition is often referred to as *countable additivity* of a probability function. However, it is best understood as a type of continuity condition on P. It is easier to see the connection between (P3) and continuity if it is expressed in terms of one of two equivalent conditions. Consider the following:

(P4) If $A_1, A_2, \ldots$, are subsets of Ω satisfying $A_1 \subset A_2 \subset \cdots$, then

$$P\left(\bigcup_{n=1}^{\infty} A_n\right) = \lim_{n \to \infty} P(A_n)$$

(P5) If $A_1, A_2, \ldots$, are subsets of Ω satisfying $A_1 \supset A_2 \supset \cdots$, then

$$P\left(\bigcap_{n=1}^{\infty} A_n\right) = \lim_{n \to \infty} P(A_n).$$

Suppose that, as in (P4), $A_1, A_2, \ldots$ is a sequence of increasing subsets of Ω. Then we may take the limit of this sequence to be the union of the A_n; that is,

$$\lim_{n \to \infty} A_n = \bigcup_{n=1}^{\infty} A_n.$$

Condition (P4) may then be written as

$$P\left(\lim_{n \to \infty} A_n\right) = \lim_{n \to \infty} P(A_n).$$

A similar interpretation applies to (P5). Thus, (P4) and (P5) may be viewed as continuity conditions on P.

The equivalence of (P3), (P4), and (P5) is established in the following theorem.

Theorem 1.1. *Consider an experiment with sample space Ω. Let P denote a function defined on subsets of Ω such that conditions (P1) and (P2) are satisfied. Then conditions (P3), (P4), and (P5) are equivalent in the sense that if any one of these conditions holds, the other two hold as well.*

Proof. First note that if $A_1, A_2, \ldots$ is an increasing sequence of subsets of Ω, then $A_1^c, A_2^c, \ldots$ is a decreasing sequence of subsets and, since, for each $k = 1, 2, \ldots$,

$$\left(\bigcup_{n=1}^{k} A_n\right)^c = \bigcap_{n=1}^{k} A_n^c,$$

$$\left(\lim_{n \to \infty} A_n\right)^c = \bigcap_{n=1}^{\infty} A_n^c = \lim_{n \to \infty} A_n^c.$$

Suppose (P5) holds. Then

$$P\left(\lim_{n\to\infty} A_n^c\right) = \lim_{n\to\infty} P(A_n^c)$$

so that

$$P\left(\lim_{n\to\infty} A_n\right) = 1 - P\left\{\left(\lim_{n\to\infty} A_n\right)^c\right\} = 1 - \lim_{n\to\infty} P(A_n^c) = \lim_{n\to\infty} P(A_n),$$

proving (P4). A similar argument may be used to show that (P4) implies (P5). Hence, it suffices to show that (P3) and (P4) are equivalent.

Suppose $A_1, A_2, \ldots$ is an increasing sequence of events. For $n = 2, 3, \ldots$, define

$$\bar{A}_n = A_n \cap A_{n-1}^c.$$

Then, for $1 < n < k$,

$$\bar{A}_n \cap \bar{A}_k = (A_n \cap A_k) \cap \left(A_{n-1}^c \cap A_{k-1}^c\right).$$

Note that, since the sequence $A_1, A_2, \ldots$ is increasing, and $n < k$,

$$A_n \cap A_k = A_n$$

and

$$A_{n-1}^c \cap A_{k-1}^c = A_{k-1}^c.$$

Hence, since $A_n \subset A_{k-1}$,

$$\bar{A}_n \cap \bar{A}_k = A_n \cap A_{k-1}^c = \emptyset.$$

Suppose $\omega \in A_k$. Then either $\omega \in A_{k-1}$ or $\omega \in A_{k-1}^c \cap A_k = \bar{A}_k$; similarly, if $\omega \in A_{k-1}$ then either $\omega \in A_{k-2}$ or $\omega \in A_1^c \cap A_{k-1} \cap A_{k-2}^c = \bar{A}_{k-1}$. Hence, ω must be an element of either one of $\bar{A}_k, \bar{A}_{k-1}, \ldots, \bar{A}_2$ or of A_1. That is,

$$A_k = A_1 \cup \bar{A}_2 \cup \bar{A}_3 \cup \cdots \cup \bar{A}_k;$$

hence, taking $\bar{A}_1 = A_1$,

$$A_k = \bigcup_{n=1}^{k} \bar{A}_n$$

and

$$\lim_{k\to\infty} A_k = \bigcup_{n=1}^{\infty} \bar{A}_n.$$

Now suppose that (P3) holds. Then

$$P(\lim_{k\to\infty} A_k) = P\left(\bigcup_{n=1}^{\infty} \bar{A}_n\right) = \sum_{n=1}^{\infty} P(\bar{A}_n) = \lim_{k\to\infty} \sum_{n=1}^{k} P(\bar{A}_n) = \lim_{k\to\infty} P(A_k),$$

proving (P4).

Now suppose that (P4) holds. Let $A_1, A_2, \ldots$ denote an arbitrary sequence of disjoint subsets of Ω and let

$$A_0 = \bigcup_{n=1}^{\infty} A_n.$$

Define

$$\tilde{A}_k = \bigcup_{n=1}^{k} A_j, \quad k = 1, 2, \ldots;$$

note that $\tilde{A}_1, \tilde{A}_2, \ldots$ is an increasing sequence and that

$$A_0 = \lim_{k \to \infty} \tilde{A}_k.$$

Hence, by (P4),

$$P(A_0) = \lim_{k \to \infty} P(\tilde{A}_k) = \lim_{k \to \infty} \sum_{n=1}^{k} P(A_n) = \sum_{n=1}^{\infty} P(A_n),$$

proving (P3). It follows that (P3) and (P4) are equivalent, proving the theorem. ■

1.3   Random Variables

Let ω denote the outcome of an experiment; that is, let ω denote an element of Ω. In many applications we are concerned primarily with certain numerical characteristics of ω, rather than with ω itself. Let $X : \Omega \to \mathcal{X}$, where $\mathcal{X}$ is a subset of $\mathbf{R}^d$ for some $d = 1, 2, \ldots$, denote a *random variable*; the set $\mathcal{X}$ is called the *range* of X or, sometimes, the *sample space of X*. For a given outcome $\omega \in \Omega$, the corresponding value of X is $x = X(\omega)$. Probabilities regarding X may be obtained from the probability function P for the original experiment. Let P_X denote a function such that for any set $A \subset \mathcal{X}$, $P_X(A)$ denotes the probability that $X \in A$. Then P_X is a probability function defined on subsets of $\mathcal{X}$ and

$$P_X(A) = P(\{\omega \in \Omega: X(\omega) \in A\}).$$

We will generally use a less formal notation in which $\text{Pr}(X \in A)$ denotes $P_X(A)$. For instance, the probability that $X \le 1$ may be written as either $\text{Pr}(X \le 1)$ or $P_X\{(-\infty, 1]\}$. In this book, we will generally focus on probabilities associated with random variables, without explicit reference to the underlying experiments and associated probability functions.

Note that since P_X defines a probability function on the subsets of $\mathcal{X}$, it must satisfy conditions (P1)–(P3). Also, the issues regarding measurability discussed in the previous section apply here as well.

When the range $\mathcal{X}$ of a random variable X is a subset of $\mathbf{R}^d$ for some $d = 1, 2, \ldots$, it is often convenient to proceed as if probability function P_X is defined on the entire space $\mathbf{R}^d$. Then the probability of any subset of $\mathcal{X}^c$ is 0 and, for any set $A \subset \mathbf{R}^d$,

$$P_X(A) \equiv \text{Pr}(X \in A) = \text{Pr}(X \in A \cap \mathcal{X}).$$

It is worth noting that some authors distinguish between random variables and random vectors, the latter term referring to random variables X for which $\mathcal{X}$ is a subset of $\mathbf{R}^d$ for $d > 1$. Here we will not make this distinction. The term *random variable* will refer to either a scalar or vector; in those cases in which it is important to distinguish between real-valued and vector random variables, the terms *real-valued random variable* and *scalar random variable* will be used to denote a random variable with $\mathcal{X} \subset \mathbf{R}$ and the term *vector random variable* and *random vector* will be used to denote a random variable with $X \subset \mathbf{R}^d$, $d > 1$. Random vectors will always be taken to be column vectors so that a d-dimensional random vector X is of the form

$$X = \begin{pmatrix} X_1 \\ X_2 \\ \vdots \\ X_d \end{pmatrix}$$

where $X_1, X_2, \ldots, X_d$ are real-valued random variables.

For convenience, when writing a d-dimensional random vector in the text, we will write $X = (X_1, \ldots, X_d)$ rather than $X = (X_1, \ldots, X_d)^T$. Also, if X and Y are both random vectors, the random vector formed by combining X and Y will be written as (X, Y), rather than the more correct, but more cumbersome, $(X^T, Y^T)^T$. We will often consider random vectors of the form (X, Y) with range $\mathcal{X} \times \mathcal{Y}$; a statement of this form should be taken to mean that X takes values in $\mathcal{X}$ and Y takes values in $\mathcal{Y}$.

Example 1.4 (*Binomial distribution*). Consider the experiment considered in Example 1.2. Recall that an element ω of Ω is of the form $(x_1, \ldots, x_n)$ where each x_j is either 0 or 1. For an element $\omega \in \Omega$, define

$$X(\omega) = \sum_{j=1}^{n} x_j.$$

Then

$$\Pr(X = 0) = P((0, 0, \ldots, 0)) = (1 - \theta)^n,$$

$$\Pr(X = 1) = P((1, 0, \ldots, 0)) + P((0, 1, 0, \ldots, 0)) + \cdots + P((0, 0, \ldots, 0, 1))$$
$$= n\theta(1 - \theta)^{n-1}.$$

It is straightforward to show that

$$\Pr(X = x) = \binom{n}{x} \theta^x (1 - \theta)^{n-x}, \quad x = 0, 1, \ldots, n;$$

X is said to have a *binomial distribution* with parameters n and θ. □

Example 1.5 (*Uniform distribution on the unit cube*). Let X denote a three-dimensional random vector with range $\mathcal{X} = (0, 1)^3$. For any subset of $A \in \mathcal{X}$, let

$$\Pr(X \in A) = \int \int \int_A dt_1 \, dt_2 \, dt_3.$$

Here the properties of the random vector X are defined without reference to any underlying experiment.

As discussed above, we may take the range of X to be $\mathbf{R}^3$. Then, for any subset $A \in \mathbf{R}^3$,

$$\Pr(X \in A) = \int \int \int_{A \cap (0,1)^3} dt_1 \, dt_2 \, dt_3.$$ □

Let X denote random variable on $\mathbf{R}^d$ with a given probability distribution. A *support* of the distribution, or, more simply, a support of X, is defined to be any set $\mathcal{X}_0 \subset \mathbf{R}^d$ such that

$$\Pr(X \in \mathcal{X}_0) = 1.$$

The *minimal support* of the distribution is the smallest closed set $\mathcal{X}_0 \subset \mathbf{R}^d$ such that

$$\Pr(X \in \mathcal{X}_0) = 1.$$

That is, the minimal support of X is a closed set $\mathcal{X}_0$ that is a support of X, and if $\mathcal{X}_1$ is another closed set that is a support of X, then $\mathcal{X}_0 \subset \mathcal{X}_1$.

The distribution of a real-valued random variable X is said to be *degenerate* if there exists a constant c such that

$$\Pr(X = c) = 1.$$

$\Pr(a^TX = c) = 1$

For a random vector X, with dimension greater than 1, the distribution of X is said to be degenerate if there exists a vector $a \neq 0$, with the same dimension as X, such that $a^T X$ is equal to a constant with probability 1. For example, a two-dimensional random vector $X = (X_1, X_2)$ has a degenerate distribution if, as in the case of a real-valued random variable, it is equal to a constant with probability 1. However, it also has a degenerate distribution if

$$\Pr(a_1 X_1 + a_2 X_2 = c) = 1$$

for some constants a_1, a_2, c. In this case, one of the components of X is redundant, in the sense that it can be expressed in terms of the other component (with probability 1).

Example 1.6 (*Polytomous random variable*). Let X denote a random variable with range

$$\mathcal{X} = \{x_1, \dots, x_m\}$$

where $x_1, \dots, x_n$ are distinct elements of $\mathbf{R}$. Assume that $\Pr(X = x_j) > 0$ for each $j = 1, \dots, m$. Any set containing $\mathcal{X}$ is a support of X; since $\mathcal{X}$ is closed in $\mathbf{R}$, it follows that the minimal support of X is simply $\mathcal{X}$. If $m = 1$ the distribution of X is degenerate; otherwise it is nondegenerate. $\square$

Example 1.7 (*Uniform distribution on the unit cube*). Let X denote the random variable defined in Example 1.5. Recall that for any $A \subset \mathbf{R}^3$,

$$\Pr(X \in A) = \iiint_{A \cap (0,1)^3} dt_1 \, dt_2 \, dt_3.$$

The minimal support of X is $[0, 1]^3$. $\square$

Example 1.8 (*Degenerate random vector*). Consider the experiment considered in Example 1.2 and used in Example 1.4 to define the binomial distribution. Recall that an element ω of Ω is of the form $(x_1, \dots, x_n)$ where each x_j is either 0 or 1. Define Y to be the two-dimensional random vector given by

$$Y(\omega) = \left(\sum_{j=1}^n x_j, 2 \sum_{j=1}^n x_j^2 \right).$$

Then

$$\Pr((2, -1)^T Y = 0) = 1.$$

Hence, Y has a degenerate distribution. $\square$

1.4   Distribution Functions

Consider a real-valued random variable X. The properties of X are described by its probability function P_X, which gives the probability that $X \in A$ for any set $A \subset \mathbf{R}$. However, it is also possible to specify the distribution of a random variable by considering $\Pr(X \in A)$ for a limited class of sets A; this approach has the advantage that the function giving such probabilities may be easier to use in computations. For instance, consider sets of the form $(-\infty, x]$, for $x \in \mathbf{R}$, so that $P_X\{(-\infty, x]\}$ gives $\Pr(X \le x)$. The *distribution function* of the distribution of X or, simply, the distribution function of X, is the function $F \equiv F_X : \mathbf{R} \to [0, 1]$ given by

$$F(x) = \Pr(X \le x), \quad -\infty < x < \infty.$$

Example* 1.9 *(Uniform distribution). Suppose that X is a real-valued random variable such that

$$\Pr(X \in A) = \int_{A \cap (0,1)} dx, \quad A \subset \mathbf{R};$$

X is said to have a uniform distribution on $(0, 1)$.

The distribution function of this distribution is given by

$$F(x) = \Pr\{X \in (-\infty, x]\} = \int_{(-\infty, x] \cap (0,1)} dx = \begin{cases} 0 & \text{if } x \le 0 \\ x & \text{if } 0 < x \le 1. \\ 1 & \text{if } x > 1 \end{cases}$$

Figure 1.1 gives a plot of F.   □

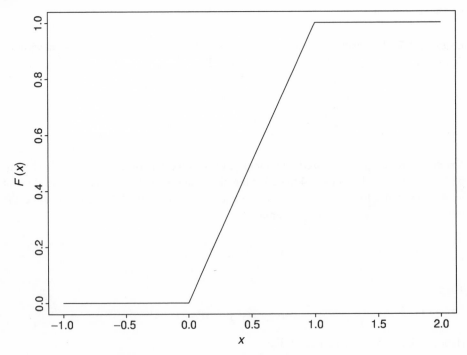

Figure 1.1. Distribution function in Example 1.9.

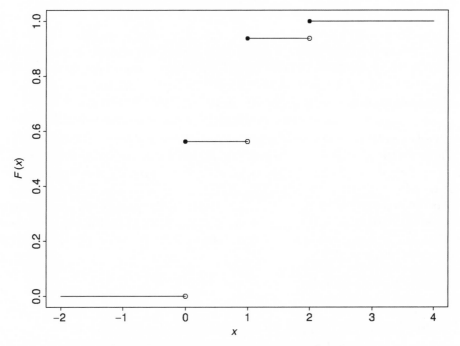

Figure 1.2. Distribution function in Example 1.10.

Note that when giving the form of a distribution function, it is convenient to only give the value of the function in the range of x for which $F(x)$ varies between 0 and 1. For instance, in the previous example, we might say that $F(x) = x$, $0 < x < 1$; in this case it is understood that $F(x) = 0$ for $x \leq 0$ and $F(x) = 1$ for $x \geq 1$.

***Example* 1.10 *(Binomial distribution)*.** Let X denote a random variable with a binomial distribution with parameters n and θ, as described in Example 1.4. Then

$$\Pr(X = x) = \binom{n}{x}\theta^x(1 - \theta)^{n-x}, \quad x = 0, 1, \ldots, n$$

and, hence, the distribution function of X is

$$F(x) = \sum_{j=0,1,\ldots;j \leq x} \binom{n}{j}\theta^j(1 - \theta)^{n-j}.$$

Thus, F is a step function, with jumps at $0, 1, 2, \ldots, n$; Figure 1.2 gives a plot of F for the case $n = 2$, $\theta = 1/4$. □

Clearly, there are some basic properties which any distribution function F must possess. For instance, as noted above, F must take values in $[0, 1]$; also, F must be nondecreasing. The properties of a distribution function are summarized in the following theorem.

Theorem 1.2. *A distribution function F of a distribution on* **R** *has the following properties:*
  (DF1) $\lim_{x \to \infty} F(x) = 1$; $\lim_{x \to -\infty} F(x) = 0$
  (DF2) If $x_1 < x_2$ *then* $F(x_1) \leq F(x_2)$
  (DF3) $\lim_{h \to 0^+} F(x + h) = F(x)$

(DF4) $\lim_{h \to 0^+} F(x - h) \equiv F(x-) = F(x) - \Pr(X = x) = \Pr(X < x)$.

Proof. Let a_n, $n = 1, 2, \ldots$ denote any increasing sequence diverging to ∞ and let A_n denote the event that $X \leq a_n$. Then $P_X(A_n) = F(a_n)$ and $A_1 \subset A_2 \subset \cdots$ with $\cup_{n=1}^{\infty} A_n$ equal to the event that $X < \infty$. It follows from (P4) that

$$\lim_{n \to \infty} F(a_n) = \Pr(X < \infty) = 1,$$

establishing the first part of (DF1); the second part follows in a similar manner.

To show (DF2), let A_1 denote the event that $X \leq x_1$ and A_2 denote the event that $x_1 < X \leq x_2$. Then A_1 and A_2 are disjoint with $F(x_1) = P_X(A_1)$ and $F(x_2) = P_X(A_1 \cup A_2) = P_X(A_1) + P_X(A_2)$, which establishes (DF2).

For (DF3) and (DF4), let a_n, $n = 1, 2, \ldots$, denote any decreasing sequence converging to 0, let A_n denote the event that $X \leq x + a_n$, let B_n denote the event that $X \leq x - a_n$, and let C_n denote the event that $x - a_n < X \leq x$. Then $A_1 \supset A_2 \supset \cdots$ and $\cap_{n=1}^{\infty} A_n$ is the event that $X \leq x$. Hence, by (P5),

$$\Pr(X \leq x) \equiv F(x) = \lim_{n \to \infty} F(x + a_n),$$

which establishes (DF3).

Finally, note that $F(x) = P_X(B_n) + P_X(C_n)$ and that $C_1 \supset C_2 \supset \cdots$ with $\cap_{n=1}^{\infty} C_n$ equal to the event that $X = x$. Hence,

$$F(x) = \lim_{n \to \infty} F(x - a_n) + \lim_{n \to \infty} P_X(C_n) = F(x-) + \Pr(X = x),$$

yielding (DF4). ∎

Thus, according to (DF2), a distribution function is nondecreasing and according to (DF3), a distribution is right-continuous.

A distribution function F gives the probability of sets of the form $(-\infty, x]$. The following result gives expressions for the probability of other types of intervals in terms of F; the proof is left as an exercise. As in Theorem 1.2, here we use the notation

$$F(x-) = \lim_{h \to 0^+} F(x - h).$$

Corollary 1.1. *Let* X *denote a real-valued random variable with distribution function* F. *Then, for* $x_1 < x_2$,
  (i) $\Pr(x_1 < X \leq x_2) = F(x_2) - F(x_1)$
  (ii) $\Pr(x_1 \leq X \leq x_2) = F(x_2) - F(x_1-)$
  (iii) $\Pr(x_1 \leq X < x_2) = F(x_2-) - F(x_1-)$
  (iv) $\Pr(x_1 < X < x_2) = F(x_2-) - F(x_1)$

Any distribution function possesses properties (DF1)–(DF4). Furthermore, properties (DF1)–(DF3) characterize a distribution function in the sense that a function having those properties must be a distribution function of some random variable.

Theorem 1.3. *If a function* $F : \mathbf{R} \to [0, 1]$ *has properties* (DF1)–(DF3), *then* F *is the distribution function of some random variable.*

Proof. Consider the experiment with $\Omega = (0, 1)$ and, suppose that, for any set $A \in \Omega$, $P(A)$ given by

$$P(A) = \int_A dx.$$

Given a function F satisfying (DF1)–(DF3), define a random variable X by

$$X(\omega) = \inf\{x \in \mathbf{R}: F(x) \geq \omega\}.$$

Then

$$\Pr(X \leq x) = P(\{\omega \in \Omega: X(\omega) \leq x\}) = P(\{\omega \in \Omega: \omega \leq F(x)\}) = F(x).$$

Hence, F is the distribution function of X. ∎

The distribution function is a useful way to describe the probability distribution of a random variable. The following theorem states that the distribution function of a random variable X completely characterizes the probability distribution of X.

Theorem 1.4. *If two random variables X_1, X_2 each have distribution function F, then X_1 and X_2 have the same probability distribution.*

A detailed proof of this result is beyond the scope of this book; see, for example, Ash (1972, Section 1.4) or Port (1994, Section 10.3).

It is not difficult, however, to give an informal explanation of why we expect such a result to hold. The goal is to show that, if X_1 and X_2 have the same distribution function, then, for 'any' set $A \subset \mathbf{R}$,

$$\Pr(X_1 \in A) = \Pr(X_2 \in A).$$

First suppose that A is an interval of the form $(a_0, a_1]$. Then

$$\Pr(X_j \in A) = F(a_1) - F(a_0), \quad j = 1, 2$$

so that $\Pr(X_1 \in A) = \Pr(X_2 \in A)$. The same is true for A^c. Now consider a second interval $B = (b_0, b_1]$. Then

$$A \cap B = \begin{cases} \emptyset & \text{if } b_0 > a_1 \text{ or } a_0 > b_1 \\ B & \text{if } a_0 \leq b_0 < b_1 \leq a_1 \\ A & \text{if } b_0 \leq a_0 < a_1 \leq b_1 \\ (a_0, b_1] & \text{if } b_1 \leq a_1 \text{ and } b_0 \leq a_0 \\ (b_0, a_1] & \text{if } a_1 \leq b_1 \text{ and } a_0 \leq b_0 \end{cases} .$$

In each case, $A \cap B$ is an interval and, hence, $\Pr(X_j \in A \cap B)$ and $\Pr(X_j \in A \cup B)$ do not depend on $j = 1, 2$. The same approach can be used for any finite collection of intervals. Hence, if a set is generated from a finite collection of intervals using set operations such as union, intersection, and complementation, then $\Pr(X_1 \in A) = \Pr(X_2 \in A)$.

However, we require that this equality holds for 'any' set A. Of course, we know that probability distibutions cannot, in general, be defined for all subsets of $\mathbf{R}$. Hence, to proceed, we must pay close attention to the class of sets A for which $\Pr(X_1 \in A)$ is defined. Essentially, the result stated above for a finite collection of intervals must be extended to a countable collection. Although this does hold, the proof is more complicated and it is useful, if not essential, to use a more sophisticated method of proof.

A number of useful properties of distribution functions follow from the fact a distribution function is nondecreasing. The following result gives one of these.

Theorem 1.5. *Let F denote the distribution function of a distribution on* **R**. *Then the set of points x at which F is not continuous is countable.*

Proof. Let D denote the set of points at which F has a discontinuity. For each positive integer m, let D_m denote the set of points x in **R** such that F has a jump of at least $1/m$ at x and let n_m denote the number of elements in D_m. Note that

$$D = \bigcup_{m=1}^{\infty} D_m$$

since

$$\lim_{x \to \infty} F(x) = 1 \quad \text{and} \quad \lim_{x \to -\infty} F(x) = 0,$$

$n_m \le m$. It follows that the number of points of discontinuity is bounded by $\sum_{m=1}^{\infty} m$. The result follows. ∎

Discrete distributions

Hence, although a distribution function is not necessarily continuous, the number of jumps must be countable; in many cases it is finite, or even 0. Let X denote a real-valued random variable with distribution function F. If F is a step function, we say that the X has a *discrete distribution* or is a *discrete random variable*.

Example 1.11 (Integer random variable). Let X denote a random variable with range $\mathcal{X} = \{1, 2, \ldots, m\}$ for some $m = 1, 2, \ldots$, and let

$$\theta_j = \Pr(X = j), \quad j = 1, \ldots, m.$$

The distribution function of X is given by

$$F(x) = \begin{cases} 0 & \text{if } x < 1 \\ \theta_1 & \text{if } 1 \le x < 2 \\ \theta_1 + \theta_2 & \text{if } 2 \le x < 3 \\ \vdots & \\ \theta_1 + \cdots + \theta_{m-1} & \text{if } m - 1 \le x < m \\ 1 & \text{if } m \le x \end{cases}$$

where $\theta_1, \ldots, \theta_m$ are constants summing to 1. Hence, F is a step function and X has a discrete distribution. □

Distribution functions for random vectors

For a random vector X taking values in $\mathbf{R}^d$, the distribution function is defined as the function $F : \mathbf{R}^d \to [0, 1]$ given by

$$F(x) = \Pr\{X \in (-\infty, x_1] \times (-\infty, x_2] \times \cdots (-\infty, x_d]\}, \quad x = (x_1, \ldots, x_d).$$

If X is written in terms of component random variables $X_1, \ldots, X_d$ each of which is real-valued, $X = (X_1, \ldots, X_d)$, then

$$F(x) = \Pr(X_1 \leq x_1, \ldots, X_d \leq x_d).$$

***Example* 1.12 *(Two-dimensional polytomous random vector)*.** Consider a two-dimensional random vector X with range

$$\mathcal{X} = \{x \in \mathbf{R}^2 \colon x = (i, j), \quad i = 1, \ldots, m; j = 1, \ldots, m\}$$

and let

$$\theta_{ij} = \Pr\{X = (i, j)\}.$$

The distribution function of X is given by

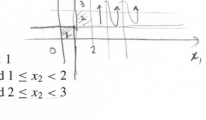

$$F(x) = \begin{cases} 0 & \text{if } x_1 < 1 \text{ or } x_2 < 1 \\ \theta_{11} & \text{if } 1 \leq x_1 < 2 \text{ and } 1 \leq x_2 < 2 \\ \theta_{11} + \theta_{12} & \text{if } 1 \leq x_1 < 2 \text{ and } 2 \leq x_2 < 3 \\ \vdots \\ \theta_{11} + \cdots + \theta_{1m} & \text{if } 1 \leq x_1 < 2 \text{ and } m \leq x_2 \\ \vdots \\ \theta_{11} + \cdots + \theta_{m1} & \text{if } m \leq x_1 \text{ and } 1 \leq x_2 < 2 \\ \theta_{11} + \cdots + \theta_{m1} + \theta_{12} + \cdots + \theta_{m2} & \text{if } m \leq x_1 \text{ and } 2 \leq x_2 < 3 \\ \vdots \\ 1 & \text{if } m \leq x_1 \text{ and } m \leq x_2 \end{cases}, \quad x = (x_1, x_2).$$

This is a two-dimensional step function. $\quad\square$

***Example* 1.13 *(Uniform distribution on the unit cube)*.** Consider the random vector X defined in Example 1.5. Recall that X has range $\mathcal{X} = (0, 1)^3$ and for any subset $A \subset \mathcal{X}$,

$$\Pr(X \in A) = \int \int \int_A dt_1 \, dt_2 \, dt_3.$$

Then X has distribution function

$$F(x) = \int_0^{x_3} \int_0^{x_2} \int_0^{x_1} dt_1 \, dt_2 \, dt_3 = x_1 x_2 x_3, \quad x = (x_1, x_2, x_3), \ 0 \leq x_j \leq 1, \ j = 1, 2, 3$$

with $F(x) = 0$ if $\min(x_1, x_2, x_3) < 0$. If $x_j > 1$ for some $j = 1, 2, 3$, then $F(x) = x_1 x_2 x_3 / x_j$; if $x_i > 1$, $x_j > 1$ for some $i, j = 1, 2, 3$, then $F(x) = x_1 x_2 x_3 / (x_i x_j)$. $\quad\square$

Like distribution functions on $\mathbf{R}$, a distribution function on $\mathbf{R}^d$ is nondecreasing and right-continuous.

***Theorem* 1.6.** *Let F denote the distribution function of a vector-valued random variable X taking values in $\mathbf{R}^d$.*

(i) *If $x = (x_1, \ldots, x_d)$ and $y = (y_1, \ldots, y_d)$ are elements of $\mathbf{R}^d$ such that $x_j \leq y_j$, $j = 1, \ldots, d$, then $F(x) \leq F(y)$.*

(ii) If $x_n = (x_{n1}, \ldots, x_{nd})$, $n = 1, 2, \ldots$ is a sequence in $\mathbf{R}^d$ such that each sequence x_{nj}, $n = 1, 2, \ldots$ is a decreasing sequence with limit x_j, $j = 1, \ldots, d$, then

$$\lim_{n \to \infty} F(x_n) = F(x).$$

Proof. Let $x = (x_1, x_2, \ldots, x_d)$ and $y = (y_1, y_2, \ldots, y_d)$ denote elements of $\mathbf{R}^d$ satisfying the condition in part (i) of the theorem. Define

$$A = (-\infty, x_1] \times \cdots \times (-\infty, x_d]$$

and

$$B = (-\infty, y_1] \times \cdots \times (-\infty, y_d].$$

Then $F(x) = \mathrm{P}_X(A)$, $F(y) = \mathrm{P}_X(B)$ and part (i) of the theorem follows from the fact that $A \subset B$.

For part (ii), define

$$A_n = (-\infty, x_{n1}] \times \cdots (-\infty, x_{nd}], \quad n = 1, 2, \ldots.$$

Then $A_1 \supset A_2 \supset \cdots$ and

$$\cap_{n=1}^{\infty} A_n = (-\infty, x_1] \times \cdots \times (-\infty, x_n].$$

The result now follows from (P5). ∎

We saw that the probability that a real-valued random variable takes values in a set $(a, b]$ can be expressed in terms of its distribution, specifically,

$$\Pr(a < X \leq b) = F(b) - F(a).$$

A similar result is available for random vectors, although the complexity of the expression increases with the dimension of the random variable. The following example illustrates the case of a two-dimensional random vector; the general case is considered in Theorem 1.7 below.

Example **1.14** *(Two-dimensional random vector)*. Let $X = (X_1, X_2)$ denote a two-dimensional random vector with distribution function F. Consider the probability

$$\Pr(a_1 < X_1 \leq b_1, \ a_2 < X_2 \leq b_2);$$

our goal is to express this probability in terms of F.

Note that

$$
\begin{aligned}
\Pr(a_1 < X_1 \leq b_1, a_2 < X_2 \leq b_2) &= \Pr(X_1 \leq b_1, a_2 < X_2 \leq b_2) \\
&\quad - \Pr(X_1 \leq a_1, a_2 < X_2 \leq b_2) \\
&= \Pr(X_1 \leq b_1, X_2 \leq b_2) - \Pr(X_1 \leq b_1, X_2 \leq a_2) \\
&\quad - \Pr(X_1 \leq a_1, X_2 \leq b_2) + \Pr(X_1 \leq a_1, X_2 \leq a_1) \\
&= F(b_1, b_2) - F(b_1, a_2) - F(a_1, b_2) + F(a_1, a_2),
\end{aligned}
$$

which yields the desired result. □

It is clear that the approach used in Example 1.14 can be extended to a random variable of arbitrary dimension. However, the statement of such a result becomes quite complicated.

Theorem 1.7. *Let F denote the distribution function of a vector-valued random variable X taking values in $\mathbf{R}^d$.*

For each $j = 1, \ldots, d$, let $-\infty < a_j < b_j < \infty$ and define the set A by

$$A = (a_1, b_1] \times \cdots \times (a_d, b_d].$$

Then

$$P_X(A) = \Delta(b - a)F(a)$$

where $a = (a_1, \ldots, a_d)$, $b = (b_1, \ldots, b_d)$, and for any arbitrary function h on $\mathbf{R}^d$,

$$\Delta(b)h(x) = \Delta_{1,b_1} \Delta_{2,b_2} \cdots \Delta_{d,b_d} h(x),$$

$$\Delta_{j,c} h(x) = h(x + ce_j) - h(x).$$

Here e_j is the jth coordinate vector in $\mathbf{R}^d$, $(0, \ldots, 0, 1, 0, \ldots, 0)$.

Proof. First note that

$$\Delta_{1,b_1-a_1} F(a) = F(b_1, a_2, \ldots, a_d) - F(a_1, a_2, \ldots, a_d)$$
$$= \Pr(a_1 < X_1 \le b_1, X_2 \le a_2, \ldots, X_d \le a_d).$$

Each of the remaining operations based on Δ_{j,b_j-a_j}, where $j = 2, \ldots, d$, concerns only the corresponding random variable X_j. Hence,

$$\Delta_{2,b_2-a_2} \Delta_{1,b_1-a_1} F(a) = \Pr(a_1 < X_1 \le b_1, a_2 < X_2 \le b_2, X_3 \le a_3, \ldots, X_d \le a_d),$$

and so on. The result follows. ∎

1.5 Quantile Functions

Consider a real-valued random variable X. The distribution function of X describes its probability distribution by giving the probability that $X \le x$ for all $x \in \mathbf{R}$. For example, if we choose an $x \in \mathbf{R}$, $F(x)$ returns the probability that X is no greater than x.

Another approach to specifying the distribution of X is to give, for a specified probability $p \in (0, 1)$, the value x_p such that $\Pr(X \le x_p) = p$. That is, instead of asking for the probability that $X \le 1$, we might ask for the point x such that $\Pr(X \le x) = .5$. One complication of this approach is that there may be many values $x_p \in \mathbf{R}$ such that $\Pr(X \le x_p) = p$ or no such value might exist. For instance, if X is a binary random variable taking the values 0 and 1 each with probability $1/2$, any value x in the interval $[0, 1)$ satisfies $\Pr(X \le x) = 1/2$ and there does not exist an $x \in \mathbf{R}$ such that $\Pr(X \le x) = 3/4$.

For a given value $p \in (0, 1)$ we define the pth *quantile* of the distribution to be

$$\inf\{z \colon F(z) \ge p\}.$$

Thus, for the binary random variable described above, the .5th quantile is 0 and the .75th quantile is 1.

The *quantile function* of the distribution or, more simply, of X, is the function Q : $(0, 1) \to \mathbf{R}$ given by

$$Q(t) = \inf\{z \colon F(z) \geq t\}.$$

The quantile function is essentially the inverse of the distribution function F; however, since F is not necessarily a one-to-one function, its inverse may not exist. The pth quantile of the distribution, as defined above, is given by $Q(p), 0 < p < 1$.

Example **1.15** *(**Integer random variable**).* Let X denote a random variable with range $\mathcal{X} = \{1, 2, \ldots, m\}$ for some $m = 1, 2, \ldots$, and let

$$\theta_j = \Pr(X = j), \quad j = 1, \ldots, m.$$

The distribution function of X is given in Example 1.11; it is a step function with jump θ_j at $x = j$.

The quantile function of X may be calculated as follows. Suppose that $t \leq \theta_1$. Then $F(x) \geq t$ provided that $x \geq 1$. Hence, $Q(t) = 1$. If $\theta_1 < t \leq \theta_1 + \theta_2$, then $F(x) \geq t$ provided that $x \geq 2$ so that $Q(t) = 2$. This procedure may be used to determine the entire function Q. It follows that

$$Q(t) = \begin{cases} 1 & \text{if } 0 < t \leq \theta_1 \\ 2 & \text{if } \theta_1 < t \leq \theta_1 + \theta_2 \\ 3 & \text{if } \theta_1 + \theta_2 < t \leq \theta_1 + \theta_2 + \theta_3 \\ \vdots & \\ m & \text{if } \theta_1 + \cdots + \theta_{m-1} < t < 1 \end{cases}$$

Figure 1.3 gives plots of F and Q for the case in which $m = 3$, $\theta_1 = 1/4$, $\theta_2 = 1/2$, and $\theta_3 = 1/4$. $\square$

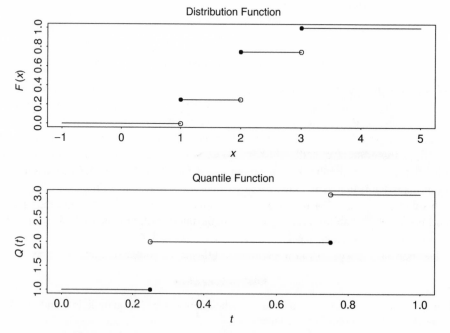

Figure 1.3. Quantile and distribution functions in Example 1.15.

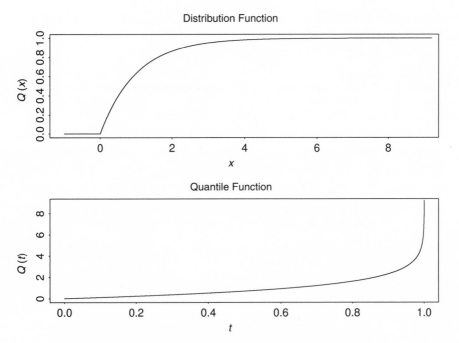

Figure 1.4. Quantile and distribution functions in Example 1.16.

Example* 1.16 *(Standard exponential distribution). Let X denote a real-valued random variable with distribution function $F(x) = 1 - \exp(-x)$, $x > 0$; this distribution is known as the *standard exponential distribution*. The quantile function of the distribution is given by $Q(t) = -\log(1 - t)$, $0 < t < 1$. Figure 1.4 gives plots of F and Q.   □

A *median* of the distribution of a real-valued random variable X is any point $m \in \mathbf{R}$ such that

$$\Pr(X \le m) \ge \frac{1}{2} \quad \text{and} \quad \Pr(X \ge m) \ge \frac{1}{2};$$

note that a median of a distribution is not, in general, unique. It may be shown that if X has quantile function Q, then $Q(.5)$ is a median of X; this problem is given as Exercise 1.20.

Example* 1.17 *(Standard exponential distribution). Let X have a standard exponential distribution as discussed in Example 1.16. Since, for any $x > 0$,

$$\Pr(X \le x) = 1 - \Pr(X \ge x)$$

and

$$\Pr(X \ge x) = \exp(-x), \quad x > 0,$$

it follows that the median of the distribution is $m = \log(2)$.   □

Example* 1.18 *(Binomial distribution). Let X denote a random variable with a binomial distribution with parameters n and θ, as described in Example 1.4. Then X is a discrete random variable with

$$\Pr(X = x) = \binom{n}{x}\theta^x(1 - \theta)^{n-x}, \quad x = 0, \ldots, n$$

where $0 < \theta < 1$. Let m_0 denote the largest positive integer for which

$$\sum_{j=0}^{m_0}\binom{n}{j}\theta^j(1 - \theta)^{n-j} \le \frac{1}{2}.$$

If

$$\sum_{j=0}^{m_0}\binom{n}{j}\theta^j(1 - \theta)^{n-j} < \frac{1}{2},$$

then the median of the distribution is $m_0 + 1$; otherwise, any value in the interval $(m_0, m_0 + 1)$ is a median of the distribution.   □

There are a number of properties that any quantile function must satisfy; for convenience, we use the convention that $Q(0) = -\infty$.

Theorem 1.8. *Consider a real-valued random variable X with distribution function F and quantile function Q. Then*

    (i) $Q(F(x)) \le x, \quad -\infty < x < \infty$
    (ii) $F(Q(t)) \ge t, \quad 0 < t < 1$
    (iii) $Q(t) \le x$ *if and only if* $F(x) \ge t$
    (iv) *If* F^{-1} *exists, then* $Q(t) = F^{-1}(t)$
    (v) *If* $t_1 < t_2$, *then* $Q(t_1) \le Q(t_2)$

Proof.   Define the set $A(t)$ by

$$A(t) = \{z : F(z) \ge t\}$$

so that $Q(t) = \inf A(t)$. Then $A[F(x)]$ clearly contains x so that $Q[F(x)] = \inf A[F(x)]$ must be no greater than x; this proves part (i). Note that if $F(x) = 0$, then $A(t) = (-\infty, x_1]$ for some x_1 so that the result continues to hold if $Q(F(x))$ is taken to be $-\infty$ in this case.

Also, for any element $x \in A(t)$, $F(x) \ge t$; clearly, this relation must hold for any sequence in $A(t)$ and, hence, must hold for the inf of the set, proving part (ii).

Suppose that, for a given x and t, $F(x) \ge t$. Then $A(t)$ contains x; hence, $Q(t) \le x$. Now suppose that $Q(t) \le x$; since F is nondecreasing, $F(Q(t)) \le F(x)$. By part (ii) of the theorem $F(Q(t)) \ge t$ so that $F(x) \ge t$, proving part (iii).

If F is invertible, then $A_t = \{x : x \ge F^{-1}(t)\}$ so that $Q(t) = F^{-1}(t)$, establishing part (iv).

Let $t_1 \le t_2$. Then $A(t_1) \supset A(t_2)$. It follows that the

$$\inf A(t_1) \le \inf A(t_2);$$

part (v) follows.   ■

***Example* 1.19** *(A piecewise-continuous distribution function).* Let X denote a real-valued random variable with distribution function given by

$$F(x) = \begin{cases} 0 & \text{if } x < 0 \\ \frac{1}{2} & \text{if } x = 0 \\ 1 - \exp(-x)/2 & \text{if } 0 < x < \infty \end{cases}.$$

Then

$$Q(t) = \begin{cases} 0 & \text{if } 0 < t \leq 1/2 \\ -\log(2(1-t)) & \text{if } 1/2 < t < 1 \end{cases}.$$

This example can be used to illustrate the results in Theorem 1.8. Note that, for $x \geq 0$,

$$Q(F(x)) = \begin{cases} 0 & \text{if } x = 0 \\ -\log(2(\exp(-x)/2)) & \text{if } 0 < x < \infty \end{cases} = x;$$

for $x < 0$, $F(x) = 0$ so that $Q(F(x)) = -\infty$. Similarly,

$$F(Q(t)) = \begin{cases} \frac{1}{2} & \text{if } t \leq 1/2 \\ 1 - \exp(\log(2(1-t)))/2 & \text{if } 1/2 < t < 1 \end{cases} = \begin{cases} \frac{1}{2} & \text{if } t \leq 1/2 \\ t & \text{if } 1/2 < t < 1 \end{cases} \geq t.$$

This illustrates parts (i) and (ii) of the theorem.

Suppose $x < 0$. Then $Q(t) \leq x$ does not hold for any value of $t > 0$ and $F(x) = 0 \geq t$ does not hold for any $t > 0$. If $x = 0$, then $Q(t) \leq x$ if and only if $t \leq 1/2$, and $F(x) = F(0) = 1/2 \geq t$ if and only if $t \leq 1/2$. Finally, if $x > 0$, $Q(t) \leq x$ if and only if

$$-\log(2(1-t)) \leq x,$$

that is, if and only if $t \leq 1 - \exp(-x)/2$, while $F(x) \geq t$ if and only if

$$1 - \exp(-x)/2 \geq t.$$

This verifies part (iii) of Theorem 1.8 for this distribution.

Part (iv) of the theorem does not apply here, while it is easy to see that part (v) holds. □

We have seen that the distribution of a random variable is characterized by its distribution function. Similarly, two random variables with the same quantile function have the same distribution.

Corollary 1.2. *Let X_1 and X_2 denote real-valued random variables with quantile functions Q_1 and Q_2, respectively. If $Q_1(t) = Q_2(t)$, $0 < t < 1$, then X_1 and X_2 have the same probability distribution.*

Proof. Let F_j denote the distribution function of X_j, $j = 1, 2$, and fix a value x_0. Then either $F_1(x_0) < F_2(x_0)$, $F_1(x_0) > F_2(x_0)$, or $F_1(x_0) = F_2(x_0)$.

First suppose that $F_1(x_0) < F_2(x_0)$. By parts (i) and (v) of Theorem 1.8,

$$Q_2(F_1(x_0)) \leq Q_2(F_2(x_0)) \leq x_0.$$

Hence, by part (iii) of Theorem 1.8, $F_2(x_0) \geq F_1(x_0)$ so that $F_1(x_0) < F_2(x_0)$ is impossible.

The same argument shows that $F_2(x_0) < F_1(x_0)$ is impossible. It follows that $F_1(x_0) = F_2(x_0)$. Since x_0 is arbitrary, it follows that $F_1 = F_2$, proving the result. ■

1.6   Density and Frequency Functions

Consider a real-valued random variable X with distribution function F and range $\mathcal{X}$. Suppose there exists a function $p : \mathbf{R} \to \mathbf{R}$ such that

$$F(x) = \int_{-\infty}^{x} p(t)\,dt, \quad -\infty < x < \infty. \tag{1.1}$$

The function p is called the *density function* of the distribution or, more simply, of X; since F is nondecreasing, p can be assumed to be nonnegative and we must have

$$\int_{-\infty}^{\infty} p(x)\,dx = 1.$$

We also assume that any density function is continuous almost everywhere and is of bounded variation, which ensures that the Riemann integral of the density function exists; see Sections A3.1.8, A.3.3.3, and A3.4.9 of Appendix 3.

In this case, it is clear that the distribution function F must be a continuous function; in fact, F is an *absolutely continuous* function. Absolute continuity is stronger than ordinary continuity; see Appendix 1 for further discussion of absolutely continuous functions. Hence, when (1.1) holds, we say that the distribution of X is an *absolutely continuous distribution*; alternatively, we say that X is an *absolutely continuous random variable*.

Conversely, if F is an absolutely continuous function, then there exists a density function p such that (1.1) holds. In many cases, the function p can be obtained from F by the fundamental theorem of calculus; see Theorem 1.9 below. It is important to note, however, that the density function of a distribution is not uniquely defined. If

$$p_1(x) = p_2(x) \quad \text{for almost all } x,$$

and

$$F(x) = \int_{-\infty}^{x} p_1(t)\,dt, \quad -\infty < x < \infty,$$

then

$$F(x) = \int_{-\infty}^{x} p_2(t)\,dt, \quad -\infty < x < \infty;$$

see Section A3.1.8 of Appendix 3 for discussion of the term "almost all." In this case, either p_1 or p_2 may be taken as the density function of the distribution. Generally, we use the version of the density that is continuous, if one exists.

The following theorem gives further details on the relationship between density and distribution functions.

Theorem 1.9. *Let F denote the distribution function of a distribution on $\mathbf{R}$.*
  (i) Suppose that F is absolutely continuous with density function p. If p is continuous at x then $F'(x)$ exists and $p(x) = F'(x)$.
 (ii) Suppose $F'(x)$ exists for all $x \in \mathbf{R}$ and

$$\int_{-\infty}^{\infty} F'(x)\,dx < \infty.$$

Then F is absolutely continuous with density function F'.

(iii) Suppose that F is absolutely continuous and there exists a function p such that

$$F'(x) = p(x) \qquad \text{for almost all } x.$$

Then p is a density function of F.

Proof. If F is absolutely continuous with density p, then

$$F(x) = \int_{-\infty}^{x} p(t)\, dt, \quad -\infty < x < \infty.$$

Hence, part (i) follows immediately from the fundamental theorem of calculus, given in Appendix 1 (Section A1.6). Part (ii) is simply a restatement of result (2) of Section A1.6.

To prove part (iii), note that if F is absolutely continuous, then there exists a function f such that

$$F(x) = \int_{-\infty}^{x} f(t)\, dt, \quad -\infty < x < \infty$$

and

$$F'(x) = f(x) \qquad \text{for almost all } x.$$

It follows that $f(x) = p(x)$ for almost all x so that

$$F(x) = \int_{-\infty}^{x} p(t)\, dt, \quad -\infty < x < \infty,$$

as well. ∎

Example 1.20 (Uniform distribution on (0, 1)). Let X denote a random variable with the uniform distribution on (0, 1), as defined in Example 1.9. Then X has distribution function

$$F(x) = \int_{0}^{x} dt, \quad 0 \le x \le 1$$

so that X has an absolutely continuous distribution with density function

$$p(x) = 1, \quad 0 \le x \le 1.$$

Note that the density function of X may also be taken to be

$$p(x) = \begin{cases} 1 & \text{if } 0 < x < 1 \\ 0 & \text{otherwise} \end{cases}.$$

☐

Example 1.21 (Distribution function satisfying a Lipschitz condition). Consider the distribution with distribution function given by

$$F(x) = \begin{cases} 0 & \text{if } x < 1 \\ (x-1)^2 & \text{if } 1 \le x \le 2 \\ 1 & \text{if } x > 2 \end{cases}.$$

We first show that there exists a constant M such that, for all $x_1, x_2 \in \mathbf{R}$,

$$|F(x_2) - F(x_1)| \le M|x_2 - x_1|.$$

This is called a *Lipschitz condition* and it implies that F is an absolutely continuous function and, hence, that part (iii) of Theorem 1.9 can be used to find the density of the distribution; see Section A1.5 of Appendix 1 for further details.

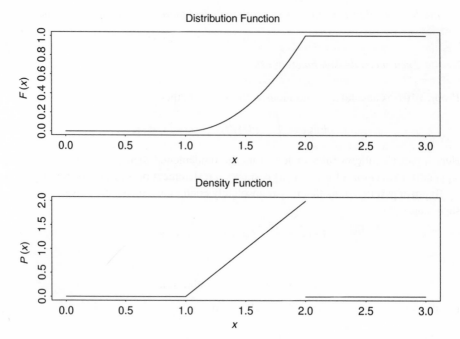

Figure 1.5. Distribution and density functions in Example 1.21.

First consider the case in which $x_1 < 1$ and $1 \leq x_2 \leq 2$; then

$$|F(x_2) - F(x_1)| = (x_2 - 1)^2 \leq |x_2 - x_1|.$$

If x_1 and x_2 are both in $[1, 2]$, then

$$|F(x_2) - F(x_1)| = \left|x_2^2 - x_1^2 + 2(x_1 - x_2)\right| \leq |x_1 + x_2 + 2||x_2 - x_1| \leq 6|x_2 - x_1|;$$

if $x_2 > 1$ and $1 < x_2 < 2$, then

$$|F(x_2) - F(x_1)| = \left|1 - (x_2 - 1)^2\right| = \left|x_2^2 - 2x_2\right| = x_2|x_2 - 2| \leq 2|x_2 - x_1|.$$

Finally, if $x_1 < 1$ and $x_2 > 2$,

$$|F(x_2) - F(x_1)| \leq 1 \leq |x_2 - x_1|.$$

Since F satisfies a Lipschitz condition, it follows that F is absolutely continuous and that the density function of the distribution is given by

$$p(x) = \begin{cases} F'(x) & \text{if } 1 < x < 2 \\ 0 & \text{otherwise} \end{cases} = \begin{cases} 2(x - 1) & \text{if } 1 < x < 2 \\ 0 & \text{otherwise} \end{cases}.$$

Figure 1.5 contains plots of F and p. □

Note that, by the properties of the Riemann integral, if X has an absolutely continuous distribution with density p, then, for small $\epsilon > 0$,

$$\Pr(x - \epsilon/2 < X < x + \epsilon/2) = \int_{x-\epsilon/2}^{x+\epsilon/2} p(t)\,dt \doteq p(x)\epsilon.$$

Hence, $p(x)$ can be viewed as being proportional to the probability that X lies in a small interval containing x; of course, such an interpretation only gives an intuitive meaning to the density function and cannot be used in formal arguments. It follows that the density function gives an indication of the relative likelihood of different possible values of X. For instance, Figure 1.5 shows that the likelihood of X taking a value x in the interval $(1, 2)$ increases as x increases.

Thus, when working with absolutely continuous distributions, density functions are often more informative than distribution functions for assessing the basic properties of a probability distribution. Of course, mathematically speaking, this statement is nonsense since the distribution function completely characterizes a probability distribution. However, for understanding the basic properties of the distribution of random variable, the density function is often more useful than the distribution function.

Example 1.22. Consider an absolutely continuous distribution with distribution function

$$F(x) = (5 - 2x)(x - 1)^2, \quad 1 < x < 2$$

and density function

$$p(x) = \begin{cases} 6(2 - x)(x - 1) & \text{if } 1 < x < 2 \\ 0 & \text{otherwise} \end{cases}.$$

Figure 1.6 gives a plot of F and p. Based on the plot of p it is clear that the most likely value of X is $3/2$ and, for $z < 1/2$, $X = 3/2 - z$ and $X = 3/2 + z$ are equally likely; these facts are difficult to discern from the plot of, or the expression for, the distribution function. The plots in Figure 1.6 can also be compared to those in Figure 1.5, which represent the distribution and density functions in Example 1.21. Based on the distribution functions,

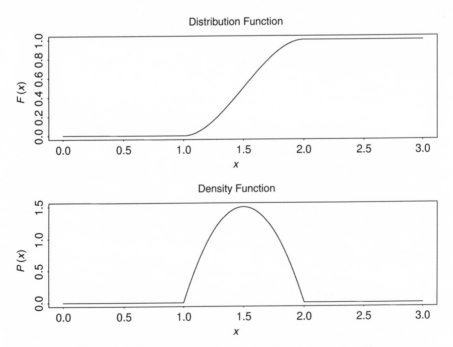

Figure 1.6. Distribution and density functions in Example 1.22.

one might conclude that the distributions are very similar; however, the density functions indicate important differences in the distributions. □

In giving expressions for density functions, it is often convenient to give the value of the density function, $p(x)$, only for those values of x for which the value is nonzero. For instance, in the previous example, the density function might be given as $p(x) = 6(2 - x)(x - 1)$, $1 < x < 2$; this statement implies that for $x \leq 1$ or $x \geq 2$, $p(x) = 0$.

Discrete distributions

A second important special case occurs when F is a step function. Suppose that F has jumps at $x_1, x_2, \ldots$. In this case,

$$F(x) = \sum_{j:x_j \leq x} p(x_j)$$

where $p(x_j)$ is the size of the jump of F at x_j. Hence,

$$p(x_j) = \Pr(X = x_j), \quad j = 1, \ldots.$$

In this case, X is a discrete random variable and the function p will be called the *frequency function* of the distribution. The set of possible values of X is given by $\mathcal{X} = \{x_1, x_2, \ldots, \}$.

Example 1.23 *(Binomial distribution)*. Let X denote the random variable defined in Example 1.4. Then $\mathcal{X} = \{0, 1, \ldots, n\}$ and

$$\Pr(X = x) = \binom{n}{x} \theta^x (1 - \theta)^{n-x}, \quad x = 0, 1, \ldots, n;$$

here $0 < \theta < 1$ is a constant. Hence, the distribution function of X is given by

$$F(x) = \sum_{j=0,1,\ldots,n; j \leq x} \binom{n}{j} \theta^j (1 - \theta)^{n-j}$$

so that X is a discrete random variable with frequency function

$$\binom{n}{x} \theta^x (1 - \theta)^{n-x}, \quad x = 0, 1, \ldots, n. \qquad \square$$

A random variable can be neither discrete nor absolutely continuous.

Example 1.24 *(A distribution function with discrete and absolutely continuous components)*. Let X denote a real-valued random variable such that, for any $A \subset \mathbf{R}$,

$$\Pr(X \in A) = \frac{1}{2} I_{\{0 \in A\}} + \frac{1}{2} \int_{A \cap (0, \infty)} \exp(-t) \, dt.$$

Thus, X is equal to 0 with probability $1/2$; if X is not equal to 0 then X is distributed like a random variable with probability function

$$\int_{A \cap (0, \infty)} \exp(-t) \, dt.$$

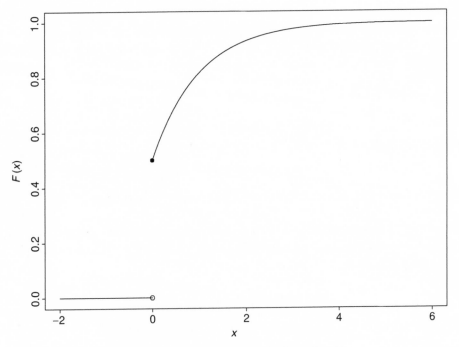

Figure 1.7. Distribution function in Example 1.24.

The distribution function of X is given by

$$F(x) = \begin{cases} 0 & \text{if } x < 0 \\ \frac{1}{2} & \text{if } x = 0 \\ 1 - \exp(-x)/2 & \text{if } 0 < x < \infty \end{cases}.$$

Recall that this distribution was considered in Example 1.19; a plot of F is given in Figure 1.7.

Note that, although F is clearly not continuous, it is continuous aside from the jump at $x = 0$ and it can be written as a weighted sum of an absolutely continuous distribution function and a distribution function based on a discrete distribution. Let

$$F_{\text{d}}(x) = \begin{cases} 0 & \text{if } x < 0 \\ 1 & \text{if } 0 \le x \end{cases}$$

and

$$F_{\text{ac}}(x) = \begin{cases} 0 & \text{if } x < 0 \\ 1 - \exp(-x) & \text{if } 0 \le x \end{cases}.$$

Note that F_{d} is a step function, F_{ac} is absolutely continuous, and

$$F = \frac{1}{2}F_{\text{d}} + \frac{1}{2}F_{\text{ac}}.$$

Hence, the distribution of X is not absolutely continuous, since F cannot be written as an integral and, since F is not a step function, the distribution of X is not discrete. In these cases, we say that X has a *mixed distribution*, with discrete and absolutely continuous components. $\square$

Random vectors

The concepts of discrete and absolutely continuous distributions can be applied to a vector-valued random variable $X = (X_1, \ldots, X_d)$ as well. If the distribution function F of X may be written

$$F(x_1, \ldots, x_d) = \int_{-\infty}^{x_d} \cdots \int_{-\infty}^{x_1} p(t) \, dt$$

for some function p on $\mathbf{R}^d$, then X is said to have an absolutely continuous distribution with density p. If the range of X is a countable set $\mathcal{X}$ with

$$p(x) = \Pr(X = x), \quad x \in \mathcal{X},$$

then X is said to have a discrete distribution with frequency function p.

***Example* 1.25** *(Uniform distribution on the unit cube).* Let X denote a three-dimensional random vector with the uniform distribution on $(0, 1)^3$, as defined in Example 1.7. Then, for any $A \in \mathbf{R}^3$

$$\Pr(X \in A) = \iiint_{A \cap (0,1)^3} dt_1 \, dt_2 \, dt_3.$$

Hence, X has an absolutely continuous distribution with density function

$$p(x) = 1, \quad x \in (0, 1)^3. \qquad \square$$

***Example* 1.26** *(A discrete random vector).* Let $X = (X_1, X_2)$ denote a two-dimensional random vector such that

$$\Pr(X = (0, 0)) = \frac{1}{2}, \quad \Pr(X = (1, 0)) = \frac{1}{4}, \quad \text{and } \Pr(X = (0, 1)) = \frac{1}{4}.$$

Then X is a discrete random variable with range $\{(0, 0), (0, 1), (1, 0)\}$ and frequency function

$$p(x_1, x_2) = \frac{1}{2} - \frac{(x_1 + x_2)}{4}, \quad x_1 = 0, 1; \quad x_2 = 0, 1. \qquad \square$$

1.7   Integration with Respect to a Distribution Function

Integrals with respect to distribution functions, that is, integrals of the form

$$\int_{\mathbf{R}^d} g(x) \, dF(x),$$

play a central role in distribution theory. For readers familiar with the general theory of integration with respect to a measure, the definition and properties of such an integral follow from noting that F defines a measure on $\mathbf{R}^d$. In this section, a brief description of such integrals is given for the case in which X is a real-valued random variable; further details and references are given in Appendix 1.

Suppose X is a real-valued random variable with distribution function F. Then we expect that

$$F(x) = \int_{-\infty}^{x} dF(t);$$

more generally, for any set $A \subset \mathbf{R}$, we should have

$$\Pr(X \in A) = \int_A dF(x).$$

It is also natural to expect that any definition of an integral satisfies certain linearity and nonnegativity conditions:

$$\int_{\mathbf{R}} [a_1 g_1(x) + a_2 g_2(x)] \, dF(x) = a_1 \int_{\mathbf{R}} g_1(x) \, dF(x) + a_2 \int_{\mathbf{R}} g_2(x) \, dF(x)$$

and

$$\int_{\mathbf{R}} g(x) \, dF(x) \geq 0 \quad \text{whenever} \quad g(x) \geq 0 \quad \text{for all} \quad x.$$

These properties, together with a continuity property, can be used to give a precise definition to integrals of the form

$$\int_{\mathbf{R}} g(x) \, dF(x);$$

see Appendix 1 for further details.

Although this type of integral can be computed for any distribution function F, the computation is particularly simple if the distribution is either absolutely continuous or discrete. Suppose X has an absolutely continuous distribution with density p. Then

$$F(x) = \int_{-\infty}^{x} dF(t) = \int_{-\infty}^{x} p(t) \, dt;$$

this result generalizes to

$$\int_{-\infty}^{\infty} g(x) \, dF(x) = \int_{-\infty}^{\infty} g(x) p(x) \, dx,$$

provided that the integrals exist. If X has discrete distribution with frequency function p, a similar result holds, so that

$$\int_{-\infty}^{\infty} g(x) \, dF(x) = \sum_{x \in \mathcal{X}} g(x) p(x),$$

provided that the sum exists.

Thus, if attention is restricted to random variables with either an absolutely continuous or discrete distribution, then

$$\int_{-\infty}^{\infty} g(x) \, dF(x) = \begin{cases} \int_{-\infty}^{\infty} g(x) p(x) \, dx & \text{if } X \text{ is absolutely continuous} \\ \sum_{x \in \mathcal{X}} g(x) p(x) & \text{if } X \text{ is discrete} \end{cases}; \qquad (1.2)$$

here p represents either the density function or the frequency function of the distribution. The case in which the distribution is neither absolutely continuous nor discrete may be viewed as an extension of these results. If X is a random vector taking values in $\mathbf{R}^d$, then (1.2) generalizes in the obvious manner:

$$\int_{-\infty}^{\infty} g(x) \, dF(x) = \begin{cases} \int_{\mathbf{R}^d} g(x) p(x) \, dx & \text{if } X \text{ is absolutely continuous} \\ \sum_{x \in \mathcal{X}} g(x) p(x) & \text{if } X \text{ is discrete} \end{cases}.$$

1.8   Expectation

Let X denote a real-valued random variable with distribution function F. The expected value of X, denoted by $E(X)$, is given by

$$E(X) = \int_{-\infty}^{\infty} x \, dF(x),$$

provided that the integral exists. It follows from the discussion in the previous section that, if X has a discrete distribution, taking the values $x_1, x_2, \ldots$ with frequency function p, then

$$E(X) = \sum_j x_j \, p(x_j).$$

If X has an absolutely continuous distribution with density p, then

$$E(X) = \int_{-\infty}^{\infty} x \, p(x) \, dx.$$

There are three possibilities for an expected value $E(X)$: $E(X) < \infty$, $E(X) = \pm\infty$, or $E(X)$ might not exist. In general, $E(X)$ fails to exist if the integral

$$\int_{-\infty}^{\infty} x \, dF(x)$$

fails to exist; see Appendix 1. Hence, the expressions for $E(X)$ given above are valid only if the corresponding sum or integral exists. If X is nonnegative, then $E(X)$ always exists, although we may have $E(X) = \infty$; in general, $E(X)$ exists and is finite provided that

$$E(|X|) = \int_{-\infty}^{\infty} |x| \, dF(x) < \infty.$$

***Example* 1.27 (*Binomial distribution*).** Let X denote a random variable with a binomial distribution with parameters n and θ, as described in Example 1.4. Then X is a discrete random variable with frequency function

$$p(x) = \binom{n}{x} \theta^x (1 - \theta)^{n-x}, \quad x = 0, \ldots, n$$

so that

$$E(X) = \sum_{x=0}^{n} x \binom{n}{x} \theta^x (1 - \theta)^{n-x} = n\theta. \qquad \square$$

***Example* 1.28 (*Pareto distribution*).** Let X denote a real-valued random variable with an absolutely continuous distribution with density function

$$p(x) = \theta x^{-(\theta+1)}, \quad x \geq 1,$$

where θ is a positive constant; this is called a *Pareto distribution with parameter* θ. Then

$$E(X) = \int_0^{\infty} x \theta x^{-(\theta+1)} \, dx = \theta \int_0^{\infty} x^{-\theta} \, dx.$$

Hence, if $\theta \leq 1$, $E(X) = \infty$; if $\theta > 1$, then

$$E(X) = \frac{\theta}{\theta - 1}. \qquad \square$$

Example 1.29 (Cauchy distribution). Let X denote a real-valued random variable with an absolutely continuous distribution with density function

$$p(x) = \frac{1}{\pi(1 + x^2)}, \quad -\infty < x < \infty;$$

this is a *standard Cauchy distribution*. If $E(X)$ exists, it must be equal to

$$\int_{-\infty}^{\infty} \frac{x}{\pi(1 + x^2)} \, dx = \int_{0}^{\infty} \frac{x}{\pi(1 + x^2)} \, dx - \int_{0}^{\infty} \frac{x}{\pi(1 + x^2)} \, dx.$$

Since

$$\int_{0}^{\infty} \frac{x}{\pi(1 + x^2)} \, dx = \infty,$$

it follows that $E(X)$ does not exist. $\square$

Now suppose that X is a random vector $X = (X_1, \ldots, X_d)$, where $X_j, j = 1, \ldots, d$, is real-valued. Then $E(X)$ is simply the vector $(E(X_1), \ldots, E(X_d))$.

Example 1.30 (Uniform distribution on the unit cube). Let $X = (X_1, X_2, X_3)$ denote a three-dimensional random vector with the uniform distribution on $(0, 1)^3$; see Examples 1.7 and 1.25. Then X has an absolutely continuous distribution with density function

$$p(x) = 1, \quad x \in (0, 1)^3.$$

It follows that

$$E(X_1) = \int_{0}^{1} x \, dx = \frac{1}{2}.$$

Similarly, $E(X_2) = E(X_3) = 1/2$. It follows that $E(X) = (1/2, 1/2, 1/2)$. $\square$

Expectation of a function of a random variable

Let X denote a random variable, possibly vector-valued, and let g denote a real-valued function defined on $\mathcal{X}$, the range of X. Let $Y = g(X)$ and let H denote the distribution function of Y. Then

$$E(Y) = \int_{-\infty}^{\infty} y \, dH(y)$$

provided that the integral exists. An important result is that we may also compute $E(Y)$ by

$$E(Y) = \int_{\mathcal{X}} g(x) \, dF(x) \tag{1.3}$$

so that the probability distribution of Y is not needed to compute its expected value.

When X has a discrete distribution, with frequency function p, proof of this result is straightforward. Let f denote the frequency function of Y. Then

$$E(Y) = \sum_{y} y f(y).$$

Note that

$$f(y) = \Pr(Y = y) = \Pr(X \in \{x: g(x) = y\}) = \sum_{x: g(x) = y} p(x).$$

Hence,

$$E(Y) = \sum_y \sum_{x:g(x)=y} yp(x) = \sum_y \sum_{x:g(x)=y} g(x)p(x).$$

Since every x value in the range of X leads to some value y in the range of Y, it follows that

$$E(Y) = \sum_x g(x)p(x).$$

In the general case, the result is simply the change-of-variable formula for integration, as discussed in Appendix 1; see, for example, Billingsley (1995, Theorem 16.13) for a proof. The result (1.3) is usually expressed without reference to the random variable Y:

$$E[g(X)] = \int_{\mathcal{X}} g(x)\,dF(x)$$

provided that the integral exists. Note that it causes no problem if $g(x)$ is undefined for $x \in A$ for some set A such that $\Pr(X \in A) = 0$; this set can simply be omitted when computing the expected value.

Example 1.31 (Standard exponential distribution). Let X denote a random variable with a standard exponential distribution; see Example 1.16. Then X has density function

$$p(x) = \exp(-x), \quad 0 < x < \infty.$$

Consider the expected value of X^r where $r > 0$ is a constant. If $E(X^r)$ exists, it is given by

$$\int_0^\infty x^r \exp(-x)\,dx,$$

which is simply the well-known gamma function evaluated at $r + 1$, $\Gamma(r + 1)$; the gamma function is discussed in detail in Section 10.2. □

Expected values of the form $E(X^r)$ for $r = 1, 2, \ldots$ are called the *moments* of the distribution or the moments of X. Thus, the moments of the standard exponential distribution are $r!$, $r = 1, 2, \ldots$. Moments will be discussed in detail in Chapter 4.

Example 1.32 (Uniform distribution on the unit cube). Let X denote a three-dimensional random vector with the uniform distribution on $(0, 1)^3$; see Example 1.30. Let $Y = X_1 X_2 X_3$, where $X = (X_1, X_2, X_3)$. Then

$$E(Y) = \int_0^1 \int_0^1 \int_0^1 x_1 x_2 x_3 \, dx_1 \, dx_2 \, dx_3 = \frac{1}{8}. \qquad \square$$

Given the correspondence between $E[g(X)]$ and integrals of the form

$$\int_{\mathcal{X}} g(x)\,dF(x),$$

many important properties of expectation may be derived directly from the corresponding properties of integrals, given in Appendix 1. Theorem 1.10 contains a number of these; the proof follows immediately from the results in Appendix 1 and, hence, it is omitted.

Theorem 1.10. *Let X denote a random variable with range* $\mathcal{X}$.

    (i) *If g is a nonnegative real-valued function on* $\mathcal{X}$, *then* $E[g(X)] \geq 0$ *and* $E[g(X)] = 0$ *if and only if* $\Pr[g(X) = 0] = 1$.

   (ii) *If g is the constant function identically equal to c then* $E[g(X)] = c$.

  (iii) *If* $g_1, g_2, \ldots, g_m$ *are real-valued functions on* $\mathcal{X}$ *such that* $E[|g_j(X)|] < \infty$, $j = 1, \ldots, m$, *then*

$$E[g_1(X) + \cdots + g_m(X)] = E[g_1(X)] + \cdots + E[g_m(X)].$$

  (iv) *Let* $g_1, g_2, \ldots$ *denote an increasing sequence of nonnegative, real-valued functions on* $\mathcal{X}$ *with limit g. Then*

$$\lim_{n \to \infty} E[g_n(X)] = E[g(X)].$$

   (v) *Let* $g_1, g_2, \ldots$ *denote a sequence of nonnegative, real-valued functions on* $\mathcal{X}$. *Then*

$$E[\liminf_{n \to \infty} g_n(X)] \leq \liminf_{n \to \infty} E[g_n(X)].$$

  (vi) *Let* $g_1, g_2, \ldots$ *denote a sequence of real-valued functions on* $\mathcal{X}$. *Suppose there exist real-valued functions g and G, defined on* $\mathcal{X}$, *such that, with probability 1,*

$$|g_n(X)| \leq G(X), \quad n = 1, 2, \ldots,$$

and

$$\lim_{n \to \infty} g_n(X) = g(X).$$

If $E[G(X)] < \infty$, *then*

$$\lim_{n \to \infty} E[g_n(X)] = E[g(X)].$$

An important property of expectation is that the expectation operator $E(\cdot)$ completely defines a probability distribution. A formal statement of this fact is given in the following theorem.

Theorem 1.11. *Let X and Y denote random variables.*

$$E[g(X)] = E[g(Y)]$$

for all bounded, continuous, real-valued functions g, if and only if X and Y have the same probability distribution.

Proof. If X and Y have the same distribution, then clearly $E[g(X)] = E[g(Y)]$ for all functions g for which the expectations exist. Since these expectations exist for bounded g, the first part of the result follows.

Now suppose that $E[g(X)] = E[g(Y)]$ for all bounded continuous g. We will show that in this case X and Y have the same distribution. Note that we may assume that X and Y have the same range, neglecting sets with probability 0, for if they do not, it is easy to construct a function g for which the expected values differ.

The proof is based on the following idea. Note that the distribution function of a random variable X can be written as the expected value of an indicator function:

$$\Pr(X \leq z) = E[I_{\{X \leq z\}}].$$

The function $g(x) = I_{\{x \leq z\}}$ is bounded, but it is not continuous; however, it can be approximated by a bounded, continuous function to arbitrary accuracy.

First suppose that X and Y are real-valued random variables. Fix a real number z and, for $\epsilon > 0$, define

$$g_\epsilon(t) \equiv g_\epsilon(t; z) = \begin{cases} 1 & \text{if } t \leq z \\ 1 - (t - z)/\epsilon & \text{if } z < t < z + \epsilon \; ; \\ 0 & \text{if } t \geq z + \epsilon \end{cases}$$

clearly g_ϵ is bounded and continuous and

$$E[g_\epsilon(X)] = F_X(z) + \int_z^{z+\epsilon} [1 - (x - z)/\epsilon] \, dF_X(x)$$

where F_X denotes the distribution function of X. Using integration-by-parts,

$$E[g_\epsilon(X)] = \frac{1}{\epsilon} \int_z^{z+\epsilon} F_X(x) \, dx.$$

Hence, for all $\epsilon > 0$,

$$\frac{1}{\epsilon} \int_z^{z+\epsilon} F_X(x) \, dx = \frac{1}{\epsilon} \int_z^{z+\epsilon} F_Y(y) \, dy$$

or, equivalently,

$$F_X(z) - F_Y(z) = \frac{1}{\epsilon} \int_z^{z+\epsilon} [F_X(x) - F_X(z)] \, dx - \frac{1}{\epsilon} \int_z^{z+\epsilon} [F_Y(y) - F_Y(z)] \, dy.$$

Since F_X and F_Y are non-decreasing,

$$|F_X(z) - F_Y(z)| = \frac{1}{\epsilon} \int_z^{z+\epsilon} [F_X(x) - F_X(z)] \, dx + \frac{1}{\epsilon} \int_z^{z+\epsilon} [F_Y(y) - F_Y(z)] \, dy,$$

and, hence, for all $\epsilon > 0$,

$$|F_X(z) - F_Y(z)| \leq [F_X(z + \epsilon) - F_X(z)] + [F_Y(z + \epsilon) - F_Y(z)].$$

Since F_X and F_Y are right-continuous, it follows that $F_X(z) = F_Y(z)$; since z is arbitrary, it follows that $F_X = F_Y$ and, hence, that X and Y have the same distribution.

The proof for the case in which X and Y are vectors is very similar. Suppose X and Y take values in a subset of $\mathbf{R}^d$. For a given value of $z \in \mathbf{R}^d$, let

$$A = (-\infty, z_1] \times \cdots \times (-\infty, z_p]$$

and $\rho(t)$ be the Euclidean distance from t to A. For $\epsilon > 0$, define

$$g_\epsilon(t) \equiv g_\epsilon(t; z) = \begin{cases} 1 & \text{if } \rho(t) = 0 \\ 1 - \rho(t)/\epsilon & \text{if } 0 < \rho(t) < \epsilon \; ; \\ 0 & \text{if } \rho(t) \geq \epsilon \end{cases}$$

clearly g is bounded; since ρ is a continuous function on $\mathbf{R}^p$, g is continuous as well.

Let $X_0 = \rho(X)$ and $Y_0 = \rho(Y)$; then X_0 and Y_0 are real-valued random variables, with distribution functions F_{X_0} and F_{Y_0}, respectively. Note that

$$E[g_\epsilon(X)] = F_X(z) + \int_0^\epsilon [1 - x/\epsilon] \, dF_{X_0}(x)$$

so that

$$F_X(z) - F_Y(z) = \int_0^\epsilon [1 - x/\epsilon] \, dF_{X_0}(x) - \int_0^\epsilon [1 - y/\epsilon] \, dF_{Y_0}(y).$$

Using integration-by-parts,

$$F_X(z) - F_Y(z) = \frac{1}{\epsilon} \int_0^\epsilon [F_{X_0}(x) - F_{X_0}(0)] \, dx + \frac{1}{\epsilon} \int_0^\epsilon [F_{Y_0}(y) - F_{Y_0}(0)] \, dy.$$

The result now follows as in the scalar random variable case. ∎

Inequalities

The following theorems give some useful inequalities regarding expectations; the proofs of Theorems 1.12–1.14 are left as exercises.

Theorem 1.12 (Cauchy-Schwarz inequality). *Let X denote a random variable with range $\mathcal{X}$ and let g_1, g_2 denote real-valued functions on $\mathcal{X}$. Then*

$$E[|g_1(X)g_2(X)|]^2 \le E[g_1(X)^2]E[g_2(X)^2]$$

with equality if and only if either $E[g_j(X)^2] = 0$ for some $j = 1, 2$ or

$$\Pr[g_1(X) = cg_2(X)] = 1$$

for some real-valued nonzero constant c.

Theorem 1.13 (Jensen's inequality). *Let X denote a real-valued random variable with range $\mathcal{X}$ and let g denote a real-valued convex function defined on some interval containing $\mathcal{X}$ such that $E(|X|) < \infty$ and $E[|g(X)|] < \infty$. Then*

$$g[E(X)] \le E[g(X)].$$

Theorem 1.14 (Markov's inequality). *Let X be a nonnegative, real-valued random variable. Then, for all $a > 0$,*

$$\Pr(X \ge a) \le \frac{1}{a} E(X).$$

Theorem 1.15 (Hölder inequality). *Let X denote a random variable with range $\mathcal{X}$ and let g_1, g_2 denote real-valued functions on $\mathcal{X}$. Let $p > 1$ and $q > 1$ denote real numbers such that $1/p + 1/q = 1$. Then*

$$E(|g_1(X)g_2(X)|) \le E(|g_1(X)|^p)^{\frac{1}{p}} E(|g_2(X)|^q)^{\frac{1}{q}}.$$

The proof is based on the following lemma.

Lemma 1.1. *Let a, b, α, β denote positive real numbers such that $\alpha + \beta = 1$.*

(i) $a^\alpha b^\beta \le \alpha a + \beta b$.

(ii) If $p > 1$ and $q > 1$ satisfy $1/p + 1/q = 1$, then

$$ab \le \frac{a^p}{p} + \frac{b^q}{q}.$$

Proof.   Consider the function $f(x) = -\log(x)$, $x > 0$. Then

$$f''(x) = \frac{1}{x^2} > 0, \quad x > 0$$

so that f is convex. It follows that

$$f(\alpha a + \beta b) \leq \alpha f(a) + \beta f(b);$$

that is,

$$-\log(\alpha a + \beta b) \leq -[\alpha \log(a) + \beta \log(b)].$$

Taking the exponential function of both sides of this inequality yields

$$\alpha a + \beta b \geq a^\alpha b^\beta,$$

proving part (i).

Consider part (ii). Let $\alpha = 1/p$ and $\beta = 1/q$. Then, by part (i) of the theorem applied to a^p and b^q,

$$ab = a^{p\alpha} b^{q\beta} \leq \alpha a^p + \beta b^q,$$

proving part (ii).   ∎

Proof of Theorem 1.15.   The result is clearly true if $E(|g_1(X)|^p) = \infty$ or $E(|g_2(X)|^q) = \infty$ so we may assume that $E(|g_1(X)|^p) < \infty$ and $E(|g_2(X)|^q) < \infty$.

If $E(|g_1(X)|^p) = 0$ then $|g_1(X)| = 0$ with probability 1, so that $E(|g_1(X)g_2(X)|) = 0$ and the result holds; similarly, the result holds if $E(|g_1(X)|^q) = 0$. Hence, assume that $E(|g_1(X)|^p) > 0$ and $E(|g_1(X)|^q) > 0$.

Applying part (ii) of Lemma 1.1 to

$$\frac{|g_1(X)|}{E(|g_1(X)|^p)^{\frac{1}{p}}} \quad \text{and} \quad \frac{|g_2(X)|}{E(|g_2(X)|^q)^{\frac{1}{q}}},$$

it follows that

$$\frac{|g_1(X)||g_2(X)|}{E(|g_1(X)|^p)^{\frac{1}{p}} E(|g_2(X)|^q)^{\frac{1}{q}}} \leq \frac{|g_1(X)|^p}{pE(|g_1(X)|^p)} + \frac{|g_2(X)|^q}{qE(|g_2(X)|^q)}.$$

Taking expectations of both sides of this inequality shows that

$$\frac{E(|g_1(X)g_2(X)|)}{E(|g_1(X)|^p)^{\frac{1}{p}} E(|g_2(X)|^q)^{\frac{1}{q}}} \leq \frac{1}{p} + \frac{1}{q} = 1;$$

the result follows.   ∎

1.9   Exercises

In problems 1 through 6, let Ω and P denote the sample space and probability function, respectively, of an experiment and let A_1, A_2, and A_3 denote events.

1.1 Show that

$$P(A_1 \cup A_2) = P(A_1) + P(A_2) - P(A_1 \cap A_2).$$

1.2 Let $A_1 \setminus A_2$ denote the elements of A_1 that are not in A_2.

(a) Suppose $A_2 \subset A_1$. Show that

$$P(A_1 \setminus A_2) = P(A_1) - P(A_2).$$

(b) Suppose that A_2 is not necessarily a subset of A_1. Does

$$P(A_1 \setminus A_2) = P(A_1) - P(A_2)$$

still hold?

1.3 Let $A_1 \triangle A_2$ denote the *symmetric difference* of A_1 and A_2, given by

$$A_1 \triangle A_2 = (A_1 \setminus A_2) \cup (A_2 \setminus A_1).$$

Give an expression for $P(A_1 \triangle A_2)$ in terms of $P(A_1)$, $P(A_2)$, and $P(A_1 \cap A_2)$.

1.4 Show that

$$P(A_1) \le P(A_1 \cap A_2) + P(A_2^c).$$

1.5 Show that

(a) $P(A_1 \cup A_2) \le P(A_1) + P(A_2)$

(b) $P(A_1 \cap A_2) \ge P(A_1) + P(A_2) - 1$.

1.6 Find an expression for $\Pr(A_1 \cup A_2 \cup A_3)$ in terms of the probabilities of A_1, A_2, and A_3 and intersections of these sets.

1.7 Show that (P3) implies (P2).

1.8 Let Ω and P denote the sample space and probability function, respectively, of an experiment and let $A_1, A_2, \ldots$ denote events. Show that

$$\Pr\left(\bigcup_{n=1}^{\infty} A_n\right) \le \sum_{n=1}^{\infty} P(A_n).$$

1.9 Consider an experiment with sample space $\Omega = [0, 1]$. Let $D(\cdot)$ denote a function defined as follows: for a given subset of Ω, A,

$$D(A) = \sup_{s,t \in A} |s - t|.$$

Is D a probability function on Ω?

1.10 Let X denote a real-valued random variable with distribution function F. Call x a *support point* of the distribution if

$$F(x + \epsilon) - F(x - \epsilon) > 0 \quad \text{for all } \epsilon > 0.$$

Let $\mathcal{X}_0$ denote the set of all support points of the distribution. Show that $\mathcal{X}_0$ is identical to the minimal support of the distribution, as defined in this chapter.

1.11 Prove Corollary 1.1.

1.12 Let X_1 and X_2 denote real-valued random variables with distribution functions F_1 and F_2, respectively. Show that, if

$$F_1(b) - F_1(a) = F_2(b) - F_2(a)$$

for all $-\infty < a < b < \infty$, then X_1 and X_2 have the same distribution.

1.13 Let X denote a real-valued random variable with distribution function F such that $F(x) = 0$ for $x \le 0$. Let

$$f(x) = \begin{cases} 1/x & \text{if } x > 0 \\ 0 & \text{otherwise} \end{cases}$$

and let $Y = f(X)$. Find the distribution function of Y in terms of F.

1.14 Let F_1 and F_2 denote distribution functions on **R**. Which of the following functions is a distribution function?

    (a) $F(x) = \alpha F_1(x) + (1 - \alpha)F_2(x)$, $x \in \mathbf{R}$, where α is a given constant, $0 \le \alpha \le 1$

    (b) $F(x) = F_1(x)F_2(x)$, $x \in \mathbf{R}$

    (c) $F(x) = 1 - F_1(-x)$, $x \in \mathbf{R}$.

1.15 Let X denote a real-valued random variable with an absolutely continuous distribution with density function

$$p(x) = \frac{2x}{(1 + x^2)^2}, \quad 0 < x < \infty.$$

Find the distribution function of X.

1.16 Let X denote a real-valued random variable with a discrete distribution with frequency function

$$p(x) = \frac{1}{2^x}, \quad x = 1, 2, \dots.$$

Find the distribution function of X.

1.17 Let $X_1, X_2, \dots, X_n$ denote independent, identically distributed random variables and let F denote the distribution function of X_1. Define

$$\hat{F}(t) = \frac{1}{n} \sum_{j=1}^{n} I_{\{X_j \le t\}}, \quad -\infty < t < \infty.$$

Hence, this is a random function on **R**. For example, if Ω denotes the sample space of the underlying experiment, then, for each $t \in \mathbf{R}$,

$$\hat{F}(t)(\omega) = \frac{1}{n} \sum_{j=1}^{n} I_{\{X_j(\omega) \le t\}}, \quad \omega \in \Omega.$$

Show that $\hat{F}(\cdot)$ is a genuine distribution function. That is, for each $\omega \in \Omega$, show that $\hat{F}(\cdot)(\omega)$ satisfies (DF1)–(DF3).

1.18 Let X denote a nonnegative, real-valued random variable with an absolutely continuous distribution. Let F denote the distribution function of X and let p denote the corresponding density function. The *hazard function* of the distribution is given by

$$H(x) = \frac{p(x)}{1 - F(x)}, \quad x > 0.$$

    (a) Give an expression for F in terms of H.

    (b) Find the distribution function corresponding to $H(x) = \lambda_0$ and $H(x) = \lambda_0 + \lambda_1 x$, where λ_0 and λ_1 are constants.

1.19 Let X denote a real-valued random variable with a Pareto distribution, as described in Example 1.28. Find the quantile function of X.

1.20 Let X denote a real-valued random variable and let Q denote the quantile function of X. Show that $Q(.5)$ is a median of X.

1.21 Let X_1 and X_2 denote real-valued random variables such that, for $j = 1, 2$, the distribution of X_j has a unique median m_j. Suppose that $\Pr(X_1 > X_2) > 1/2$. Does it follow that $m_1 \ge m_2$?

1.22 Let F_1 and F_2 denote distribution functions for absolutely continuous distributions on the real line and let p_1 and p_2 denote the corresponding density functions. Which of the following functions is a density function?

    (a) $\alpha p_1(\alpha x)$ where $\alpha > 0$

    (b) $p_1^\alpha p_2^{1-\alpha}$ where $0 \le \alpha \le 1$

    (c) $\alpha p_1 + (1 - \alpha)p_2$ where $0 \le \alpha \le 1$

1.23 Prove the Cauchy-Schwarz inequality.

Hint: $E\{[g_1(X) - tg_2(X)]^2\} \geq 0$ for all t. Find the minimum of this expression.

1.24 Prove Jensen's inequality.

Hint: For a convex function g, show that there exists a constant c such that

$$g(x) \geq g(E(X)) + c(x - E(X))$$

for all $x \in \mathbf{R}$.

1.25 Prove Markov's inequality.

1.26 Let X denote a real-valued random variable with distribution function F; assume that $E(|X|) < \infty$. Show that

$$E(X) = \int_0^\infty (1 - F(x)) \, dx - \int_{-\infty}^0 F(x) \, dx$$

$$= \int_0^\infty [1 - F(x) - F(-x)] \, dx.$$

1.27 Let X denote a real-valued random variable with distribution function F. Find the distribution functions of $|X|$ and X^+, where

$$X^+ = \begin{cases} X & \text{if } X > 0 \\ 0 & \text{otherwise} \end{cases}.$$

1.28 Let $\mathcal{L}_2$ denote the set of real-valued random variables X satisfying $E(X^2) < \infty$. Show that $\mathcal{L}_2$ is a linear space: if X_1 and X_2 are elements of $\mathcal{L}_2$ and a and b are scalar constants, then $aX_1 + bX_2 \in \mathcal{L}_2$.

1.29 Consider the space of random variables $\mathcal{L}_2$ described in Exercise 1.28. Let 0 denote the random variable identically equal to 0 and, for $X \in \mathcal{L}_2$, write $X = 0$ if $\Pr(X = 0) = 1$. Define a function $|| \cdot ||$ on $\mathcal{L}_2$ as follows: for $X \in \mathcal{L}_2$, $||X||^2 = E(X^2)$. Show that $|| \cdot ||$ defines a *norm* on $\mathcal{L}_2$: for all X_1 and X_2 in $\mathcal{L}_2$ and all scalar constants a,

(a) $||X_1|| \geq 0$ and $||X_1|| = 0$ if and only if $X_1 = 0$

(b) $||X_1 + X_2|| \leq ||X_1|| + ||X_2||$

(c) $||aX|| = |a| \, ||X||$

1.30 Let X denote a real-valued random variable and suppose that the distribution of X is symmetric about 0; that is, suppose that X and $-X$ have the same distribution. Show that, for $r = 1, 3, \ldots,$ $E(X^r) = 0$ provided that $E(X^r)$ exists.

1.31 Let X be real-valued random variable with a discrete distribution with frequency function

$$p(x) = \lambda^x \exp(-\lambda)/x!, \quad x = 0, 1, 2, \ldots$$

where $\lambda > 0$; this is a *Poisson distribution* with parameter λ. Find $E(X)$.

1.32 Let X denote a real-valued random variable with an absolutely continuous distribution with density $\alpha x^{\alpha-1}, 0 < x < 1$. Find $E[X^r]$.

1.33 Let X denote a real-valued random variable with quantile function Q and assume that $E(|X|) < \infty$. Show that

$$E(X) = \int_0^1 Q(t) \, dt. \tag{1.3}$$

Let g denote a function defined on the range of X such that $E[|g(X)|] < \infty$. Find an expression for $E[g(X)]$ similar to (1.4).

1.34 Let X denote a real-valued, non-negative random variable with quantile function Q; assume that $E(X) < \infty$. Fix $0 < p < 1$ and let $x_p = Q(p)$ denote the pth quantile of the distribution. Show that

$$x_p \leq \frac{1}{1-p} E(X).$$

1.35 Let X denote a real-valued random variable with an absolutely continuous distribution with density function

$$p(x) = \frac{1}{\sqrt{(2\pi)}} \exp\left(-\frac{1}{2}x^2\right), \quad -\infty < x < \infty.$$

Let $g : \mathbf{R} \to \mathbf{R}$ denote a differentiable function such that $E[|g'(X)|] < \infty$. Show that

$$E[g'(X)] = E[Xg(X)].$$

1.10   Suggestions for Further Reading

The topics covered in this chapter are standard topics in probability theory and are covered in many books on probability and statistics. See, for example, Ash (1972), Billingsley (1995), Karr (1993), and Port (1994) for rigorous discussion of these topics. Capinski and Kopp (2004) has a particularly accessible, yet rigorous, treatment of measure theory. Casella and Berger (2002), Ross (1995), Snell (1988), and Woodroofe (1975) contain good introductory treatments. Theorem 1.11 is based on Theorem 1.2 of Billingsley (1968).

2

Conditional Distributions and Expectation

2.1 Introduction

Consider an experiment with sample space Ω and let P denote a probability function on Ω so that a given event $A \subset \Omega$ has a probability $P(A)$. Now suppose we are told that a certain event B has occurred. This information affects our probabilities for all other events since now we should only consider those sample points ω that are in B; hence, the probability $P(A)$ must be updated to the conditional probability $P(A|B)$. From elementary probability theory, we know that

$$P(A|B) = \frac{P(A \cap B)}{P(B)},$$

provided that $P(B) > 0$.

In a similar manner, we can consider conditional probabilities based on random variables. Let (X, Y) denote a random vector. Then the conditional probability that $X \in A$ given $Y \in B$ is given by

$$Pr(X \in A|Y \in B) = \frac{Pr(X \in A \cap Y \in B)}{Pr(Y \in B)}$$

provided that $Pr(Y \in B) > 0$.

In this chapter, we extend these ideas in order to define the conditional distribution and conditional expectation of one random variable given another. Conditioning of this type represents the introduction of additional information into a probability model and, thus, plays a central role in many areas of statistics, including estimation theory, prediction, and the analysis of models for dependent data.

2.2 Marginal Distributions and Independence

Consider a random vector of the form (X, Y), where each of X and Y may be a vector and suppose that the range of (X, Y) is of the form $\mathcal{X} \times \mathcal{Y}$ so that $X \in \mathcal{X}$ and $Y \in \mathcal{Y}$. The probability distribution of X when considered alone, called the *marginal distribution of X*, is given by

$$Pr(X \in A) = Pr(X \in A, Y \in \mathcal{Y}), \quad A \subset \mathcal{X}.$$

Let F denote the distribution function of (X, Y). Then

$$Pr(X \in A) = \int_{A \times \mathcal{Y}} dF(x, y).$$

Let F_X denote the distribution function of the marginal distribution of X. Clearly, F_X and F are related. For instance, for any $A \subset \mathcal{X}$,

$$\Pr(X \in A) = \int_A dF_X(x).$$

Hence, F_X must satisfy

$$\int_A dF_X(x) = \int_{A \times \mathcal{Y}} dF(x, y)$$

for all $A \subset \mathcal{X}$.

The cases in which (X, Y) has either an absolutely continuous or a discrete distribution are particularly easy to handle.

Lemma 2.1. *Consider a random vector (X, Y), where X is d-dimensional and Y is q-dimensional and let $\mathcal{X} \times \mathcal{Y}$ denote the range of (X, Y).*

(i) *Let $F(x_1, \ldots, x_d, y_1, \ldots, y_q)$ denote the distribution function of (X, Y). Then X has distribution function*

$$F_X(x_1, \ldots, x_d) = F(x_1, \ldots, x_d, \infty, \ldots, \infty)$$
$$\equiv \lim_{y_1 \to \infty} \cdots \lim_{y_q \to \infty} F(x_1, \ldots, x_d, y_1, \ldots, y_q).$$

(ii) *If (X, Y) has an absolutely continuous distribution with density function $p(x, y)$ then the marginal distribution of X is absolutely continuous with density function*

$$p_X(x) = \int_{\mathbf{R}^q} p(x, y) dy, x \in \mathbf{R}^d$$

(iii) *If (X, Y) has a discrete distribution with frequency function $p(x, y)$, then the marginal distribution of X is discrete with frequency function of X given by*

$$p_X(x) = \sum_{y \in \mathcal{Y}} p(x, y), x \in \mathcal{X}.$$

Proof.   Part (i) follows from the fact that

$$\Pr(X_1 \leq x_1, \ldots, X_d \leq x_d) = \Pr(X_1 \leq x_1, \ldots, X_d \leq x_d, Y_1 < \infty, \ldots, Y_q < \infty).$$

Let A be a subset of the range of X. Then, by Fubini's Theorem (see Appendix 1),

$$\Pr(X \in A) = \int_A \int_{\mathbf{R}^q} p(x, y) \, dy \, dx \equiv \int_A p_X(x) \, dx,$$

proving part (ii). Part (iii) follows in a similar manner. ∎

Example 2.1 (Bivariate distribution). Suppose that X and Y are both real-valued and that (X, Y) has an absolutely continuous distribution with density function

$$p(x, y) = 6(1 - x - y), \quad x > 0, \ y > 0, \ x + y < 1.$$

Then the marginal density function of X is given by

$$p_X(x) = 6 \int_0^\infty (1 - x - y)\mathrm{I}_{\{x+y<1\}}\, dy$$

$$= 6 \int_0^{1-x} (1 - x - y)\, dy = 3(1 - x)^2, \quad 0 < x < 1.$$

Since $p(x, y)$ is symmetric in x and y, the marginal density of Y has the same form. $\quad\square$

Example 2.2 (Multinomial distribution). Let $X = (X_1, \ldots, X_m)$ denote a random vector with a discrete distribution with frequency function

$$p(x_1, \ldots, x_m) = \binom{n}{x_1, x_2, \ldots, x_m} \theta_1^{x_1} \theta_2^{x_2} \cdots \theta_m^{x_m},$$

for $x_j = 0, 1, \ldots, n$, $j = 1, 2, \ldots, m$, $\sum_{j=1}^m x_j = n$; here $\theta_1, \ldots, \theta_m$ are nonnegative constants satisfying $\theta_1 + \cdots + \theta_m = 1$. This is called the *multinomial distribution* with parameters n and $(\theta_1, \ldots, \theta_m)$. Note that

$$\binom{n}{x_1, x_2, \ldots, x_m} = \frac{n!}{x_1! x_2! \cdots x_m!}.$$

Consider the marginal distribution of X_1. This distribution has frequency function

$$p_{X_1}(x_1)$$

$$= \sum_{(x_2,\ldots,x_m):\sum_{j=2}^m x_j = n - x_1}^{n} p(x_1, \ldots, x_{m-1}, j)$$

$$= \binom{n}{x_1} \theta_1^{x_1} \sum_{(x_2,\ldots,x_m):\sum_{j=2}^m x_j = n - x_1}^{n} \binom{n - x_1}{x_2, \ldots, x_m} \theta_2^{x_2} \cdots \theta_m^{x_m}$$

$$= \binom{n}{x_1} \theta_1^{x_1}(1 - \theta_1)^{n - x_1} \sum_{(x_2,\ldots,x_m):\sum_{j=2}^m x_j = n - x_1}^{n} \binom{n - x_1}{x_2, \ldots, x_m} \left(\frac{\theta_2}{1 - \theta_1}\right)^{x_2} \cdots \left(\frac{\theta_m}{1 - \theta_1}\right)^{x_m}$$

$$= \binom{n}{x_1} \theta_1^{x_1}(1 - \theta_1)^{n - x_1}.$$

Hence, the marginal distribution of X_1 is a binomial distribution with parameters n and θ_1. $\quad\square$

Example 2.3 (A distribution that is neither discrete nor absolutely continuous). Let (X, Y) denote a two-dimensional random vector with range $(0, \infty) \times \{1, 2\}$ such that, for any set $A \subset (0, \infty)$ and $y = 1, 2$,

$$\Pr(X \in A, \ Y = y) = \frac{1}{2} \int_A y \exp(-yx)\, dx.$$

Thus, the distribution of (X, Y) is neither discrete nor absolutely continuous.

The marginal distribution of X has distribution function

$$F_X(x) = \Pr(X \le x) = \Pr(X \le x, \ Y = 1) + \Pr(X \le x, \ Y = 2)$$

$$= 1 - \frac{1}{2}[\exp(-x) + \exp(-2x)], \quad 0 < x < \infty.$$

This is an absolutely continuous distribution with density function

$$\frac{1}{2}[\exp(-x) + 2\exp(-2x)].$$

Since, for $y = 1, 2$,

$$\Pr(Y = y) = \frac{1}{2}\int_0^\infty y\exp(-yx)\,dx = \frac{1}{2},$$

it follows that Y has a discrete distribution with frequency function $1/2$, $y = 1, 2$. $\square$

Independence

Consider a random vector (X, Y) with range $\mathcal{X} \times \mathcal{Y}$. We say X and Y are independent if for any $A \subset \mathcal{X}$ and $B \subset \mathcal{Y}$, the events $X \in A$ and $Y \in B$ are independent events in the usual sense of elementary probability theory; that is, if

$$\Pr(X \in A, Y \in B) = \Pr(X \in A)\Pr(Y \in B).$$

Independence may easily be characterized in terms of either distribution functions or expected values.

Theorem 2.1. *Let (X, Y) denote a random vector with range $\mathcal{X} \times \mathcal{Y}$ and distribution function F. Let F_X and F_Y denote the marginal distribution functions of X and Y, respectively.*

(i) X and Y are independent if and only if for all x, y

$$F(x, y) = F_X(x)F_Y(y).$$

(ii) X and Y are independent if and only if for all bounded functions $g_1 : \mathcal{X} \to \mathbf{R}$ and $g_2 : \mathcal{Y} \to \mathbf{R}$

$$\mathrm{E}[g_1(X)g_2(Y)] = \mathrm{E}[g_1(X)]\mathrm{E}[g_2(Y)].$$

Proof. Suppose X and Y are independent. Let m denote the dimension of X and let n denote the dimension of Y. Fix $x = (x_1, \ldots, x_m)$ and $y = (y_1, \ldots, y_n)$; let

$$A = (-\infty, x_1] \times \cdots \times (-\infty, x_m]$$

and

$$B = (-\infty, y_1] \times \cdots \times (-\infty, y_n]$$

so that

$$F(x, y) = \Pr(X \in A, Y \in B), \quad F_X(x) = \Pr(X \in A), \quad \text{and} \quad F_Y(y) = \Pr(Y \in B).$$

Then

$$F(x, y) = \Pr(X \in A, Y \in B) = \Pr(X \in A)\Pr(Y \in B) = F_X(x)F_Y(y).$$

Now suppose $F(x, y) = F_X(x)F_Y(y)$. Since $F_X(x)F_Y(y)$ is the distribution function of a random variable (X_1, Y_1) such that X_1 and Y_1 are independent with marginal distribution functions F_X and F_Y, respectively, it follows that (X, Y) has the same distribution as (X_1, Y_1); that is, X and Y are independent. This proves part (i).

If X and Y are independent,

$$E[g_1(X)g_2(Y)] = \int_{\mathcal{X} \times \mathcal{Y}} g_1(x)g_2(y)\, dF(x,y) = \int_{\mathcal{X} \times \mathcal{Y}} g_1(x)g_2(y)\, dF_X(x)\, dF_Y(y)$$

$$= E[g_1(X)]E[g_2(Y)].$$

Conversely, suppose that $E[g_1(X)g_2(Y)] = E[g_1(X)]E[g_2(Y)]$ for all bounded, real-valued, g_1, g_2. Let d_1 denote the dimension of X and let d_2 denote the dimension of Y. Then, for a given $x \in \mathbf{R}^{d_1}$ and a given $y \in \mathbf{R}^{d_2}$, let g_1 denote the indicator function for the set

$$(-\infty, x_1] \times \cdots \times (-\infty, x_{d_1}]$$

and let g_2 denote the indicator function for the set

$$(-\infty, y_1] \times \cdots \times (-\infty, y_{d_2}];$$

here $x = (x_1, \ldots, x_{d_1})$ and $y = (y_1, \ldots, y_{d_2})$. Since $E[g_1(X)g_2(Y)] = E[g_1(X)]E[g_2(Y)]$, it follows that $F(x, y) = F_X(x)F_Y(y)$. Since x and y are arbitrary, X and Y are independent; this proves part (ii).   ∎

For the case in which the distribution of (X, Y) is either absolutely continuous or discrete, it is straightforward to characterize independence in terms of either the density functions or frequency functions of X and Y. The formal result is given in the following corollary to Theorem 2.1; the proof is left as an exercise.

Corollary 2.1.

  (i) *Suppose (X, Y) has an absolutely continuous distribution with density function p and let p_X and p_Y denote the marginal density functions of X and Y, respectively. X and Y are independent if and only if*

$$p(x, y) = p_X(x)\, p_Y(y)$$

    for almost all x, y.

  (ii) *Suppose (X, Y) has a discrete distribution with frequency function p and let p_X and p_Y denote the marginal frequency functions of X and*

$$p(x, y) = p_X(x)p_Y(y)$$

    for all x, y.

***Example* 2.4 (*Bivariate distribution*).** Consider the distribution considered in Example 2.1. The random vector (X, Y) has an absolutely continuous distribution with density function

$$p(x, y) = 6(1 - x - y), \quad x > 0, \quad y > 0, \quad x + y < 1$$

and the marginal density of X is

$$p_X(x) = 3(1 - x)^2, \quad 0 < x < 1;$$

the same argument used to derive p_X may be used to show that the marginal density of Y is also

$$p_Y(y) = 3(1 - y)^2, \quad 0 < y < 1.$$

Clearly, $p \neq p_X p_Y$ so that X and Y are not independent.   □

Independence of a sequence of random variables

Independence of a sequence of random variables may be defined in a similar manner. Consider a sequence of random variables $X_1, X_2, \ldots, X_n$ any of which may be vector-valued, with ranges $\mathcal{X}_1, \mathcal{X}_2, \ldots$, respectively; we may view these random variables as the components of a random vector. We say $X_1, X_2, \ldots, X_n$ are independent if for any sets $A_1, A_2, \ldots, A_n$, $A_j \subset \mathcal{X}_j$, $j = 1, \ldots, n$, the events $X_1 \in A_1, \ldots, X_n \in A_n$ are independent so that

$$\Pr(X_1 \in A_1, \ldots, X_n \in A_n) = \Pr(X_1 \in A_1) \cdots \Pr(X_n \in A_n).$$

Theorem 2.2 gives analogues of Theorem 2.1 and Corollary 2.1 in this setting; the proof is left as an exercise.

Theorem 2.2. *Let $X_1, \ldots, X_n$ denote a sequence of random variables and let F denote the distribution function of $(X_1, \ldots, X_n)$. For each $j = 1, \ldots, n$, let $\mathcal{X}_j$ and F_j denote the range and marginal distribution function, respectively, of X_j.*

(i) *$X_1, \ldots, X_n$ are independent if and only if for all $x_1, \ldots, x_n$ with $x_j \in \mathbf{R}^{d_j}$, $d_j = \dim(\mathcal{X}_j)$,*

$$F(x_1, \ldots, x_n) = F_1(x_1) \cdots F_n(x_n).$$

(ii) *$X_1, X_2, \ldots, X_n$ are independent if and only if for any sequence of bounded, real-valued functions $g_1, g_2, \ldots, g_n$, $g_j : \mathcal{X}_j \to \mathbf{R}$, $j = 1, \ldots, n$,*

$$E[g_1(X_1)g_2(X_2) \cdots g_n(X_n)] = E[g_1(X_1)] \cdots E[g_n(X_n)].$$

(iii) *Suppose $(X_1, \ldots, X_n)$ has an absolutely continuous distribution with density function p. Let p_j denote the marginal density function of X_j, $j = 1, \ldots, n$. Then $X_1, \ldots, X_n$ are independent if and only if*

$$p(x_1, \ldots, x_n) = p_1(x_1) \cdots p_n(x_n)$$

for almost all $x_1, x_2, \ldots, x_n$.

(iv) *Suppose $(X_1, \ldots, X_n)$ has a discrete distribution with frequency function f. Let p_j denote the marginal frequency function of X_j, $j = 1, \ldots, n$. Then $X_1, \ldots, X_n$ are independent if and only if*

$$p(x_1, \ldots, x_n) = p_1(x_1) \cdots p_n(x_n)$$

for all $x_1, x_2, \ldots, x_n$.

Example 2.5 (Uniform distribution on the unit cube). Let $X = (X_1, X_2, X_3)$ denote a three-dimensional random vector with the uniform distribution on $(0, 1)^3$; see Examples 1.7 and 1.25. Then X has an absolutely continuous distribution with density function

$$p(x_1, x_2, x_3) = 1, \quad x_j \in (0, 1), \quad j = 1, 2, 3.$$

The marginal density of X_1 is given by

$$p_1(x_1) = \int_0^1 \int_0^1 dx_2 \, dx_3 = 1, \quad 0 < x_1 < 1.$$

Clearly, X_2 and X_3 have the same marginal density. It follows that X_1, X_2, X_3 are independent. $\square$

***Example* 2.6 (*Multinomial distribution*).** Let $X = (X_1, \ldots, X_m)$ denote a random vector with a multinomial distribution, as in Example 2.2. Then X has frequency function

$$p(x_1, \ldots, x_m) = \binom{n}{x_1, x_2, \ldots, x_m} \theta_1^{x_1} \theta_2^{x_2} \cdots \theta_m^{x_m},$$

for $x_j = 0, 1, \ldots, n$, $j = 1, \ldots, m$, $\sum_{j=1}^{m} x_j = n$; here $\theta_1, \ldots, \theta_m$ are nonnegative constants satisfying $\theta_1 + \cdots + \theta_m = 1$.

According to Example 2.2, for $j = 1, \ldots, m$, X_j has a binomial distribution with parameters n and θ_j so that X_j has frequency function

$$\binom{n}{x_j} \theta_j^{x_j} (1 - \theta_j)^{n - x_j}, \quad x_j = 0, 1, \ldots, n.$$

Suppose there exists a $j = 1, 2, \ldots, m$ such that $0 < \theta_j < 1$; it then follows from part (iv) of Theorem 2.2 that $X_1, X_2, \ldots, X_m$ are not independent. This is most easily seen by noting that $\Pr(X_j = 0) > 0$ for all $j = 1, 2, \ldots, m$, while

$$\Pr(X_1 = X_2 = \cdots = X_m = 0) = 0.$$

If all θ_j are either 0 or 1 then $X_1, \ldots, X_m$ are independent. To see this, suppose that $\theta_1 = 1$ and $\theta_2 = \cdots = \theta_m = 0$. Then, with probability 1, $X_1 = n$ and $X_2 = \cdots = X_m = 0$. Hence,

$$E[g_1(X_1) \cdots g_m(X_m)] = g_1(n) g_2(0) \cdots g_m(0) = E[g_1(X_1)] \cdots E[g_m(X_m)]$$

and independence follows from part (ii) of Theorem 2.2. $\quad\square$

Random variables $X_1, X_2, \ldots, X_n$ are said to be *independent and identically distributed* if, in addition to being independent, each X_j has the same marginal distribution. Thus, in Example 2.5, X_1, X_2, X_3 are independent and identically distributed. The assumption of independent identically distributed random variables is often used in the specification of the distribution of a vector $(X_1, X_2, \ldots, X_n)$.

***Example* 2.7 (*Independent standard exponential random variables*).** Let $X_1, X_2, \ldots,$ X_n denote independent, identically distributed, real-valued random variables such that each X_j has a standard exponential distribution; see Example 1.16. Then the vector $(X_1, \ldots, X_n)$ has an absolutely continuous distribution with density function

$$p(x_1, \ldots, x_n) = \prod_{j=1}^{n} \exp(-x_j) = \exp\left(-\sum_{j=1}^{n} x_j\right), \quad x_j > 0, \ j = 1, \ldots, n. \quad\square$$

It is often necessary to refer to infinite sequences of random variables, particularly in the development of certain large-sample approximations. An important result, beyond the scope of this book, is that such a sequence can be defined in a logically consistent manner. See, for example, Feller (1971, Chapter IV) or Billingsley (1995, Section 36). As might be expected, technical issues, such as measurability of sets, become much more difficult in this setting. An infinite sequence of random variables $X_1, X_2, \ldots$ is said to be independent if each finite subset of $\{X_1, X_2, \ldots\}$ is independent.

2.3   Conditional Distributions

Consider random variables X and Y. Suppose that Y is a discrete random variable taking the values 0 and 1 with probabilities θ and $1 - \theta$, respectively, where $0 < \theta < 1$. From elementary probability theory we know that the conditional probability that $X \in A$ given that $Y = y$ is given by

$$\Pr(X \in A | Y = y) = \frac{\Pr(X \in A, Y = y)}{\Pr(Y = y)}, \tag{2.1}$$

provided that $y = 0, 1$ so that $\Pr(Y = y) > 0$. Hence, for any set A, the conditional probability function $\Pr(X \in A | Y = y)$ satisfies the equation

$$\begin{aligned}
\Pr(X \in A) &= \Pr(X \in A, Y = 0) + \Pr(X \in A, Y = 1) \\
&= \Pr(X \in A | Y = 0)\Pr(Y = 0) + \Pr(X \in A | Y = 1)\Pr(Y = 1) \\
&= \int_{-\infty}^{\infty} \Pr(X \in A | Y = y) \, dF_Y(y).
\end{aligned}$$

Furthermore, for any subset B of $\{0, 1\}$,

$$\begin{aligned}
\Pr(X \in A, \ Y \in B) &= \sum_{y \in B} \Pr(X \in A, \ Y = y) = \sum_{y \in B} \Pr(X \in A | Y = y)\Pr(Y = y) \\
&= \int_B \Pr(X \in A | Y = y) \, dF_Y(y). \tag{2.2}
\end{aligned}$$

Now suppose that Y has an absolutely continuous distribution and consider $\Pr(X \in A | Y = y)$. If the distribution of Y is absolutely continuous, then $\Pr(Y = y) = 0$ for all y so that (2.1) cannot be used as a definition of $\Pr(X \in A | Y = y)$. Instead, we use a definition based on a generalization of (2.2).

Let (X, Y) denote a random vector, where X and Y may each be vectors, and let $\mathcal{X} \times \mathcal{Y}$ denote the range of (X, Y). In general, the conditional distribution of X given $Y = y$ is a function $q(A, y)$, defined for subsets $A \subset \mathcal{X}$ and elements $y \in \mathcal{Y}$ such that for $B \subset \mathcal{Y}$

$$\Pr(X \in A, Y \in B) = \int_B q(A, y) \, dF_Y(y) \tag{2.3}$$

where F_Y denotes the marginal distribution function of Y and such that for each fixed $y \in \mathcal{Y}$, $q(\cdot, y)$ defines a probability distribution on $\mathcal{X}$. The quantity $q(A, y)$ will be denoted by $\Pr(X \in A | Y = y)$.

***Example* 2.8 (*Two-dimensional discrete random variable*).** Let (X, Y) denote a two-dimensional discrete random variable with range

$$\{1, 2, \ldots, m\} \times \{1, 2, \ldots, n\}.$$

For each $i = 1, 2, \ldots, m$ let

$$q_i(y) = \Pr(X = i | Y = y).$$

Then, according to (2.3), $q_1(y), \ldots, q_m(y)$ must satisfy

$$\Pr(X = i, Y = j) = q_i(j)\Pr(Y = j)$$

for each $i = 1, \ldots, m$ and $j = 1, \ldots, n$.

Hence, if $\Pr(Y = j) > 0$, then

$$\Pr(X = i | Y = j) = \frac{\Pr(X = i, Y = j)}{\Pr(Y = j)};$$

if $\Pr(Y = j) = 0$, $\Pr(X = i | Y = j)$ may be taken to have any finite value. □

***Example* 2.9 (*Independent random variables*).** Consider the random vector (X, Y) where X and Y may each be a vector. If X and Y are independent, then, F, the distribution function of (X, Y), may be written

$$F(x, y) = F_X(x) F_Y(y)$$

for all x, y, where F_X denotes the distribution function of the marginal distribution of X and F_Y denotes the distribution function of the marginal distribution of Y.

Hence, the conditional distribution of X given $Y = y$, $q(\cdot, y)$ must satisfy

$$\int_B \int_A dF_X(x) \, dF_Y(y) = \int_B q(A, y) \, dF_Y(y).$$

Clearly, this equation is satisfied by

$P_1(X \in A) = P_0(X \in A | Y = y) \neq$ $\qquad q(A, y) = \int_A dF_X(x)$

so that

$$\Pr(X \in A | Y = y) = \int_A dF_X(x) = \Pr(X \in A). \qquad \text{by independence}$$

Two important issues are the existence and uniqueness of conditional probability distributions. Note that, for fixed A, if B satisfies $\Pr(Y \in B) = 0$, then $\Pr(X \in A, Y \in B) = 0$. The *Radon-Nikodym Theorem* now guarantees the existence of a function $q(A, \cdot)$ satisfying (2.3). Furthermore, it may be shown that this function may be constructed in such a way that $q(\cdot, y)$ defines a probability distribution on $\mathcal{X}$ for each y. Thus, a conditional probability distribution always exists. Formal proofs of these results are quite difficult and are beyond the scope of this book; see, for example, Billingsley (1995, Chapter 6) for a detailed discussion of the technical issues involved.

If, for a given set A, $q_1(A, \cdot)$ and $q_2(A, \cdot)$ satisfy

$$\int_B q_1(A, y) \, dF_Y(y) = \int_B q_2(A, y) \, dF_Y(y)$$

for all $B \subset \mathcal{Y}$ and $q_1(A, y) = \Pr(X \in A | Y = y)$, then $q_2(A, y) = \Pr(X \in A | Y = y)$ as well. In this case, $q_1(A, y)$ and $q_2(A, y)$ are said to be two *versions* of the conditional probability. The following result shows that, while conditional probabilities are not unique, they are essentially unique.

***Lemma* 2.2.** *Let* (X, Y) *denote a random vector with range* $\mathcal{X} \times \mathcal{Y}$ *and let* $q_1(\cdot, y)$ *and* $q_2(\cdot, y)$ *denote two versions of the conditional probability distribution of* X *given* $Y = y$. *For a given set* $A \subset \mathcal{X}$, *let*

$$\mathcal{Y}_0 = \{y \in \mathcal{Y} : q_1(A, y) \neq q_2(A, y)\}.$$

Then $\Pr(Y \in \mathcal{Y}_0) = 0$.

The conclusion of this lemma can be stated as follows: for any $A \subset \mathcal{X}$, $q_1(A, y) = q_2(A, y)$ *for almost all* y (F_Y). As in the statement of the lemma, this means that the set of y for which $q_1(A, y) = q_2(A, y)$ does not hold has probability 0 under the distribution given by F_Y. Alternatively, we may write that $q_1(A, \cdot) = q_2(A, \cdot)$ *almost everywhere* (F_Y), or, more simply, $q_1(A, \cdot) = q_2(A, \cdot)$ *a.e.* (F_Y).

Proof of Lemma 2.2.   Fix a set $A \subset \mathcal{X}$. For $n = 1, 2, \ldots$, define

$$\mathcal{Y}_n = \{y \in \mathcal{Y}: |q_1(A, y) - q_2(A, y)| \geq 1/n\}.$$

Note that $\mathcal{Y}_1 \subset \mathcal{Y}_2 \cdots$ and

$$\bigcup_{n=1}^{\infty} \mathcal{Y}_n = \mathcal{Y}_0.$$

For each $n = 1, 2, \ldots$, let

$$B_n = \{y \in \mathcal{Y}: q_1(A, y) - q_2(A, y) \geq 1/n\}$$

and

$$C_n = \{y \in \mathcal{Y}: q_2(A, y) - q_1(A, y) \geq 1/n\}$$

so that $\mathcal{Y}_n = B_n \cup C_n$, $n = 1, 2, \ldots$.

Fix n. Since both q_1 and q_2 satisfy (2.3),

$$0 = \int_{B_n} q_1(A, y) \, dF_Y(y) - \int_{B_n} q_2(A, y) \, dF_Y(y) \geq \frac{1}{n} \int_{B_n} dF_Y(y) = \frac{1}{n} \Pr(Y \in B_n)$$

so that $\Pr(Y \in B_n) = 0$. Similarly, $\Pr(Y \in C_n) = 0$. Hence, for each $n = 1, 2, \ldots$, $\Pr(Y \in \mathcal{Y}_n) = 0$.

By condition (P4) on probability distributions, given in Chapter 1, together with the fact that $\mathcal{Y}_0 = Y_1 \cup Y_2 \cup \cdots$,

$$\Pr(Y \in \mathcal{Y}_0) = \lim_{n \to \infty} \Pr(Y \in \mathcal{Y}_n) = 0,$$

proving the result.   ∎

Here we will refer to *the* conditional probability, with the understanding that there may be another version of the conditional probability that is equal to the first for y in a set of probability 1.

It is important to note that, when Y has an absolutely continuous distribution, conditional probabilities of the form $\Pr(X \in A | Y = y_0)$ for a specific value $y_0 \in \mathcal{Y}$ are not well defined, except as a function $q(A, y)$ satisfying (2.3) evaluated at $y = y_0$, and this fact can sometimes cause difficulties.

For instance, suppose we wish to determine $\Pr(X \in A | (X, Y) \in B)$, for some sets A and B, $A \subset \mathcal{X}$ and $B \subset \mathcal{X} \times \mathcal{Y}$, where $\Pr((X, Y) \in B) = 0$. Suppose further that the event $(X, Y) \in B$ can be described in terms of two different random variables W and Z, each of which is a function of (X, Y); that is, suppose there exist functions W and Z and values w_0 and z_0 in the ranges of W and Z, respectively, such that

$$\{(x, y) \in \mathcal{X} \times \mathcal{Y}: W(x, y) = w_0\} = \{(x, y) \in \mathcal{X} \times \mathcal{Y}: Z(x, y) = z_0\} = B.$$

Let $q_W(A, w) = \Pr(X \in A|W = w)$ and $q_Z(A, z) = \Pr(X \in A|Z = z)$. Then $\Pr(X \in A|(X, Y) \in B)$ is given by both $q_W(A, w_0)$ and $q_Z(A, z_0)$; however, there is no guarantee that $q_W(A, w_0) = q_Z(A, z_0)$ so that these two approaches could yield different results. This possibility is illustrated in the following example.

Example 2.10. Let X and Y denote independent random variables such that

$$\Pr(X = 1) = \Pr(X = c) = \frac{1}{2},$$

for some constant $c > 1$, and Y has an absolutely continuous distribution with density

$$p(y) = \frac{1}{2}, \quad -1 < y < 1.$$

Let $Z = XY$. Note that the events $Z = 0$ and $Y = 0$ are identical; that is, $Z = 0$ if and only if $Y = 0$. However, it will be shown that

$$\Pr(X = 1|Z = 0) \neq \Pr(X = 1|Y = 0).$$

Using (2.3), for $z \in \mathbf{R}$,

$$\Pr(Z \le z) = \frac{1}{2}\Pr(Z \le z|X = 1) + \frac{1}{2}\Pr(Z \le z|X = c).$$

Since the events $X = 1$ and $X = c$ both have nonzero probability, we know from elementary probability theory that, for $x = 1, c$,

$$\Pr(Z \le z|X = x) = 2\Pr(XY \le z \cap X = x) = 2\Pr(Y \le z/x \cap X = x)$$
$$= 2\Pr(Y \le z/x)\Pr(X = x) = \Pr(Y \le z/x).$$

It follows that, for $z > -c$,

$$\Pr(Z \le z) = \frac{1}{4}\int_{-1}^{z/c} dy + \frac{1}{4}\int_{-1}^{z} dy$$

so that Z has an absolutely continuous distribution with density

$$p_Z(z) = \frac{1}{4c}I_{\{|z|<c\}} + \frac{1}{4}I_{\{|z|<1\}}.$$

Define

$$h(z) = \begin{cases} 0 & \text{if } |z| \ge 1 \\ c/(c+1) & \text{if } |z| < 1 \end{cases}. \tag{2.4}$$

It will be shown that $h(z) = \Pr(X = 1|Z = z)$. To do this, first note that, for $B \subset \mathbf{R}$,

$$\Pr(X = 1 \cap Z \in B) = \Pr(X = 1 \cap Y \in B) = \frac{1}{4}\int_{B \cap (-1, 1)} dz.$$

Using (2.4),

$$\int_B h(z)\, dF_Z(z) = \int_B \frac{c}{c+1}I_{\{|z|<1\}}p_Z(z)\, dz = \frac{1}{4}\int_{B \cap (-1, 1)} dz$$

so that (2.3) is satisfied and, hence, $\Pr(X = 1|Z = z) = h(z)$ as claimed.

Consider $\Pr(X = 1|Y = 0)$. Since X and Y are independent, it follows from Example 2.9 that

$$\Pr(X = 1|Y = 0) = \Pr(X = 1) = \frac{1}{2}.$$

Note that the event $Y = 0$ is equivalent to the event $Z = 0$. Using (2.4),

$$\Pr(X = 1|Z = 0) = \frac{c}{c+1}.$$

Thus, two different answers are obtained, even though the events $Y = 0$ and $Z = 0$ are identical.

Some insight into this discrepency can be obtained by considering the conditioning events $|Y| < \epsilon$ and $|Z| < \epsilon$, for some small ϵ, in place of the events $Y = 0$ and $Z = 0$, respectively. Note that the events $|Y| < \epsilon$ and $|Z| < \epsilon$ are not equivalent. Suppose that c is a large number. Then, $|Z| < \epsilon$ strongly suggests that X is 1 so that we would expect $\Pr(X = 1| |Z| < \epsilon)$ to be close to 1. In fact, a formal calculation shows that, for any $0 < \epsilon < 1$,

$$\Pr(X = 1| |Z| < \epsilon) = \frac{c}{c+1},$$

while, by the independence of X and Y,

$$\Pr(X = 1| |Y| < \epsilon) = \Pr(X = 1) = \frac{1}{2},$$

results which are in agreement with the values for $\Pr(X = 1|Z = 0)$ and $\Pr(X = 1|Y = 0)$ obtained above.   □

Conditional distribution functions and densities

Since $\Pr(X \in A|Y = y)$ defines a probability distribution for X, for each y, there exists a distribution function $F_{X|Y}(x|y)$ such that

$$\Pr(X \in A|Y = y) = \int_A dF_{X|Y}(x|y);$$

the distribution function $F_{X|Y}(\cdot|y)$ is called the conditional distribution function of X given $Y = y$. By (2.2), F, the distribution function of (X, Y), F_Y, the marginal distribution function of Y, and $F_{X|Y}$ are related by

$$F(x, y) = F_{X|Y}(x|y) F_Y(y) \qquad \text{for all } x, y.$$

If the conditional distribution of X given $Y = y$ is absolutely continuous, then

$$\Pr(X \in A|Y = y) = \int_A p_{X|Y}(x|y)\,dx$$

where $p_{X|Y}(\cdot|y)$ denotes the conditional density of X given $Y = y$. If the conditional distribution of X given $Y = y$ is discrete, then

$$\Pr(X \in A|Y = y) = \sum_{x \in A} p_{X|Y}(x|y)$$

where $p_{X|Y}(\cdot|y)$ denotes the conditional frequency function of X given $Y = y$.

Theorem 2.3. *Consider a random vector* (X, Y) *where either or both of X and Y may be vectors.*

(i) Suppose that (X, Y) *has an absolutely continuous distribution with density p and let* p_Y *denote the marginal density function of Y. Then the conditional distribution of X given* $Y = y$ *is absolutely continuous with density* $p_{X|Y}$ *given by*

$$p_{X|Y}(x|y) = \frac{p(x, y)}{p_Y(y)}$$

provided that $p_Y(y) > 0$; *if* $p_Y(y) = 0$, $p_{X|Y}(x|y)$ *may be taken to be any finite value.*

(ii) Suppose that (X, Y) *has a discrete distribution with frequency function p and let* p_Y *denote the marginal frequency function of Y. Then the conditional distribution of X given* $Y = y$ *is discrete with frequency function* $p_{X|Y}(x|y)$, *given by*

$$p_{X|Y}(x|y) = \frac{p(x, y)}{p_Y(y)}$$

provided that $p_Y(y) > 0$; *if* $p_Y(y) = 0$, $p_{X|Y}(x|y)$ *may be taken to have any finite value.*

Proof.   Consider case (i) in which the distribution of (X, Y) is absolutely continuous. Then $\Pr(X \in A | Y = y)$ must satisfy

$$\Pr(X \in A, Y \in B) = \int_B \Pr(X \in A | Y = y) p_Y(y) \, dy.$$

For y satisfying $p_Y(y) > 0$, let

$$q(A, y) = \int_A \frac{p(x, y)}{p_Y(y)} \, dx.$$

Then

$$\int_B q(A, y) p_Y(y) \, dy = \int_B \int_A p(x, y) \, dx \, dy = \Pr(X \in A, Y \in B)$$

so that

$$\Pr(X \in A | Y = y) = \int_A \frac{p(x, y)}{p_Y(y)} \, dx.$$

Note that the value of $q(A, y)$ for those y satisfying $p_Y(y) = 0$ is irrelevant. Clearly this distribution is absolutely continuous with density $p_{X|Y}$ as given in the theorem.

The result for the discrete case follows along similar lines.   ∎

Example 2.11 (Bivariate distribution). Consider the distribution considered in Example 2.1. The random vector (X, Y) has an absolutely continuous distribution with density function

$$p(x, y) = 6(1 - x - y), \quad x > 0, \ y > 0, \ x + y < 1$$

and the marginal density of Y is

$$p_Y(y) = 3(1 - y)^2, \quad 0 < y < 1.$$

Hence, the conditional distribution of X given $Y = y$ is absolutely continuous with density function

$$p_{X|Y}(x|y) = 2\frac{1 - x - y}{(1 - y)^2}, \quad 0 < x < 1 - y$$

where $0 < y < 1$. □

Example* 2.12 *(Trinomial distribution). Let (X_1, X_2, X_3) be a random vector with a multinomial distribution with parameters n and $(\theta_1, \theta_2, \theta_3)$; see Example 2.2. This is sometimes called a *trinomial* distribution. It was shown in Example 2.2 that the marginal distribution of X_1 is binomial with parameters n and θ_1.

Hence, (X_1, X_2, X_3) has frequency function

$$p(x_1, x_2) = \binom{n}{x_1, x_2, x_3}\theta_1^{x_1}\theta_2^{x_2}\theta_3^{x_3}$$

and the marginal frequency function of X_1 is given by

$$p_{X_1}(x_1) = \binom{n}{x_1}\theta_1^{x_1}(1 - \theta_1)^{n-x_1}.$$

It follows that the conditional distribution of (X_2, X_3) given $X_1 = x_1$ is discrete with frequency function

$$p_{X_2,X_3|X_1}(x_2, x_3|x_1) = \frac{\binom{n}{x_1,x_2}\theta_1^{x_1}\theta_2^{x_2}\theta_3^{x_3}}{\binom{n}{x_1}\theta_1^{x_1}(1 - \theta_1)^{n-x_1}} = \binom{n - x_1}{x_2, x_3}\frac{\theta_2^{x_2}\theta_3^{x_3)}}{(\theta_2 + \theta_3)^{n-x_1}},$$

where $x_2, x_3 = 0, \ldots, n - x_1$ with $x_2 + x_3 = n - x_1$, for $x_1 = 0, \ldots, n$; recall that $\theta_1 + \theta_2 + \theta_3 = 1$.

That is, the conditional distribution of (X_2, X_3) given $X_1 = x_1$ is multinomial with parameters $n - x_1$ and $(\theta_2/(\theta_2 + \theta_3), \theta_3/(\theta_2 + \theta_3))$. Alternatively, we can say X_2 has a binomial distribution with parameters $n - x_1$ and $\theta_2/(\theta_2 + \theta_3)$ with $X_3 = n - x_1 - X_2$. □

Example* 2.13 *(A mixed distribution). Let (X, Y) denote a two-dimensional random vector with the distribution described in Example 2.3. Recall that this distribution is neither absolutely continuous nor discrete.

First consider the conditional distribution of X given $Y = y$. Recall that for $A \subset \mathbf{R}^+$ and $y = 1, 2$,

$$\Pr(X \in A, \; Y = y) = \frac{1}{2}\int_A y\exp(-yx)\,dx.$$

Note that

$$\Pr(X \leq x, \; Y = y) = \frac{1}{2}[1 - \exp(-yx)], \quad x > 0, \; y = 1, 2$$

so that, for $y = 1, 2$,

$$\Pr(X \leq x|Y = y) = 1 - \exp(-yx), \quad x > 0.$$

Since

$$\Pr(X \le x | Y = y) = \int_0^x y \exp(-yx) \, dx,$$

it follows that, for $y = 1, 2$, the conditional distribution of X given $Y = y$ is absolutely continuous with density function $y \exp(-yx)$, $x > 0$.

It is worth noting that this same result may be obtained by the following informal method. Since, for $A \subset \mathcal{Y}$ and $y = 1, 2$,

$$\Pr(X \in A, \ Y = y) = \frac{1}{2} \int_A y \exp(-yx) \, dx,$$

the function

$$\frac{1}{2} y \exp(-yx)$$

plays the role of a density function for (X, Y) with the understanding that we must integrate with respect to x and sum with respect to y. Since the marginal distribution of Y is discrete with frequency function $1/2$, $y = 1, 2$, the conditional density function of X given $Y = y$ is $y \exp(-yx)$, $x > 0$.

Now consider the conditional distribution of Y given $X = x$. Recall that the marginal distribution of X is absolutely continuous with density function

$$\frac{1}{2} [\exp(-x) + 2 \exp(-2x)], \quad x > 0.$$

Using the informal method described above, the conditional distribution of Y given $X = x$, $x > 0$, has frequency function

$$\frac{y \exp(-yx)}{\exp(-x) + 2 \exp(-2x)}, \quad y = 1, 2.$$

It is easy to verify that this result is correct by noting that

$$\Pr(Y = y, \ X \in A) = \frac{1}{2} \int_A y \exp(-yx) \, dx$$

$$= \int_A \frac{y \exp(-yx)}{\exp(-x) + 2 \exp(-2x)} p_X(x) \, dx, \quad y = 1, 2.$$

Hence, the conditional distribution of Y given $X = x$ is discrete with

$$\Pr(Y = 1 | X = x) = 1 - \Pr(Y = 2 | X = x) = \frac{1}{1 + 2 \exp(-x)}$$

for $x > 0$. $\square$

2.4   Conditional Expectation

Let (X, Y) denote a random vector with range $\mathcal{X} \times \mathcal{Y}$ and let $F_{X|Y}(\cdot | y)$ denote the distribution function corresponding to the conditional distribution of X given $Y = y$. For a function $g : X \to \mathbf{R}$ such that $\mathrm{E}[|g(X)|] < \infty$, $\mathrm{E}[g(X) | Y = y]$ may be defined by

$$\mathrm{E}[g(X) | Y = y] = \int_{\mathcal{X}} g(x) \, dF_{X|Y}(x | y).$$

It is sometimes convenient to define $E[g(X)|Y = y]$ directly, without reference to $F_{X|Y}(\cdot|y)$. First consider the case in which $g(x) = I_{\{x \in A\}}$ for some set $A \subset \mathcal{X}$. Then $E[g(X)|Y = y] = \Pr[X \in A|Y = y]$ so that $E[g(X)|Y = y]$ satisfies the equation

$$E[g(X)I_{\{Y \in B\}}] = \int_B E[g(X)|Y = y] \, dF_Y(y)$$

for all $B \subset \mathcal{Y}$.

This definition can be extended to an arbitrary function g. Suppose that $g : X \to \mathbf{R}$ satisfies $E[|g(X)|] < \infty$. We define $E[g(X)|Y = y]$ to be any function of y satisfying

$$E[g(X)I_{\{Y \in B\}}] = \int_B E[g(X)|Y = y] \, dF_Y(y) \tag{2.5}$$

for all sets $B \subset \mathcal{Y}$. The issues regarding existence and uniqueness are essentially the same as they are for conditional probabilities. If $E[|g(X)|] < \infty$, then the Radon-Nikodym Theorem guarantees existence of the conditional expected value. Conditional expected values are not unique, but any two versions of $E[g(X)|Y = y]$ differ only for y in a set of probability 0.

Let $F_{X|Y}(\cdot|y)$ denote the conditional distribution function of X given $Y = y$ and consider

$$h(y) = \int_{\mathcal{X}} g(x) \, dF_{X|Y}(x|y). \quad = E\big[g(X)\,|\,Y=y\big]$$

Then

$$E\big[E[g(x)|Y]\big] = \int_B h(y) \, dF_Y(y) = \int_B \left[\int_{\mathcal{X}} g(x) \, dF_{X|Y}(x|y)\right] dF_Y(y) = \int_{\mathcal{X} \times \mathcal{Y}} I_{\{y \in B\}} g(x) \, dF_{X,Y}(x, y)$$
$$= E[g(X)I_{\{Y \in B\}}].$$

Hence, one choice for $E[g(X)|Y = y]$ is given

$$E[g(X)|Y = y] = \int g(x) \, dF_{X|Y}(x|y);$$

that is, the two approaches to defining $E[g(X)|Y = y]$ considered here are in agreement. Generally speaking, the expression based on $F_{X|Y}(\cdot|y)$ is more convenient for computing conditional expected values for a given distribution, while the definition based on (2.5) is more convenient for establishing general properties of conditional expected values.

***Example* 2.14** (*Bivariate distribution*). Let (X, Y) denote a two-dimensional random vector with the distribution described in Example 2.1. This distribution is absolutely continuous with density function

$$p(x, y) = 6(1 - x - y), \quad x > 0, \ y > 0, \ x + y < 1;$$

it was shown in Example 2.11 that the conditional distribution of X given $Y = y$ is absolutely continuous with density function

$$p_{X|Y}(x|y) = 2\frac{1 - x - y}{(1 - y)^2}, \quad 0 < x < 1 - y$$

where $0 < y < 1$. It follows that

$$E[X|Y = y] = \frac{2}{(1 - y)^2} \int_0^{1-y} x(1 - x - y) \, dx = \frac{1}{3}(1 - y), \quad 0 < y < 1. \qquad \square$$

***Example* 2.15** *(A mixed distribution).* Let (X, Y) denote a two-dimensional random vector with the distribution described in Example 2.3 and considered further in Example 2.13.

Recall that the conditional distribution of X given $Y = y$ is absolutely continuous with density function $y \exp(-yx)$, $x > 0$. It follows

$$E(X|Y = y) = 1/y.$$

The following theorem gives several properties of conditional expected values. These follow immediately from the properties of integrals, as described in Appendix 1, and, hence, the proof is left as an exercise. In describing these results, we write that a property holds for "almost all y (F_Y)" if the set of $y \in \mathcal{Y}$ for which the property does not hold has probability 0 under F_Y.

Theorem 2.4. *Let (X, Y) denote a random vector with range $\mathcal{X} \times \mathcal{Y}$; note that X and Y may each be vectors. Let $g_1, \ldots, g_m$ denote a real-valued functions defined on $\mathcal{X}$ such that $E[|g_j(X)|] < \infty$, $j = 1, \ldots, m$. Then*

(i) If g_1 is nonnegative, then

$$E[g_1(X)|Y = y] \geq 0 \qquad \text{for almost all } y \ (F_Y).$$

(ii) If g_1 is constant, $g_1(x) \equiv c$, then

$$E[g_1(X)|Y = y] = c \qquad \text{for almost all } y \ (F_Y).$$

(iii) For almost all y (F_Y),

$$E[g_1(X) + \cdots + g_m(X)|Y = y] = E[g_1(X)|Y = y] + \cdots + E[g_m(X)|Y = y].$$

Note that $E[g(X)|Y = y]$ is a function of y, which we may denote, for example, by $f(y)$. It is often convenient to consider the random variable $f(Y)$, which we denote by $E[g(X)|Y]$ and call the conditional expected value of $g(X)$ given Y. This random variable is a function of Y, yet it retains some of the properties of $g(X)$. According to (2.5), $E[g(X)|Y]$ is any function of Y satisfying

$$E\{g(X)I_{\{Y \in B\}}\} = E\{E[g(X)|Y]I_{\{Y \in B\}}\} \qquad \text{for all } B \subset \mathcal{Y}. \tag{2.6}$$

The following result gives a number of useful properties of conditional expected values.

Theorem 2.5. *Let (X, Y) denote a random vector with range $\mathcal{X} \times \mathcal{Y}$, let $T : \mathcal{Y} \to \mathcal{T}$ denote a function on $\mathcal{Y}$, let g denote a real-valued function on $\mathcal{X}$ such that $E[|g(X)|] < \infty$, and let h denote a real-valued function on $\mathcal{Y}$ such that $E[|g(X)h(Y)|] < \infty$. Then*

(i) $E\{E[g(X)|Y]\} = E[g(X)]$
(ii) $E[g(X)h(Y)|Y] = h(Y)E[g(X)|Y]$ with probability 1
(iii) $E[g(X)|Y, T(Y)] = E[g(X)|Y]$ with probability 1
(iv) $E[g(X)|T(Y)] = E\{E[g(X)|Y]|T(Y)\}$ with probability 1
(v) $E[g(X)|T(Y)] = E\{E[g(X)|T(Y)]|Y\}$ with probability 1

Proof. Part (i) follows immediately from Equation (2.6) with B taken to be $\mathcal{Y}$.

Let $f(y) = \mathrm{E}[g(X)|Y = y]$; then, for almost all y (F_Y),

$$f(y) = \int_{\mathcal{X}} g(x)\, dF_{X|Y}(x|y).$$

Hence, for any $B \subset \mathcal{Y}$,

$$\int_B h(y)\mathrm{E}[g(X)|Y = y]\, dF_Y(y) = \int_B h(y) \int_{\mathcal{X}} g(x)\, dF_{X|Y}(x|y)\, dF_Y(y)$$

$$= \int\int_{\mathcal{X}\times\mathcal{Y}} \mathrm{I}_{\{y \in B\}} g(x)h(y)\, dF_{X,Y}(x, y)$$

$$= \mathrm{E}[\mathrm{I}_{\{Y \in B\}} g(X)h(Y)]$$

so that (2.6) is satisfied; part (ii) follows.

By (2.6), for any $B \subset \mathcal{Y}$,

$$\mathrm{E}\{\mathrm{E}[g(X)|Y, T(Y)]\mathrm{I}_{\{Y \in B,\ T(Y) \in T\}}\} = \mathrm{E}[g(X)\mathrm{I}_{\{Y \in B,\ T(Y) \in T\}}] = \mathrm{E}[g(X)\mathrm{I}_{\{Y \in B\}}];$$

part (iii) now follows from (2.6).

Let $\bar{g}(Y) = \mathrm{E}[g(X)|Y]$. Then, by (2.6),

$$\mathrm{E}\{\mathrm{E}[\bar{g}(Y)|h(Y)]\mathrm{I}_{\{h(Y) \in B_0\}}\} = \mathrm{E}[\bar{g}(Y)\mathrm{I}_{\{h(Y) \in A\}}]$$

for any subset A of the range of $h(y)$. Let $B \subset \mathcal{Y}$ denote a set satisfying

$$\mathrm{I}_{\{h(Y) \in A\}} = \mathrm{I}_{\{Y \in B\}} \quad \text{with probability } 1.$$

Then, by (2.6),

$$\mathrm{E}[\bar{g}(Y)\mathrm{I}_{\{h(Y) \in A\}}] = \mathrm{E}\{\mathrm{E}[g(X)|Y]\mathrm{I}_{\{Y \in B\}}\} = \mathrm{E}[g(X)\mathrm{I}_{\{Y \in B\}}] = \mathrm{E}[g(X)\mathrm{I}_{\{h(Y) \in A\}}].$$

That is, for all subsets A of the range of $h(\cdot)$,

$$\mathrm{E}\{\mathrm{E}[\bar{g}(Y)|h(Y)]\mathrm{I}_{\{h(Y) \in A\}}\} = \mathrm{E}[g(X)\mathrm{I}_{\{h(Y) \in A\}}].$$

It now follows from (2.6) that

$$\mathrm{E}[\bar{g}(Y)|h(Y)] = \mathrm{E}[g(X)|h(Y)],$$

proving part (iv).

Note that $\mathrm{E}[g(X)|h(Y)]$ is a function of Y such that

$$\mathrm{E}\{|\mathrm{E}[g(X)|h(Y)]|\} \le \mathrm{E}\{\mathrm{E}[|g(X)|\,|h(Y)]\} = \mathrm{E}[|g(X)|] < \infty;$$

part (v) now follows from part (iii). ∎

Example 2.16. Let $Y = (Y_1, Y_2)$ where Y_1, Y_2 are independent, real-valued random variables, each with a uniform distribution on $(0, 1)$. Let X denote a real-valued random variable such that the conditional distribution of X given $Y = y$ has an absolutely continuous distribution with density

$$p_{X|Y}(x|y) = \frac{1}{y_1 + y_2}\exp\{-x/(y_1 + y_2)\}, \quad x > 0$$

where $y = (y_1, y_2)$ and $y_1 > 0, y_2 > 0$.

It is straightforward to show that

$$E(X|(Y_1, Y_2)) = Y_1 + Y_2$$

and, since each of Y_1, Y_2 has expected value $1/2$, it follows from part (i) of Theorem 2.5 that $E(X) = 1$.

Now consider $E(X|Y_1)$. Since $E(X|(Y_1, Y_2)) = Y_1 + Y_2$, it follows from part (iv) of Theorem 2.5 that

$$E(X|Y_1) = E(Y_1 + Y_2|Y_1) = Y_1 + E(Y_2) = Y_1 + 1/2. \qquad \square$$

The following characterization of conditional expected values is often useful.

Theorem 2.6. *Let (X, Y) denote a random vector with range $\mathcal{X} \times \mathcal{Y}$ and let g denote a real-valued function on $\mathcal{X}$.*

(i) Suppose that $E[|g(X)|] < \infty$ and let Z denote a real-valued function of Y such that $E(|Z|) < \infty$. If

$$Z = E[g(X)|Y] \quad \text{with probability 1}$$

then

$$E[Zh(Y)] = E[g(X)h(Y)] \tag{2.7}$$

for all functions $h : \mathcal{Y} \to \mathbf{R}$ such that

$$E[|h(Y)|] < \infty \quad \text{and} \quad E[|g(X)h(Y)|] < \infty.$$

(ii) If Z is a function of Y such that (2.7) holds for all bounded functions $h : \mathcal{Y} \to \mathbf{R}$ then

$$Z = E[g(X)|Y] \quad \text{with probability 1}.$$

Proof.   Suppose $Z = E[g(X)|Y]$ with probability 1. Let h be a real-valued function on $\mathcal{Y}$ such that $E[|h(Y)|] < \infty$ and $E[|g(X)h(Y)|] < \infty$. Then, since

$$\{Z - E[g(X)|Y]\}h(Y) = 0 \quad \text{with probability } 1,$$

$$E\{(Z - E[g(X)|Y])h(Y)\} = 0.$$

It follows from Theorem 2.5 that

$$E\{E[g(X)|Y]h(Y)\} = E[g(X)h(Y)] < \infty,$$

so that

$$E[Zh(Y)] = E\{E[g(X)|Y]h(Y)\} = E[g(X)h(Y)],$$

proving the first part of the theorem.

Now suppose that (2.7) holds for all bounded functions $h : \mathcal{Y} \to \mathbf{R}$. Let $B \subset \mathcal{Y}$. Since $h(y) = I_{\{y \in B\}}$ is a bounded, real-valued function on $\mathcal{Y}$, it follows that (2.6) holds for any B. Part (ii) follows.   ∎

Conditional expectation as an approximation

Conditional expectation may be viewed as the solution to the following approximation problem. Let X and Y denote random variables, which may be vectors. Let $g(X)$ denote a real-valued function of X and suppose that we wish to approximate the random variable $g(X)$ by a real-valued function of Y. Suppose further that we decide to measure the quality of an approximation $h(Y)$ by $\mathrm{E}[(g(X) - h(Y))^2]$. Then the best approximation in this sense is given by $h(Y) = \mathrm{E}[g(X)|Y]$. This idea is frequently used in the context of statistical forecasting in which X represents a random variable that is, as of yet, unobserved, while Y represents the information currently available. A formal statement of this result is given in the following corollary to Theorem 2.6.

Corollary 2.2. Let (X, Y) denote a random vector with range $\mathcal{X} \times \mathcal{Y}$ and let g denote a real-valued function on $\mathcal{X}$ such that $\mathrm{E}[g(X)^2] < \infty$. Let $Z = \mathrm{E}[g(X)|Y]$. Then, for any real-valued function h on $\mathcal{Y}$ such that $\mathrm{E}[h(Y)^2] < \infty$,

$$\mathrm{E}[(h(Y) - g(X))^2] \geq \mathrm{E}[(Z - g(X))^2]$$

with equality if and only if $h(Y) = Z$ with probability 1.

Proof. Note that

$$\mathrm{E}[(h(Y) - g(X))^2] = \mathrm{E}[(h(Y) - Z + Z - g(X))^2]$$
$$= \mathrm{E}[(h(Y) - Z)^2] + \mathrm{E}[(Z - g(X))^2] + 2\mathrm{E}\{(h(Y) - Z)(Z - g(X))\}.$$

Since Z is a function of Y, $h(Y) - Z$ is a function of Y. Furthermore,

$$\mathrm{E}[|h(Y) - Z|] \leq \mathrm{E}[|h(Y)|] + \mathrm{E}[|Z|] < \infty$$

and

$$\mathrm{E}[|g(X)| \, |h(Y) - Z|] \leq \mathrm{E}[g(X)^2]^{\frac{1}{2}} \mathrm{E}[|h(Y) - Z|^2]^{\frac{1}{2}}$$
$$\leq \mathrm{E}[|g(X)|^2]^{\frac{1}{2}} \{2\mathrm{E}[|h(Y)|^2] + 2\mathrm{E}[|Z|^2]\}^{\frac{1}{2}} < \infty,$$

using the fact that

$$\mathrm{E}[|Z|^2] = \mathrm{E}\{\mathrm{E}[g(X)|Y]^2\} \leq \mathrm{E}\{\mathrm{E}[g(X)^2|Y]\} = \mathrm{E}[g(X)^2] < \infty.$$

Hence, by Theorem 2.6,

$$\mathrm{E}\{(h(Y) - Z)Z\} = \mathrm{E}\{(h(Y) - Z)g(X)\}$$

so that

$$\mathrm{E}[(h(Y) - g(X))^2] = \mathrm{E}[(h(Y) - Z)^2] + \mathrm{E}[(Z - g(X))^2].$$

It follows that

$$\mathrm{E}[(g(X) - h(Y))^2] \geq \mathrm{E}[(g(X) - Z)^2]$$

with equality if and only if $\mathrm{E}[(h(Y) - Z)^2] = 0$, which occurs if and only if $h(Y) = Z$ with probability 1. ∎

Example 2.17 (Independent random variables). Let Y denote a real-valued random variable with $\mathrm{E}(Y^2) < \infty$ and let X denote a random vector such that X and Y are independent.

Note that $E(Y|X) = \mu$, where $\mu = E(Y)$. Then, according to Corollary 2.2, for any real-valued function h of Y,

$$E[(h(X) - Y)^2] \le E[(Y - \mu)^2];$$

that is, the best approximation to Y among all functions of X is simply the constant function $h(X) = \mu$. $\quad\Box$

Example 2.18. Let X and Y denote real-valued random variables such that $E(X^4) < \infty$ and $E(Y^4) < \infty$. Suppose that $Y = X + Z$ where Z is a real-valued random variable with $E(Z) = 0$ and $E(Z^2) = 1$; assume that X and Z are independent.

Then, according to Corollary 2.2, the best approximation to Y among functions of X is

$$E(Y|X) = E(X + Z|X) = X.$$

The best approximation to Y^2 among functions of X is

$$\begin{aligned}
E(Y|X) &= E[(X + Z)^2|X] = E(X^2|X) + E(2XZ|X) + E(Z^2|X) \\
&= X^2 + 2XE(Z|X) + E(Z^2) \\
&= X^2 + 1.
\end{aligned}$$

Hence, although the best approximation to Y is X, the best approximation to Y^2 is $X^2 + 1$, not X^2. This is due to the criterion used to evaluate approximations. $\quad\Box$

2.5 Exchangeability

Recall that random variables $X_1, X_2, \ldots, X_n$ are independent and identically distributed if they are independent and each X_j has the same marginal distribution. An infinite sequence of random variables $X_1, X_2, \ldots$ is independent and identically distributed if each finite subset is.

Exchangeability provides a useful generalization of this concept. Recall that a permutation of $(1, 2, \ldots, n)$ is a rearrangement of the form $(i_1, \ldots, i_n)$ such that each $1 \le i_j \le n$ is an integer and $i_j \ne i_k$ for $j \ne k$. Real-valued random variables $X_1, \ldots, X_n$ are said to have an *exchangeable distribution* or, more simply, to be *exchangeable* if the distribution of $(X_1, \ldots, X_n)$ is the same as the distribution of $(X_{i_1}, \ldots, X_{i_n})$ for any permutation $(i_1, i_2, \ldots, i_n)$ of $(1, 2, \ldots, n)$.

As noted above, the simplest example of exchangeability is the case of independent, identically distributed random variables. A formal statement of this is given in the following theorem; the proof is straightforward and is left as an exercise.

Theorem 2.7. *Suppose* $X_1, \ldots, X_n$ *are independent identically distributed random variables. Then* $X_1, \ldots, X_n$ *are exchangeable.*

Example 2.19 (Bivariate distribution). Consider the distribution considered in Examples 2.1 and 2.11. The random vector (X, Y) has an absolutely continuous distribution with density function

$$p(x, y) = 6(1 - x - y), \quad x > 0, \ y > 0, \ x + y < 1.$$

Let g denote a bounded, real-valued function defined on $(0, 1)^2$. Then

$$E[g(X, Y)] = \int_0^1 \int_0^1 g(x, y)6(1 - x - y)I_{\{x+y<1\}} \, dx \, dy$$

$$= \int_0^1 \int_0^1 g(y, x)6(1 - y - x)I_{\{y+x<1\}} \, dy \, dx = E[g(Y, X)].$$

It follows that (Y, X) has the same distribution as (X, Y) so that X, Y are exchangeable. □

***Example* 2.20.** Let (X_1, X_2) denote a two-dimensional random vector with an absolutely continuous distribution with density function

$$p(x, y) = \frac{1}{x_1}, \quad 0 < x_2 < x_1 < 1.$$

Note that $X_2 < X_1$ with probability 1. Let $(Y_1, Y_2) = (X_2, X_1)$. Then

$$\Pr(Y_2 < Y_1) = 0.$$

It follows that (Y_1, Y_2) does not have the same distribution as (X_1, X_2); that is, (X_2, X_1) does not have the same distribution as (X_1, X_2). It follows that the distribution of (X_1, X_2) is not exchangeable. □

Suppose that $X_1, X_2, \ldots, X_n$ are exchangeable random variables. Then the distribution of $(X_1, X_2, \ldots, X_n)$ is the same as the distribution of $(X_2, X_1, \ldots, X_n)$. In this case,

$$\Pr(X_1 \le x, X_2 < \infty, \ldots, X_n < \infty) = \Pr(X_2 \le x, X_1 < \infty, \ldots, X_n < \infty).$$

That is, the marginal distribution of X_1 is the same as the marginal distribution of X_2; it follows that each X_j has the same marginal distribution. This result may be generalized as follows.

***Theorem* 2.8.** *Suppose $X_1, \ldots, X_n$ are exchangeable real-valued random variables. Let m denote a positive integer less than or equal to n and let $t_1, \ldots, t_m$ denote distinct elements of $\{1, 2, \ldots, n\}$. Then the distribution of $(X_{t_1}, \ldots, X_{t_m})$ does not depend on the choice of $t_1, t_2, \ldots, t_m$.*

Proof. Fix m and let $t_1, \ldots, t_m$ and $r_1, \ldots, r_m$ denote two sets of distinct elements from $\{1, 2, \ldots, n\}$. Then we may find $t_{m+1}, \ldots, t_n$ in $\{1, \ldots, n\}$ such that $(t_1, \ldots, t_n)$ is a permutation of $(1, 2, \ldots, n)$; similarly, suppose that $(r_1, \ldots, r_n)$ is a permutation of $(1, 2, \ldots, n)$. Then $(X_{t_1}, \ldots, X_{t_n})$ and $(X_{r_1}, \ldots, X_{r_n})$ have the same distribution. Hence,

$$\Pr(X_{t_1} \le x_1, \ldots, X_{t_m} \le x_m, X_{t_{m+1}} < \infty, \ldots, X_{t_n} < \infty)$$
$$= \Pr(X_{r_1} \le x_1, \ldots, X_{r_m} \le x_m, X_{r_{m+1}} < \infty, \ldots, X_{r_n} < \infty);$$

the result follows. ■

Thus, exchangeable random variables $X_1, \ldots, X_n$ are identically distributed and any two subsets of $X_1, \ldots, X_n$ of the same size are also identically distributed. However, exchangeable random variables are generally not independent.

Let $\mathcal{T}$ denote a subset of the set of all permutations of $(1, \ldots, n)$. A function $h : \mathbf{R}^n \to \mathbf{R}^m$ is said to be *invariant with respect to* $\mathcal{T}$ if for any permutation $\tau \in \mathcal{T}, h(x) = h(\tau x)$. Here τ is of the form $\tau = (i_1, \ldots, i_n)$ and $\tau x = (x_{i_1}, x_{i_2}, \ldots, x_{i_n})$. If $\mathcal{T}$ is the set of all permutations, h is said to be *permutation invariant*.

Example 2.21 (Sample mean). Let h denote the function on $\mathbf{R}^n$ given by

$$h(x) = \frac{1}{n} \sum_{j=1}^{n} x_j, \quad x = (x_1, \ldots, x_n).$$

Since changing the order of $(x_1, \ldots, x_n)$ does not change the sum, this function is permutation invariant. $\square$

Theorem 2.9. *Suppose $X_1, \ldots, X_n$ are exchangeable real-valued random variables and h is invariant with respect to $\mathcal{T}$ for some set of permutations $\mathcal{T}$. Let g denote a real-valued function on the range of $X = (X_1, \ldots, X_n)$ such that $\mathrm{E}[|g(X)|] < \infty$. Then*

(i) The distribution of $(g(\tau X), h(X))$ is the same for all $\tau \in \mathcal{T}$.

(ii) $\mathrm{E}[g(X)|h(X)] = \mathrm{E}[g(\tau X)|h(X)]$, with probability 1, for all $\tau \in \mathcal{T}$.

Proof. Since $X_1, \ldots, X_n$ are exchangeable, the distribution of $(g(\tau X), h(\tau X))$ is the same for any permutation τ. Part (i) now follows from the fact that, for $\tau \in \mathcal{T}, h(\tau X) = h(X)$ with probability 1.

By Theorem 2.6, part (ii) follows provided that, for any bounded, real-valued function f on the range of h,

$$\mathrm{E}[\mathrm{E}[g(\tau X)|h(X)]f(h(X))] = \mathrm{E}[g(X)f(h(X))].$$

Since

$$\mathrm{E}[\mathrm{E}[g(\tau X)|h(X)]f(h(X))] = \mathrm{E}[\mathrm{E}[g(\tau X)f(h(X))|h(X)]] = \mathrm{E}[g(\tau X)f(h(X))],$$

part (ii) follows provided that

$$\mathrm{E}[g(\tau X)f(h(X))] = \mathrm{E}[g(X)f(h(X))];$$

the result now follows from part (i). ∎

Example 2.22 (Conditioning on the sum of random variables). Let $X_1, X_2, \ldots, X_n$ denote exchangeable, real-valued random variables such that

$$\mathrm{E}[|X_j|] < \infty, \quad j = 1, \ldots, n,$$

and let $S = \sum_{j=1}^{n} X_j$. Since S is a permutation invariant function of $X_1, \ldots, X_n$, it follows from Theorem 2.9 that $\mathrm{E}[X_j|S]$ does not depend on j. This fact, together with the fact that

$$S = \mathrm{E}(S|S) = \mathrm{E}[\sum_{j=1}^{n} X_j|S] = \sum_{j=1}^{n} \mathrm{E}[X_j|S]$$

shows that $E(X_j|S) = S/n$. $\square$

2.6   Martingales

Consider a sequence of real-valued random variables $\{X_1, X_2, \ldots\}$, such that $E(|X_n|) < \infty$ for all $n = 1, 2, \ldots$. The sequence $\{X_1, X_2, \ldots\}$ is said to be a *martingale* if for any $n = 1, 2, \ldots$,

$$E(X_{n+1}|X_1, \ldots, X_n) = X_n$$

with probability 1.

A martingale may be viewed as the fortunes of a gambler playing a sequence of fair games with X_n denoting the fortune of the gambler after the nth game, $n = 2, 3, \ldots$, and X_1 representing the initial fortune. Since the games are fair, the conditional expected value of X_{n+1}, the fortune at stage $n + 1$, given the current fortune X_n, is X_n.

Example 2.23 *(Sums of independent random variables).* Let $Y_1, Y_2, \ldots$ denote a sequence of independent random variables each with mean 0. Define

$$X_n = Y_1 + \cdots + Y_n, \quad n = 1, 2, \ldots$$

Then

$$E(X_{n+1}|X_1, \ldots, X_n) = E(X_n|X_1, \ldots, X_n) + E(Y_{n+1}|X_1, \ldots, X_n).$$

Clearly, $E(X_n|X_1, \ldots, X_n) = X_n$ and, since $(X_1, \ldots, X_n)$ is a function of $(Y_1, \ldots, Y_n)$, $E(Y_{n+1}|X_1, \ldots, X_n) = 0$. It follows that $\{X_1, X_2, \ldots\}$ is a martingale. □

Example 2.24 *(Polya's urn scheme).* Consider an urn containing b black and r red balls. A ball is randomly drawn from the urn and c balls of the color drawn are added to the urn. Let X_n denote the proportion of black balls after the nth draw. Hence, $X_0 = b/(r + b)$,

$$\Pr[X_1 = (b + c)/(r + b + c)] = 1 - \Pr[X_1 = b/(r + b + c)] = \frac{b}{r + b},$$

and so on.

Let $Y_n = 1$ if the nth draw is a black ball and 0 otherwise. Clearly,

$$\Pr(Y_{n+1} = 1|X_1, \ldots, X_n) = X_n.$$

After n draws, there are $r + b + nc$ balls in the urn and the number of black balls is given by $(r + b + nc)X_n$. Hence,

$$X_{n+1} = \frac{(r + b + nc)X_n + Y_{n+1}c}{r + b + (n + 1)c}$$

so that

$$E[X_{n+1}|X_1, \ldots, X_n] = \frac{(r + b + nc)X_n + X_n c}{r + b + (n + 1)c} = X_n;$$

it follows that $\{X_1, X_2, \ldots\}$ is a martingale. □

Using the interpretation of a martingale in terms of fair games, it is clear that if the gambler has fortune c after n games, then the gamblers expected fortune after a number of additional games will still be c. A formal statement of this result for martingales is given in the following theorem.

Theorem 2.10. *Let* $\{X_1, X_2, \ldots\}$ *be a martingale and let n and m be positive integers. If* $n < m$, *then*

$$E(X_m|X_1, \ldots, X_n) = X_n$$

with probability 1.

Proof. Since $\{X_1, X_2, \ldots\}$ is a martingale,

$$E[X_{n+1}|X_1, \ldots, X_n] = X_n \quad \text{with probability } 1.$$

Note that

$$E[X_{n+2}|X_1, \ldots, X_n] = E[E(X_{n+2}|X_1, \ldots, X_n, X_{n+1})|X_1, \ldots, X_n]$$
$$= E[X_{n+1}|X_1, \ldots, X_n] = X_n$$

with probability 1. Similarly,

$$E(X_{n+3}|X_1, \ldots, X_n) = E[E(X_{n+3}|X_1, \ldots, X_n, X_{n+1}, X_{n+2}|X_1, \ldots, X_n]$$
$$= E(X_{n+2}|X_1, \ldots, X_n) = X_n$$

with probability 1. Continuing this argument yields the result. ∎

The martingale properties of a sequence $X_1, X_2, \ldots$ can also be described in terms of the differences

$$D_n = X_n - X_{n-1}, \quad n = 1, 2, \ldots$$

where $X_0 = 0$. Note that, for each $n = 1, 2, \ldots, (X_1, \ldots, X_n)$ is a one-to-one function of $(D_1, \ldots, D_n)$ since

$$X_m = D_1 + \cdots + D_m, \quad m = 1, 2, \ldots, n.$$

Suppose that $\{X_1, X_2, \ldots\}$ is a martingale. Then, by Theorem 2.5,

$$E\{D_{n+1}|D_1, \ldots, D_n\} = E\{E(D_{n+1}|X_1, \ldots, X_n)|D_1, \ldots, D_n\}$$
$$= E\{E(X_{n+1} - X_n|X_1, \ldots, X_n)|D_1, \ldots, D_n\} = 0, \quad n = 1, 2, \ldots.$$

A sequence of real-valued random variables $D_1, D_2, \ldots$ satisfying

$$E\{D_{n+1}|D_1, \ldots, D_n\} = 0, \quad n = 1, 2, \ldots$$

is said to be a sequence of *martingale differences*.

As noted above, if $X_1, X_2, \ldots$ is a martingale, then X_n can be interpreted as the fortune of a gambler after a series of fair games. In the same manner, if $D_1, D_2, \ldots$ is a martingale difference sequence, then D_n can be interpreted as the amount won by the gambler on the nth game.

Example 2.25 (Gambling systems). Suppose that a gambler plays a series of fair games with outcomes $D_1, D_2, \ldots$ such that, if the gambler places a bet B_n on the nth game, her winnings are $B_n D_n$. For each n, the bet B_n is a function of $D_1, \ldots, D_{n-1}$, the outcomes of the first $n-1$ games, but, of course, B_n cannot depend on $D_n, \ldots$. We take $B_1 = 1$ so that initial fortune of the gambler is given by D_1. The random variables $B_1, B_2, \ldots$ describe how the gambler uses the information provided by the game to construct her series of bets and is called a *gambling system*. The gambler's fortune after n games is thus

$$W_{n+1} = B_1 D_1 + \cdots + B_{n+1} D_{n+1}.$$

Then, using the fact that $D_1, D_2, \ldots$ is a martingale difference sequence, and assuming that $B_2, B_3, \ldots$ are bounded,

$$\begin{aligned} E(W_{n+1}|D_1, \ldots, D_n) &= D_1 + B_2 D_2 + \cdots + B_n D_n + E(B_{n+1} D_{n+1}|D_1, \ldots, D_n) \\ &= D_1 + B_2 D_2 + \cdots + B_n D_n + B_{n+1} E(D_{n+1}|D_1, \ldots, D_n) \\ &= D_1 + B_2 D_2 + \cdots + B_n D_n \equiv W_n. \end{aligned}$$

It follows that $E(W_{n+1}) = E(W_n)$, $n = 1, 2, \ldots$ so that

$$E(W_1) = E(W_2) = \cdots;$$

that is, the expected fortune of the gambler after n games is always equal to the initial fortune. Thus, a gambling system of the type described here cannot convert a fair game into one that is advantageous to the gambler. $\square$

2.7 Exercises

2.1 Let X and Y denote real-valued random variables such that the distribution of (X, Y) is absolutely continuous with density function $p(x, y)$. Suppose that there exist real-valued nonnegative functions g and h such that

$$\int_{-\infty}^{\infty} g(x)\,dx < \infty \quad \text{and} \quad \int_{-\infty}^{\infty} h(y)\,dy < \infty \qquad \text{bounded.}$$

and

$$p(x, y) = g(x)h(y), \quad -\infty < x < \infty, \ -\infty < y < \infty.$$

Does it follow that X and Y are independent?

2.2 Let X and Y denote independent random vectors with ranges $\mathcal{X}$ and $\mathcal{Y}$, respectively. Consider functions $f : \mathcal{X} \to \mathbf{R}$ and $g : \mathcal{Y} \to \mathbf{R}$. Does it follow that $f(X)$ and $g(Y)$ are independent random variables?

2.3 Let $X_1, X_2, \ldots, X_m$ denote real-valued random variables. Suppose that for each $n = 1, 2, \ldots, m$, $(X_1, \ldots, X_{n-1})$ and X_n are independent. Does it follow that $X_1, \ldots, X_m$ are independent?

2.4 Let X and Y denote real-valued random variables such that the distribution of (X, Y) is absolutely continuous with density function

$$p(x, y) = \frac{1}{x^3 y^2}, \quad x > 1, \ y > 1/x.$$

Find the density functions of the marginal distributions of X and Y.

2.5 Let X and Y denote real-valued random variables such that the distribution of (X, Y) is discrete with frequency function

$$p(x, y) = \frac{1}{2e} \frac{e^{-1} + 2^{-(x+y)}}{x! y!}, \quad x, y = 0, 1, \ldots.$$

Find the frequency functions of the marginal distributions of X and Y.

2.6   Prove Corollary 2.1.

2.7   Prove Theorem 2.2.

2.8   Let (X, Y) denote a random vector with an absolutely continuous distribution with density function p and let p_X and p_Y denote the marginal densities of X and Y, respectively. Let $p_{X|Y}(\cdot|y)$ denote the density of the conditional distribution of X given $Y = y$ and let $p_{Y|X}(\cdot|x)$ denote the density of the conditional distribution of Y given $X = x$. Show that

$$p_{X|Y}(x|y) = \frac{p_{Y|X}(y|x)p_X(x)}{p_Y(y)}$$

provided that $p_Y(y) > 0$. This is Bayes Theorem for density functions.

2.9   Consider a sequence of random variables $X_1, X_2, \ldots, X_n$ which each take the values 0 and 1. Assume that

$$\Pr(X_j = 1) = 1 - \Pr(X_j = 0) = \phi, \quad j = 1, \ldots, n$$

where $0 < \phi < 1$ and that

$$\Pr(X_j = 1|X_{j-1} = 1) = \lambda, \quad j = 2, \ldots, n.$$

(a) Find $\Pr(X_j = 0|X_{j-1} = 1)$, $\Pr(X_j = 1|X_{j-1} = 0)$, $\Pr(X_j = 0|X_{j-1} = 0)$.

(b) Find the requirements on λ so that this describes a valid probability distribution for $X_1, \ldots, X_n$.

2.10 Let X and Y denote real-valued random variables such that the distribution of (X, Y) is absolutely continuous with density function p and let p_X denote the marginal density function of X. Suppose that there exists a point x_0 such that $p_X(x_0) > 0$, p_X is continuous at x_0, and for almost all y, $p(\cdot, y)$ is continuous at x_0. Let A denote a subset of $\mathbf{R}$. For each $\epsilon > 0$, let

$$d(\epsilon) = \Pr(Y \in A|x_0 \le X \le x_0 + \epsilon].$$

Show that

$$\Pr[Y \in A|X = x_0] = \lim_{\epsilon \to 0} d(\epsilon).$$

2.11 Let (X, Y) denote a random vector with the distribution described in Exercise 2.4. Find the density function of the conditional distribution of X given $Y = y$ and of the conditional distribution of Y given $X = x$.

2.12 Let X denote a real-valued random variable with range $\mathcal{X}$, such that $\mathrm{E}(|X|) < \infty$. Let $A_1, \ldots, A_n$ denote disjoint subsets of $\mathcal{X}$. Show that

$$\mathrm{E}(X) = \sum_{j=1}^{n} \mathrm{E}(X|X \in A_j)\Pr(X \in A_j).$$

2.13 Let X denote a real-valued random variable with an absolutely continuous distribution with distribution function F and density p. For $c \ge 0$, find an expression for $\Pr(X > 0||X| = c)$.

2.14 Let X, Y, and Z denote random variables, possibly vector-valued. Let $\mathcal{X}$ denote the range of X and let $\mathcal{Y}$ denote the range of Y. X and Y are said to be *conditionally independent given Z* if, for any $A \subset \mathcal{X}$ and $B \subset \mathcal{Y}$,

$$\Pr(X \in A, \ Y \in B|Z) = \Pr(X \in A|Z)\Pr(Y \in B|Z)$$

with probability 1.

(a) Suppose that X and Y are conditionally independent given Z and that Y and Z are independent. Does it follow that X and Z are independent?

(b) Suppose that X and Y are conditionally independent given Z and that X and Z are conditionally independent given Y. Does it follow that X and Y are independent?

2.15 Let X and Y denote random vectors with ranges $\mathcal{X}$ and $\mathcal{Y}$, respectively. Show that, if X and Y are independent, then

$$E[g(X)|Y] = E[g(X)]$$

for any function $g : \mathcal{X} \to \mathbf{R}$ such that $E[|g(X)|] < \infty$.

Does the converse to this result hold? That is, suppose that

$$E[g(X)|Y] = E[g(X)]$$

for any function $g : \mathcal{X} \to \mathbf{R}$ such that $E[|g(X)|] < \infty$. Does it follow that X and Y are independent?

2.16 Prove Theorem 2.4.

2.17 Let X, Y and Z denote real-valued random variables such that (X, Y) and Z are independent. Assume that $E(|Y|) < \infty$. Does it follow that

$$E(Y|X, Z) = E(Y|X)?$$

2.18 Let X denote a nonnegative, real-valued random variable. The *expected residual life* function of X is given by

$$R(x) = E(X - x|X \geq x), \quad x > 0.$$

Let F denote the distribution function of X.

(a) Find an expression for R in terms of the integral

$$\int_x^\infty F(t)\, dt.$$

(b) Find an expression for F in terms of R.

(c) Let X_1 and X_2 denote nonnegative, real-valued random variables with distribution functions F_1 and F_2 and expected residual life functions R_1 and R_2. If

$$R_1(x) = R_2(x), \quad x > 0$$

does it follow that

$$F_1(x) = F_2(x), \quad -\infty < x < \infty?$$

2.19 Let $\mathcal{L}_2$ denote the linear space of random variables X such that $E(X^2) < \infty$, as described in Exercises 1.28 and 1.29. Let X_1, X_2 denote elements of $\mathcal{L}_2$; we say that X_1 and X_2 are orthogonal, written $X_1 \perp X_2$, if $E[X_1 X_2] = 0$.

Let Z denote a given element of $\mathcal{L}_2$ and let $\mathcal{L}_2(Z)$ denote the elements of $\mathcal{L}_2$ that are functions of Z. For a given random variable $Y \in \mathcal{L}_2$, let $P_Z Y$ denote the projection of Y onto $\mathcal{L}_2(Z)$, defined to be the element of $\mathcal{L}_2(Z)$ such that $Y - P_Z Y$ is orthogonal to all elements of $\mathcal{L}_2(Z)$. Show that $P_Z Y = E(Y|Z)$.

2.20 Let $X_1, X_2,$ and Z denote independent, real-valued random variables. Assume that

$$\Pr(Z = 0) = 1 - \Pr(Z = 1) = \alpha$$

for some $0 < \alpha < 1$. Define

$$Y = \begin{cases} X_1 & \text{if } Z = 0 \\ X_2 & \text{if } Z = 1 \end{cases}.$$

(a) Suppose that $E(X_1)$ and $E(X_2)$ exist. Does it follow that $E(Y)$ exists?

(b) Assume that $E(|X_1|) < \infty$ and $E(|X_2|) < \infty$. Find $E(Y|X_1)$.

2.21 Let (X, Y) denote a two-dimensional random vector with an absolutely continuous distribution with density function

$$p(x, y) = \frac{1}{y} \exp(-y), \quad 0 < x < y < \infty.$$

Find $E(X^r | Y)$ for $r = 1, 2, \ldots$.

2.22 For some $n = 1, 2, \ldots$, let $Y_1, Y_2, \ldots, Y_{n+1}$ denote independent, identically distributed, real-valued random variables. Define

$$X_j = Y_j Y_{j+1}, \quad j = 1, \ldots, n.$$

(a) Are $X_1, X_2, \ldots, X_n$ independent?

(b) Are $X_1, X_2, \ldots, X_n$ exchangeable?

2.23 Let X and Y denote real-valued exchangeable random variables. Find $\Pr(X \le Y)$.

2.24 Prove Theorem 2.7.

2.25 Let $X_0, X_1, \ldots, X_n$ denote independent, identically distributed real-valued random variables. For each definition of $Y_1, \ldots, Y_n$ given below, state whether or not $Y_1, \ldots, Y_n$ are exchangeable and justify your answer.

(a) $Y_j = X_j - X_0, j = 1, \ldots, n$.

(b) $Y_j = X_j - X_{j-1}, j = 1, \ldots, n$.

(c) $Y_j = X_j - \bar{X}, j = 1, \ldots, n$ where $\bar{X} = \sum_{j=1}^{n} X_j / n$.

(d) $Y_j = (j/n)X_j + (1 - j/n)X_0, j = 1, \ldots, n$.

2.26 Let $Y_1, Y_2, \ldots$ denote independent, identically distributed nonnegative random variables with $E(Y_1) = 1$. For each $n = 1, 2, \ldots$, let

$$X_n = Y_1 \cdots Y_n.$$

Is $\{X_1, X_2, \ldots\}$ a martingale?

2.27 Let $\{X_1, X_2, \ldots\}$ denote a martingale. Show that

$$E(X_1) = E(X_2) = \cdots.$$

Exercises 2.28 and 2.29 use the following definition. A sequence of real-valued random variables $\{X_1, X_2, \ldots\}$ such that $E[|X_n|] < \infty, n = 1, 2, \ldots$, is said to be a *submartingale* if, for each $n = 1, 2, \ldots$,

$$E[X_{n+1} | X_1, \ldots, X_n] \ge X_n$$

with probability 1.

2.28 Show that if $\{X_1, X_2, \ldots\}$ is a submartingale, then $\{X_1, X_2, \ldots\}$ is a martingale if and only if

$$E(X_1) = E(X_2) = \cdots.$$

2.29 Let $\{X_1, X_2, \ldots\}$ denote a martingale. Show that $\{|X_1|, |X_2|, \ldots\}$ is a submartingale.

2.8 Suggestions for Further Reading

Conditional distributions and expectation are standard topics in probability theory. A mathematically rigorous treatment of these topics requires measure-theoretic probability theory; see, for example, Billingsley (1995, Chapter 6) and Port (1994, Chapter 14). For readers without the necessary background for these references, Parzen (1962, Chapter 2), Ross (1985, Chapter 3), Snell (1988, Chapter 4), and Woodroofe (1975, Chapter 10) give more elementary, but still very useful, discussions of conditioning.

Exercise 2.19 briefly considers an approach to conditional expectation based on projections in spaces of square-integrable random variables. This approach is developed more fully in Karr (1993, Chapter 8).

Exchangeable random variables are considered in Port (1994, Chapter 15). Schervish (1995, Chapter 1) discusses the relevance of this concept to statistical inference.

Martingales play an important role in probability and statistics. The definition of a martingale used in this chapter is a special case of a more general, and more useful, definition. Let $X_1, X_2, \ldots$ and $Y_1, Y_2, \ldots$ denote sequences of random variables and suppose that, for each $n = 1, 2, \ldots, X_n$ is a function of $Y_1, Y_2, \ldots, Y_n$. The sequence $(X_1, X_2, \ldots)$ is said to be a martingale with respect to $(Y_1, Y_2, \ldots)$ if

$$\mathrm{E}(X_{n+1}|Y_1, \ldots, Y_n) = X_n, \quad n = 1, 2, \ldots.$$

Thus, the definition used in this chapter is a special case in which Y_n is taken to be $X_n, n = 1, 2, \ldots.$ See, for example, Billingsley (1995, Section 35), Karr (1993, Chapter 9), Port (1994, Chapter 17), and Woodroofe (1975, Chapter 12) for further discussion of martingales.

3

Characteristic Functions

3.1 Introduction

The properties of a random variable may be described by its distribution function or, in some cases, by its density or frequency function. In Section 1.8 it was shown that expectations of the form $E[g(X)]$ for all bounded, continuous, real-valued functions g completely determine the distribution of X (Theorem 1.11). However, the entire set of all bounded, continuous real-valued functions is not needed to characterize the distribution of a random variable in this way.

Let X denote a real-valued random variable with distribution function F. For each $t \in \mathbf{R}$, let g_t denote a function on the range of X such that $E[|g_t(X)|] < \infty$. Then the function

$$W(t) = E[g_t(X)] = \int_{-\infty}^{\infty} g_t(X) \, dF(x), \quad t \in \mathbf{R},$$

gives the expected values of all functions of the form g_t. If the set of functions $\mathcal{G} = \{g_t : t \in \mathbf{R}\}$ is chosen appropriately, then function W will completely characterize the distribution of X, and certain features of F will be reflected in W. In fact, we have already seen one simple example of this with the distribution function, in which $g_t(x) = I_{\{x \le t\}}$.

A function such as W is called an *integral transform* of F; the properties of an integral transform will depend on the properties of the class of functions $\mathcal{G}$. In this chapter, we consider a particular integral transform, the *characteristic function*. Two other integral transforms, the Laplace transform and the moment-generating function, are discussed in Chapter 4.

The characteristic function of the distribution of a random variable X, or, more simply, the characteristic function of X, is defined as

$$\varphi(t) \equiv \varphi_X(t) = E[\exp(itX)] \equiv \int_{-\infty}^{\infty} \exp(itx) \, dF(x), \quad -\infty < t < \infty,$$

where $\exp(itx)$ is a complex number; writing

$$\exp(itx) = \cos(tx) + i \, \sin(tx),$$

as described in Appendix 2, we may write $\varphi(t) = u(t) + iv(t)$ where

$$u(t) = \int_{-\infty}^{\infty} \cos(tx) \, dF(x) \quad \text{and} \quad v(t) = \int_{-\infty}^{\infty} \sin(tx) \, dF(x). \tag{3.1}$$

Characteristic Functions

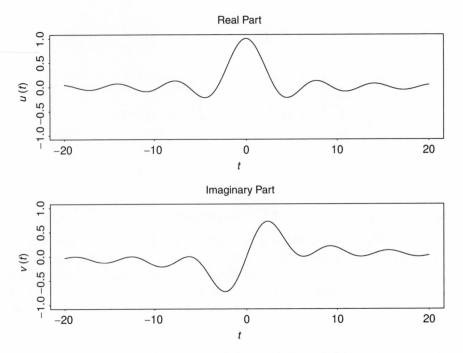

Figure 3.1. Characteristic function in Example 3.1.

Thus, a characteristic function may be viewed as two real-valued functions, $u(\cdot)$ and $v(\cdot)$, the real and imaginary parts of $\varphi(\cdot)$, respectively. Note that, since cos and sin are bounded functions, the characteristic function of a random variable always exists.

Example 3.1 *(Uniform distribution on the unit interval).* Let X denote a real-valued random variable with distribution function

$$F(x) = x, \quad 0 \le x \le 1.$$

The characteristic function of this distribution is given by

$$\varphi(t) = \int_0^1 \exp(itx)\,dx = \frac{\exp(it) - 1}{it} = \frac{\sin(t)}{t} + i\frac{1 - \cos(t)}{t}, \quad t \in \mathbf{R}.$$

Plots of the real and imaginary parts of φ are given in Figure 3.1. □

Example 3.2 *(Standard normal distribution).* Let Z denote a real-valued random variable with an absolutely continuous distribution with density function

$$p(z) = \frac{1}{\sqrt{(2\pi)}} \exp\left(-\frac{1}{2}z^2\right), \quad -\infty < z < \infty;$$

this is called the *standard normal distribution*. The characteristic function of this distribution is given by

$$\varphi(t) = \frac{1}{\sqrt{(2\pi)}} \int_{-\infty}^{\infty} \exp(itz) \exp\left(-\frac{1}{2}z^2\right) dz = \exp\left(-\frac{1}{2}t^2\right), \quad t \in \mathbf{R}.$$

Thus, aside from a constant, the characteristic function of the standard normal distribution is the same as its density function.   □

Example 3.3 (*Binomial distribution*). Let X denote a real-valued random variable with a discrete distribution with frequency function

$$p(x) = \binom{n}{x} \theta^x (1 - \theta)^{n-x}, \quad x = 0, \dots, n;$$

here θ and n are fixed constants, with θ taking values in the interval $(0, 1)$ and n taking values in the set $\{1, 2, \dots\}$. This is a binomial distribution with parameters n and θ. The characteristic function of this distribution is given by

$$\begin{aligned}
\varphi(t) &= \sum_{x=0}^{n} \exp(itx) \binom{n}{x} \theta^x (1 - \theta)^{n-x} \\
&= (1 - \theta)^n \sum_{x=0}^{n} \binom{n}{x} \left(\frac{\theta \exp(itx)}{1 - \theta} \right)^x \\
&= [1 - \theta + \theta \exp(it)]^n.
\end{aligned}$$

Plots of the real and imaginary parts of φ for the case $n = 3, \theta = 1/2$ are given in Figure 3.2.   □

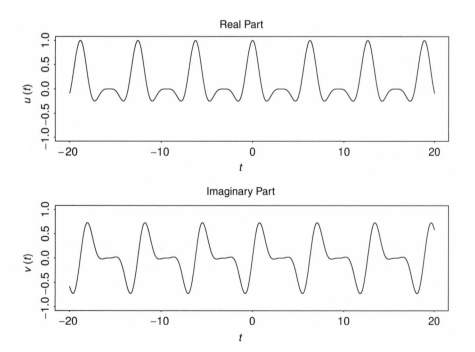

Figure 3.2. Characteristic function in Example 3.3.

***Example* 3.4 (*Gamma distribution*).** Let X denote a real-valued random variable with an absolutely continuous distribution with density function

$$f(x) = \frac{\beta^\alpha}{\Gamma(\alpha)} x^{\alpha-1} \exp(-\beta x), \quad x > 0$$

where α and β are nonnegative constants. This is called a *gamma distribution with parameters α and β*.

The characteristic function of this distribution is given by

$$\varphi(t) = \int_0^\infty \exp(itx) \frac{\beta^\alpha}{\Gamma(\alpha)} x^{\alpha-1} \exp(-\beta x)\, dx$$

$$= \frac{\beta^\alpha}{\Gamma(\alpha)} \int_0^\infty x^{\alpha-1} \exp(-(\beta - it)x) = \frac{\beta^\alpha}{(\beta - it)^\alpha}, \quad -\infty < t < \infty. \qquad \square$$

The characteristic function has a number of useful properties that make it convenient for deriving many important results. The main drawback of the characteristic function is that it requires some results in complex analysis. However, the complex analysis required is relatively straightforward and the advantages provided by the use of characteristic functions far outweigh this minor inconvenience. For readers unfamiliar with complex analysis, a brief summary is given in Appendix 2.

3.2 Basic Properties

Characteristic functions have a number of basic properties.

***Theorem* 3.1.** *Let $\varphi(\cdot)$ denote the characteristic function of a distribution on* **R**. *Then*
  (i) *φ is a continuous function*
  (ii) *$|\varphi(t)| \leq 1$ for all $t \in$ **R***
  (iii) *Let X denote a real-valued random variable with characteristic function φ, let a, b denote real-valued constants, and let $Y = aX + b$. Then φ_Y, the characteristic function of Y, is given by*

$$\varphi_Y(t) = \exp(ibt)\varphi(at).$$

  (iv) *u is an even function and v is an odd function, where u and v are given by (3.1).*

Proof. Note that

$$|\varphi(t + h) - \varphi(t)| \leq \int_{-\infty}^\infty |\exp\{ix(t + h)\} - \exp\{itx\}|\, dF_X(x)$$

$$\leq \int_{-\infty}^\infty |\exp\{ixh\} - 1|\, dF_X(x).$$

Note that $|\exp\{ixh\} - 1|$ is a real-valued function bounded by 2. Hence, the continuity of φ follows from the Dominated Convergence Theorem (see Appendix 1), using the fact that $\exp(ixh)$ is continuous at $h = 0$. This establishes part (i).

Part (ii) follows from the fact that

$$|\varphi(t)| \leq \int_{-\infty}^\infty |\exp\{itx\}|\, dF_X(x) \leq 1.$$

Parts (iii) and (iv) are immediate. ∎

Example* 3.5 *(Uniform distribution on the unit interval). In Example 3.1 it is shown that the characteristic function of this distribution is given by

$$\varphi(t) = \int_0^1 \exp(itx)\, dx = \frac{\exp(it) - 1}{it}.$$

Hence, part (ii) of Theorem 3.1 implies that for all $-\infty < t < \infty$,

$$|\exp(it) - 1| \leq |t|.$$

This is a useful result in complex analysis; see Appendix 2. □

Example* 3.6 *(Normal distribution). Let Z denote a real-valued random variable with a standard normal distribution and let μ and σ be real-valued constants with $\sigma > 0$. Define a random variable X by

$$X = \sigma Z + \mu.$$

The distribution of X is called a *normal distribution with parameters μ and σ*.

Recall that the characteristic function of Z is $\exp(-t^2/2)$; according to part (iii) of Theorem 3.1, the characteristic function of X is

$$\exp(i\mu t) \exp\left(-\frac{1}{2}\sigma^2 t^2\right) = \exp\left(-\frac{\sigma^2}{2}t^2 + i\mu t\right).$$

□

Uniqueness and inversion of characteristic functions

The characteristic function is essentially the Fourier transform used in mathematical analysis. Let g denote a function of bounded variation such that

$$\int_{-\infty}^{\infty} |g(x)|\, dx < \infty.$$

The *Fourier transform* of g is given by

$$G(t) = \frac{1}{\sqrt{(2\pi)}} \int_{-\infty}^{\infty} g(x) \exp\{itx\}\, dx, \quad \infty < t < \infty.$$

The following result shows that it is possible to recover a function from its Fourier transform.

***Theorem* 3.2.** *Let G denote the Fourier transform of a function g, which is of bounded variation.*

(i) $G(t) \to 0$ as $t \to \pm\infty$.

(ii) Suppose x_0 is a continuity point of g. Then

$$g(x_0) = \frac{1}{\sqrt{(2\pi)}} \lim_{T \to \infty} \int_{-T}^{T} G(t) \exp\{-itx_0\}\, dt.$$

Proof. The proof of this theorem uses the Riemann-Lebesgue Lemma (Section A3.4.10) along with the result in Section A3.4.11 of Appendix 3.

Note that part (i) follows provided that

$$\lim_{t \to \infty} \int_{-\infty}^{\infty} g(x)\, \sin(tx)\, dx = 0.$$

and that

$$\lim_{t \to \infty} \int_{-\infty}^{\infty} g(x) \cos(tx) \, dx = 0.$$

These follow immediately from the Riemann-Lebesgue Lemma.

Now consider part (ii). Note that it suffices to show that

$$g(x_0) = \frac{1}{2\pi} \lim_{T \to \infty} \int_{-T}^{T} \int_{-\infty}^{\infty} g(x) \cos(t(x - x_0)) \, dx \, dt \qquad (3.2)$$

and

$$\lim_{T \to \infty} \int_{-T}^{T} \int_{-\infty}^{\infty} g(x) \sin(t(x - x_0)) \, dx \, dt = 0. \qquad (3.3)$$

Consider (3.3). Changing the order of integration,

$$\int_{-T}^{T} \int_{-\infty}^{\infty} g(x) \sin(t(x - x_0)) \, dx \, dt = \int_{-\infty}^{\infty} g(x) \int_{-T}^{T} \sin(t(x - x_0)) \, dt \, dx.$$

Equation (3.3) now follows from the fact that, for $T > 0$,

$$\int_{-T}^{T} \sin(t(x - x_0)) \, dt = 0.$$

Now consider (3.2). Again, changing the order of integration, and using the change-of-variable $u = x - x_0$,

$$\int_{-T}^{T} \int_{-\infty}^{\infty} g(x) \cos(t(x - x_0)) \, dx \, dt = \int_{-\infty}^{\infty} g(x) \int_{-T}^{T} \cos(t(x - x_0)) \, dt \, dx$$

$$= \int_{-\infty}^{\infty} g(u + x_0) \frac{\sin(Tu)}{u} \, du.$$

It now follows from Section A3.4.11 that

$$\lim_{T \to \infty} \int_{-\infty}^{\infty} g(u + x_0) \frac{\sin(Tu)}{u} \, du = g(x_0). \quad \blacksquare$$

The following theorem applies this result to characteristic functions and shows that the distribution function of a real-valued random variable may be obtained from its characteristic function, at least at continuity points of the distribution function.

Theorem 3.3. *Let X denote a real-valued random variable with distribution function F and characteristic function φ. If F is continuous at $x_0, x_1, x_0 < x_1$, then*

$$F(x_1) - F(x_0) = \frac{1}{2\pi} \lim_{T \to \infty} \int_{-T}^{T} \frac{\exp\{-itx_0\} - \exp\{-itx_1\}}{it} \varphi(t) \, dt.$$

Proof. Fix x. Define

$$h(y) = F(x + y) - F(y).$$

Then h is of bounded variation and for any $a < b$

$$\int_a^b |h(y)|\,dy = \int_a^b F(x+y)\,dy - \int_a^b F(y)\,dy$$

$$= \int_b^{b+x} F(z)\,dz - \int_a^{a+x} F(z)\,dz$$

$$\le [F(b+x) - F(a)]x \le x.$$

Hence,

$$\int_{-\infty}^{\infty} |h(y)|\,dy < \infty.$$

It follows from the Theorem 3.2 that

$$h(y) = \frac{1}{2\pi} \lim_{T\to\infty} \int_{-T}^{T} \exp\{-ity\} \left\{ \int_{-\infty}^{\infty} h(z)\exp(itz) \right\}\,dz\,dy$$

provided that h is continuous at y; note that, in this expression, the integral with respect to dz is simply the Fourier transform of h. Consider the integral

$$\int_{-\infty}^{\infty} h(z)\exp\{itz\}\,dz = \frac{1}{it} \int_{-\infty}^{\infty} h(z)\,d\exp\{itz\}.$$

Using integration-by-parts, this integral is equal to

$$-\frac{1}{it} \int_{-\infty}^{\infty} \exp\{itz\}\,dh(z) + \frac{\exp\{itz\}}{it} h(z) \Big|_{-\infty}^{\infty}.$$

Note that $\exp\{itz\}$ is bounded,

$$\lim_{z\to\infty} h(z) = \lim_{z\to-\infty} h(z) = 0,$$

and

$$\int_{-\infty}^{\infty} \exp\{itz\}\,dh(z) = \int_{-\infty}^{\infty} \exp\{itz\}\,dF(x+z) - \int_{-\infty}^{\infty} \exp\{itz\}\,dF(z)$$

$$= \varphi(t)[\exp\{-itx\} - 1].$$

Hence,

$$\int_{-\infty}^{\infty} h(z)\exp(itz)\,dz = \varphi(t)\frac{1 - \exp(-itx)}{it},$$

so that

$$h(y) = \frac{1}{2\pi} \lim_{T\to\infty} \int_{-T}^{T} \frac{\exp\{-ity\} - \exp\{-it(x+y)\}}{it} \varphi(t)\,dt,$$

provided that h is continuous at y, which holds provided that F is continuous at x and $x + y$. Choosing $y = x_0$ and $x = x_1 - x_0$ yields the result. ∎

Thus, given the characteristic function of a random variable X we may determine differences of F, the distribution function of X, of the form $F(x_1) - F(x_0)$ for continuity points x_0, x_1. However, since set of points at which F_X is discontinuous is countable, and F is right-continuous, the characteristic function determines the entire distribution of X. The details are given in the following corollary to Theorem 3.3.

***Corollary* 3.1.** *Let X and Y denote real-value random variables with characteristic functions φ_X and φ_Y, respectively. X and Y have the same distribution if and only if*

$$\varphi_X(t) = \varphi_Y(t), \quad -\infty < t < \infty. \tag{3.4}$$

Proof. Clearly, if X and Y have the same distribution then they have the same characteristic function.

Now suppose that (3.4) holds; let F_X denote the distribution function of X and let F_Y denote the distribution function of Y. It follows from Theorem 3.3 that if a and b are continuity points of both F_X and F_Y, then

$$F_X(b) - F_X(a) = F_Y(b) - F_Y(a).$$

Let a_n, $n = 1, 2, \ldots$, denote a sequence of continuity points of both F_X and F_Y such that a_n diverges to $-\infty$ as $n \to \infty$. Note that, since the points at which either F_X or F_Y is not continuous is countable, such a sequence must exist. Then

$$F_X(b) - F_X(a_n) = F_Y(b) - F_Y(a_n), \quad n = 1, 2, \ldots$$

so that

$$F_X(b) - F_Y(b) = \lim_{n \to \infty} F_X(a_n) - F_Y(a_n) = 0.$$

Hence, $F_X(b)$ and $F_Y(b)$ are equal for any point b that is a continuity point of both F_X and F_Y.

Now suppose at least one of F_X and F_Y is not continuous at b. Let b_n, $n = 1, 2, \ldots$, denote a sequence of continuity points of F_X and F_Y decreasing to b. Then

$$F_X(b_n) = F_Y(b_n), \quad n = 1, 2, \ldots$$

and, by the right-continuity of F_X and F_Y,

$$F_X(b) = F_Y(b),$$

proving the result. ∎

Characteristic function of a sum

The following result illustrates one of the main advantages of working with characteristic functions rather than with distribution functions or density functions. The proof is straightforward and is left as an exercise.

***Theorem* 3.4.** *Let X and Y denote independent, real-valued random variables with characteristic functions φ_X and φ_Y, respectively. Let φ_{X+Y} denote the characteristic function of the random variable $X + Y$. Then*

$$\varphi_{X+Y}(t) = \varphi_X(t)\varphi_Y(t), \quad t \in \mathbf{R}.$$

The result given in Theorem 3.4 clearly extends to a sequence of n independent random variables and, hence, gives one method for determining the distribution of a sum $X_1 + \cdots + X_n$ of random variables $X_1, \ldots, X_n$; other methods will be discussed in Chapter 7.

Example 3.7 (Chi-squared distribution). Let Z denote a random variable with a standard normal distribution and consider the distribution of Z^2. This distribution has characteristic function

$$\int_{-\infty}^{\infty} \exp(itz^2) \frac{1}{\sqrt{(2\pi)}} \exp\left(-\frac{1}{2}z^2\right) dz = \frac{1}{\sqrt{(2\pi)}} \int_{-\infty}^{\infty} \exp\left[-\frac{1}{2}(1-2it)z^2\right] dz$$

$$= \frac{1}{(1-2it)^{\frac{1}{2}}}.$$

Now consider independent standard normal random variables $Z_1, Z_2, \cdots, Z_n$ and let $X = Z_1^2 + \cdots + Z_n^2$. By Theorem 3.4, the characteristic function of X is

$$\varphi(t) = \frac{1}{(1-2it)^{\frac{n}{2}}}.$$

Comparing this result to the characteristic function derived in Example 3.4 shows that X has a gamma distribution with parameters $\alpha = n/2$ and $\beta = 1/2$. This special case of the gamma distribution is called the *chi-squared distribution with n degrees of freedom*; note that this distribution is defined for any positive value of n, not just integer values. ☐

Example 3.8 (Sum of binomial random variables). Let X_1 and X_2 denote independent random variables such that, for $j = 1, 2$, X_j has a binomial distribution with parameters n_j and θ_j. Recall from Example 3.3 that X_j has characteristic function

$$\varphi_j(t) = [1 - \theta_j + \theta_j \exp(it)]^{n_j}, \quad j = 1, 2.$$

Let $X = X_1 + X_2$. Then X has characteristic function

$$\varphi(t) = [1 - \theta_1 + \theta_1 \exp(it)]^{n_1}[1 - \theta_2 + \theta_2 \exp(it)]^{n_2}.$$

Hence, if $\theta_1 = \theta_2$, then X also has a binomial distribution. ☐

An expansion for characteristic functions

It is well known that the exponential function of a real-valued argument, $\exp(x)$, can be expanded in a power series in x:

$$\exp(x) = \sum_{j=0}^{\infty} \frac{x^j}{j!}.$$

The same result holds for complex arguments, so that

$$\exp(itx) = \sum_{j=0}^{\infty} \frac{(itx)^j}{j!};$$

see Appendix 2 for further discussion. Thus, the characteristic function of a random variable X can be expanded in power series whose coefficients involve expected values of the form $E(X^m), m = 0, 1, \ldots$.

This fact can be used to show that the existence of $E(X^m), m = 1, 2, \ldots$, is related to the smoothness of the characteristic function at 0; in particular, if $E(|X|^m) < \infty$, then φ_X is m-times differentiable at 0. The converse to this result is also useful, but it applies only to moments, and derivatives, of even order. Specifically, if φ_X is $2m$-times differentiable at 0 then $E(X^{2m}) < \infty$. The details are given in the following theorem.

Theorem 3.5. *Let X denote a real-valued random variable with characterstic function $\varphi(\cdot)$. If, for some $m = 1, 2, \ldots$, $\mathrm{E}(X^m)$ exists and is finite, then*

$$\varphi(t) = 1 + \sum_{j=1}^{m} \frac{(it)^j}{j!} \mathrm{E}(X^j) + o(t^m) \quad \text{as } t \to 0$$

and, hence, φ is m-times differentiable at 0 with

$$\varphi^{(j)}(0) = (i)^j \mathrm{E}(X^j), \quad j = 1, \ldots, m.$$

If, for some $m = 1, 2, \ldots$, $\varphi^{(2m)}(0)$ exists, then $\mathrm{E}(X^{2m}) < \infty$.

Proof. Assume that $\mathrm{E}(X^m)$ exists and is finite for some $m = 1, 2, \ldots$. By Lemma A2.1 in Appendix 2, we may write

$$\exp\{it\} = \sum_{j=0}^{m} \frac{(it)^j}{j!} + R_m(t)$$

where

$$|R_m(t)| \le \min\{|t|^{m+1}/(m+1)!, \ 2|t|^m/m!\}.$$

Hence,

$$\varphi(t) = \mathrm{E}[\exp\{itX\}] = 1 + \sum_{j=1}^{m} \frac{(it)^j \mathrm{E}(X^j)}{j!} + \mathrm{E}[R_m(tX)].$$

It remains to be shown that

$$\frac{|\mathrm{E}[R_m(tX)]|}{|t|^m} \to 0 \quad \text{as } t \to 0.$$

We know that $|\mathrm{E}[R_m(tX)]| \le \mathrm{E}[|R_m(tX)|]$ and that

$$\frac{m!|R_m(tX)|}{|t|^m} \le \min\left\{ \frac{|t||X|^{m+1}}{(m+1)}, 2|X|^m \right\}.$$

Let

$$M_t(x) = \min\left\{ \frac{|t||x|^{m+1}}{(m+1)}, 2|x|^m \right\} = \begin{cases} 2|x|^m & \text{if } |x| \ge 2(m+1)/|t| \\ |t||x|^{m+1}/(m+1) & \text{if } |x| \le 2(m+1)/|t| \end{cases}.$$

Hence, it suffices to show that

$$\mathrm{E}[M_t(X)] \to 0 \quad \text{as } t \to 0.$$

Note that for each x, $M_t(x) \to 0$ as $t \to 0$. Furthermore,

$$|M_t(x)| \le 2|x|^m$$

and, under the conditions of the theorem, $\mathrm{E}(2|X|^m) < \infty$. It now follows from the Dominated Convergence Theorem (see Appendix 1) that

$$\mathrm{E}[M_t(X)] \to 0 \quad \text{as } t \to 0.$$

This proves the first part of the theorem.

Now suppose that $\varphi^{(2)}(0)$ exists. Writing $\varphi^{(2)}(0)$ in terms of the first derivative φ' and then writing φ' in terms of φ, we have

$$\varphi^{(2)}(0) = \lim_{h \to 0} \frac{\varphi(h) - 2\varphi(0) + \varphi(-h)}{h^2}.$$

Now,

$$\varphi(h) - 2\varphi(0) + \varphi(-h) = E[\exp(ihX) - 2 + \exp(-ihX)]$$
$$= E\{[\exp(ihX/2) - \exp(-ihX/2)]^2\} = E\{[2i \ \sin(hX/2)]^2\}.$$

Hence,

$$\varphi^{(2)}(0) = \lim_{h \to 0} E\left\{\frac{[i \ \sin(hX/2)]^2}{(hX/2)^2} X^2\right\}$$

and

$$|\varphi^{(2)}(0)| = \lim_{h \to 0} E\left\{\frac{\sin(hX/2)^2}{(hX/2)^2} X^2\right\}.$$

By Fatou's Lemma,

$$\lim_{h \to 0} E\left\{\frac{\sin(hX/2)^2}{(hX/2)^2} X^2\right\} \geq E\left\{\liminf_{h \to 0} \frac{\sin(hX/2)^2}{(hX/2)^2} X^2\right\}$$

and since

$$\lim_{t \to 0} \frac{\sin(t)}{t} = 1$$

we have that

$$E(X^2) \leq |\varphi^{(2)}(0)|$$

so that $E(X^2) < \infty$.

The general case is similar, but a bit more complicated. Suppose that $\varphi^{(2m)}(0)$ exists for some $m = 1, 2, \dots$. Since

$$\varphi^{(2m)}(0) = \lim_{h \to 0} \frac{\varphi^{(2m-2)}(h) - 2\varphi^{(2m-2)}(0) + \varphi^{(2m-2)}(-h)}{h^2},$$

it may be shown by induction that

$$\varphi^{(2m)}(0) = \lim_{h \to 0} \frac{1}{h^{2m}} \sum_{j=0}^{2m} (-1)^j \binom{2m}{j} \varphi((j - m)h).$$

Furthermore,

$$\sum_{j=0}^{2m} (-1)^j \binom{2m}{j} \varphi((j - m)h) = E\left\{\sum_{j=0}^{2m} (-1)^j \binom{2m}{j} \exp[i(j - m)hX]\right\}$$
$$= E\{[\exp(ihX/2) - \exp(-ihX/2)]^{2m}\}$$
$$= E\{[2i \ \sin(hX/2)]^{2m}\}.$$

The remainder of the proof is essentially the same as the $m = 1$ case considered above. Since

$$\varphi^{(2m)}(0) = \lim_{h \to 0} E \left\{ \frac{[i \, \sin(hX/2)]^{2m}}{(hX/2)^{2m}} X^{2m} \right\}$$

and

$$|\varphi^{(2m)}(0)| = \lim_{h \to 0} E \left\{ \frac{\sin(hX/2)^{2m}}{(hX/2)^{2m}} X^{2m} \right\},$$

and by Fatou's Lemma (see Appendix 1),

$$\lim_{h \to 0} E \left\{ \frac{\sin(hX/2)^{2m}}{(hX/2)^{2m}} X^{2m} \right\} \ge E(X^{2m}),$$

it follows that

$$E(X^{2m}) \le |\varphi^{(2m)}(0)|$$

so that $E(X^{2m}) < \infty$. ∎

Example 3.9 *(Normal distribution).* Let X denote a random variable with a normal distribution with parameters μ and σ; see Example 3.6. This distribution has characteristic function

$$\varphi(t) = \exp \left(-\frac{\sigma^2}{2} t^2 + i \mu t \right).$$

Clearly, $\varphi(t)$ is m-times differentiable for any $m = 1, 2, \ldots$ so that $E(X^r)$ exists for all $r = 1, 2, \ldots$. It is straightforward to show that

$$\varphi'(0) = \mu \quad \text{and} \quad \varphi''(0) = -(\mu^2 + \sigma^2)$$

so that

$$E(X) = \mu \quad \text{and} \quad E(X^2) = \mu^2 + \sigma^2. \qquad \square$$

Example 3.10 *(Gamma distribution).* Consider a gamma distribution with parameters α and β, as in Example 3.4; recall that this distribution has characteristic function

$$\varphi(t) = \frac{\beta^\alpha}{(\beta - it)^\alpha}, \quad t \in \mathbf{R}.$$

Since the density function $p(x)$ decreases exponentially fast as $x \to \infty$, it is straightforward to show that $E(X^m)$ exists for any m and, hence, the moments may be determined by differentiating φ.

Note that

$$\varphi^{(m)}(t) = -(-i)^m [\alpha(\alpha + 1) \cdots (\alpha + m - 1)] \frac{\beta^\alpha}{(\beta - it)^{\alpha + m}}$$

so that

$$\varphi^{(m)}(0) = -(-i)^m [\alpha(\alpha + 1) \cdots (\alpha + m - 1)] \frac{1}{\beta^m}$$

and, hence, that

$$E(X^m) = \frac{\alpha(\alpha + 1) \cdots (\alpha + m - 1)}{\beta^m}.$$ □

Random vectors

Characteristic functions may also be defined for vector-valued random variables. Let X denote a random vector taking values in $\mathbf{R}^d$. The characteristic function of X is given by

$$\varphi(t) \equiv \varphi_X(t) = E[\exp(it^T X)], \quad t \in \mathbf{R}^d.$$

Many of the basic results regarding characteristic functions generalize in a natural way to the vector case. Several of these are given in the following theorem; the proof is left as an exercise.

Theorem 3.6. *Let* $\varphi(\cdot)$ *denote the characteristic function of a random variable taking values in* $\mathbf{R}^d$. *Then*

(i) φ *is a continuous function*

(ii) $|\varphi(t)| \leq 1, t \in \mathbf{R}^d$

(iii) *Let* X *denote a d-dimensional random vector with characteristic functin* φ_X. *Let* A *denote an* $r \times d$ *matrix, let* b *denote a d-dimensional vector of constants, and let* $Y = AX + b$. *Then* φ_Y, *the characteristic function of* Y, *satisfies* $\varphi_Y(t) = \exp(it^T b)\varphi_X(A^T t)$.

(iv) *Let* X *and* Y *denote independent random variables, each taking values in* $\mathbf{R}^d$, *with characteristic functions* φ_X *and* φ_Y, *respectively. Then*

$$\varphi_{X+Y}(t) = \varphi_X(t)\varphi_Y(t).$$

As in the case of a real-valued random variable, the characteristic function of a random vector completely determines its distribution. This result is stated without proof in the theorem below; see, for example, Port (1994, Section 51.1) for a proof.

Theorem 3.7. *Let* X *and* Y *denote random variables, taking values in* $\mathbf{R}^d$, *with characteristic functions* φ_X *and* φ_Y, *respectively. X and Y have the same distribution if and only if*

$$\varphi_X(t) = \varphi_Y(t), \quad t \in \mathbf{R}^d.$$

There is a very useful corollary to Theorem 3.7. It essentially reduces the problem of determining the distribution of a random vector to the problem of determining the distribution of all linear functions of the random vector, a problem that can be handled using methods for real-valued random variables; the proof is left as an exercise.

Corollary 3.2. *Let* X *and* Y *denote random vectors, each taking values in* $\mathbf{R}^d$. *X and Y have the same distribution if and only if* $a^T X$ *and* $a^T Y$ *have the same distribution for any* $a \in \mathbf{R}^d$.

Another useful corollary to Theorem 3.7 is that it is possible to determine if two random vectors are independent by considering their characteristic function; again, the proof is left as an exercise.

Corollary 3.3. *Let X denote a random vector taking values in $\mathbf{R}^d$ and let $X = (X_1, X_2)$ where X_1 takes values in $\mathbf{R}^{d_1}$ and X_2 takes values in $\mathbf{R}^{d_2}$. Let φ denote the characteristic function of X, let φ_1 denote the characteristic function of X_1, and let φ_2 denote the characteristic function of X_2. Then X_1 and X_2 are independent if and only if*

$$\varphi(t) = \varphi_1(t_1)\varphi(t_2) \quad \text{for all } t = (t_1, t_2), \quad t_1 \in \mathbf{R}^{d_1}, \ t_2 \in \mathbf{R}^{d_2}.$$

Example 3.11 (*Multinomial distribution*). Consider a multinomial distribution with parameters n and $(\theta_1, \ldots, \theta_m)$, $\sum_{j=1}^{m} \theta_j = 1$. Recall that this is a discrete distribution with frequency function

$$p(x_1, \ldots, x_m) = \binom{n}{x_1, x_2, \ldots, x_m} \theta_1^{x_1} \theta_2^{x_2} \cdots \theta_m^{x_m}$$

for $x_j = 0, 1, \ldots, n$, $j = 1, \ldots, m$, such that $\sum_{j=1}^{m} x_j = n$; see Example 2.2.

The characteristic function of the distribution is given by

$$\varphi(t) = \sum_{\mathcal{X}} \exp(it_1 x_1 + \cdots + it_m x_m) p(x_1, \ldots, x_m)$$

where the sum is over all

$$(x_1, \ldots, x_m) \in \mathcal{X} = \left\{ (x_1, \ldots, x_m) \in \{0, 1, \ldots\}^m : \sum_{j=1}^{m} x_j = n \right\}.$$

Hence,

$$\varphi(t) = \sum_{\mathcal{X}} \binom{n}{x_1, x_2, \ldots, x_m} [\exp(it_1)\theta_1]^{x_1} \cdots [\exp(it_m)\theta_m]^{x_m}$$

$$= \left[\sum_{j=1}^{m} \exp(it_j)\theta_j \right]^n \sum_{\mathcal{X}} \binom{n}{x_1, x_2, \ldots, x_m}$$

$$\times \left[\frac{\exp(it_1)\theta_1}{\sum_{j=1}^{m} \exp(it_j)\theta_j} \right]^{x_1} \cdots \left[\frac{\exp(it_m)\theta_m}{\sum_{j=1}^{m} \exp(it_j)\theta_j} \right]^{x_m}$$

$$= \left[\sum_{j=1}^{m} \exp(it_j)\theta_j \right]^n$$

where $t = (t_1, \ldots, t_m)$.

Using Theorems 3.6 and 3.7, it follows immediately from this result that the sum of r independent identically distributed multinomial random variables with $n = 1$ has a multinomial distribution with $n = r$. □

3.3 Further Properties of Characteristic Functions

We have seen that the characteristic function of a random variable completely determines its distribution. Thus, it is not surprising that properties of the distribution of X are reflected in the properties of its characteristic function. In this section, we consider several of these properties for the case in which X is a real-valued random variable.

In Theorem 3.5, it was shown that the existence of expected values of the form $E(X^m)$, for $m = 1, 2, \ldots$, is related to the smoothness of the characteristic function of X at 0. Note that finiteness of $E(|X|^m)$ depends on the rates at which $F(x) \to 1$ as $x \to \infty$ and $F(x) \to 0$ as $x \to -\infty$; for instance, if X has an absolutely continuous distribution with density p, then $E(|X|^m) < \infty$ requires that $|x|^m p(x) \to 0$ as $x \to \pm\infty$. That is, the behavior of the distribution function of X at $\pm\infty$ is related to the smoothness of its characteristic function at 0.

The following result shows that the behavior of $|\varphi(t)|$ for large t is similarly related to the smoothness of the distribution function of X.

Theorem 3.8. *Consider a probability distribution on the real line with characteristic function φ. If*

$$\int_{-\infty}^{\infty} |\varphi(t)| \, dt < \infty$$

then the distribution is absolutely continuous. Furthermore, the density function of the distribution is given by

$$p(x) = \frac{1}{2\pi} \int_{-\infty}^{\infty} \exp(-itx)\varphi(t) \, dt, \quad x \in \mathbf{R}.$$

Proof. By Lemma A2.1, $|\exp(it) - 1| \le |t|$ so that for any x_0, x_1,

$$|\exp(-itx_0) - \exp(-itx_1)| = |\exp(-itx_0)| \, |1 - \exp\{-it(x_1 - x_0)\}| \le |t||x_1 - x_0|. \tag{3.5}$$

Hence, for any $T > 0$,

$$\left| \lim_{T \to \infty} \int_{-T}^{T} \frac{\exp\{-itx_0\} - \exp\{-itx_1\}}{it} \varphi(t) \, dt \right| \le |x_1 - x_0| \int_{-\infty}^{\infty} |\varphi(t)| \, dt. \tag{3.6}$$

Let F denote the distribution function of the distribution; it follows from Theorem 3.3, along with Equation (3.6), that, for any $x_0 < x_1$ at which F is continuous,

$$F(x_1) - F(x_0) = \frac{1}{2\pi} \lim_{T \to \infty} \int_{-T}^{T} \frac{\exp\{-itx_0\} - \exp\{-itx_1\}}{it} \varphi(t) \, dt$$

$$= \frac{1}{2\pi} \int_{-\infty}^{\infty} \frac{\exp\{-itx_0\} - \exp\{-itx_1\}}{it} \varphi(t) \, dt.$$

Since

$$|\exp(-itx_0) - \exp(-itx_1)| \le |t| \, |x_1 - x_0|,$$

it follows that

$$|F(x_1) - F(x_0)| \le \frac{1}{2\pi} \int_{-\infty}^{\infty} |\varphi(t)| \, dt \, |x_1 - x_0| \le M \, |x_1 - x_0|$$

for some constant M.

Now let x and y, $y < x$, be arbitrary points in $\mathbf{R}$, i.e., not necessarily continuity points of F. There exist continuity points of F, x_0 and x_1, such that $x_0 \le y < x \le x_1$ and

$$|x_1 - x_0| \le 2|x - y|.$$

Then

$$|F(x) - F(y)| \leq |F(x_1) - F(x_0)| \leq M|x_1 - x_0| \leq 2M|x - y|.$$

That is, F satisfies a Lipschitz condition and, hence, is absolutely continuous; see Appendix 1.

Fix x and h. By Theorem 3.3,

$$\frac{F(x+h) - F(x)}{h} = \frac{1}{2\pi} \int_{-\infty}^{\infty} G_h(t)\varphi(t)\,dt$$

where

$$G_h(t) = \frac{\exp\{-itx\} - \exp\{-it(x+h)\}}{ith}.$$

It follows from (3.5) that, for all h, $|G_h(t)| \leq 1$ so that, by the Dominated Convergence Theorem (see Appendix 1),

$$\lim_{h \to 0} \frac{F(x+h) - F(x)}{h} = \frac{1}{2\pi} \int_{-\infty}^{\infty} G_0(t)\varphi(t)\,dt$$

where

$$G_0(t) = \lim_{h \to 0} G_h(t) = \exp\{-itx\} \lim_{h \to 0} \frac{1 - \exp\{-ith\}}{ith}.$$

By Lemma A2.1,

$$1 - \exp(-ith) = ith + R_2(th)$$

where $|R_2(th)| \leq (th)^2/2$. Hence,

$$\lim_{h \to 0} \frac{1 - \exp(-ith)}{ith} = 1$$

so that $G_0(t) = \exp(-itx)$ and

$$\lim_{h \to 0} \frac{F(x+h) - F(x)}{h} = \frac{1}{2\pi} \int_{-\infty}^{\infty} \exp(-itx)\varphi(t)\,dt.$$

Hence, F is differentiable with derivative

$$F'(x) = \frac{1}{2\pi} \int_{-\infty}^{\infty} \exp(-itx)\varphi(t)\,dt.$$

The result now follows from Theorem 1.9. ∎

Hence, for those cases in which the integral of the modulus of the characteristic function is finite, Theorem 3.8 gives an alternative to Theorem 3.3 for obtaining the distribution of a random variable from its characteristic function. When this condition is satisfied, calculation of the integral in Theorem 3.8 is often easier than calculation of the integral appearing in Theorem 3.3.

***Example* 3.12 (*Normal distribution*).** Let X denote a random variable with a normal distribution with parameters μ and σ, as described in Example 3.6. The characteristic function of this distribution function is

$$\varphi(t) = \exp\left(-\frac{\sigma^2}{2}t^2 + i\mu t\right), \quad \infty < t < \infty.$$

Note that

$$\int_{-\infty}^{\infty} |\varphi(t)| \le \int_{-\infty}^{\infty} \exp\left(-\frac{\sigma^2}{2}t^2\right) dt = \frac{\sqrt{(2\pi)}}{\sigma};$$

hence, by Theorem 3.8, the distribution of X has density function

$p(x)$

$$= \frac{1}{2\pi} \int_{-\infty}^{\infty} \exp\left\{-\frac{\sigma^2}{2}t^2 - i(x - \mu)t\right\} dt$$

$$= \frac{1}{2\pi} \int_{-\infty}^{\infty} \exp\left\{-\frac{\sigma^2}{2}[t^2 - 2i(x - \mu)t/\sigma^2 - (x - \mu)^2/\sigma^4]\right\} dt \exp\left\{-\frac{1}{2\sigma^2}(x - \mu)^2\right\}$$

$$= \frac{1}{\sigma\sqrt{(2\pi)}} \exp\left\{-\frac{1}{2\sigma^2}(x - \mu)^2\right\}, \quad -\infty < x < \infty.$$

It follows that this is the density of the normal distribution with parameters μ and σ.   □

Example 3.13 (Ratio of normal random variables). Let Z_1 and Z_2 denote independent scalar random variables such that each Z_j, $j = 1, 2$, has a standard normal distribution. Define a random variable X by $X = Z_1/Z_2$. The characteristic function of this distribution is given by

$$\varphi(t) = \mathrm{E}[\exp(itZ_1/Z_2)] = \mathrm{E}\{\mathrm{E}[\exp(itZ_1/Z_2)|Z_2]\}, \quad -\infty < t < \infty.$$

Since the characteristic function of Z_1 is $\exp(-t^2/2)$ and Z_1 and Z_2 are independent,

$$\varphi(t) = \mathrm{E}\left[\exp\left(\left(-\frac{1}{2}t^2\right)/Z_2^2\right)\right] = \int_{-\infty}^{\infty} \frac{1}{\sqrt{(2\pi)}} \exp\left\{-\frac{1}{2}(t^2/z^2 + z^2)\right\} dz$$

$$= \exp(-|t|), \quad -\infty < t < \infty.$$

The density of this distribution may be obtained using Theorem 3.8. Since

$$\int_{-\infty}^{\infty} \exp(-|t|) dt = 2,$$

it follows that the distribution of X has density function

$$p(x) = \frac{1}{2\pi} \int_{-\infty}^{\infty} \exp(-itx) \exp(-|t|) dt$$

$$= \frac{1}{2\pi} \int_{0}^{\infty} \exp(-itx) \exp(-t) dt + \frac{1}{2\pi} \int_{0}^{\infty} \exp(itx) \exp(-t) dt$$

$$= \frac{1}{2\pi}\left[\frac{1}{1 + ix} + \frac{1}{1 - ix}\right]$$

$$= \frac{1}{\pi}\frac{1}{1 + x^2}, \quad -\infty < x < \infty.$$

Hence, X has a standard Cauchy distribution; see Example 1.29.   □

In Theorem 3.8 it was shown that if $|\varphi(t)| \to 0$ fast enough as $|t| \to \infty$ for

$$\int_{-\infty}^{\infty} |\varphi(t)|\, dt$$

to be finite, then the distribution is absolutely continuous. The following theorem gives a partial converse to this: if X has an absolutely continuous distribution, then the characteristic function of X approaches 0 at $\pm\infty$. Furthermore, the smoothness of p is related to the rate at which the characteristic function approaches 0.

Theorem 3.9. *Let X denote a real-valued random variable with characteristic function φ. If the distribution of X is absolutely continuous with density function p then*

$$\varphi(t) \to 0 \quad as \quad |t| \to \infty. \tag{3.7}$$

If p is k-times differentiable with

$$\int_{-\infty}^{\infty} |p^{(k)}(x)|\, dx < \infty,$$

then

$$|\varphi(t)| = o(|t|^{-k}) \quad as \quad |t| \to \infty.$$

Proof. If X has an absolutely continuous distribution with density p, then the characteristic function of X is given by

$$\varphi_X(t) = \int_{-\infty}^{\infty} \exp(itx) p(x)\, dx.$$

Hence, (3.7) follows directly from Theorem 3.2.

Suppose p is differentiable. Then, using integration-by-parts,

$$\varphi_X(t) = \frac{1}{it} \exp(itx) p(x) \Big|_{-\infty}^{\infty} - \frac{1}{it} \int_{-\infty}^{\infty} \exp(itx) p'(x)\, dx.$$

Clearly, $p(x)$ must approach 0 at $\pm\infty$ and, since $\exp(itx)$ is bounded,

$$\varphi_X(t) = -\frac{1}{it} \int_{-\infty}^{\infty} \exp(itx) p'(x)\, dx.$$

If p' satisfies

$$\int_{-\infty}^{\infty} |p'(x)|\, dx < \infty,$$

then Theorem 3.2 applies to p' so that

$$\left| \int_{-\infty}^{\infty} \exp(itx) p'(x)\, dx \right| \to 0 \quad as \quad |t| \to \infty.$$

Hence,

$$|\varphi_X(t)| = o(|t|^{-1}) \quad as \quad |t| \to \infty.$$

The results for the higher-order derivatives follow in a similar manner. ■

Example 3.14 *(Uniform distribution on the unit interval).* Consider the uniform distribution on the unit interval; the characteristic function of this distribution is given by

$$\varphi(t) = \int_0^1 \exp(itx)\,dx = \frac{\exp(it) - 1}{it} = \frac{\sin(t)}{t} - i\left(\frac{\cos(t) - 1}{t}\right), \quad -\infty < t < \infty.$$

Hence,

$$|\varphi(t)|^2 = \frac{\sin(t)^2 + (\cos(t) - 1)^2}{t^2} = O(|t|^{-2}) \quad \text{as} \quad |t| \to \infty$$

so that the density of this distribution is differentiable, at most, one time. Here $p(x) = 1$ if $0 < x < 1$ and $p(x) = 0$ otherwise so that p is not differentiable.   □

Example 3.15 *(Binomial distribution).* Consider a binomial distribution with parameters n and θ. The characteristic function of this distribution is given by

$$\varphi(t) = [1 - \theta + \theta \exp(it)]^n$$

so that

$$|\varphi(t)| = [(1 - \theta)^2 + 2\theta(1 - \theta)\cos(t) + \theta^2]^{\frac{n}{2}}.$$

It is easy to see that

$$\liminf_{|t|\to\infty} |\varphi(t)| = |2\theta - 1|^n \quad \text{and} \quad \limsup_{|t|\to\infty} |\varphi(t)| = 1$$

so that $\varphi(t)$ does not have a limit as $|t| \to 0$; see Figure 3.2 for a graphical illustration of this fact. It follows that the binomial distribution is not absolutely continuous, which, of course, is obvious from its definition.   □

Symmetric distributions

The distribution of X is said to be *symmetric* about a point x_0 if $X - x_0$ and $-(X - x_0)$ have the same distribution. Note that this implies that

$$F(x_0 + x) = 1 - F((x_0 - x)-), \quad -\infty < x < \infty,$$

where F denotes the distribution function of X.

The following theorem shows that the imaginary part of the characteristic function of a distribution symmetric about 0 is 0; the proof is left as an exercise.

Theorem 3.10. *Let X denote a real-valued random variable with characteristic function φ. The distribution of X is symmetric about 0 if and only if $\varphi(t)$ is real for all $-\infty < t < \infty$.*

Example 3.16 *(Symmetrization of a distribution).* Let X denote a real-valued random variable with characteristic function φ. Let X_1, X_2 denote independent random variables, such that X_1 and X_2 each have the same distribution as X, and let $Y = X_1 - X_2$. Then Y has characteristic function

$$\varphi_Y(t) = \mathrm{E}[\exp(itX_1)\exp(-itX_2)] = \varphi(t)\varphi(-t)$$
$$= |\varphi(t)|^2, \quad -\infty < t < \infty$$

Hence, the distribution of Y is symmetric about 0; the distribution of Y is called the *symmetrization* of the distribution of X.

For instance, suppose that X has a uniform distribution on the interval $(0, 1)$. Then Y has characteristic function

$$\varphi_Y(t) = \frac{\sin(t)^2 + (\cos(t) - 1)^2}{t^2}, \quad -\infty < t < \infty;$$

see Example 3.14. □

Lattice distributions

Let X denote a real-valued discrete random variable with range

$$\mathcal{X} = \{x_1, x_2, \ldots\}$$

where $x_1 < x_2 < \cdots$ and $\Pr(X = x_j) > 0$, $j = 1, 2, \ldots$. The distribution of X is said to be a *lattice distribution* if there exists a constant b such that, for any j and k, $x_j - x_k$ is a multiple of b. This occurs if and only if $\mathcal{X}$ is a subset of the set

$$\{a + bj, \quad j = 0, \pm 1, \pm 2, \ldots\}$$

for some constant a; b is said to be a *span* of the distribution. A span b is said to be a *maximal* span if $b \geq b_1$ for any span b_1.

Stated another way, X has a lattice distribution if and only if there is a linear function of X that is integer-valued.

***Example* 3.17 (*Binomial distribution*).** Consider a binomial distribution with parameters n and θ. Recall that range of this distribution is $\{0, 1, \ldots, n\}$ with frequency function

$$p(x) = \binom{n}{x} \theta^x (1 - \theta)^{n-x}, \quad x = 0, 1, \ldots, n.$$

Hence, this is a lattice distribution with maximal span 1. □

***Example* 3.18 (*A discrete distribution that is nonlattice*).** Let X denote a real-valued random variable such that the range of X is $\mathcal{X} = \{0, 1, \sqrt{2}\}$. Suppose this is a lattice distribution. Then there exist integers m and n and a constant b such that

$$\sqrt{2} = mb \quad \text{and} \quad 1 = nb.$$

It follows that $\sqrt{2} = m/n$ for some integers m, n. Since $\sqrt{2}$ is irrational, this is impossible. It follows that the distribution of X is non-lattice.

More generally, if X has range $\mathcal{X} = \{0, 1, c\}$ for some $c > 0$, then X has a lattice distribution if and only if c is rational. □

The characteristic function of a lattice distribution has some special properties.

***Theorem* 3.11.**

(i) *The distribution of X is a lattice distribution if and only if its characteristic function φ satisfies $|\varphi(t)| = 1$ for some $t \neq 0$. Furthermore, if X has a lattice distribution, then $|\varphi(t)| = 1$ if and only if $2\pi/t$ is a span of the distribution.*

(ii) *Let X denote a random variable with a lattice distribution and let φ denote its characteristic function. A constant b is a maximal span of the distribution if and only if $\varphi(2\pi/b) = 1$ and*

$$|\varphi(t)| < 1 \quad for\ all\ \ 0 < |t| < 2\pi/b.$$

(iii) *Let X denote a random variable with range $\mathcal{X}$ that is a subset of*

$$\{a + bj, \ \ j = 0, \pm 1, \pm 2, \ldots\}$$

and let φ denote the characteristic function of the distribution. Then, for all $-\infty < t < \infty$,

$$\varphi(t) = \exp\left(-2\pi k \frac{a}{b} i\right) \varphi\left(t + 2\pi \frac{k}{b}\right), \quad k = 0, \pm 1, \pm 2, \ldots.$$

Thus, if $a = 0$, the characteristic function is periodic.

Proof.   Suppose X has a lattice distribution. Then the characteristic function of X is of the form

$$\varphi(t) = \exp\{ita\} \sum_{j=-\infty}^{\infty} \exp\{itjb\} p_j$$

where the p_j are nonnegative and sum to 1. Hence,

$$\varphi(2\pi/b) = \exp\{i2\pi a/b\} \sum_{j=-\infty}^{\infty} \exp\{i2\pi j\} p_j = \exp\{i2\pi a/b\}$$

so that

$$|\varphi(2\pi/b)| = 1.$$

Now suppose that $|\varphi(t)| = 1$ for some $t \neq 0$. Writing

$$\varphi(t) = y_1 + iy_2,$$

we must have $y_1^2 + y_2^2 = 1$. Let F denote the distribution function of the distribution. Then

$$\int \exp\{itx\} dF(x) = \exp\{itz\}$$

for some real number z. It follows that

$$\int [\exp\{itz\} - \exp\{itx\}] dF(x) = 0$$

and, hence, that

$$\int [\cos(tz) - \cos(tx)] dF(x) = \int [1 - \cos(t(x - z))] dF(x) = 0.$$

Note that $1 - \cos(t(x - z))$ is nonnegative and continuous in x. Hence, F must be discontinuous with mass points at the zeros of $1 - \cos(t(x - z))$. It follows that the mass points of the distribution must be of the form $z + 2\pi j/t$, for $j = 0, \pm 1, \ldots$. This proves part (i) of the theorem.

Let

$$t^* = \inf\{t \in \mathbf{R} \colon t > 0 \quad \text{and} \quad |\varphi(t)| = 1\}.$$

Then $2\pi/t^*$ is the maximal span of the distribution and

$$|\varphi(t)| < 1 \quad \text{for } |t| < t^*.$$

That is, the maximal span $b = 2\pi/t^*$ satisfies

$$|\varphi(2\pi/b)| = 1$$

and

$$|\varphi(t)| < 1 \quad \text{for } |t| < 2\pi/b,$$

proving part (ii).

Let $p_j = \Pr(X = a + bj)$, $j = 0, \pm 1, \pm 2, \ldots$. Then

$$\varphi(t) = \sum_{j=-\infty}^{\infty} p_j \exp\{it(a + bj)\}$$

and

$$\varphi(t + 2\pi k/b) = \sum_{j=-\infty}^{\infty} p_j \exp\{it(a + bj)\} \exp\{i2\pi k(a + bj)/b\}$$

$$= \exp(i2\pi ka/b) \sum_{j=-\infty}^{\infty} p_j \exp\{it(a + bj)\} \exp(i2\pi kj).$$

Part (iii) of the theorem now follows from the fact that

$$\exp(i2\pi kj) = \cos(kj2\pi) + i \, \sin(kj2\pi) = 1. \quad \blacksquare$$

Example 3.19 (Binomial distribution). Consider a binomial distribution with parameters n and θ. The characteristic function of this distribution is given by

$$\varphi(t) = [1 - \theta + \theta \exp(it)]^n$$

so that

$$|\varphi(t)| = [(1 - \theta)^2 + 2\theta(1 - \theta) \cos(t) + \theta^2]^{\frac{n}{2}}.$$

Hence, $|\varphi(t)| = 1$ if and only if $\cos(t) = 1$; that is, $|\varphi(t)| = 1$ if and only if $t = 2\pi j$ for some integer j.

Thus, according to part (1) of Theorem 3.11, the binomial distribution is a lattice distribution. According to part (ii) of the theorem, the maximal span is 1. $\quad \square$

3.4   Exercises

3.1 Let X denote a real-valued random variable with a discrete distribution with frequency function

$$p(x) = \theta(1 - \theta)^x, \quad x = 0, 1, \ldots;$$

where θ is a fixed constant, $0 < \theta < 1$. Find the characteristic function of X.

3.2 Let X denote a real-valued random variable with an absolutely continuous distribution with density function

$$p(x) = \frac{1}{2} \exp(-|x|), \quad -\infty < x < \infty.$$

This is the standard Laplace distribution. Find the characteristic function of X.

3.3 Let X denote a real-valued random variable with a uniform distribution on the interval (a, b), $b > a$. That is, X has an absolutely continuous distribution with density function

$$p(x) = \frac{1}{b - a}, \quad a < x < b;$$

here a and b are fixed constants. Find the characteristic function of X.

3.4 Let X denote a real-valued random variable with distribution function F and characteristic function φ. Suppose that φ satisfies the following condition:

$$\lim_{T \to \infty} \frac{1}{T} \int_{-T}^{T} \varphi(t) \, dt = 2.$$

Based on this, what can you conclude about the distribution of X. Be as specific as possible.

3.5 Let X_1 and X_2 denote independent real-valued random variables with distribution functions F_1, F_2, and characteristic functions φ_1, φ_2, respectively. Let Y denote a random variable such that $X_1, X_2,$ and Y are independent and

$$\Pr(Y = 0) = 1 - \Pr(Y = 1) = \alpha$$

for some $0 < \alpha < 1$. Define

$$Z = \begin{cases} X_1 & \text{if } Y = 0 \\ X_2 & \text{if } Y = 1 \end{cases}.$$

Find the characteristic function of Z in terms of φ_1, φ_2, and α.

3.6 Let X denote a real-valued random variable with characteristic function φ. Suppose that

$$|\varphi(1)| = \varphi(\pi) = 1.$$

Describe the distribution of X; be as specific as possible.

3.7 Prove Theorem 3.4.

3.8 Let X_1 and X_2 denote independent random variables each with a standard normal distribution and let $Y = X_1 X_2$. Find the characteristic function and density function of Y.

3.9 Prove Theorem 3.6.

3.10 Let X_1 and X_2 denote independent random variables, each with the standard Laplace distribution; see Exercise 3.2. Let $Y = X_1 + X_2$. Find the characteristic function and density function of Y.

3.11 Prove Corollary 3.2.

3.12 Let X denote a real-valued random variable with characteristic function φ. Suppose that g is a real-valued function on $\mathbf{R}$ that has the representation

$$g(x) = \int_{-\infty}^{\infty} G(t) \exp(itx) \, dt$$

for some function G satisfying

$$\int_{-\infty}^{\infty} |G(t)| \, dt < \infty.$$

Show that

$$E[g(X)] = \int_{-\infty}^{\infty} G(t)\varphi(t)\,dt.$$

3.13 Consider a distribution on the real line with characteristic function

$$\varphi(t) = (1 - t^2/2)\exp(-t^2/4), \quad -\infty < t < \infty.$$

Show that this distribution is absolutely continuous and find the density function of the distribution.

3.14 Let $\varphi_1, \ldots, \varphi_n$ denote characteristic functions for distributions on the real line. Let $a_1, \ldots, a_n$ denote nonnegative constants such that $a_1 + \cdots + a_n = 1$. Show that

$$\varphi(t) = \sum_{j=1}^{n} a_j \varphi_j(t), \quad -\infty < t < \infty$$

is also a characteristic function.

3.15 Let X and Y denote independent, real-valued random variables each with the same marginal distribution and let φ denote the characteristic function of that distribution. Consider the random vector $(X + Y, X - Y)$ and let φ_0 denote its characteristic function. Show that

$$\varphi_0((t_1, t_2)) = \varphi(t_1 + t_2)\varphi(t_1 + t_2), \quad (t_1, t_2) \in \mathbf{R}^2.$$

3.16 Consider an absolutely continuous distribution on the real line with density function p. Suppose that p is piece-wise continuous with a jump at x_0, $-\infty < x_0 < \infty$. Show that

$$\int_{-\infty}^{\infty} |\varphi(t)|\,dt = \infty,$$

where φ denotes the characteristic function of the distribution.

3.17 Suppose that X is a real-valued random variable. Suppose that there exists a constant $M > 0$ such that the support of X lies entirely in the interval $[-M, M]$. Let φ denote the characteristic function of X. Show that φ is infinitely differentiable at 0.

3.18 Prove Corollary 3.3.

3.19 Let X denote a real-valued random variable with characteristic function

$$\varphi(t) = \frac{1}{2}[\cos(t) + \cos(t\pi)], \quad -\infty < t < \infty.$$

    (a) Is the distribution of X absolutely continuous?

    (b) Does there exist an r such that $E(X^r)$ either does not exist or is infinite?

3.20 Let φ denote the characteristic function of the distribution with distribution function

$$F(x) = \begin{cases} 0 & \text{if } x < 0 \\ 1 - \exp(-x), & \text{if } 0 \le x < \infty \end{cases}.$$

Show that this distribution is absolutely continuous and that

$$\int_{-\infty}^{\infty} |\varphi(t)|\,dt = \infty.$$

Thus, the converse to Theorem 3.8 does not hold.

3.21 Find the density function of the distribution with characteristic function

$$\varphi(t) = \begin{cases} 1 - |t| & \text{if } |t| \le 1 \\ 0 & \text{otherwise} \end{cases}.$$

3.22 Let the random variable Z be defined as in Exercise 3.5.

    (a) Suppose that, for each $j = 1, 2$, the distribution of X_j is symmetric about 0. Is the distribution of Z symmetric about 0?

    (b) Suppose that X_1 and X_2 each have a lattice distribution. Does Z have a lattice distribution?

3.23 Let X denote a real-valued random variable with an absolutely continuous distribution. Suppose that the distribution of X is symmetric about 0. Show that there exists a density function p for the distribution satisfying

$$p(x) = p(-x) \qquad \text{for all } x.$$

3.24 Let X denote a d-dimensional random vector with characteristic function φ. Show that X has a degenerate distribution if and only if there exists an $a \in \mathbf{R}^d$ such that

$$|\varphi(a^T t)| = 1 \qquad \text{for all } t \in \mathbf{R}^d.$$

3.25 Prove Theorem 3.10.

3.5 Suggestions for Further Reading

A comprehensive reference for characteristic functions is Lukacs (1960); see also Billingsley (1995, Section 26), Feller (1971, Chapter XV), Karr (1993, Chapter 6), Port (1994, Chapter 51), Stuart and Ord (1994, Chapter 4). See Apostol (1974, Chapter 11) for further details regarding Fourier transforms. Lattice distributions are discussed in detail in Feller (1968). Symmetrization is discussed in Feller (1971, Section V.5).

4

Moments and Cumulants

4.1 Introduction

Let X denote a real-valued random variable. As noted in Chapter 1, expected values of the form $E(X^r)$, for $r = 1, 2, \ldots$, are called the *moments* of X. The moments of a random variable give one way to describe its properties and, in many cases, the sequence of moments uniquely determines the distribution of the random variable. The most commonly used moment is the first moment, $E(X)$, called the *mean* of the distribution or the mean of X. In elementary statistics courses, the mean of a distribution is often described as a "measure of central tendency."

Let $\mu = E(X)$. The *central moments* of X are the moments of $X - \mu$. The most commonly used central moment is the second central moment, $E[(X - \mu)^2]$, called the *variance*. The variance is a measure of the dispersion of the distribution around its mean.

In this chapter, we consider properties of moments, along with associated quantities such as moment-generating functions and *cumulants*, certain functions of the moments that have many useful properties.

4.2 Moments and Central Moments

Let X denote a real-valued random variable. It is important to note that, for a given value of r, $E(X^r)$ may be infinite, or may not exist. As with any function of X, if

$$E(|X^r|) \equiv E(|X|^r) < \infty,$$

then $E(X^r)$ exists and is finite. If, for some r, $E(|X|^r) < \infty$, then,

$$E(|X|^j) < \infty, \quad j = 1, 2, \ldots, r.$$

This follows from Jensen's inequality, using the fact that the function t^m, $t > 0$, is convex for $m \geq 1$; then,

$$E(|X|^j)^{\frac{r}{j}} \leq E[(|X|^j)^{\frac{r}{j}}] = E(|X|^r) < \infty.$$

so that $E(|X|^j) < \infty$.

***Example* 4.1 (*Standard exponential distribution*).** In Example 1.31 it was shown that if X has a standard exponential distribution, then

$$E(X^r) = \Gamma(r + 1), \quad r > 0.$$

Hence, the moments of X are $r!, r = 1, 2, \ldots$. □

Example* 4.2 *(Cauchy distribution). Let X denote a random variable with a standard Cauchy distribution; see Example 1.29. The same argument used in that example to show that $E(X)$ does not exist may be used to show that $E(X^r)$ does not exist when r is an odd integer and $E(X^r) = \infty$ when r is an even integer. $\square$

Central moments

Let X denote a real-valued random variable such that $E(|X|) < \infty$ and let $\mu = E(X)$. As noted earlier, the central moments of X are the moments of the random variable $X - \mu$. Clearly, the first central moment of X, $E(X - \mu)$, is 0. The second central moment of X, $E[(X - \mu)^2]$, is called the *variance* of X, which we will often denote by $\text{Var}(X)$. The *standard deviation* of X, defined to be the (positive) square root of its variance, $\sqrt{\text{Var}(x)}$, is also often used. Note that, if μ exists, $\text{Var}(X)$ always exists, but it may be infinite. The following theorem gives some simple properties of the variance. The proof is straightforward and left as an exercise.

Theorem 4.1. *Let X denote a real-valued random variable such that $E(X^2) < \infty$. Then*
  (i) $\text{Var}(X) < \infty$
  (ii) $\text{Var}(X) = E(X^2) - \mu^2$ *where* $\mu = E(X)$
  (iii) Let a, b denote real-valued constants and let $Y = aX + b$. Then $\text{Var}(Y) = a^2 \text{Var}(X)$.
  (iv) For any $c \in \mathbf{R}$, $E[(X - c)^2] \geq \text{Var}(X)$ with equality if and only if $c = \mu$.
  (v) For any $c \in \mathbf{R}$,

$$\Pr\{|X - \mu| \geq c\} \leq \frac{1}{c^2} \text{Var}(X).$$

Part (v) of Theorem 4.1 is known as *Chebyshev's inequality*.

Example* 4.3 *(Standard exponential distribution). Suppose that X has a standard exponential distribution. Then, according to Example 4.1, $E(X) = 1$ and $E(X^2) = 2$. Hence, $\text{Var}(X) = 1$. $\square$

Since $(X - \mu)^r, r = 3, 4, \ldots$, may be expanded in powers of X and μ, clearly the central moments may be written as functions of the standard moments. In particular, the central moment of order r is a function of $E(X^j)$ for $j = 1, 2, \ldots, r$ so that if $E(|X|^r) < \infty$, then the central moment of order r also exists and is finite.

Example* 4.4 *(Third central moment). Let X denote a random variable such that $E(|X|^3) < \infty$. Since

$$(X - \mu)^3 = X^3 - 3\mu X^2 + 3\mu^2 X - \mu^3,$$

it follows that

$$E[(X - \mu)^3] = E(X^3) - 3\mu E(X^2) + 2\mu^3. \qquad \square$$

Moments of random vectors

Let X and Y denote real-valued random variables such that $E(X^2) < \infty$ and $E(Y^2) < \infty$. In addition to the individual moments of X and Y, $E(X)$, $E(Y)$, $E(X^2)$, $E(Y^2)$, and so on, we may also consider the moments and central moments of the random vector (X, Y), which are called the *joint moments* and *joint central moments*, respectively, of (X, Y); the terms *product moments* and *central product moments* are also used.

The joint moment of (X, Y) of order (r, s) is defined to be $E(X^r Y^s)$, given that the expected value exists. Similarly, if $\mu_X = E(X)$ and $\mu_Y = E(Y)$, the joint central moment of order (r, s) is defined to be

$$E[(X - \mu_X)^r (Y - \mu_Y)^s].$$

The most commonly used joint moment or joint central moment is the central moment of order $(1, 1)$, generally known as the *covariance* of X and Y. The covariance of X and Y will be denoted by $\text{Cov}(X, Y)$ and is given by

$$\text{Cov}(X, Y) = E[(X - \mu_X)(Y - \mu_Y)] = E(XY) - \mu_X \mu_Y.$$

Note that the $\text{Cov}(X, Y) = \text{Cov}(Y, X)$ and that $\text{Cov}(X, X) = \text{Var}(X)$. It follows from Theorem 2.1 that if X and Y are independent, then $\text{Cov}(X, Y) = 0$.

The covariance arises in a natural way in computing the variance of a sum of random variables. The result is given in the following theorem, whose proof is left as an exercise.

Theorem 4.2. *Let X and Y denote real-valued random variables such that $E(X^2) < \infty$ and $E(Y^2) < \infty$. Then*

$$\text{Var}(X + Y) = \text{Var}(X) + \text{Var}(Y) + 2\,\text{Cov}(X, Y)$$

and, for any real-valued constants a, b,

$$\text{Cov}(aX + b, Y) = a\,\text{Cov}(X, Y).$$

The results of Theorem 4.2 are easily extended to the case of several random variables; again, the proof is left as an exercise.

Corollary 4.1. *Let $Y, X_1, \ldots, X_n$ denote real-valued random variables such that*

$$E(Y^2) < \infty, \quad E\left(X_j^2\right) < \infty, \qquad j = 1, \ldots, n.$$

Then

$$\text{Cov}\left(Y, \sum_{j=1}^{n} X_j\right) = \sum_{j=1}^{n} \text{Cov}(Y, X_j)$$

and

$$\text{Var}\left(\sum_{j=1}^{n} X_j\right) = \sum_{j=1}^{n} \text{Var}(X_j) + 2\sum_{i<j} \text{Cov}(X_i, X_j).$$

Correlation

The correlation of X and Y, which we denote by $\rho(X, Y)$, is defined by

$$\rho(X, Y) = \frac{\text{Cov}(X, Y)}{[\text{Var}(X)\,\text{Var}(Y)]^{\frac{1}{2}}},$$

provided that $\text{Var}(X) > 0$ and $\text{Var}(Y) > 0$. The correlation is simply the covariance of the random variables X and Y standardized to have unit variance. Hence, if a, b, c, d are constants, then the correlation of $aX + b$ and $cY + d$ is the same as the correlation of X and Y.

The covariance and correlation are measures of the strength of the linear relationship between X and Y; the correlation has the advantage of being easier to interpret, because of the following result.

Theorem 4.3. *Let X and Y denote real-valued random variables satisfying $\text{E}(X^2) < \infty$ and $\text{E}(Y^2) < \infty$. Assume that $\text{Var}(X) > 0$ and $\text{Var}(Y) > 0$. Then*

(i) $\rho(X, Y)^2 \le 1$

(ii) $\rho(X, Y)^2 = 1$ *if and only if there exist real-valued constants a, b such that*

$$\Pr(Y = aX + b) = 1.$$

(iii) $\rho(X, Y) = 0$ *if and only if, for any real-valued constants a, b,*

$$\text{E}\{[Y - (aX + b)]^2\} \ge \text{Var}(Y).$$

Proof. Let $Z = (X, Y)$, $g_1(Z) = X - \mu_X$, and $g_2(Z) = Y - \mu_Y$, where $\mu_X = \text{E}(X)$ and $\mu_Y = \text{E}(Y)$. Then, by the Cauchy-Schwarz inequality,

$$\text{E}[g_1(Z)g_2(Z)]^2 \le \text{E}[g_1(Z)^2]\,\text{E}[g_2(Z)^2].$$

That is,

$$\text{E}[(X - \mu_X)(Y - \mu_Y)]^2 \le \text{Var}(X)\,\text{Var}(Y);$$

part (i) follows.

The condition $\rho(X, Y)^2 = 1$ is equivalent to equality in the Cauchy-Schwarz inequality; under the conditions of the theorem, this occurs if and only there exists a constant c such that

$$\Pr(Y - \mu_Y = c(X - \mu_X)) = 1.$$

This proves part (ii).

By part (iv) of Theorem 4.1, for any constants a and b,

$$\text{E}\{[Y - (aX + b)]^2\} \ge \text{E}\{[Y - \mu_Y - a(X - \mu_X)]^2\}$$
$$= \text{Var}(Y) + a^2\,\text{Var}(X) - 2a\,\text{Cov}(Y, X).$$

If $\rho(X, Y) = 0$, then $\text{Cov}(Y, X) = 0$ so that, by Theorem 4.1,

$$\text{E}\{[Y - (aX + b)]^2\} \ge \text{Var}(Y) + a^2\,\text{Var}(X) \ge \text{Var}(Y).$$

Now suppose that, for any constants a, b,

$$\text{E}\{[Y - (aX + b)]^2\} \ge \text{Var}(Y).$$

Let $\sigma_X^2 = \text{Var}(X)$ and $\sigma_Y^2 = \text{Var}(Y)$. Taking

$$a = \frac{\sigma_Y}{\sigma_X}\rho(X, Y)$$

and

$$b = \mu_Y - \frac{\sigma_Y}{\sigma_X}\rho(X, Y)\mu_X,$$

we have that

$$\text{Var}\left(Y - \frac{\sigma_Y}{\sigma_X}\rho(X, Y)X\right) = \text{Var}(Y) + \rho(X, Y)^2\,\text{Var}(Y) - 2\frac{\sigma_Y}{\sigma_X}\rho(X, Y)\,\text{Cov}(X, Y)$$

$$= (1 - \rho(X, Y)^2)\text{Var}(Y).$$

Therefore, we must have $\rho(X, Y) = 0$, proving part (iii). ■

According to Theorem 4.3, $0 \leq \rho(X, Y)^2 \leq 1$ with $\rho(X, Y)^2 = 1$ if and only if Y is, with probability 1, a linear function of X. Part (iii) of the theorem states that $\rho(X, Y)^2 = 0$ if and only if the linear function of X that best predicts Y in the sense of the criterion $\text{E}\{[Y - (aX + b)]^2\}$ is the function with $a = 0$ and $b = \text{E}(Y)$; that is, X is of no help in predicting Y, at least if we restrict attention to linear functions of X. The restriction to linear functions is crucial, as the following example illustrates.

***Example* 4.5 (*Laplace distribution*).** Let X denote a random variable with an absolutely continuous distribution with density function

$$p(x) = \frac{1}{2}\,\exp\{-|x|\}, \quad -\infty < x < \infty;$$

this distribution is called the *Laplace distribution*. Note that $\text{E}(X) = 0$, $\text{E}(X^2) = 2$, and $\text{E}(X^3) = 0$.

Let $Y = X^2$. Then

$$\text{Cov}(Y, X) = \text{E}[(Y - 2)X] = \text{E}[X^3 - 2X] = 0$$

so that $\rho(Y, X) = 0$. Hence, linear functions of X are not useful for predicting Y. However, there are clearly nonlinear functions of X that are useful for predicting Y; in particular, X^2 yields Y exactly. □

Covariance matrices

Joint moments and joint central moments for sets of more than two real-valued random variables may be defined in a similar manner. For instance, the joint moment of $(X_1, \ldots, X_d)$ of order $(i_1, i_2, \ldots, i_d)$ is given by

$$\text{E}[X_1^{i_1} \cdots X_d^{i_d}]$$

provided that the expectation exists. Such moments involving three or more random variables arise only occasionally and we will not consider them here.

Let X denote a d-dimensional random vector and write $X = (X_1, X_2, \ldots, X_d)$, where $X_1, \ldots, X_d$ are real-valued. We are often interested in the set of all covariances of pairs of (X_i, X_j). Let

$$\sigma_{ij} = \text{Cov}(X_i, X_j), \quad i, j = 1, \ldots, d.$$

It is often convenient to work with these covariances in matrix form. Hence, let Σ denote the $d \times d$ matrix with (i, j)th element given by σ_{ij}; note that this matrix is symmetric. The matrix Σ is known as the *covariance matrix* of X. The following theorem gives some basic properties of covariance matrices; the proof is left as an exercise.

Theorem 4.4. *Let X denote a d-dimensional random vector such that* $\text{E}(X^T X) < \infty$ *and let Σ denote the covariance matrix of X. Then*

  (i) $\text{Var}(a^T X) = a^T \Sigma a, \quad a \in \mathbf{R}^d$

  (ii) Σ *is nonnegative definite*

4.3 Laplace Transforms and Moment-Generating Functions

Let X denote a real-valued random variable. Consider the expected value $\text{E}[\exp\{tX\}]$ where t is a given number, $-\infty < t < \infty$. Since $\exp\{tX\}$ is a nonnegative random variable, this expected value always exists, although it may be infinite.

The expected value of $\exp\{tX\}$ is closely related to the characteristic function of X, the expected value of $\exp\{itX\}$. However, there is an important difference between the functions $\exp\{tx\}$ and $\exp\{itx\}$. Although $\exp\{itx\}$ is bounded, with $|\exp\{itx\}| = 1$, the function $\exp\{tx\}$ is unbounded for any nonzero t and grows very fast as either $x \to \infty$ or $x \to -\infty$. Hence, for many random variables, the set of values of t for which $\text{E}\{\exp(tX)\}$ is finite is quite small.

Suppose there exists number $\delta > 0$ such that $\text{E}[\exp\{tX\}] < \infty$ for $|t| < \delta$. In this case, we say that X, or, more precisely, the distribution of X, has *moment-generating function*

$$M_X(t) = \text{E}[\exp\{tX\}], \quad |t| < \delta.$$

As noted above, it is not unusual for a random variable to not have a moment-generating function.

Laplace transforms

The situation is a little better if X is nonnegative. In that case, we know that $\text{E}\{\exp(tX)\} < \infty$ for all $t \le 0$. Hence, we may define a function

$$\mathcal{L}(t) = \text{E}\{\exp(-tX)\}, \quad t \ge 0;$$

we will refer to this function as the *Laplace transform* of the distribution or, more simply, the Laplace transform of X.

Example 4.6 *(Gamma distribution).* Consider the gamma distribution with parameters α and β, as discussed in Example 3.4. The Laplace transform of this distribution is given by

$$\mathcal{L}(t) = \int_0^\infty \exp(-tx) \frac{\beta^\alpha}{\Gamma(\alpha)} x^{\alpha-1} \exp(-\beta x)\, dx = \frac{\beta^\alpha}{(\beta + t)^\alpha}, \quad t \ge 0. \qquad \square$$

***Example* 4.7 (*Inverse gamma distribution*).** Let X denote a scalar random variable with an absolutely continuous distribution with density function

$$x^{-3} \exp(-1/x), \quad x > 0;$$

this is an example of an inverse gamma distribution. The Laplace transform of this distribution is given by

$$\mathcal{L}(t) = \int_0^\infty \exp(-tx) \underbrace{x^{-3} \exp(-1/x)}_{dF(x)} \, dx = 2t K_2(2\sqrt{t}), \quad t \geq 0.$$

Here K_2 denotes the modified Bessel function of order 2; see, for example, Temme (1996). □

As might be expected, the properties of the Laplace transform of X are similar to those of the characteristic function of X; in particular, if two random variables have the same Laplace transform, then they have the same distribution.

Theorem 4.5. *Let X and Y denote real-valued, nonnegative random variables. If $\mathcal{L}_X(t) = \mathcal{L}_Y(t)$ for all $t > 0$, then X and Y have the same distribution.*

Proof. Let $X_0 = \exp\{-X\}$ and $Y_0 = \exp\{-Y\}$. Then X_0 and Y_0 are random variables taking values in the interval $[0, 1]$. Since $\mathcal{L}_X(t) = \mathcal{L}_Y(t)$, it follows that $E[X_0^t] = E[Y_0^t]$ for all $t > 0$; in particular, this holds for $t = 1, 2, \ldots$. Hence, for any polynomial g, $E[g(X_0)] = E[g(Y_0)]$.

From the Weierstrass Approximation Theorem (see Appendix 3), we know that any continuous function on $[0, 1]$ may be approximated to arbitrary accuracy by a polynomial. More formally, let h denote a bounded, continuous function on $[0, 1]$. Given $\epsilon > 0$, there exists a polynomial g_ϵ such that

$$\sup_{z \in [0,1]} |h(z) - g_\epsilon(z)| \leq \epsilon.$$

Then

$$\left| E[h(X_0) - h(Y_0)] - E[g_\epsilon(X_0) - g_\epsilon(Y_0)] \right| = \left| E[h(X_0) - g_\epsilon(X_0)] - E[h(Y_0) - g_\epsilon(Y_0)] \right|$$
$$\leq E[|h(X_0) - g_\epsilon(X_0)|] + E[|h(Y_0) - g_\epsilon(Y_0)|]$$
$$\leq 2\epsilon.$$

Since $E[g_\epsilon(X_0) - g_\epsilon(Y_0)] = 0$,

$$\left| E[h(X_0)] - E[h(Y_0)] \right| \leq 2\epsilon$$

and, since ϵ is arbitrary, it follows that $E[h(X_0)] = E[h(Y_0)]$ for any bounded continuous function h. It follows from Theorem 1.14 that X_0 and Y_0 have the same distribution. That is, for any bounded continuous function f, $E[f(X_0)] = E[f(Y_0)]$. Let g denote a bounded, continuous, real-valued function on the range of X and Y. Since $X = -\log(X_0)$ and $Y = -\log(Y_0)$, $g(X) = f(X_0)$ where $f(t) = g(-\log(t))$, $0 < t < 1$. Since g is bounded and continuous, it follows that f is bounded and continuous; it follows that

$$E[g(X)] = E[f(X_0)] = E[f(Y_0)] = E[g(Y)]$$

so that X and Y have the same distribution. ■

Let X denote a real-valued random variable with Laplace transform $\mathcal{L}$. The Laplace transform has the property that moments of X may be obtained from the derivatives of $\mathcal{L}(t)$ at $t = 0$. Note that since $\mathcal{L}(t)$ is defined only for $t \geq 0$, $\mathcal{L}'(0)$, $\mathcal{L}''(0)$, and so on will refer to the right-hand derivatives of $\mathcal{L}(t)$ at $t = 0$; for example,

$$\mathcal{L}'(0) = \lim_{h \to 0^+} \frac{\mathcal{L}(h) - \mathcal{L}(0)}{h}.$$

Theorem 4.6. *Let X denote a real-valued, nonnegative random variable and let $\mathcal{L}$ denote its Laplace transform. Suppose that, for some $m = 1, 2, \ldots$, $E[X^m] < \infty$. Then $\mathcal{L}^{(m)}(0)$ exists and*

$$E[X^m] = (-1)^m \mathcal{L}^{(m)}(0).$$

Conversely, if $\mathcal{L}^{(m)}(0)$ exists, then $E(X^m) < \infty$.

Proof.   We will consider only the case $m = 1$; the general case follows along similar lines. Note that, by the mean-value theorem, for all h, x, there exists a $q \equiv q(x, h)$, $0 \leq q \leq h$, such that

$$\exp(-hx) - 1 = -x \exp(-q(x, h)x)h. \tag{4.1}$$

Hence,

$$\lim_{h \to 0} q(x, h) = 0$$

and, for all $h > 0$ and all x,

$$\exp\{-q(x, h)h\} \leq 1.$$

By (4.1), the existence of $\mathcal{L}'_X(0)$ is related to the existence of

$$\lim_{h \to 0^+} \int_0^\infty -x \exp\{-q(x, h)x\} \, dF(x).$$

Suppose that $\mathcal{L}'(0)$ exists. Then the limits

$$\lim_{h \to 0^+} \int_0^\infty \frac{\exp\{-hx\} - 1}{h} \, dF(x) = \lim_{h \to 0^+} \int_0^\infty -x \exp\{-q(x, h)x\} \, dF(x)$$

exist and are finite. By Fatou's Lemma (see Appendix 1),

$$\lim_{h \to 0^+} \int_0^\infty x \exp\{-q(x, h)x\} \, dF(x) \geq \int_0^\infty x \liminf_{h \to 0^+} \exp\{-q(x, h)x\} \, dF(x)$$

$$= \int_0^\infty x \, dF(x),$$

so that $E(X) < \infty$.

Suppose $E(X) < \infty$. Since, for $h \geq 0$, $x \exp(-hx) \leq x$ and, for all $x \geq 0$,

$$x \exp(-hx) \to x \quad \text{as } h \to 0$$

by the Dominated Convergence Theorem (see Appendix 1),

$$\lim_{h \to 0^+} \int_0^\infty -x \exp\{-hx\} \, dF(x) = -E(X).$$

Hence, $\mathcal{L}'_X(0)$ exists and is equal to $-E(X)$. ∎

***Example* 4.8 (*Inverse gamma distribution*).** Consider the inverse gamma distribution considered in Example 4.7. The Laplace transform of this distribution is given by

$$\mathcal{L}(t) = 2t K_2(2\sqrt{t})$$

where K_2 denotes a modified Bessel function. The following properties of the modified Bessel functions are useful; see, for example, Temme (1996) for further details.

The derivative of K_ν satisfies

$$K_\nu'(t) = -K_{\nu-1}(t) - \frac{\nu}{t} K_\nu(t).$$

The function K_ν for $\nu = 0, 1, 2$ has the following behavior near 0:

$$K_\nu(t) \sim \begin{cases} 2/t^2 & \text{if } \nu = 2 \\ 1/t & \text{if } \nu = 1 \\ -\log(t) & \text{if } \nu = 0 \end{cases} \qquad \text{as } t \to 0.$$

Using these results it is easy to show that

$$\mathcal{L}'(t) = -2\sqrt{t} K_1(2\sqrt{t})$$

and that $\mathcal{L}'(0) = -1$. Hence, $\mathrm{E}(X) = 1$. Similarly,

$$\mathcal{L}''(t) = 2K_0(2\sqrt{t}) \sim \log(1/t) \quad \text{as } t \to 0$$

so that $\mathcal{L}''(0)$ does not exist. It follows that $E(X^2)$ is not finite. □

The Laplace transform of a sum of independent random variables is easily determined from the Laplace transforms of the individual random variables. This result is stated in the following theorem; the proof is left as an exercise.

***Theorem* 4.7.** *Let X and Y denote independent, real-valued nonnegative random variables with Laplace tranforms $\mathcal{L}_X$ and $\mathcal{L}_Y$, respectively. Let $\mathcal{L}_{X+Y}$ denote the Laplace transform of the random variable $X + Y$. Then*

$$\mathcal{L}_{X+Y}(t) = \mathcal{L}_X(t)\mathcal{L}_Y(t), \quad t \geq 0.$$

***Example* 4.9 (*Gamma distribution*).** Let X_1 and X_2 denote independent random variables such that, for $j = 1, 2$, X_j has a gamma distribution with parameters α_j and β_j. Let $\mathcal{L}_j$ denote the Laplace transform of X_j, $j = 1, 2$. Then

$$\mathcal{L}_j(t) = \frac{\beta_j^{\alpha_j}}{(\beta_j + t)^{\alpha_j}}, \quad j = 1, 2.$$

Let $X = X_1 + X_2$. The Laplace transform of X is therefore given by

$$\mathcal{L}(t) = \frac{\beta_1^{\alpha_1} \beta_2^{\alpha_2}}{(\beta_1 + t)^{\alpha_1} (\beta_2 + t)^{\alpha_2}};$$

see Example 4.6. It follows that X has a gamma distribution if and only if $\beta_1 = \beta_2$. □

Moment-generating functions

The main drawback of Laplace transforms is that they apply only to nonnegative random variables. The same idea used to define the Laplace transform can be applied to a random variable with range **R**, yielding the moment-generating function; however, as noted earlier, moment-generating functions do not always exist.

Let X denote a real-valued random variable and suppose there exists a number $\delta > 0$ such that $E[\exp\{tX\}] < \infty$ for $|t| < \delta$. In this case, we say that X has *moment-generating function*

$$M(t) \equiv M_X(t) = E[\exp\{tX\}], \quad |t| < \delta;$$

δ is known as the *radius of convergence* of M_X. The moment-generating function is closely related to the characteristic function of X and, if X is nonnegative, to the Laplace transform of X.

Example **4.10** *(Poisson distribution).* Let X denote a discrete random variable taking values in the set $\{0, 1, 2, \ldots\}$ and let

$$p(x) = \lambda^x \exp(-\lambda)/x!, \quad x = 0, 1, 2, \ldots$$

denote the frequency function of the distribution, where $\lambda > 0$. This is a *Poisson distribution with parameter* λ.

Note that, for any value of t,

$$E[\exp(tX)] = \sum_{x=0}^{\infty} \exp(tx)\lambda^x \exp(-\lambda)/x!$$

$$= \sum_{x=0}^{\infty} [\exp(t)\lambda]^x \exp(-\lambda)/x! = \exp\{[\exp(t) - 1]\lambda\}.$$

Hence, the moment-generating function of this distribution exists and is given by

$$M(t) = \exp\{[\exp(t) - 1]\lambda\}, \quad -\infty < t < \infty. \qquad \square$$

Example **4.11** *(Exponential distribution).* Let X denote a random variable with a standard exponential distribution. Recall that this distribution is absolutely continuous with density function

$$p(x) = \exp(-x) \quad x > 0.$$

Note that

$$E[\exp(tX)] = \int_0^{\infty} \exp(tx) \exp(-tx)\,dx;$$

clearly, for any $t > 1$, $E[\exp(tX)] = \infty$; hence, the moment-generating function of this distribution is given by

$$M(t) = \frac{1}{1-t}, \quad |t| < 1.$$

This function can be compared with the Laplace transform of the distribution, which can be obtained from Example 4.6 by taking $\alpha = \beta = 1$:

$$\mathcal{L}(t) = \frac{1}{1+t}, \quad t \geq 0. \qquad \boxtimes$$

***Example* 4.12 (*Log-normal distribution*).** Let X denote a real-valued random variable with an absolutely continuous distribution with density function

$$p(x) = \frac{1}{x\sqrt{(2\pi)}} \exp\{-\frac{1}{2}[\log(x)]^2\}, \quad x > 0;$$

this is an example of a *log-normal distribution.* Consider the integral

$$\int_0^\infty \exp(tx)\frac{1}{x\sqrt{(2\pi)}} \exp\left\{-\frac{1}{2}[\log(x)]^2\right\} dx$$

$$= \frac{1}{\sqrt{(2\pi)}} \int_0^\infty \exp\{tx - [\log(x)]^2/2 - \log(x)\} dx.$$

Since for any $t > 0$,

$$tx - [\log(x)]^2/2 - \log(x) \sim tx \quad \text{as } x \to \infty,$$

it follows that $E[\exp(tx)] = \infty$ for all $t > 0$ and, hence, the moment-generating function of this distribution does not exist. □

When moment-generating functions exist, they have many of the important properties possessed by characteristic functions and Laplace transforms. The following theorem shows that the moment-generating function can be expanded in a power series expansion and that moments of the distribution can be obtained by differentiation.

***Theorem* 4.8.** *Let X denote a real-valued random variable with moment-generating function $M_X(t)$, $|t| < \delta$, for some $\delta > 0$. Then $E[X^n]$ exists and is finite for all $n = 1, 2, \ldots$ and*

$$M_X(t) = \sum_{n=0}^\infty t^n \, E\,(X^n)/n!, \quad |t| < \delta.$$

Furthermore,

$$E(X^n) = M_X^{(n)}(0), \quad n = 1, 2, \ldots.$$

Proof. Choose $t \neq 0$ in the interval $(-\delta, \delta)$. Then $E[\exp\{tX\}]$ and $E[\exp\{-tX\}]$ are both finite. Hence,

$$E[\exp\{|tX|\}] = E[\exp(tX)I_{\{tX>0\}}] + E[\exp(-tX)I_{\{tX\leq 0\}}]$$
$$\leq E[\exp\{tX\}] + E[\exp\{-tX\}] < \infty.$$

Note that, since

$$\exp\{|tX|\} = \sum_{j=0}^\infty \frac{|tX|^j}{j!}, \quad \geq \frac{|tX|^m}{n!}$$

$$|X|^n \leq \frac{n!}{|t|^n} \exp\{|tX|\}, \quad n = 0, 1, 2, \ldots.$$

It follows that $E[|X|^n] < \infty$ for $n = 1, 2, \ldots$ and, hence, that $E[X^n]$ exists and is finite for $n = 1, 2, \ldots$.

For fixed t, define

$$f_n(x) = \sum_{j=0}^{n} (tx)^j / j!.$$

Note that for each fixed x,

$$f_n(x) \to \exp\{tx\} \quad \text{as} \quad n \to \infty.$$

Also,

$$|f_n(x)| \le \left| \sum_{j=0}^{n} (tx)^j / j! \right| \le \sum_{j=0}^{n} |tx|^j / j! \le \exp\{|tx|\}$$

and, for $|t| < \delta$,

$$E[\exp(|tX|)] < \infty.$$

Hence, by the Dominated Convergence Theorem (see Appendix 1),

$$\lim_{n \to \infty} E[f_n(X)] = \sum_{j=0}^{\infty} t^j E(X^j) / j! = M_X(t), \quad |t| < \delta.$$

That is,

$$M_X(t) = \sum_{n=0}^{\infty} t^n E(X^n) / n!, \quad |t| < \delta.$$

The remaining part of the theorem now follows from the fact that a power series may be differentiated term-by-term within its radius of convergence (see Appendix 3). ■

Example* 4.13 *(Poisson distribution). Let X denote a random variable with a Poisson distribution with parameter λ; see Example 4.10. Recall that the moment-generating function of this distribution is given by

$$M(t) = \exp\{[\exp(t) - 1]\lambda\}, \quad -\infty < t < \infty.$$

Note that

$$M'(t) = M(t) \exp(t)\lambda,$$

$$M''(t) = M(t) \exp(2t)\lambda^2 + M(t) \exp(t)\lambda$$

and

$$M'''(t) = M(t)[\lambda \exp(t) + 3(\lambda \exp(t))^2 + (\lambda \exp(t))^3].$$

It follows that

$$E(X) = M'(0) = \lambda, \qquad E(X^2) = M''(0) = \lambda^2 + \lambda$$

and

$$E(X^3) = \lambda + 3\lambda^2 + \lambda^3.$$

In particular, X has mean λ and variance λ. □

Example* 4.14 *(Log-normal distribution). Let X denote a random variable with the log-normal distribution considered in Example 4.12; recall that the moment-generating function of this distribution does not exist. Note that

$$
\begin{aligned}
\mathrm{E}(X^r) &= \int_0^\infty x^r \frac{1}{x\sqrt{(2\pi)}} \exp\left\{-\frac{1}{2}[\log(x)]^2\right\} dx \\
&= \int_{-\infty}^\infty \exp(rt) \frac{1}{\sqrt{(2\pi)}} \exp\left(-\frac{1}{2}t^2\right) dt \\
&= \exp\left(\frac{1}{2}r^2\right).
\end{aligned}
$$

Hence, all the moments of the distribution exist and are finite, even though the moment-generating function does not exist. That is, the converse to the first part of Theorem 4.8 does not hold.   □

If two moment-generating functions agree in a neighborhood of 0, then they represent the same distribution; a formal statement of this result is given in the following theorem.

Theorem 4.9. *Let X and Y denote real-valued random variables with moment-generating functions $M_X(t)$, $|t| < \delta_X$, and $M_Y(t)$, $|t| < \delta_Y$, respectively. X and Y have the same distribution if and only if there exists a $\delta > 0$ such that*

$$
M_X(t) = M_Y(t), \quad |t| < \delta.
$$

Proof.   Note that, since $M_X(t)$ and $M_Y(t)$ agree in a neighborhood of 0,

$$
\mathrm{E}(X^j) = \mathrm{E}(Y^j), \quad j = 1, 2, \ldots.
$$

Since, for $|t| < \delta$,

$$
\mathrm{E}(\exp\{t|X|\}) \le \mathrm{E}(\exp\{tX\}) + \mathrm{E}(\exp\{-tX\}) < \infty,
$$

it follows that moment-generating function of $|X|$ exists and, hence, all moments of $|X|$ exist. Let

$$
\gamma_j = \mathrm{E}(|X|^j), \quad j = 1, 2, \ldots.
$$

Since

$$
\sum_{j=0}^\infty \gamma_j t^j / j! < \infty \quad \text{for } |t| < \delta,
$$

it follows that

$$
\lim_{j\to\infty} \frac{\gamma_j t^j}{j!} = 0, \quad |t| < \delta.
$$

By Lemma A2.1,

$$
\left| \exp\{ihx\} - \sum_{j=0}^n (ihx)^j / j! \right| \le \frac{|hx|^{n+1}}{(n+1)!}.
$$

Hence,

$$\left| \exp\{itx\} \left[\exp\{ihx\} - \sum_{j=0}^{n} (ihx)^j / j! \right] \right| \le \frac{|hx|^{n+1}}{(n+1)!}$$

and

$$\left| \varphi_X(t+h) - \sum_{j=0}^{n} \frac{h^j}{j!} \int_{-\infty}^{\infty} (ix)^j \exp\{itx\} \, dF_X(x) \right| \le \frac{|h|^{n+1} \gamma_{n+1}}{(n+1)!},$$

where φ_X denotes the characteristic function of X.

Note that

$$\int_{-\infty}^{\infty} (ix)^k \exp\{itx\} \, dF_X(x) = \varphi_X^{(k)}(t).$$

Hence,

$$\left| \varphi_X(t+h) - \sum_{j=0}^{n} \frac{h^j}{j!} \varphi_X^{(j)}(t) \right| \le \frac{|h|^{n+1} \gamma_{n+1}}{(n+1)!}, \quad n = 1, 2, \ldots.$$

It follows that

$$\varphi_X(t+h) = \sum_{j=0}^{\infty} \frac{h^j}{j!} \varphi_X^{(j)}(t), \quad |h| < \delta. \tag{4.2}$$

Applying the same argument to Y shows that φ_Y, the characteristic function of Y, satisfies

$$\varphi_Y(t+h) = \sum_{j=0}^{\infty} \frac{h^j}{j!} \varphi_Y^{(j)}(t), \quad |h| < \delta. \tag{4.3}$$

Taking $t = 0$ and using the fact that

$$\varphi_X^{(k)}(0) = E(X^k) = E(Y^k) = \varphi_Y^{(k)}(0),$$

it follows that

$$\varphi_X(t) = \varphi_Y(t), \quad |t| < \delta$$

and also that

$$\varphi_X^{(k)}(t) = \varphi_Y^{(k)}(t), \quad k = 1, 2, \ldots, \quad |t| < \delta.$$

Taking $t = \delta/2$ in (4.2) and (4.3) shows that

$$\varphi_X(\delta/2 + h) = \sum_{j=0}^{\infty} \frac{h^j}{j!} \varphi_X^{(j)}(\delta/2), \quad |h| < \delta$$

and

$$\varphi_Y(\delta/2 + h) = \sum_{j=0}^{\infty} \frac{h^j}{j!} \varphi_Y^{(j)}(\delta/2), \quad |h| < \delta.$$

Since

$$\varphi_X^{(k)}(\delta/2) = \varphi_Y^{(k)}(\delta/2), \quad k = 1, 2, \ldots,$$

it follows that

$$\varphi_X(\delta/2 + h) = \varphi_Y(\delta/2 + h), \quad |h| < \delta.$$

The same argument may be used with $-\delta/2$ so that

$$\varphi_X(-\delta/2 + h) = \varphi_Y(-\delta/2 + h), \quad |h| < \delta$$

and, hence, that

$$\varphi_X(t) = \varphi_Y(t), \quad |t| < \frac{3\delta}{2}.$$

Using this same argument with δ and $-\delta$, it can be shown that

$$\varphi_X(t) = \varphi_Y(t), \quad |t| < 2\delta.$$

Continuing in this way shows that

$$\varphi_X(t) = \varphi_Y(t), \quad |t| < \frac{r\delta}{2}$$

for any $r = 1, 2, \ldots$ and, hence, that

$$\varphi_X(t) = \varphi_Y(t), \quad -\infty < t < \infty.$$

It now follows from Corollary 3.1 that X and Y have the same distribution. ∎

Hence, in establishing that two random variables have the same distribution, there is a slight difference between moment-generating function and characteristic functions. For the distributions of X and Y to be the same, $\varphi_X(t)$ and $\varphi_Y(t)$ must be equal for all $t \in \mathbf{R}$, while $M_X(t)$ and $M_Y(t)$ only need to be equal for all t in some neighborhood of 0.

Theorem 4.9 is often used in conjunction with the following results to determine the distribution of a function of random variables. The first of these results shows that there is a simple relationship between the moment-generating function of a linear function of a random variable and the moment-generating function of the random variable itself; the proof is left as an exercise.

Theorem 4.10. *Let X denote a real-valued random variable with moment-generating function $M_X(t)$, $|t| < \delta$. Let a and b denote real-valued constants and let $Y = a + bX$. Then the moment-generating function of Y is given by*

$$M_Y(t) = \exp(at)M_X(bt), \quad |t| < \delta/|b|.$$

Like characteristic functions and Laplace transforms, the moment-generating function of a sum of independent random variables is simply the product of the individual moment-generating functions. This result is given in the following theorem; the proof is left as an exercise.

Theorem 4.11. *Let X and Y denote independent, real-valued random variables; let $M_X(t)$, $|t| < \delta_X$, denote the moment-generating function of X and let $M_Y(t)$, $|t| < \delta_Y$, denote the moment-generating function of Y. Let $M_{X+Y}(t)$ denote the moment-generating function of the random variable $X + Y$. Then*

$$M_{X+Y}(t) = M_X(t)M_Y(t), \quad |t| < \min(\delta_X, \delta_Y)$$

where $M_{X+Y}(t)$ denotes the moment-generating function of the random variable $X + Y$.

Example* 4.15 *(Poisson distribution). Let X_1 and X_2 denote independent random variables such that, for $j = 1, 2$, X_j has a Poisson distribution with mean λ_j; see Example 4.10. Then X_j has moment-generating function

$$M_j(t) = \exp\{[\exp(t) - 1]\lambda_j\}, \quad -\infty < t < \infty.$$

Let $X = X_1 + X_2$. By Theorem 4.11, the moment-generating function of X is given by

$$M(t) = M_1(t)M_2(t) = \exp\{[\exp(t) - 1](\lambda_1 + \lambda_2)\}, \quad -\infty < t < \infty.$$

Note that this is the moment-generating function of a Poisson distribution with mean $\lambda_1 + \lambda_2$. Thus, the sum of two independent Poisson random variables also has a Poisson distribution. □

Example* 4.16 *(Sample mean of normal random variables). Let Z denote a random variable with standard normal distribution; then the moment-generating function of Z is given by

$$M(t) = \int_{-\infty}^{\infty} \exp(tz) \frac{1}{\sqrt{(2\pi)}} \exp\left(-\frac{1}{2}z^2\right) dx = \exp(t^2/2), \quad -\infty < t < \infty.$$

Let μ and σ denote real-valued constants, $\sigma > 0$, and let X denote a random variable with a normal distribution with mean μ and standard deviation σ. Recall that X has the same distribution as $\mu + \sigma Z$; see Example 3.6. According to Theorem 4.10, the moment-generating function of X is given by

$$M_X(t) = \exp(\mu t) \exp(\sigma^2 t^2/2), \quad -\infty < t < \infty.$$

Let $X_1, X_2, \ldots, X_n$ denote independent, identically distributed random variables, each with the same distribution as X. Then, by repeated application of Theorem 4.11, $\sum_{j=1}^{n} X_j$ has moment-generating function

$$\exp(n\mu t) \exp(n\sigma^2 t^2/2), \quad -\infty < t < \infty$$

and, by Theorem 4.10, the sample mean $\bar{X} = \sum_{j=1}^{n} X_j/n$ has moment-generating function

$$M_{\bar{X}}(t) = \exp(\mu t) \exp[(\sigma^2/n)t^2/2], \quad -\infty < t < \infty.$$

Comparing this to M_X above, we see that $\bar{X}$ has a normal distribution with mean μ and standard deviation $\sigma/\sqrt{n}$. □

Moment-generating functions for random vectors

Moment-generating functions are defined for random vectors in a manner that is analogous to the definition of a characteristic function for a random vector. Let X denote a d-dimensional random vector and let t denote an element of $\mathbf{R}^d$. If there exists a $\delta > 0$ such that

$$E(\exp\{t^T X\}) < \infty \quad \text{for all } ||t|| < \delta,$$

then the moment-generating function of X exists and is given by

$$M_X(t) = E(\exp\{t^T X\}), \quad t \in \mathbf{R}^d, \quad ||t|| < \delta;$$

as in the case of a real-valued random variable, δ is known as the radius of convergence of M_X.

Many of the properties of a moment-generating function for a real-valued random variable extend to the vector case. Several of these properties are given in the following theorem; the proof is left as an exercise.

Theorem 4.12. *Let X and Y denote d-dimensional random vectors with moment-generating functions M_X and M_Y, and radii of convergence δ_X and δ_Y, respectively.*

(i) *Let A denote an $m \times d$ matrix of real numbers and let b denote an element of $\mathbf{R}^d$. Then M_{AX+b}, the moment-generating function of $AX + b$, satisfies*

$$M_{AX+b}(t) = \exp\{t^T b\} M_X(A^T t), \quad ||t|| < \delta_A$$

for some $\delta_A > 0$, possibly depending on A.

(ii) *If X and Y are independent then M_{X+Y}, the moment-generating function of $X + Y$, exists and is given by*

$$M_{X+Y}(t) = M_X(t) M_Y(t), \quad ||t|| < \min(\delta_X, \delta_Y).$$

(iii) *X and Y have the same distribution if and only if there exists a $\delta > 0$ such that*

$$M_X(t) = M_Y(t) \qquad \text{for all } ||t|| < \delta.$$

As is the case with the characteristic function, the following result shows that the moment-generating function can be used to establish the independence of two random vectors; the proof is left as an exercise.

Corollary 4.2. *Let X denote a random vector taking values in $\mathbf{R}^d$ and let $X = (X_1, X_2)$ where X_1 takes values in $\mathbf{R}^{d_1}$ and X_2 takes values in $\mathbf{R}^{d_2}$. Let M denote the moment-generating function of X with radius of convergence δ, let M_1 denote the moment-generating function of X_1 with radius of convergence δ_1, and let M_2 denote the moment-generating function of X_2 with radius of convergence δ_2. X_1 and X_2 are independent if and only if there exists a $\delta_0 > 0$ such that for all $t = (t_1, t_2)$, $t_1 \in \mathbf{R}^{d_1}$, $t_2 \in \mathbf{R}^{d_2}$, $||t|| < \delta_0$,*

$$M(t) = M_1(t_1) M(t_2).$$

4.4 Cumulants

Although moments provide a convenient summary of the properties of a random variable, they are not always easy to work with. For instance, let X denote a real-valued random variable and let a, b denote constants. Then the relationship between the moments of X and those of $aX + b$ can be quite complicated. Similarly, if Y is a real-valued random variable such that X and Y are independent, then the moments of $X + Y$ do not have a simple relationship to the moments of X and Y.

Suppose that X and Y have moment-generating functions M_X and M_Y, respectively. Some insight into the properties described above can be gained by viewing moments of a random variable as derivatives of its moment-generating function at 0, rather than as integrals with respect to a distribution function. Since the moment-generating function of

$aX + b$ is given by $\exp(bt)M_X(at)$, it is clear that repeated differentiation of this expression with respect to t will lead to a fairly complicated expression; specifically, by Leibnitz's rule for differentiation (see Appendix 3), the nth moment of $aX + b$ is given by

$$\sum_{j=0}^{n} \binom{n}{j} b^{n-j} a^j E(X^j), \quad n = 0, 1, \ldots.$$

Similarly, the moment-generating of $X + Y$ is given by $M_X(t)M_Y(t)$ and differentiating this function can lead to a complicated result.

In both cases, the situation is simplified if, instead of using the moment-generating function itself, we use the log of the moment-generating function. For instance, if M_{aX+b} denotes the moment-generating function of $aX + b$, then

$$\log M_{aX+b}(t) = bt + \log M_X(at)$$

and the derivatives of $\log M_{aX+b}(t)$ at $t = 0$ have a relatively simple relationship to the derivatives of $\log M_X(at)$. Of course, these derivatives no longer represent the moments of the distribution, although they will be functions of the moments; these functions of the moments are called the *cumulants* of the distribution. In this section, the basic theory of cumulants is presented.

Let X denote a real-valued random variable with moment-generating function $M_X(t)$, $|t| < \delta$. The *cumulant-generating function* of X is defined as

$$K_X(t) = \log M_X(t), \quad |t| < \delta.$$

Since, by Theorem 4.8, M_X has a power series expansion for t near 0, the cumulant-generating function may be expanded

$$K_X(t) = \sum_{j=1}^{\infty} \frac{\kappa_j}{j!} t^j, \quad |t| < \delta$$

where $\kappa_1, \kappa_2, \ldots$ are constants that depend on the distribution of X. These constants are called the *cumulants* of the distribution or, more simply, the cumulants of X. The constant κ_j will be called the jth cumulant of X; we may also write the jth cumulant of X as $\kappa_j(X)$.

Hence, the cumulants may be obtained by differentiation of K_X, in the same way that the moments of a distribution may be obtained by differentiation of the moment-generating function:

$$\kappa_j = \frac{d^j}{dt^j} K_X(t)\Big|_{t=0}, \quad j = 1, 2, \ldots.$$

Example 4.17 (Standard normal distribution). Let Z denote a random variable with a standard normal distribution. It is straightforward to show that the moment-generating function of this distribution is given by

$$M_Z(t) = \int_{-\infty}^{\infty} \exp(tz) \frac{1}{\sqrt{(2\pi)}} \exp\left(-\frac{1}{2}z^2\right) dz = \exp(t^2/2), \quad -\infty < t < \infty$$

and, hence, the cumulant-generating function is given by

$$K_Z(t) = \frac{1}{2}t^2, \quad -\infty < t < \infty.$$

It follows that $\kappa_1 = 0$, $\kappa_2 = 1$, and $\kappa_j = 0$, $j = 3, 4, \ldots.$

Now let X denote a random variable with a normal distribution with parameters μ and $\sigma > 0$. Then X has the same distribution as $\sigma Z + \mu$; see Example 3.6. It follows that X has moment-generating function

$$M_X(t) = \exp(\mu t) M_Z(\sigma t) = \exp(\mu t) \exp(\sigma^2 t^2/2), \quad -\infty < t < \infty$$

and, hence, the cumulant-generating function of X is

$$K_X(t) = \mu t + \frac{1}{2}\sigma^2 t^2, \quad -\infty < t < \infty.$$

The cumulants of X are given by $\kappa_1(X) = \mu$, $\kappa_2(X) = \sigma^2$, and $\kappa_j(X) = 0$, $j = 3, 4, \ldots$; the distribution of X is often described as a normal distribution with mean μ and standard deviation σ. $\square$

Example 4.18 (Poisson distribution). Let X denote a random variable with a Poisson distribution with parameter λ; see Example 4.10. Here

$$M_X(t) = \exp\{[\exp(t) - 1]\lambda\}, \quad -\infty < t < \infty$$

so that

$$K_X(t) = [\exp(t) - 1]\lambda, \quad -\infty < t < \infty.$$

It follows that all cumulants of this distribution are equal to λ. $\square$

Example 4.19 (Laplace distribution). Let X denote a random variable with a standard Laplace distribution; see Example 4.5. This distribution is absolutely continuous with density function

$$p(x) = \frac{1}{2}\exp\{-|x|\}, \quad -\infty < x < \infty.$$

Hence, the moment-generating function of the distribution is given by

$$M_X(t) = \frac{1}{1-t^2}, \quad |t| < 1$$

and the cumulant-generating function is given by

$$K_X(t) = -\log(1 - t^2), \quad |t| < 1.$$

It follows that $\kappa_1 = 0$, $\kappa_2 = 2$, $\kappa_3 = 0$, and $\kappa_4 = 12$. $\square$

Since

$$\boxed{M_X(t) = \sum_{j=0}^{\infty} t^j \frac{E(X^j)}{j!}, \quad |t| < \delta,}$$

$$K_X(t) = \log M_X(t) = \log\left[1 + \sum_{j=1}^{\infty} t^j E(X^j)/j!\right].$$

Hence, it is clear that cumulants are functions of moments. The exact relationship may be obtained by expanding

$$\log\left[1 + \sum_{j=1}^{\infty} t^j E(X^j)/j!\right]$$

in a power series in t and then equating terms with an expansion for the cumulant generating function. For example,

$$\log\left[1 + \sum_{j=1}^{\infty} t^j E(X^j)/j!\right] = E(X)t + \frac{1}{2!}[E(X^2) - E(X)^2]t^2$$

$$+ \frac{1}{3!}[E(X^3) - E(X)E(X^2) - 2(E(X^2))$$

$$- E(X)^2)E(X)]t^3 + \cdots$$

so that

$$\kappa_1 = E(X), \quad \kappa_2 = E(X^2) - E(X)^2, \quad \kappa_3 = E(X^3) - E(X)E(X^2) - 2[E(X^2) - E(X)^2]E(X).$$

Hence, the first cumulant is the mean of X, the second cumulant is the variance of X, and, with a little algebra, it may be shown that the third cumulant is $E[(X - E(X))^3]$, often called the _skewness_ of X.

The general form of the relationship between moments and cumulants is based on the relationship between the coefficients of a power series and the coefficients in an expansion of the log of that power series. Consider a function

$$\alpha(t) = \sum_{j=0}^{\infty} \frac{\alpha_j}{j!} t^j$$

defined for t in a neighborhood of 0. Suppose that $\alpha(t) > 0$ for t near 0 and write

$$\beta(t) = \log \alpha(t) = \sum_{j=0}^{\infty} \frac{\beta_j}{j!} t^j.$$

Clearly, the coefficients $\beta_1, \beta_2, \ldots$ are functions of $\alpha_1, \alpha_2, \ldots$. The following result can be used to determine an expression for α_r in terms of $\beta_1, \ldots, \beta_r$; conversely, an expression for β_r can be given in terms of $\alpha_1, \ldots, \alpha_r$.

Lemma 4.1. _Define the functions $\alpha(t)$ and $\beta(t)$ and the coefficients $\alpha_1, \alpha_2, \ldots$ and $\beta_1, \beta_2, \ldots$ as above. Then $\alpha_0 = \exp(\beta_0)$ and_

$$\alpha_{r+1} = \sum_{j=0}^{r} \binom{r}{j} \beta_{j+1} \alpha_{r-j}, \quad r = 0, 1, \ldots.$$

Proof. Note that

$$\alpha_j = \frac{d^j}{dt^j} \alpha(t)\Big|_{t=0} \quad \text{and} \quad \beta_j = \frac{d^j}{dt^j} \beta(t)\Big|_{t=0}, \quad j = 0, 1, \ldots.$$

The result relating α_0 and β_0 follows immediately.

Since $\alpha(t) = \exp\{\beta(t)\}$,

$$\alpha'(t) = \exp\{\beta(t)\}\beta'(t) = \alpha(t)\beta'(t).$$

Hence, by Leibnitz's rule,

$$\frac{d^r}{dt^r}\alpha'(t) = \sum_{j=0}^{r}\binom{r}{j}\frac{d^j}{dt^j}\beta'(t)\frac{d^{r-j}}{dt^{r-j}}\alpha(t).$$

The result now follows by evaluating both sides of this expression at $t = 0$. ∎

Hence, applying this result to the moment-generating and cumulant-generating functions yields a general formula relating moments to cumulants. In this context, $\alpha_1, \alpha_2, \ldots$ are the moments and $\beta_1, \beta_2, \ldots$ are the cumulants. It follows that

$$E(X^{r+1}) = \sum_{j=0}^{r}\binom{r}{j}\kappa_{j+1}E(X^{r-j}), \quad r = 0, 1, \ldots.$$

An important consequence of this result is that κ_r is a function of $E(X), \ldots, E(X^r)$.

Lemma 4.1 can also be used to derive an expression for central moments in terms of cumulants by interpreting $\alpha_1, \alpha_2, \ldots$ as the central moments and $\beta_1, \beta_2, \ldots$, under the assumption that $\alpha_1 = \beta_1 = 0$. Hence,

$$E[(X - \mu)^2] = \kappa_2, \qquad E[(X - \mu)^3] = \kappa_3$$

and

$$E[(X - \mu)^4] = \kappa_4 + 3\kappa_2^2.$$

The approach to cumulants taken thus far in this section requires the existence of the moment-generating function of X. A more general approach may be based on the characteristic function. Suppose that X has characteristic function $\varphi_X(t)$ and that $E(X^m)$ exists and is finite. Then, by Theorem 3.5, $\varphi_X^{(m)}(0)$ exists and, hence, the mth derivative of $\log\varphi_X^{(m)}(t)$ at $t = 0$ exists. We may define the jth cumulant of X, $1 \le j \le m$, by

$$\kappa_j = \frac{1}{(i)^j}\frac{d^j}{dt^j}\left.\log\varphi_X(t)\right|_{t=0}.$$

Of course, if the cumulant-generating function X exists, it is important to confirm that the definition of cumulants based on the characteristic function agrees with the definition based on the cumulant-generating function. This fact is established by the following lemma.

Lemma 4.2. *Let X denote a random variable with moment-generating function M_X and characteristic function φ_X. Then, for any $m = 1, 2, \ldots$,*

$$\frac{d^m}{dt^m}\left.\log M_X(t)\right|_{t=0} = \frac{1}{(i)^m}\frac{d^m}{dt^m}\left.\log\varphi_X(t)\right|_{t=0}.$$

Proof. Fix m. Since the mth moment of X exists we may write

$$M_X(t) = 1 + \sum_{j=1}^{m}t^j E(X^j)/j! + o(t^m) \quad \text{as } t \to 0$$

and

$$\varphi_X(t) = 1 + \sum_{j=1}^{m}(it)^j E(X^j)/j! + o(t^m) \quad \text{as } t \to 0.$$

Hence, we may write

$$\log M_X(t) = h(t; E(X), \ldots, E(X^m)) + o(t^m)$$

for some function h. Since $\varphi_X(t)$ has the same expansion as $M_X(t)$, except with it replacing t, we must have

$$\log \varphi_X(t) = h(it; E(X), \ldots, E(X^m)) + o(t^m).$$

It follows that

$$\frac{d^m}{dt^m} \log M_X(t)\Big|_{t=0} = \frac{d^m}{d(it)^m} \log \varphi_X(t)\Big|_{t=0}.$$

The result now follows from the fact that

$$\frac{d^m}{dt^m} \log \varphi_X(t)\Big|_{t=0} = i^m \frac{d^m}{d(it)^m} \log \varphi_X(t)\Big|_{t=0}. \quad\blacksquare$$

Therefore, the log of the characteristic function has an expansion in terms of the cumulants that is similar to the expansion of the characterstic function itself in terms of moments.

Theorem 4.13. *Let X denote a real-valued random variable and let φ_X denote the characteristic function of X. If $E(|X|^m) < \infty$, then*

cumulant of characteristic func.

$$\log(\varphi_X(t)) = \sum_{j=1}^{m}(it)^j \kappa_j/j! + o(t^m) \quad as\ t \to 0$$

where $\kappa_1, \kappa_2, \ldots, \kappa_m$ denote the cumulants of X.
If, for some $m = 1, 2, \ldots, \varphi^{(2m)}(0)$ exists then $\kappa_1, \kappa_2, \ldots, \kappa_{2m}$ all exist and are finite.

Proof. We have seen that if $E(X^m)$ exists and is finite, then

$$\varphi(t) = 1 + \sum_{j=1}^{m} \frac{(it)^j}{j!} E(X^j) + o(t^m) \quad as\ t \to 0.$$

Since, for a complex number z,

$$\log(1 + z) = \sum_{j=1}^{d}(-1)^j z^j/j + o(|z|^d) \quad as\ |z| \to 0,$$

for any $d = 1, 2, \ldots$, it follows that $\log(\varphi(t))$ may be expanded in a series of the form

$$\log(\varphi(t)) = \sum_{j=1}^{m}(it)^j c_j/j! + o(t^m) \quad as\ t \to 0,$$

for some constants $c_1, c_2, \ldots, c_m$. Using the relationship between cumulants and moments, it follows that these constants must be the cumulants; that is, $c_j = \kappa_j$, proving the first part of the theorem.

The second part of the theorem follows from the fact that the existence of $\varphi^{(2m)}(0)$ implies that all moments of order less than or equal to $2m$ exist and are finite. Since each κ_j, $j = 1, \ldots, 2m$, is a function of these moments, the result follows. $\blacksquare$

Although cumulants may be calculated directly from the characteristic function, there are relatively few cases in which the moment-generating function of a distribution does not exist, while the characteristic function is easily calculated and easy to differentiate. It is often simpler to calculate the cumulants by calculating the moments of the distribution and then using the relationship between cumulants and moments.

***Example* 4.20 (*Log-normal distribution*).** Let X denote a random variable with the log-normal distribution considered in Example 4.14. Recall that, although the moment-generating function of this distribution does not exist, all moments do exist and are given by

$$\mathrm{E}(X^r) = \exp\left(\frac{1}{2}r^2\right), \quad r = 1, 2, \ldots.$$

Hence,

$$\kappa_1 = \exp(1/2), \ \ \kappa_2 = \exp(2) - \exp(1), \ \ \kappa_3 = \exp(9/2) - 3\exp(5/2) + 2\exp(3/2),$$

and so on. □

Earlier in this section, the relationship between the the cumulants of a linear function of a random variable and the cumulants of the random variable itself was described. The following theorem gives a formal statement of this relationship for the more general case in which the moment-generating function does not necessarily exist.

Theorem 4.14. *Let X denote a real-valued random variable with mth cumulant $\kappa_m(X)$ for some $m = 1, 2, \ldots$ and let $Y = aX + b$ for some constants a, b. Then the mth cumulant of Y, denoted by $\kappa_m(Y)$, is given by*

$$\kappa_m(Y) = \begin{cases} a\kappa_1(X) + b & \text{if } m = 1 \\ a^m\kappa_m(X) & \text{if } m = 2, 3, \ldots \end{cases}$$

[Expectation]

[variance···]

Proof. Let φ_X and φ_Y denote the characteristic functions of X and Y, respectively. We have seen that $\varphi_Y(t) = \exp(ibt)\varphi_X(at)$. Hence,

$$\log(\varphi_Y(t)) = ibt + \log(\varphi_X(at)).$$

If $\mathrm{E}(|X|^m) < \infty$ then, by Theorem 4.13,

$$\log(\varphi_X(t)) = \sum_{j=1}^m (it)^j \kappa_j(X)/j! + o(t^m) \quad \text{as} \ \ t \to 0;$$

it follows that

$$\log(\varphi_Y(t)) = ibt + \sum_{j=1}^m (ita)^j \kappa_j(X)/j! + o(t^m)$$

$$= (it)(a\kappa_1(X) + b) + \sum_{j=2}^m (it)^j a^j \kappa_j(X)/j! + o(t^m).$$

The result now follows from Theorem 4.13. ∎

***Example* 4.21 *(Laplace distribution)*.** Let X denote a random variable with a standard Laplace distribution; see Example 4.19. Let μ and $\sigma > 0$ denote constants and let $Y = \sigma X + \mu$; the distribution of Y is called a Laplace distribution with location parameter μ and scale parameter σ.

Using the results in Example 4.19, together with Theorem 4.14, the first four cumulants of this distribution are μ, $2\sigma^2$, 0, and $12\sigma^4$. $\quad\square$

***Example* 4.22 *(Standardized cumulants)*.** Consider a random variable X with cumulants $\kappa_1, \kappa_2, \ldots$ and consider the standardized variable $Y = (X - \kappa_1)/\sqrt{\kappa_2}$. The cumulants of Y are given by $0, 1, \kappa_j/\kappa_2^{\frac{j}{2}}, j = 3, 4, \ldots$. The cumulants of Y of order 3 and greater are sometimes called the *standardized cumulants* of X and are dimensionless quantities. They are often denoted by $\rho_3, \rho_4, \ldots$ so that

$$\rho_j(X) = \kappa_j(X)/\kappa_2(X)^{\frac{j}{2}}, \quad j = 3, 4, \ldots. \qquad\square$$

We have seen that if X and Y are independent random variables, then the characteristic function of $X + Y$ satisfies

$$\varphi_{X+Y}(t) = \varphi_X(t)\varphi_Y(t);$$

hence,

$$\log \varphi_{X+Y}(t) = \log \varphi_X(t) + \log \varphi_Y(t).$$

Since the cumulants are simply the coefficients in the expansion of the log of the characteristic function, it follows that the jth cumulant of $X + Y$ will be the sum of the jth cumulant of X and the jth cumulant of Y.

Theorem 4.15. *Let X and Y denote independent real-valued random variables with mth cumulants $\kappa_m(X)$ and $\kappa_m(Y)$, respectively, and let $\kappa_m(X + Y)$ denote the mth cumulant of $X + Y$. Then*

$$\kappa_m(X + Y) = \kappa_m(X) + \kappa_m(Y).$$

Proof. We have seen that $\varphi_{X+Y}(t) = \varphi_X(t)\varphi_Y(t)$. Hence, by Theorem 4.13,

$$\log(\varphi_{X+Y}(t)) = \log(\varphi_X(t)) + \log(\varphi_Y(t)) = \sum_{j=1}^{m}(it)^j(\kappa_j(X) + \kappa_j(Y))/j! + o(t^m).$$

The result now follows from noting that

$$\log \varphi_{X+Y}(t) = \sum_{j=1}^{m}(it)^j \kappa_j(X + Y)/j! + o(t^m)$$

as $t \to 0$. $\quad\blacksquare$

***Example* 4.23 *(Independent identically distributed random variables)*.** Let $X_1, X_2, \ldots,$ X_n denote independent, identically distributed scalar random variables and let $\kappa_1, \kappa_2, \ldots$ denote the cumulants of X_1. Let $S = \sum_{j=1}^{n} X_j$. Then

$$\kappa_j(S) = n\kappa_j, \quad j = 1, 2, \ldots.$$

The standardized cumulants satisfy

$$\rho_j(S) = \rho_j/n^{\frac{(j-1)}{2}}, \quad j = 3, 4, \ldots$$

where $\rho_3, \rho_4, \ldots$ denote the standardized cumulants of X_1. $\quad\square$

Cumulants of a random vector

Let $X = (X_1, \ldots, X_d)$ denote a d-dimensional random vector. Joint cumulants of elements of X may be defined using the same approach used to define the cumulants of a real-valued random variable. For simplicity, we consider only the case in which the moment-generating function of X exists. However, as in the case of a real-valued random variable, the same results may be obtained provided only that moments of a certain order exist.

Let M denote the moment-generating function of X with radius of convergence $\delta > 0$. Then the cumulant-generating function of X is given by $K(t) = \log M(t)$ and the *joint cumulant* of order $(i_1, \ldots, i_d)$, where the i_j are nonnegative integers, is given by

$$\kappa_{i_1 \cdots i_d} = \frac{\partial^{i_1 + \cdots + i_d}}{\partial t_1^{i_1} \cdots \partial t_d^{i_d}} K(t) \Big|_{t=0};$$

here $t = (t_1, \ldots, t_d)$. Although this definition may be used to define joint cumulants of arbitrary order, the most commonly used joint cumulants are those in which $i_1 + \cdots + i_d = 2$, for example, $\kappa_{110\cdots0}$, $\kappa_{1010\cdots0}$, and so on.

The following result gives some basic properties of joint cumulants.

Theorem 4.16. *Let* $X = (X_1, \ldots, X_d)$ *denote a d-dimensional random vector with cumulant-generating function K.*

    (i) Fix $1 \le j \le d$ and assume that

$$\sum_{k \ne j} i_k = 0.$$

    Then the joint cumulant of order $(i_1, \ldots, i_d)$ is the i_jth cumulant of X_j.

  (ii) Suppose that, for some $1 \le j < k \le d, i_j = i_k = 1$ and $i_1 + \cdots + i_d = 2$. Then the joint cumulant of order $(i_1, \ldots, i_d)$ is the covariance of X_{i_j} and X_{i_k}.

  (iii) Suppose that, for $1 \le j < k \le d$, X_j and X_k are independent. Then any joint cumulant of order $(i_1, \ldots, i_d)$ where $i_j > 0$ and $i_k > 0$ is 0.

  (iv) Let Y denote a d-dimensional random variable such that all cumulants of Y exist and assume that X and Y are independent. Let $\kappa_{i_1\cdots i_d}(X)$, $\kappa_{i_1\cdots i_d}(Y)$, and $\kappa_{i_1\cdots i_d}(X + Y)$ denote the cumulant of order $(i_1, \ldots, i_d)$ of X, Y, and X + Y, respectively. Then

$$\kappa_{i_1\cdots i_d}(X + Y) = \kappa_{i_1\cdots i_d}(X) + \kappa_{i_1\cdots i_d}(Y).$$

Proof.   Consider part (i); without loss of generality we may assume that $j = 1$. Let K_1 denote the cumulant-generating function of X_1. Then

$$K_1(t_1) = K((t_1, 0, \ldots, 0)).$$

Part (i) of the theorem now follows.

Suppose that $d = 2$. Let a and b denote constants and let $Z = aX_1 + bX_2$. Then the cumulant-generating function of Z is given by

$$K_Z(s) = K((as, bs)).$$

It follows that the second cumulant of Z, $\mathrm{Var}(Z)$, is given by

$$
\begin{aligned}
\mathrm{Var}(Z) &= \frac{\partial^2}{\partial s^2} K_X((as, bs))\Big|_{s=0} \\
&= a^2 \frac{\partial^2}{\partial t_1^2} K_X(t_1, 0)\Big|_{t_1=0} + b^2 \frac{\partial^2}{\partial t_2^2} K_X(0, t_2)\Big|_{t_2=0} + 2ab \frac{\partial^2}{\partial t_1 \partial t_2} K_X(t_1, t_2)\Big|_{t=0}
\end{aligned}
$$

Hence, by part (i) of the theorem,

$$\mathrm{Var}(Z) = a^2 \, \mathrm{Var}(X_1) + b^2 \, \mathrm{Var}(X_2) + 2ab\kappa_{11};$$

it follows that $\kappa_{11} = \mathrm{Cov}(X_1, X_2)$.

Now consider the case of general d; without loss of generality we may assume that $j = 1$ and $k = 2$. Part (ii) of the theorem now follows from an argument analogous to the one used in the proof of part (i): the cumulant-generating function of (X_1, X_2) is given by $K_X((t_1, t_2, 0, \ldots, 0))$ so that, from the result above, $\mathrm{Cov}(X_1, X_2) = \kappa_{110\cdots0}$.

Consider part (iii). Without loss of generality we may take $j = 1$ and $k = z$. Let K_1 denote the cumulant-generating function of X_1 and let K_2 denote the cumulant-generating function of $(X_2, \ldots, X_d)$. Then

$$K(t) = K_1(t_1) + K_2(\bar{t})$$

where $t = (t_1, \ldots, t_d)$ and $\bar{t} = (t_2, \ldots, t_d)$. It follows that

$$\frac{\partial K}{\partial t_1 \partial t_2}(t) = 0,$$

proving the result.

The proof of part (iv) follows from the same argument used in the scalar random variable case (Theorem 4.15). ∎

Example 4.24   (*Multinomial distribution*). Let $X = (X_1, \ldots, X_m)$ denote a random vector with a multinomial distribution, as in Example 2.2. The frequency function of the distribution is given by

$$p(x_1, \ldots, x_m) = \binom{n}{x_1, x_2, \ldots, x_m} \theta_1^{x_1} \theta_2^{x_2} \cdots \theta_m^{x_m},$$

for $x_j = 0, 1, \ldots, n$, $j = 1, \ldots, m$, $\sum_{j=1}^m x_j = n$; here $\theta_1, \ldots, \theta_m$ are nonnegative constants satisfying $\theta_1 + \cdots + \theta_m = 1$.

For $t = (t_1, \ldots, t_m)$,

$$\mathrm{E}\left[\exp\left(\sum_{j=1}^m t_j X_j\right)\right] = \sum_{x_1, \ldots, x_{m+1}} \binom{m}{x_1, \ldots, x_m} \prod_{j=1}^m \exp(t_j x_j) \theta_j^{x_j}$$

where the sum is over all nonnegative integers $x_1, x_2, \ldots, x_m$ summing to n. Writing

$$\prod_{j=1}^{m} \exp(t_j x_j)\theta_j^{x_j} = \left[\sum_{j=1}^{m} \exp(t_j)\theta_j\right]^n \prod_{j=1}^{m} \left(\frac{\exp(t_j)\theta_j}{\sum_{j=1}^{m} \exp(t_j)\theta_j}\right)^{x_j},$$

it follows from the properties of the multinomial distribution that the moment-generating function of X is

$$M_X(t) = \left[\sum_{j=1}^{m} \exp(t_j)\theta_j\right]^n = \left[\sum_{j=1}^{m} \exp(t_j)\theta_j\right]^n, \quad t = (t_1, \ldots, t_m) \in \mathbf{R}^m.$$

The cumulant-generating function is therefore given by

$$K_X(t) = n \, \log \left[\sum_{j=1}^{m} \exp(t_j)\theta_j\right].$$

It follows that, for $j = 1, \ldots, m$,

$$\mathrm{E}(X_j) = n\theta_j, \qquad \mathrm{Var}(X_j) = n\theta_j(1 - \theta_j)$$

and, for $j, k = 1, \ldots, m$,

$$\mathrm{Cov}(X_j, X_k) = -n\theta_j\theta_k.$$

Thus, the covariance matrix of X is the $m \times m$ matrix with (j, k)th element given by

$$\sigma_{jk} = \begin{cases} n\theta_j(1 - \theta_j) & \text{if } j = k \\ -n\theta_j\theta_k & \text{if } j \neq k \end{cases}. \qquad \qquad \square$$

4.5   Moments and Cumulants of the Sample Mean

Let $X_1, X_2, \ldots, X_n$ denote independent, identically distributed, real-valued random variables. Let

$$\bar{X}_n = \frac{1}{n} \sum_{j=1}^{n} X_j \qquad K_1 = \frac{d}{dt}(\log M(t))\Big|_{t=0}$$

denote the sample mean based on $X_1, X_2, \ldots, X_n$.

In this section, we consider the moments and cumulants of $\bar{X}_n$. First consider the cumulants. Let $\kappa_1, \kappa_2, \ldots$ denote the cumulants of X_1 and let $\kappa_1(\bar{X}_n), \kappa_2(\bar{X}_n), \ldots$ denote the cumulants of $\bar{X}_n$. Using Theorems 4.14 and 4.15, it follows that

$$\kappa_j(\bar{X}_n) = \frac{1}{n^{j-1}}\kappa_j, \quad j = 1, 2, \ldots. \qquad (4.4)$$

For convenience, here we are assuming that all cumulants of X_1 exist; however, it is clear that the results only require existence of the cumulants up to a given order. For instance, $\kappa_2(\bar{X}_n) = \kappa_2/n$ holds only provided that κ_2 exists.

To obtain results for the moments of $\bar{X}_n$, we can use the expressions relating moments and cumulants; see Lemma 4.1. Then

$$\mathrm{E}(\bar{X}_n) = \mathrm{E}(X_1) \quad \text{and} \quad \mathrm{E}\left(\bar{X}_n^2\right) = \frac{n-1}{n}\mathrm{E}(X_1)^2 + \frac{1}{n}\mathrm{E}\left(X_1^2\right).$$

Another approach that is often useful is to use the fact that, for $r = 2, 3, \ldots,$

$$\bar{X}_n^r = \frac{1}{n^r} \left[\sum_{j=1}^n X_j \right]^r = \frac{1}{n^r} \sum_{j_1=1}^n \cdots \sum_{j_r=1}^n X_{j_1} X_{j_2} \cdots X_{j_r}$$

$(a_1 + a_2)^2 = a_1^2 + 2a_1 a_2 + a_2^2$

$= \sum_{i=1}^2 \sum_{j=1}^2 a_i a_j$

and then take the expected value of resulting sum. For instance, consider $E(\bar{X}_n^3)$. Since

$$\bar{X}_n^3 = \frac{1}{n^3} \sum_{i=1}^n \sum_{j=1}^n \sum_{k=1}^n X_i X_j X_k,$$

$$E\left(\bar{X}_n^3\right) = \frac{1}{n^3} \sum_{i=1}^n \sum_{j=1}^n \sum_{k=1}^n E(X_i X_j X_k).$$

In order to evaluate

$$\sum_{i,j,k=1}^n E(X_i X_j X_k),$$

we must keep track of the number of terms in the sum in which i, j, k are unique, the number in which exactly two indices are equal, and so on. It is straightforward to show that of the n^3 terms in

$$\sum_{i=1}^n \sum_{j=1}^n \sum_{k=1}^n X_i X_j X_k,$$

n terms have all indices the same, $3n(n-1)$ terms have exactly two indices the same, and in $n(n-1)(n-2)$ terms all indices are unique. Hence,

$$\sum_{i,j,k=1}^n E(X_i X_j X_k) = nE\left(X_1^3\right) + 3n(n-1)E(X_1)E\left(X_1^2\right) + n(n-1)(n-2)E(X_1)^3.$$

It follows that

$$E\left(\bar{X}_n^3\right) = \frac{(n-1)(n-2)}{n^2}E(X_1)^3 + \frac{3(n-1)}{n^2}E(X_1)E\left(X_1^2\right) + \frac{1}{n^2}E\left(X_1^3\right).$$

The same approach can be used for any moment of $\bar{X}_n$, although the algebra becomes tedious very quickly. The following theorem gives expressions for the first four moments; the proof is left as an exercise.

Theorem 4.17. *Let $X_1, X_2, \ldots, X_n$ denote independent, identically distributed, real-valued random variables such that $E(X_1^4) < \infty$ and let*

$$E\left(\bar{X}_n\right) = \frac{1}{n} \sum_{j=1}^n X_j.$$

$= \frac{1}{n} \sum_{j \geq 1}^n E(X_j) = \frac{1}{n} \cdot n E(X_j)$

$= E(X_1)$

The moments of $\bar{X}_n$ satisfy

$$E(\bar{X}_n) = E(X_1), \quad E\left(\bar{X}_n^2\right) = \frac{n-1}{n}E(X_1)^2 + \frac{1}{n}E\left(X_1^2\right) = E(X_1)^2 + \frac{1}{n}\left[E\left(X_1^2\right) - E(X_1)^2\right],$$

$$E\left(\bar{X}_n^3\right) = \frac{(n-1)(n-2)}{n^2}E(X_1)^3 + \frac{3(n-1)}{n^2}E(X_1)E\left(X_1^2\right) + \frac{E(X_1^3)}{n^2}$$

$$= E(X_1)^3 + \frac{3}{n}\left[E(X_1)E\left(X_1^2\right) - E(X_1)^3\right] + \frac{1}{n^2}\left[2E(X_1)^3 - 3E(X_1)E\left(X_1^2\right) + E\left(X_1^3\right)\right],$$

and

$$\text{E}\left(\bar{X}_n^4\right) = \frac{(n-1)(n-2)(n-3)}{n^3}\text{E}(X_1)^4 + \frac{4(n-1)}{n^3}\text{E}(X_1)\text{E}\left(X_1^3\right)$$

$$+ \frac{6(n-1)(n-2)}{n^3}\text{E}(X_1)^2\text{E}\left(X_1^2\right) + \frac{3(n-1)}{n^3}\text{E}\left(X_1^2\right)^2 + \frac{1}{n^3}\text{E}\left(X_1^4\right)$$

$$= \text{E}(X_1)^4 + \frac{6}{n}\left[\text{E}(X_1)^2\text{E}\left(X_1^2\right) - \text{E}(X_1)^4\right]$$

$$+ \frac{1}{n^2}\left[11\text{E}(X_1)^4 - 18\text{E}(X_1)^2\text{E}\left(X_1^2\right) + 4\text{E}(X_1)\text{E}\left(X_1^3\right) + 3\text{E}\left(X_1^2\right)^2\right]$$

$$+ \frac{1}{n^3}\left[\text{E}\left(X_1^4\right) + 12\text{E}(X_1)^2\text{E}\left(X_1^2\right) - 4\text{E}(X_1)\text{E}\left(X_1^3\right) - 3\text{E}\left(X_1^2\right)^2 - 6\text{E}(X_1)^4\right].$$

Example 4.25 (Standard exponential distribution). Let $X_1, X_2, \ldots, X_n$ denote independent, identically distributed random variables, each with a standard exponential distribution. The cumulant-generating function of the standard exponential distribution is $-\log(1-t)$, $|t| < 1$; see Example 4.11. Hence, the cumulants of the distribution are given by $\kappa_r = (r-1)!$, $r = 1, 2, \ldots$. It follows from (4.4) that the cumulants of $\bar{X}_n = \sum_{j=1}^{n} X_j/n$ are given by

$$\kappa_r(\bar{X}_n) = \frac{1}{n^{r-1}}(r-1)!, \quad r = 1, 2, \ldots.$$

The moments of the standard exponential distribution are given by

$$\text{E}\left(X_1^r\right) = \int_0^{\infty} x^r \exp(-x)\,dx = r!, \quad r = 1, 2, \ldots.$$

Hence, the first four moments of $\bar{X}_n$ are given by $\text{E}(\bar{X}_n) = 1$,

$$\text{E}\left(\bar{X}_n^2\right) = 1 + \frac{1}{n}, \quad \text{E}\left(\bar{X}_n^3\right) = 1 + \frac{3}{n} + \frac{2}{n^2},$$

and

$$\text{E}\left(\bar{X}_n^4\right) = 1 + \frac{6}{n} + \frac{11}{n^2} + \frac{6}{n^3}. \qquad \square$$

Expressions for moments and cumulants of a sample mean can also be applied to sample moments of the form

$$\frac{1}{n}\sum_{j=1}^{n} X_j^m, \quad m = 1, 2, \ldots$$

by simply redefining the cumulants and moments in the theorem as the cumulants and moments, respectively, of X_1^m. This is generally simpler to carry out with the moments, since the moments of X_1^m are given by $\text{E}(X_1^m)$, $\text{E}(X_1^{2m})$, ...

Example 4.26 (Standard exponential distribution). As in Example 4.25, let X_1, $X_2, \ldots, X_n$ denote independent, identically distributed, standard exponential random variables and consider the cumulants of

$$T_n = \frac{1}{n}\sum_{j=1}^{n} X_j^2.$$

It follows from the results in Example 4.25 that

$$E\left(X_1^{2r}\right) = (2r)!, \quad r = 1, 2, \ldots.$$

Hence, the first four moments of T_n are given by $E(T_n) = 2$,

$$E\left(T_n^2\right) = 4 + \frac{20}{n}, \qquad E\left(T_n^3\right) = 8 + \frac{120}{n} + \frac{592}{n^2}$$

and

$$E\left(T_n^4\right) = 16 + \frac{480}{n} + \frac{5936}{n^2} + \frac{31584}{n^3}. \qquad \square$$

Central moments of $\bar{X}_n$

Results analogous to those given in Theorem 4.17 for the moments can be obtained for the central moments by taking $E(\bar{X}_1) = 0$ and then interpreting $E(X_1^2)$, $E(X_1^3)$, and $E(X_1^4)$ in Theorem 4.17 as central moments. The resulting expressions are given in the following corollary; the proof is left as an exercise.

Corollary 4.3. *Let $X_1, X_2, \ldots, X_n$ denote independent, identically distributed, real-valued random variables and let*

$$\bar{X}_n = \frac{1}{n} \sum_{j=1}^{n} X_j.$$

Assume that $E[X_1^4] < \infty$. Let $\mu = E(X_1)$, and let μ_2, μ_3, μ_4 denote the second, third, and fourth central moments, respectively, of X_1. Let $\mu_2(\bar{X}_n), \mu_3(\bar{X}_n),$ and $\mu_4(\bar{X}_n)$ denote the second, third, and fourth central moments, respectively, of $\bar{X}_n$. Then $E(\bar{X}_n) = \mu$,

$$\mu_2(\bar{X}_n) = \frac{1}{n}\mu_2, \qquad \mu_3(\bar{X}_n) = \frac{1}{n^2}\mu_3$$

and

$$\mu_4(\bar{X}_n) = \frac{3(n-1)}{n^3}\mu_2^2 + \frac{1}{n^3}\mu_4 = \frac{3}{n^2}\mu_2^2 + \frac{1}{n^3}\left(\mu_4 - 3\mu_2^2\right).$$

Example 4.27 (Standard exponential distribution). As in Examples 4.25 and 4.26, let $X_1, X_2, \ldots, X_n$ denote independent, identically distributed, standard exponential random variables. It is straightforward to show that the first four central moments of the standard exponential distribution are $0, 1, 2, 9$; these may be obtained using the expressions for central moments in terms of cumulants, given in Section 4.4. It follows from Corollary 4.3 that the first four central moments of $\bar{X}_n$ are $0, 1/n, 2/n^2$, and

$$\frac{3}{n^2} + \frac{6}{n^3},$$

respectively. $\square$

We can see from Corollary 4.3 that, as k increases, the order of $E[(\bar{X}_n - \mu)^k]$ as $n \to \infty$ is a nondecreasing power of $1/n$:

$$E[(\bar{X}_n - \mu)^2] = O\left(\frac{1}{n}\right), \quad E[(\bar{X}_n - \mu)^3] = O\left(\frac{1}{n^2}\right), \quad E[(\bar{X}_n - \mu)^4] = O\left(\frac{1}{n^2}\right).$$

The following theorem gives a generalization of these results.

Theorem 4.18. *Let $X_1, X_2, \ldots, X_n$ denote independent, identically distributed, real-valued random variables such that all moments of X_1 exist and are finite. Let $\mu = E(X_1)$ and*

$$\bar{X}_n = \frac{1}{n} \sum_{j=1}^{n} X_j.$$

Then, for $k = 1, 2, \ldots$,

$$E[(\bar{X}_n - \mu)^{2k-1}] = O\left(\frac{1}{n^k}\right) \quad and \quad E[(\bar{X}_n - \mu)^{2k}] = O\left(\frac{1}{n^k}\right) \qquad as \ n \to \infty.$$

Proof. The proof is by induction. For $k = 1$, the result follows immediately from Theorem 4.17. Assume that the result holds for $k = 1, 2, \ldots, m$. For each $j = 1, 2, \ldots$, let

$$\bar{\mu}_j = E[(\bar{X}_n - \mu)^j].$$

Note that, applying Lemma 4.1 to the moment- and cumulant-generating functions of $\bar{X}_n - \mu$, the cumulants and central moments of $\bar{X}_n$ are related by

$$\bar{\mu}_{r+1} = \sum_{j=0}^{r} \binom{r}{j} \kappa_{j+1}(\bar{X}_n)\bar{\mu}_{r-j}, \quad r = 0, 1, \ldots.$$

Since, by (4.4), $\kappa_{j+1}(\bar{X}_n) = O(1/n^j)$, and taking $\bar{\mu}_0 = 1$,

$$\bar{\mu}_{r+1} = \sum_{j=0}^{r} \bar{\mu}_{r-j} O\left(\frac{1}{n^j}\right).$$

Consider $r = 2m$. Then, since the theorem is assumed to hold for $k = 1, 2, \ldots, m$,

$$\bar{\mu}_{2m-j} = \begin{cases} O(\frac{1}{n^{m-(j-1)/2}}) & \text{if } j = 1, 3, \ldots, 2m-1 \\ O(\frac{1}{n^{m-j/2}}) & \text{if } j = 0, 2, 4, \ldots, 2m \end{cases}.$$

Hence,

$$\bar{\mu}_{2m+1} = O\left(\frac{1}{n^m}\right) + O\left(\frac{1}{n^{m+1}}\right) + O\left(\frac{1}{n^{m+1}}\right) + \cdots + O\left(\frac{1}{n^{2m}}\right) = O\left(\frac{1}{n^m}\right)$$

as $n \to \infty$.

Now consider $r = 2m + 1$. Then

$$\bar{\mu}_{2m+2} = O\left(\frac{1}{n^m}\right) + O\left(\frac{1}{n^{m+1}}\right) + \cdots + O\left(\frac{1}{n^{2m+1}}\right) = O\left(\frac{1}{n^m}\right)$$

as $n \to \infty$.

It follows that the result holds for $k = m + 1$, proving the theorem. ∎

4.6 Conditional Moments and Cumulants

Let X denote a real-valued random variable and let Y denote a random variable which may be a vector. The conditional moments and cumulants of X given $Y = y$ are simply the moments and cumulants, respectively, of the conditional distribution of X given $Y = y$;

substituting the random variable Y for y yields the conditional moments and cumulants of X given Y. In this section, we consider the relationship between the conditional moments and cumulants of X given Y and the unconditional moments and cumulants of X. As we will see, this is one area in which it is easier to work with moments than cumulants.

Suppose that $E(|X|) < \infty$. We have seen (Theorem 2.5) that $E(X) = E[E(X|Y)]$. The same result holds for any moment of X, provided that it exists. Suppose that, for some r, $E(|X|^r) < \infty$; then $E(X^r) = E[E(X^r|Y)]$.

Now consider cumulants; for simplicity, suppose that the cumulant-generating function of X, K, and the conditional cumulant-generating function of X given $Y = y$, $K(\cdot, y)$, both exist. Then, for any integer $m = 1, 2, \ldots,$

$$K(t) = \sum_{j=1}^{m} \frac{t^j}{j!} \kappa_j + o(t^m) \quad \text{as } t \to 0$$

where $\kappa_1, \kappa_2, \ldots$ denote the (unconditional) cumulants of X. Similarly,

$$K(t, y) = \sum_{j=1}^{m} \frac{t^j}{j!} \kappa_j(y) + o(t^m) \quad \text{as } t \to 0$$

where $\kappa_1(y), \kappa_2(y), \ldots$ denote the conditional cumulants of X given $Y = y$. The conditional cumulants of X given Y are then given by $\kappa_1(Y), \kappa_2(Y), \ldots$. Given the indirect way in which cumulants are defined, the relationship between conditional and unconditional cumulants is not as simple as the relationship between conditional and unconditional moments.

For the low-order cumulants, the simplest approach is to rewrite the cumulants in terms of moments and use the relationship between conditional and unconditional moments. For instance, since the first cumulant is simply the mean of the distribution, we have already seen that

$$\kappa_1 = E[\kappa_1(Y)]. \qquad \text{Cov}(\kappa_1 y) = E\left(\text{Var}(x, y \mid z)\right)$$

For the second cumulant, the variance, note that

$$\begin{aligned} \kappa_2 &= E(X^2) - E(X)^2 = E[E(X^2|Y)] - E[E(X|Y)]^2 \\ &= E[E(X^2|Y)] - E[E(X|Y)^2] + E[E(X|Y)^2] - E[E(X|Y)]^2 \\ &= E[\text{Var}(X|Y)] + \text{Var}[E(X|Y)]. \end{aligned}$$

We now consider a general approach that can be used to relate conditional and unconditional cumulants. The basic idea is that the conditional and unconditional cumulant-generating functions are related by the fact that $K(t) = \log E[\exp\{K(t, Y)\}]$. As $t \to 0$,

$$K(t) = \log E\left[\exp\left\{ \sum_{j=1}^{m} t^j \kappa_j(Y)/j! \right\} \right] + o(t^m). \tag{4.5}$$

Note that $\kappa_1(Y), \ldots, \kappa_m(Y)$ are random variables; let $K_m(t_1, \ldots, t_m)$ denote the cumulant-generating function of the random vector $(\kappa_1(Y), \ldots, \kappa_m(Y))$. Then, by (4.5),

$$K(t) = \log K_m(t, t^2/2, \ldots, t^m/m!) + o(t^m) \quad \text{as } t \to 0.$$

This result can now be used to express the unconditional cumulants in terms of the conditional ones. Although a general expression relating the conditional and unconditional cumulants may be given, here we simply outline the approach that may be used.

Consider κ_1. Note that

$$K'(t) = \sum_{j=1}^{m} \frac{\partial}{\partial t_j} K_m(t, t^2/2, \ldots, t^m/m!) \frac{t^{j-1}}{(j-1)!}. \tag{4.6}$$

Let $\bar{\kappa}_{i_1 \cdots i_m}$ denote the joint cumulant of order $(i_1, \ldots, i_m)$ of $(\kappa_1(Y), \ldots, \kappa_m(Y))$; we will use the convention that trailing 0s in the subscript of $\bar{\kappa}$ will be omitted so that, for example, $\bar{\kappa}_{10 \cdots 0}$ will be written $\bar{\kappa}_1$. Then evaluating (4.6) at $t = 0$ shows that

$$\kappa_1 = K'(0) = \bar{\kappa}_1;$$

that is, the first cumulant of X is the first cumulant of $\kappa_1(Y)$. Of course, this is simply the result that

$$E(X) = E[E(X|Y)].$$

Now consider the second cumulant of X. We may use the same approach as that used above; the calculation is simplified if we keep in mind that any term in the expansion of $K''(t)$ in terms of the derivatives of K_m that includes a nonzero power of t will be 0 when evaluated at $t = 0$. Hence, when differentiating the expression in (4.5), we only need to consider

$$\frac{d}{dt} \left\{ \frac{\partial}{\partial t_1} K_m(t, t^2/2, \ldots, t^m/m!) + \frac{\partial}{\partial t_2} K_m(t, t^2/2, \ldots, t^m/m!)t \right\} \Bigg|_{t=0}$$

$$= \frac{\partial^2}{\partial t_1^2} K_m(t, t^2/2, \ldots, t^m/m!) \Bigg|_{t=0} + \frac{\partial}{\partial t_2} K_m(t, t^2/2, \ldots, t^m/m!) \Bigg|_{t=0}.$$

It follows that

$$\kappa_2 \equiv K''(0) = \bar{\kappa}_2 + \bar{\kappa}_{01};$$

that is,

$$\text{Var}(X) = \text{Var}[E(X|Y)] + \text{Var}[E(X|Y)].$$

The expressions for the higher-order cumulants follow in a similar manner. We may obtain $K'''(0)$ by taking the second derivative of

$$\frac{\partial}{\partial t_1} K_m(t, t^2/2, \ldots, t^m/m!) + \frac{\partial}{\partial t_2} K_m(t, t^2/2, \ldots, t^m/m!)t$$

$$+ \frac{\partial}{\partial t_3} K_m(t, t^2/2, \ldots, t^m/m!)t^2/2$$

at $t = 0$. The result is

$$\kappa_3 = \bar{\kappa}_3 + 3\bar{\kappa}_{11} + \bar{\kappa}_{001}.$$

Example 4.28 (Poisson random variable with random mean). Let X denote a Poisson random variable with mean Y and suppose that Y is a random variable with a standard exponential distribution. That is, the conditional distribution of X given Y is Poisson with mean Y and the marginal distribution of Y is a standard exponential distribution.

It follows from Example 4.13 that

$$E(X|Y) = Y, \quad E(X^2|Y) = Y + Y^2, \quad \text{and} \quad E(X^3|Y) = Y + 3Y^2 + Y^3.$$

Since $E(Y^r) = r!$, it follows that

$$E(X) = 1, \quad E(X^2) = 3, \quad \text{and} \quad E(X^3) = 13.$$

From Example 4.18, we know that all conditional cumulants of X given Y are equal to Y. Hence,

$$\text{Var}(X) = E[\text{Var}(X|Y)] + \text{Var}[E(X|Y)] = 2.$$

To determine κ_3, the third cumulant of X, we need $\bar{\kappa}_3$, the third cumulant of $E(X|Y) = Y$, $\bar{\kappa}_{11}$, the covariance of $E(X|Y)$ and $\text{Var}(X|Y)$, that is, the variance of Y, and $\bar{\kappa}_{001}$, the expected value of $\kappa_3(Y) = Y$.

Letting $\gamma_1, \gamma_2, \ldots$ denote the cumulants of the standard exponential distribution, it follows that

$$\kappa_3 = \gamma_3 + 2\gamma_2 + \gamma_1.$$

According to Example 4.11, the cumulant-generating function of the standard exponential distribution is $-\log(1 - t)$. Hence,

$$\gamma_1 = 1, \quad \gamma_2 = 1, \quad \text{and} \quad \gamma_3 = 2.$$

It follows that $\kappa_3 = 6$. $\quad\square$

4.7 Exercises

4.1 Let X denote a real-valued random variable with an absolutely continuous distribution with density function

$$p(x) = \frac{\beta^\alpha}{\Gamma(\alpha)} x^{\alpha-1} \exp\{-\beta x\}, \quad x > 0,$$

where $\alpha > 0$ and $\beta > 0$; this is a gamma distribution. Find a general expression for the moments of X.

4.2 Let X denote a real-valued random variable with an absolutely continuous distribution with density function

$$p(x) = \frac{\Gamma(\alpha + \beta)}{\Gamma(\alpha)\Gamma(\beta)} x^{\alpha-1}(1 - x)^{\beta-1}, \quad 0 < x < 1,$$

where $\alpha > 0$ and $\beta > 0$; this is a beta distribution. Find a general expression for the moments of X.

4.3 Prove Theorem 4.1.

4.4 Prove Theorem 4.2.

4.5 Prove Corollary 4.1.

4.6 Prove Theorem 4.4.

4.7 Let X denote a d-dimensional random vector with covariance matrix Σ satisfying $|\Sigma| < \infty$. Show that X has a nondegenerate distribution if and only if Σ is positive definite.

4.8 Let X and Y denote real-valued random variables such that X has mean μ_X and standard deviation σ_X, Y has mean μ_Y and standard deviation σ_Y, and X and Y have correlation ρ.

(a) Find the value of $\beta \in \mathbf{R}$ that minimizes $\text{Var}(Y - \beta X)$.

(b) Find the values of $\beta \in \mathbf{R}$ such that Y and $Y - \beta X$ are uncorrelated.

(c) Find the values of $\beta \in \mathbf{R}$ such that X and $Y - \beta X$ are uncorrelated.

(d) Find conditions under which, for some β, $Y - \beta X$ is uncorrelated with both X and Y.

(e) Suppose that $E(Y|X) = \alpha + \beta X$ for some constants α, β. Express α and β in terms of $\mu_X, \mu_Y, \sigma_X, \sigma_Y, \rho$.

4.9 Let X and Y denote real-valued random variables such that $E(X^2) < \infty$ and $E(Y^2) < \infty$. Suppose that $E[X|Y] = 0$. Does it follow that $\rho(X, Y) = 0$?

4.10 Let X and Y denote real-valued identically distributed random variables such that $E(X^2) < \infty$. Give conditions under which

$$\rho(X, X + Y)^2 \geq \rho(X, Y)^2.$$

4.11 Let X and Y denote real-valued random variables such that $E(X^2) < \infty$ and $E(Y^2) < \infty$ and let ρ denote the correlation of X and Y. Find the values of ρ for which

$$E[(X - Y)^2] \geq \text{Var}(X).$$

4.12 Let X denote a nonnegative, real-valued random variable; let F denote the distribution function of X and let $\mathcal{L}$ denote the Laplace transform of X. Show that

$$\mathcal{L}(t) = t \int_0^\infty \exp(-tx) F(x) \, dx, \quad t \geq 0.$$

4.13 Prove Theorem 4.7.

4.14 Let X denote a nonnegative, real-valued random variable and let $\mathcal{L}(t)$ denote the Laplace transform of X. Show that

$$(-1)^n \frac{d^n}{dt^n} \mathcal{L}(t) \geq 0, \quad t \geq 0.$$

A function with this property is said to be *completely monotone*.

4.15 Let X denote a random variable with frequency function

$$p(x) = \theta(1 - \theta)^x, \quad x = 0, 1, 2, \ldots$$

where $0 < \theta < 1$.

Find the moment-generating function of X and the first three moments.

4.16 Let X denote a real-valued random variable and suppose that, for some $r = 1, 2, \ldots$, $E(|X|^r) = \infty$. Does it follow that $E(|X|^m) = \infty$ for all $m = r + 1, r + 2, \ldots$?

4.17 Prove Theorem 4.10.

4.18 Prove Theorem 4.11.

4.19 Let X denote a random variable with the distribution given in Exercise 4.15. Find the cumulant-generating function of X and the first three cumulants.

4.20 Let X denote a random variable with a gamma distribution, as in Exercise 4.1. Find the cumulant-generating function of X and the first three cumulants.

4.21 Let Y be a real-valued random variable with distribution function F and moment-generating function $M(t)$, $|t| < \delta$, where $\delta > 0$ is chosen to be as large as possible. Define

$$\beta = \inf\{M(t) \colon 0 \leq t < \delta\}.$$

Suppose that there exists a unique real number $\tau \in (0, \delta)$ such that $M(\tau) = \beta$.

(a) Show that $\Pr(Y \geq 0) \leq \beta$.

(b) Show that $M'(\tau) = 0$.

(c) Let

$$G(x) = \frac{1}{\beta} \int_{-\infty}^{x} \exp(\tau y) \, dF(y), \quad -\infty < x < \infty.$$

Note that G is a distribution function on $\mathbf{R}$ and let X denote a random variable with distribution function G. Find the moment-generating function of X.

(d) Find $\mathrm{E}(X)$.

4.22 Let X denote a real-valued random variable with distribution function F. Let $K(t)$, $|t| < \delta$, $\delta > 0$, denote the cumulant-generating function of X; assume that δ is chosen to be as large as possible.

Let $a \geq 0$ be a fixed, real-valued constant and define a function

$$K_a(t) = K(t) - at, \quad t \geq 0.$$

Define

$$\rho_a = \inf\{K_a(t): 0 \leq t < \delta\}.$$

(a) Calculate ρ_a, as a function of a, for the standard normal distribution and for the Poisson distribution with mean 1.

(b) Show that

$$\Pr(X \geq a) \leq \exp(\rho_a).$$

(c) Let $X_1, X_2, \ldots, X_n$ denote independent random variables, each with the same distribution as X. Obtain a bound for

$$\Pr\left(\frac{X_1 + \cdots + X_n}{n} \geq a\right)$$

that generalizes the result given in part (b).

4.23 Let X denote a real-valued random variable with moment-generating function $M_X(t)$, $|t| < \delta$, $\delta > 0$. Suppose that the distribution of X is symmetric about 0; that is, suppose that X and $-X$ have the same distribution. Find $\kappa_j(X)$, $j = 1, 3, 5, \ldots$.

4.24 Consider a distribution on the real line with moment-generating function $M(t)$, $|t| < \delta$, $\delta > 0$ and cumulants $\kappa_1, \kappa_2, \ldots$. Suppose that $\mathrm{E}(X^r) = 0$, $r = 1, 3, 5, \ldots$. Show that

$$\kappa_1 = \kappa_3 = \cdots = 0.$$

Does the converse hold? That is, suppose that all cumulants of odd order are 0. Does it follow that all moments of odd order are 0?

4.25 Let X and Y denote real-valued random variables and assume that the moment-generating function of (X, Y) exists. Write

$$M(t_1, t_2) = \mathrm{E}[\exp(t_1 X + t_2 Y)], \quad (t_1^2 + t_2^2)^{\frac{1}{2}} < \delta,$$

$K(t_1, t_2) = \log M(t_1, t_2)$, and let κ_{ij}, $i, j = 0, 1, \ldots$, denote the joint cumulants of (X, Y). Let $S = X_1 + X_2$ and let K_S denote the cumulant-generating function of S.

(a) Show that

$$K_S(t) = K(t, t), \quad |t| \leq \delta/\sqrt{2}.$$

(b) Let $\kappa_j(S)$, $j = 1, 2, \ldots$, denote the cumulants of S. Show that

$$\kappa_2(S) = \kappa_{20} + 2\kappa_{11} + \kappa_{02}.$$

(c) Derive a general expression for $\kappa_j(S)$ in terms of κ_{ik}, $i, k = 1, 2, \ldots$.

4.26 Let X and Y denote discrete random variables, with ranges $\mathcal{X}$ and $\mathcal{Y}$, respectively. Suppose that $\mathcal{X}$ and $\mathcal{Y}$ each contain m distinct elements of $\mathbf{R}$, for some $m = 1, 2, \ldots$; assume that, for each $x \in \mathcal{X}$ and each $y \in \mathcal{Y}$,

$$\Pr(X = x) > 0 \quad \text{and} \quad \Pr(Y = y) > 0.$$

Suppose that

$$\mathrm{E}(X^j) = \mathrm{E}(Y^j), \quad j = 1, 2, \ldots, m.$$

Does it follow that X and Y have the same distribution? Why or why not?

4.27 Let (X, Y) denote a two-dimensional random vector with joint cumulants κ_{ij}, $i, j = 1, 2, \ldots$. Given an expression for $\mathrm{Var}(XY)$ in terms of the κ_{ij}.

4.28 Let $X_1, X_2, \ldots$ denote a sequence of real-valued random variables such that $X_1, X_2, \ldots$ are exchangeable and $(X_1, X_2, \ldots)$ is a martingale. Find the correlation of X_i and X_j, $i, j = 1, 2, \ldots$.

4.29 For each $n = 1, 2, \ldots$, let $(X_1, Y_1), (X_2, Y_2), \ldots, (X_n, Y_n)$ denote independent, identically distributed, two-dimensional random vectors with joint cumulants κ_{ij}, $i, j = 1, 2, \ldots$. Let $\bar{X} = \sum_{j=1}^n X_j/n$ and $\bar{Y} = \sum_{j=1}^n Y_j/n$. Find $r = 1, 2, \ldots$ such that $\mathrm{Var}(\bar{X}\bar{Y}) = O(n^{-r})$ as $n \to \infty$.

4.30 Let $(X_1, Y_1), \ldots, (X_n, Y_n)$ denote independent, identically distributed random vectors such that, for each j, X_j and Y_j are real-valued; assume that all moments of (X_1, Y_1) exist and are finite. Let

$$\bar{X} = \frac{1}{n} \sum_{j=1}^n X_j \quad \text{and} \quad \bar{Y} = \frac{1}{n} \sum_{j=1}^n Y_j.$$

(a) Express $\mathrm{E}(\bar{X}\bar{Y})$ and $\mathrm{E}(\bar{X}^2\bar{Y})$ in terms of the moments of (X_1, Y_1).

(b) Express the cumulants of $(\bar{X}, \bar{Y})$ of orders $(1, 1)$ and $(2, 1)$ in terms of the cumulants of (X_1, Y_1).

4.31 Let X and Y denote real-valued random variables. Let $\kappa_1(Y), \kappa_2(Y), \ldots$ denote the cumulants of the conditional distribution of X given Y and let $\kappa_1, \kappa_2, \ldots$ denote the cumulants of the marginal distribution of X.

(a) Show that $\mathrm{E}[\kappa_1(Y)] \le \kappa_1$ and $\mathrm{E}[\kappa_2(Y)] \le \kappa_2$.

(b) Does the same result hold for $\kappa_3(Y)$ and κ_3? That is, is it true that $\mathrm{E}[\kappa_3(Y)] \le \kappa_3$?

4.32 Let X, Y, and Z denote real-valued random variables such that $\mathrm{E}(X^2)$, $\mathrm{E}(Y^2)$, and $\mathrm{E}(Z^2)$ are all finite. Find an expression for $\mathrm{Cov}(X, Y)$ in terms of $\mathrm{Cov}(X, Y|Z)$, $\mathrm{E}(X|Z)$, and $\mathrm{E}(Y|Z)$.

4.33 Let $X_1, \ldots, X_n$ denote real-valued, exchangeable random variables such that $\mathrm{E}(X_1^2) < \infty$. Let $S = \sum X_j$. For $1 \le i < j \le n$, find the conditional correlation of X_i and X_j given S.

4.8 Suggestions for Further Reading

Moments and central moments, particularly the mean and variance, are discussed in nearly every book on probability. Laplace transforms are considered in detail in Feller (1971, Chapters XIII and XIV); see also Billingsley (1995, Section 22) and Port (1994, Chapter 50). Laplace transforms are often used in nonprobabilistic contexts; see, for example, Apostol (1974, Chapter 11) and Widder (1971). Moment-generating functions are discussed in Port (1994, Chapter 56); see Lukacs (1960) for a detailed discussion of the relationship between characteristic functions and moment-generating functions.

Stuart and Ord (1994, Chapter 3) gives a comprehensive discussion of cumulants; in particular, this reference contains extensive tables relating cumulants to moments and central moments. Another excellent reference on cumulants is McCullagh (1987) which emphasizes the case of vector-valued random variables and the properties of cumulants under transformations of the random variable.

Cramér (1946, Chapter 27) gives many results on the moments, central moments, and cumulants of the sample mean; similar results are also given for the sample variance, a topic that is not considered here. Conditional cumulants are discussed in McCullagh (1987, Section 2.9); the approach used in Section 4.6 to relate unconditional and conditional cumulants is based on Brillinger (1969).

5

Parametric Families of Distributions

5.1 Introduction

Statistical inference proceeds by modeling data as the observed values of certain random variables; these observed values are then used to draw conclusions about the process that generated the data. Let Y denote a random variable with probability distribution P. The function P is typically unknown and the goal is to draw conclusions about P on the basis of observing $Y = y$. The starting point for such an analysis is generally the specification of a *model* for the data. A model consists of a set of possible distributions $\mathcal{P}$ such that we are willing to proceed as if P is an element of $\mathcal{P}$.

Thus, in addition to the properties of the individual distributions in $\mathcal{P}$, the properties of the family itself are of interest; it is these properties that we consider in this chapter.

5.2 Parameters and Identifiability

Consider a family $\mathcal{P}$ of probability distributions. A *parameterization* of $\mathcal{P}$ is a mapping from a *parameter space* Θ to the set $\mathcal{P}$ so that $\mathcal{P}$ may be represented

$$\mathcal{P} = \{P(\cdot\,;\theta)\colon \theta \in \Theta\}.$$

Hence, corresponding to any statement regarding the elements P of $\mathcal{P}$ is an equivalent statement regarding the elements θ of Θ.

Example 5.1 (Normal distributions). Let $\mathcal{P}$ denote the set of all normal distributions with finite mean and nonnegative variance. For $\theta = (\mu, \sigma)$, let $P(\cdot\,;\theta)$ be the normal distribution with mean μ and standard deviation σ and take $\Theta = \mathbf{R} \times \mathbf{R}^+$. Then $\mathcal{P}$ may be written

$$\mathcal{P} = \{P(\cdot\,;\theta)\colon \theta \in \Theta\}.$$

Let $\mathcal{P}_0$ denote the subset of $\mathcal{P}$ consisting of those normal distributions with mean 0. Then $\mathcal{P}_0$ consists of those elements of $\mathcal{P}$ of the form $P(\cdot\,;\theta)$ with $\theta = (0, \sigma), \sigma > 0$. $\square$

Example 5.2 (Distributions with median 0). Let $\mathcal{P}$ denote the set of all probability distributions on the real line such that 0 is a median of the distribution. An element P of $\mathcal{P}$ is given by

$$P(A; F) = \int_A dF(x)$$

where F is a distribution function on the real line satisfying $F(0) \geq 1/2$ and $F(0-) \geq 1/2$. Let $\mathcal{F}$ denote the set of all nondecreasing, right-continuous functions on **R** satisfying

$$\lim_{x \to -\infty} F(x) = 0, \quad \lim_{x \to \infty} F(x) = 1, \quad F(0) \geq 1/2, \quad \text{and} \quad F(0-) \geq 1/2.$$

Then the elements of P may be written $P(\cdot; F)$ where $F \in \mathcal{F}$ so that $\mathcal{F}$ is the parameter space for the model. $\square$

In parametric statistical inference, the set of possible distributions is assumed to be parameterized by a finite-dimensional parameter so that Θ is a subset of finite-dimensional Euclidean space; such a model is said to be a *parametric* model. Models that are not parametric are said to be *nonparametric*. The model described in Example 5.1 is a parametric model, with Θ a subset of $\mathbf{R}^2$; the model described in Example 5.2 is nonparametric. Here we will focus on parametric models.

In a parametric model for a random variable Y with parameter θ, all quantities based on the probability distribution of Y will depend on the value of θ under consideration. When we wish to emphasize this we will include the parameter in the notation of these quantities; for instance, we will write probabilities as $\Pr(\cdot; \theta)$ and expectations as $E(\cdot; \theta)$.

***Example* 5.3** *(Normal distributions).* Consider a random variable Y with a normal distribution with mean μ and standard deviation σ, where $-\infty < \mu < \infty$ and $\sigma > 0$, and let $\theta = (\mu, \sigma)$. Then

$$E(Y; \theta) = \mu, \qquad E(Y^2; \theta) = \mu^2 + \sigma^2$$

and the characteristic function of Y is given by

$$\varphi(t; \theta) = \exp\left(-\frac{\sigma^2}{2}t^2 + \mu i t\right), \quad -\infty < t < \infty. \qquad \square$$

Although in the discussion above models have been described in terms of probability distributions, we may equivalently describe the model in terms of distribution functions, density or frequency functions when these exist, or even characteristic functions. In many cases, either all the distributions in $\mathcal{P}$ are absolutely continuous or all are discrete with the same minimal support. In these cases, we describe the model in terms of either the density functions or the frequency functions; such a function is called the *model function* of the model. Furthermore, we will generally describe such a model informally, by specifying the model function and the parameter space, without explicit construction of the set $\mathcal{P}$.

***Example* 5.4** *(Normal distributions).* Consider the set of normal distributions considered in Examples 5.1 and 5.3. A more informal way of describing this set is the following:

consider a random variable Y with a normal distribution with mean μ and standard deviation σ, where $-\infty < \mu < \infty$ and $\sigma > 0$. The model function of the model is

$$p(y;\theta) = \frac{1}{\sigma\sqrt{(2\pi)}} \exp\left\{-\frac{1}{2\sigma^2}(y-\mu)^2\right\}, \quad -\infty < y < \infty,$$

where $\theta = (\mu, \sigma) \in \mathbf{R} \times \mathbf{R}^+$.   □

A parameterization of a model is not unique; for example, given a model with parameter θ, the model may also be parameterized by any one-to-one function of θ. Selection of a parameterization is arbitrary, although in certain cases a particular parameterization may be more useful if, for example, it simplifies the interpretation of the results.

Example 5.5 (Poisson distribution). Let X denote a random variable with a Poisson distribution with mean $\lambda > 0$; the model function is therefore given by

$$p(x;\lambda) = \lambda^x \exp(-\lambda)/x!, \quad x = 0, 1, \ldots.$$

Let $\mathcal{P}$ denote the set of all such Poisson distributions with mean $\lambda > 0$.

If X represents the number of "arrivals" observed in a given unit of time, then λ represents the mean arrival rate measured in arrivals per unit of time. We could also parameterize the model in terms of $\theta = 1/\lambda$, which has the interpretation as the mean time between arrivals. The set $\mathcal{P}$ could be described as the set of all Poisson distributions with $\theta > 0$. The model function in terms of θ is given by

$$f(x;\theta) = \theta^{-x} \exp(-1/\theta)/x!, \quad x = 0, 1, \ldots.$$

A statistical analysis could be based on either parameterization.   □

Identifiability

One requirement of a parameterization is that it be *identifiable*; that is, there must be exactly one value of $\theta \in \Theta$ corresponding to each element of $\mathcal{P}$. Stated another way, a parameterization $P(\cdot;\theta)$, $\theta \in \Theta$, of $\mathcal{P}$ is identifiable if $\theta_1 \neq \theta_2$ implies that $P(\cdot;\theta_1) \neq P(\cdot;\theta_2)$. The condition for identifiability may also be expressed in terms of the model function; however, when doing so, it is important to keep in mind that absolutely continuous distributions whose density functions are equal almost everywhere are, in fact, the same distribution. For instance, if two density functions differ at only a finite number of points, the densities represent the same distribution.

Example 5.6 (Binomial distribution with a random index). Let X denote a random variable with a binomial distribution with parameters n and $\eta, 0 < \eta < 1$, and suppose that n is itself a random variable with a Poisson distribution with mean $\lambda > 0$. Take $\theta = (\eta, \lambda)$ with $\Theta = (0, 1) \times \mathbf{R}^+$.

Consider a model for X. Given n, X has a binomial distribution so that

$$\Pr(X = x|n) = \binom{n}{x}\eta^x(1-\eta)^{n-x}, \quad x = 0, 1, \ldots, n.$$

$$\Rightarrow \Pr(X=x) = \sum_{n=1}^{\infty} \binom{n}{x}\eta^x(1-\eta)^{n-x} = \frac{(1-\eta)^n}{(1-\eta)^x}$$

$$= \Pr(X=x, n=1) + \Pr(X=x, n=2) + \ldots$$

n X independent

Joint probability

It follows that, for any $x = 0, 1, \ldots,$

$$
\begin{aligned}
\Pr(X = x) &= \left(\frac{\eta}{1 - \eta}\right)^x \exp(-\lambda) \sum_{n=0}^{\infty} \binom{n}{x} (1 - \eta)^n \lambda^n / n! \\
&= \left(\frac{\eta}{1 - \eta}\right)^x \frac{\exp(-\lambda)}{x!} \sum_{n=x}^{\infty} (1 - \eta)^n \lambda^n / (n - x)! \\
&= \left(\frac{\eta}{1 - \eta}\right)^x \exp(-\lambda)(1 - \eta)^x \lambda^x \sum_{j=0}^{\infty} (1 - \eta)^j \lambda^j / j! \\
&= (\eta\lambda)^x \frac{\exp(-\lambda)}{x!} \exp\{(1 - \eta)\lambda\} \\
&= (\eta\lambda)^x \exp(-\eta\lambda) / x!
\end{aligned}
$$

so that X has a Poisson distribution with mean $\eta\lambda$.

Hence, the model function is given by

$$
p(x; \theta) = (\eta\lambda)^x \exp(-\eta\lambda)/x!, \quad x = 0, 1, \ldots.
$$

Since the distribution of X depends on $\theta = (\eta, \lambda)$ only through $\eta\lambda$, the parameterization given by θ is not identifiable; that is, we may have $(\eta_1, \lambda_1) \neq (\eta_2, \lambda_2)$ yet $\eta_1 \lambda_1 = \eta_2 \lambda_2$.

Suppose that instead we parameterize the model in terms of $\psi = \eta\lambda$ with parameter space $\mathbf{R}^+$. The model function in terms of this parameterization is given by

$$
\psi^x \exp(-\psi)/x!, \quad x = 0, 1, \ldots
$$

and it is straightforward to show that this parameterization is identifiable.   □

Statistical models are often based on independence. For instance, we may have independent identically distributed random variables $X_1, X_2, \ldots, X_n$ such that X_1 has an absolutely continuous distribution with density $p_1(\cdot; \theta)$ where $\theta \in \Theta$. Then the model function for the model for $(X_1, \ldots, X_n)$ is given by

$$
p(x_1, \ldots, x_n; \theta) = \prod_{j=1}^{n} p_1(x_j; \theta);
$$

a similar result holds for discrete distributions. More generally, the random variables $X_1, X_2, \ldots, X_n$ may be independent, but not identically distributed.

Example 5.7 (Normal distributions). Let $X_1, \ldots, X_n$ denote independent identically distributed random variables, each with a normal distribution with mean μ, $-\infty < \mu < \infty$ and standard deviation $\sigma > 0$. The model function is therefore given by

$$
p(x; \theta) = \frac{1}{\sigma^n (2\pi)^{\frac{n}{2}}} \exp\left\{ -\frac{1}{2\sigma^2} \sum_{j=1}^{n} (x_j - \mu)^2 \right\}, \quad x = (x_1, \ldots, x_n) \in \mathbf{R}^n;
$$

here $\theta = (\mu, \sigma)$ and $\Theta = \mathbf{R} \times \mathbf{R}^+$.

Now suppose that $X_1, \ldots, X_n$ are independent, but not identically distributed; specifically, for each $j = 1, 2, \ldots, n$, let X_j have a normal distribution with mean βt_j and standard deviation $\sigma > 0$, where $t_1, \ldots, t_n$ are fixed constants and β and σ are parameters. The model function is then given by

$$p(x;\theta) = \frac{1}{\sigma^n(2\pi)^{\frac{n}{2}}} \exp\left\{-\frac{1}{2\sigma^2}\sum_{j=1}^n (x_j - \beta t_j)^2\right\}, \quad x = (x_1,\ldots,x_n) \in \mathbf{R}^n;$$

here $\theta = (\beta,\sigma)$ and $\Theta = \mathbf{R} \times \mathbf{R}^+$. □

Likelihood ratios

Likelihood ratios play an important role in statistical inference. Consider a parametric model for a random variable X with model function $p(\cdot;\theta)$, $\theta \in \Theta$. A function of X of the form

$$\frac{p(X;\theta_1)}{p(X;\theta_0)},$$

where $\theta_0, \theta_1 \in \Theta$, is called a likelihood ratio. For cases in which the true parameter value θ is unknown, the ratio $p(x;\theta_1)/p(x;\theta_0)$ may be used as a measure of the strength of the evidence supporting $\theta = \theta_1$ versus $\theta = \theta_0$, based on the observation of $X = x$.

Note that

$$\mathrm{E}\left[\frac{p(X;\theta_1)}{p(X;\theta_0)};\theta_0\right] = 1$$

for all $\theta_1 \in \Theta$. To see this, suppose that X has an absolutely continuous distribution with density function $p(\cdot;\theta_0)$. Then

$$\mathrm{E}\left[\frac{p(X;\theta_1)}{p(X;\theta_0)};\theta_0\right] = \int_{-\infty}^{\infty} \frac{p(x;\theta_1)}{p(x;\theta_0)} p(x;\theta_0)\,dx$$

$$= \int_{-\infty}^{\infty} p(x;\theta_1)\,dx = 1.$$

A similar result holds for frequency functions.

Another important property of likelihood ratios is given in the following example.

Example 5.8 *(Martingale property of likelihood ratios).* Consider a sequence of real-valued random variables $Y_1, Y_2, \ldots$; in particular, we are interested in the case in which $Y_1, Y_2, \ldots$ are not independent, although the following analysis also applies in the case of independence. Suppose that, for each $n = 1, 2, \ldots$, distribution of $Y_1, Y_2, \ldots, Y_n$ is absolutely continuous with density function $p_n(\cdot;\theta)$ where θ is a parameter taking values in a set Θ. We assume that for each $\theta \in \Theta$ the density functions $p_n(\cdot;\theta)$, $n = 1, 2, \ldots$, are consistent in the following sense. Fix n. For any $m < n$ the marginal density of $(Y_1, \ldots, Y_m)$ based on $p_n(\cdot;\theta)$ is equal to $p_m(\cdot;\theta)$. That is,

$$p_m(y_1,\ldots,y_m;\theta) = \int_{-\infty}^{\infty}\cdots\int_{-\infty}^{\infty} p_n(y_1,\ldots,y_m,y_{m+1},\ldots,y_n;\theta)\,dy_{m+1}\cdots dy_n, \quad \theta \in \Theta.$$

Let θ_0 and θ_1 denote distinct elements of Θ and define

$$X_n = \frac{p_n(Y_1,\ldots,Y_n;\theta_1)}{p_n(Y_1,\ldots,Y_n;\theta_0)}, \quad n = 1, 2, \ldots$$

where $Y_1, Y_2, \ldots$ are distributed according to the distribution with parameter θ_0. Note that the event $p_n(Y_1,\ldots,Y_n;\theta_0) = 0$ has probability 0 and, hence, may be ignored.

Then

$$E(X_{n+1}|Y_1, \ldots, Y_n; \theta_0) = \int_{-\infty}^{\infty} \frac{p_{n+1}(Y_1, \ldots, Y_n, y; \theta_1)}{p_{n+1}(Y_1, \ldots, Y_n, y; \theta_0)} \frac{p_{n+1}(Y_1, \ldots, Y_n, y; \theta_0)}{p_n(Y_1, \ldots, Y_n; \theta_0)} \, dy$$

$$= \int_{-\infty}^{\infty} \frac{p_{n+1}(Y_1, \ldots, Y_n, y; \theta_1)}{p_n(Y_1, \ldots, Y_n; \theta_0)} \, dy$$

$$= \frac{p_n(Y_1, \ldots, Y_n; \theta_1)}{p_n(Y_1, \ldots, Y_n; \theta_0)}$$

$$= X_n.$$

Since $(X_1, \ldots, X_n)$ is a function of $(Y_1, \ldots, Y_n)$,

$$E(X_{n+1}|X_1, \ldots, X_n; \theta_0) = E[E(X_{n+1}|X_1, \ldots, X_n, Y_1, \ldots, Y_n; \theta_0)|X_1, \ldots, X_n; \theta_0]$$

$$= E[E(X_{n+1}|Y_1, \ldots, Y_n; \theta_0)|X_1, \ldots, X_n; \theta_0]$$

$$= E[X_n|X_1, \ldots, X_n; \theta_0] = X_n.$$

It follows that $X_1, X_2, \ldots$ is a martingale. $\square$

5.3 Exponential Family Models

Many frequently used families of distributions have a common structure. Consider a family of disributions on $\mathbf{R}^d$, $\{P(\cdot; \theta): \theta \in \Theta\}$, such that each distribution in the family is either absolutely continuous or discrete with support not depending on θ. For each θ, let $p(\cdot; \theta)$ denote either the density function or frequency function corresponding to $P(\cdot; \theta)$. The family of distributions is said to be an *m-parameter exponential family* if each $p(\cdot; \theta)$ may be written

$$p(y; \theta) = \exp\{c(\theta)^T T(y) - A(\theta)\} h(y), \quad y \in \mathcal{Y} \tag{5.1}$$

where $\mathcal{Y} \subset \mathbf{R}^d$, $c: \Theta \to \mathbf{R}^m$, $T: \mathcal{Y} \to \mathbf{R}^m$, $A: \Theta \to \mathbf{R}$, and $h: \mathcal{Y} \to \mathbf{R}^+$. It is important to note that the representation (5.1) is not unique; for example, we may replace $c(\theta)$ by $c(\theta)/2$ and $T(y)$ by $2T(y)$.

Example 5.9 (Normal distributions). Let Y denote a random variable with a normal distribution with mean μ, $-\infty < \mu < \infty$ and standard deviation $\sigma > 0$; then Y has density function

$$\frac{1}{\sigma \sqrt{(2\pi)}} \exp\left\{-\frac{1}{2\sigma^2}(y - \mu)^2\right\}, \quad -\infty < y < \infty.$$

Hence, $\theta = (\mu, \sigma)$ and $\Theta = \mathbf{R} \times \mathbf{R}^+$. This density may be written

$$\exp\left\{-\frac{1}{2\sigma^2}y^2 + \frac{\mu}{\sigma^2}y - \frac{1}{2}\frac{\mu^2}{\sigma^2} - \log\sigma\right\} \frac{1}{(2\pi)^{\frac{1}{2}}}, \quad y \in \mathbf{R}.$$

This is of the form (5.1) with $T(y) = (y^2, y)$,

$$c(\theta) = \left(-\frac{1}{2\sigma^2}, \frac{\mu}{\sigma^2}\right), \quad \theta = (\mu, \sigma),$$

$A(\theta) = \mu^2/(2\sigma^2) - \log\sigma$, $h(y) = (2\pi)^{-\frac{1}{2}}$, and $\mathcal{Y} = \mathbf{R}$. Hence, this is a two-parameter exponential family distribution. $\square$

Example 5.10 (Poisson distributions). As in Example 5.5, let $\mathcal{P}$ denote the set of all Poisson distributions with mean $\lambda > 0$; the model function is therefore given by

$$p(x; \lambda) = \lambda^x \exp(-\lambda)/x!, \quad x = 0, 1, \ldots.$$

This can be written

$$\exp\{x \log(\lambda) - \lambda\} \frac{1}{x!}, \quad x \in \{0, 1, 2, \ldots\};$$

hence, this is a one-parameter exponential family with $c(\lambda) = \log(\lambda)$, $T(x) = x$, $A(\lambda) = \lambda$, $h(x) = 1/x!$, and $\mathcal{X} = \{0, 1, 2, \ldots\}$. □

One important consequence of the exponential family structure is that if $Y_1, Y_2, \ldots, Y_n$ are independent random variables such that the marginal distribution of each Y_j has model function of the form (5.1), then the model function for $Y = (Y_1, \ldots, Y_n)$ is also of the form (5.1). The situation is particularly simple if $Y_1, \ldots, Y_n$ are identically distributed.

Example 5.11 (Exponential distributions). Let Y denote a random variable with density function

$$\frac{1}{\theta} \exp(-y/\theta), \quad y > 0$$

where $\theta > 0$; this is an exponential distribution with mean θ. This density function may be written in the form (5.1) with $T(y) = y$, $c(\theta) = -1/\theta$, $A(\theta) = \log \theta$, and $h(y) = 1$. Hence, this is a one-parameter exponential family distribution.

Now suppose that $Y = (Y_1, \ldots, Y_n)$ where $Y_1, \ldots, Y_n$ are independent random variables, each with an exponential distribution with mean θ. Then the model function for Y is given by

$$\frac{1}{\theta^n} \exp\left(-\frac{1}{\theta} \sum_{j=1}^{n} y_j\right).$$

This is of the form (5.1) with $T(y) = \sum_{j=1}^{n} y_j$, $c(\theta) = -1/\theta$, $A(\theta) = n \log \theta$, and $h(y) = 1$; it follows that the distribution of Y is also a one-parameter exponential family distribution. □

Natural parameters

It is often convenient to reparameterize the models in order to simplify the structure of the exponential family representation. For instance, consider the reparameterization $\eta = c(\theta)$ so that the model function (5.1) becomes

$$\exp\{\eta^T T(y) - A[\theta(\eta)]\} h(y), \quad y \in \mathcal{Y}.$$

Writing $k(\eta)$ for $A[\theta(\eta)]$, the model function has the form

$$\exp\{\eta^T T(y) - k(\eta)\} h(y), \quad y \in \mathcal{Y}; \tag{5.2}$$

the parameter space of the model is given by

$$\mathcal{H}_0 = \{\eta \in \mathbf{R}^m : \eta = c(\theta), \ \theta \in \Theta\}.$$

The model function (5.2) is called the *canonical form* of the model function and the parameter η is called the *natural parameter* of the exponential family distribution. Note that the function k can be obtained from T, h, and $\mathcal{Y}$. For instance, if the distribution is absolutely continuous, we must have

$$\int_{\mathcal{Y}} \exp\{\eta^T T(y) - k(\eta)\} h(y)\, dy = 1, \quad \eta \in \mathcal{H}_0$$

so that

$$k(\eta) = \log \int_{\mathcal{Y}} \exp\{\eta^T T(y)\} h(y)\, dy.$$

The set

$$\mathcal{H} = \{\eta \in \mathbf{R}^m : \int_{\mathcal{Y}} \exp\{\eta^T T(y)\} h(y)\, dy < \infty\}$$

is the largest set in $\mathbf{R}^m$ for which (5.2) defines a valid probability density function; it is called the *natural parameter space*. A similar analysis, in which integrals are replaced by sums, holds in the discrete case.

Consider an exponential family of distributions with model function of the form

$$\exp\{\eta^T T(y) - k(\eta)\} h(y), \quad y \in \mathcal{Y},$$

where $\eta \in \mathcal{H}_0$ and $\mathcal{H}_0$ is a subset of the natural parameter space. In order to use this family of distributions as a statistical model, it is important that the parameter η is identifiable. The following result shows that this holds provided that the distribution of $T(Y)$ corresponding to some $\eta_0 \in \mathcal{H}$ is nondegenerate; in this case, we say that the *rank* of the exponential family is m, the dimension of T.

Lemma 5.1. *Consider an m-dimensional exponential family with model function*

$$\exp\{\eta^T T(y) - k(\eta)\} h(y), \quad y \in \mathcal{Y},$$

where $\eta \in \mathcal{H}_0 \subset \mathcal{H}$. The parameter η is identifiable if and only if $T(Y)$ has a nondegenerate distribution under some $\eta_0 \in \mathcal{H}$.

Proof.   We consider the case in which the distribution is absolutely continuous; the argument for the discrete case is similar. The parameter η is not identifiable if and only if there exist $\eta_1, \eta_2 \in \mathcal{H}_0$ such that

$$\int_A \exp\{\eta_1^T T(y) - k(\eta_1)\} h(y)\, dy = \int_A \exp\{\eta_2^T T(y) - k(\eta_2)\} h(y)\, dy$$

for all $A \subset \mathcal{Y}$. That is, if and only if

$$\int_A \exp\{(\eta_1 - \eta_0)^T T(y) - [k(\eta_1) - k(\eta_0)]\} \exp\{\eta_0^T T(y) - k(\eta_0)\} h(y)\, dy$$

$$= \int_A \exp\{(\eta_2 - \eta_0)^T T(y) - [k(\eta_2) - k(\eta_0)]\} \exp\{\eta_0^T T(y) - k(\eta_0)\} h(y)\, dy$$

where η_0 is an arbitrary element of $\mathcal{H}$.

This is true if and only if

$$\exp\{\eta_1^T T(Y) - k(\eta_1)\} = \exp\{\eta_2^T T(Y) - k(\eta_2)\}$$

with probability 1 under the distribution of Y with parameter η_0. It follows that η is not identifiable if and only if, with probability 1 under the distribution with parameter η_0,

$$(\eta_1 - \eta_2)^T T(Y) = k(\eta_2) - k(\eta_1).$$

That is, η is not identifiable if and only if $T(Y)$ has a degenerate distribution under the distribution with parameter η_0. Since η_0 is arbitrary, it follows that η is not identifiable if and only if $T(Y)$ has a degenerate distribution under all $\eta_0 \in \mathcal{H}$. Equivalently, η is identifiable if and only if $T(Y)$ has a nondegenerate distribution for some $\eta_0 \in \mathcal{H}$. ∎

An important property of the natural parameter space $\mathcal{H}$ is that it is convex; also, the function k is a convex function.

Theorem 5.1. *Consider an m-dimensional exponential family of probability distributions*

$$\exp\{\eta^T T(y) - k(\eta)\} h(y), \quad y \in \mathcal{Y}.$$

Then the natural parameter space $\mathcal{H}$ is a convex set and k is a convex function.

Proof. We give the proof for the case in which the distribution is absolutely continuous. Let η_1 and η_2 denote elements of $\mathcal{H}$ and let $0 < t < 1$. By the Hölder inequality,

$$\int_{\mathcal{Y}} \exp\left\{\left[t\eta_1^T + (1-t)\eta_2^T\right] T(y)\right\} h(y)\, dy$$

$$= \int_{\mathcal{Y}} \exp\left\{t\eta_1^T T(y)\right\} \exp\left\{(1-t)\eta_2^T T(y)\right\} h(y)\, dy$$

$$= \int_{Y} \exp\left\{\eta_1^T T(y)\right\}^t \exp\left\{\eta_2^T T(y)\right\}^{(1-t)} h(y)\, dy$$

$$\leq \left[\int_{\mathcal{Y}} \exp\left\{\eta_1^T T(y)\right\} h(y)\, dy\right]^t \left[\int_{\mathcal{Y}} \exp\left\{\eta_2^T T(y)\right\} h(y)\, dy\right]^{(1-t)} < \infty.$$

It follows that $t\eta_1 + (1-t)\eta_2 \in \mathcal{H}$ and, hence, that $\mathcal{H}$ is convex. Furthermore,

$$\exp\{k(t\eta_1 + (1-t)\eta_2)\} \leq \exp\{tk(\eta_1) + (1-t)k(\eta_2)\},$$

proving that k is a convex function. ∎

The function k is called the *cumulant function* of the family. This terminology is based on the fact that if the natural parameter space is open set, in which case the exponential family is said to be *regular*, then the cumulant-generating function of $T(Y)$ may be written in terms of k.

Theorem 5.2. *Let Y denote a random variable with model function of the form*

$$\exp\{\eta^T T(y) - k(\eta)\} h(y), \quad y \in \mathcal{Y},$$

where $\eta \in \mathcal{H}$ and $\mathcal{H}$ is an open set. Then the cumulant-generating function of $T(Y)$ under the distribution with parameter $\eta \in \mathcal{H}$ is given by

$$K_T(t; \eta) = k(\eta + t) - k(\eta), \quad t \in \mathbf{R}^m, \quad \|t\| < \delta$$

for some $\delta > 0$.

Proof. The proof is given for the case in which the distribution is absolutely continuous. Let M denote the moment-generating function of $T(Y)$. Then

$$M(t) = \mathrm{E}[\exp\{t^T T(Y)\}]$$
$$= \int_{\mathcal{Y}} \exp\{t^T T(y)\} \exp\{\eta^T T(y) - k(\eta)\} h(y) \, dy$$
$$= \int_{\mathcal{Y}} \exp\{(t + \eta)^T T(y) - k(\eta)\} h(y) \, dy.$$

For sufficiently small $\|t\|$, $t + \eta \in \mathcal{H}$. Then, by definition of the function k,

$$M(t) = \exp\{k(t + \eta) - k(\eta)\},$$

proving the result. ∎

Example 5.12 *(Poisson distributions).* Consider the family of Poisson distributions described in Example 5.10. Recall that the model function is given by

$$p(x; \lambda) = \lambda^x \exp(-\lambda)/x!, \quad x = 0, 1, \dots$$

which can be written

$$\exp\{x \log(\lambda) - \lambda\} \frac{1}{x!}, \quad x \in \{0, 1, 2, \dots\}.$$

Hence, the natural parameter is $\eta = \log(\lambda)$, the natural parameter space is $\mathcal{H} = \mathbf{R}$, and the cumulant function is $k(\eta) = \exp(\eta)$. It follows that the cumulant-generating function of X is

$$\exp(t + \eta) - \exp(\eta), \quad t \in \mathbf{R}.$$

In terms of the original parameter λ, this can be written

$$\lambda[\exp(t) - 1], \quad t \in \mathbf{R}. \qquad \square$$

Example 5.13 *(Exponential distributions).* Consider the family of exponential distributions described in Example 5.11. Recall that the model function is given by

$$\frac{1}{\theta} \exp(-y/\theta), \quad y > 0,$$

where $\theta > 0$; this may be written

$$\exp\{\eta y + \log(-\eta)\}, \quad y > 0$$

where $-\infty < \eta < 0$. It follows that the cumulant-generating function of Y is given by

$$\log(t - \eta) - \log(-\eta), \quad |t| < -\eta.$$

In terms of the original parameter θ, this can be written

$$\log(\theta t + 1), \quad |t| < \frac{1}{\theta}. \qquad \square$$

Some distribution theory for exponential families

The importance of exponential family distributions lies in the way in which the parameter of the model interacts with the argument of the density or frequency function in the model function. For instance, if $p(y; \theta)$ is of the form (5.1) and θ_0 and θ_1 are two elements of the parameter space, then $\log[p(y; \theta_1)/p(y; \theta_0)]$ is a linear function of $T(y)$ with coefficients depending on θ_0, θ_1:

$$\log \frac{p(y; \theta_1)}{p(y; \theta_0)} = A(\theta_0) - A(\theta_1) + [c(\theta_1) - c(\theta_0)]^T T(y).$$

This type of structure simplifies certain aspects of the distribution theory of the model, particularly those aspects concerned with how the distributions change under changes in parameter values. The following lemma gives a relationship between expectations under two different parameter values.

Lemma 5.2. *Let Y denote a random variable with model function of the form*

$$\exp\{\eta^T T(y) - k(\eta)\} h(y), \quad y \in \mathcal{Y},$$

where $\eta \in \mathcal{H}$.

Fix $\eta_0 \in \mathcal{H}$ and let $g : \mathcal{Y} \to \mathbf{R}$. Then

$$E[g(Y); \eta] = \exp\{k(\eta_0) - k(\eta)\} E[g(Y) \exp\{(\eta - \eta_0)^T T(Y)\}; \eta_0]$$

for any $\eta \in \mathcal{H}$ such that

$$E[|g(Y)|; \eta] < \infty.$$

Proof. The proof is given for the case in which Y has an absolutely continuous distribution. Suppose $E[|g(Y)|; \eta] < \infty$; then the integral

$$\int_{\mathcal{Y}} g(y) p(y; \eta) \, dy$$

exists and is finite. Note that

$$\int_Y g(y) p(y; \eta) \, dy = \int_{\mathcal{Y}} g(y) \frac{p(y; \eta)}{p(y; \eta_0)} p(y; \eta_0) \, dy = E\left[g(Y) \frac{p(Y; \eta)}{p(Y; \eta_0)}; \eta_0 \right];$$

The result now follows from the fact that

$$\frac{p(Y; \eta)}{p(Y; \eta_0)} = \exp\{k(\eta_0) - k(\eta)\} \exp\{(\eta - \eta_0)^T T(Y)\}. \qquad \blacksquare$$

Consider a random variable Y with model function of the form

$$\exp\{c(\theta)^T T(y) - A(\theta)\} h(y);$$

this function can be written as the product of two terms, the term given by the exponential function and $h(y)$. Note that only the first of these terms depends on θ and that term depends on y only through $T(y)$. This suggests that, in some sense, the dependence of the distribution of Y on θ is primarily through the dependence of the distribution of $T(Y)$ on θ. The following two theorems give some formal expressions of this idea.

Theorem 5.3. *Let Y denote a random variable with model function of the form*

$$\exp\{c(\theta)^T T(y) - A(\theta)\} h(y), \quad y \in \mathcal{Y},$$

where $\theta \in \Theta$. Then the conditional distribution of Y given $T(Y)$ does not depend on θ.

Proof. Let $\eta = c(\theta)$, $\mathcal{H}$ denote the natural parameter space of the model, and

$$\mathcal{H}_0 = \{\eta \in \mathcal{H} : \eta = c(\theta), \theta \in \Theta\}.$$

Then the model function for this model can be written

$$\exp\{\eta^T T(y) - k(\eta)\} h(y), \quad y \in \mathcal{Y},$$

where $\eta \in \mathcal{H}_0$. Hence, it suffices to show that the conditional distribution of Y given $T(Y)$, based on this model, with the parameter space enlarged to $\mathcal{H}$, does not depend on η.

We prove this result by showing that for any bounded, real-valued function g on $\mathcal{Y}$, $\mathrm{E}[g(Y)|T; \eta]$ does not depend on η.

Fix $\eta_0 \in \mathcal{H}$. The idea of the proof is that the random variable

$$Z = \mathrm{E}[g(Y)|T; \eta_0]$$

satisfies

$$\mathrm{E}[Zh(T); \eta] = \mathrm{E}[g(Y)h(T); \eta]$$

for any $\eta \in \mathcal{H}$, for all bounded functions h of T. Hence, by Theorem 2.6,

$$Z = \mathrm{E}[g(Y)|T; \eta].$$

That is, for all $\eta_0, \eta \in \mathcal{H}$,

$$\mathrm{E}[g(Y)|T; \eta] = \mathrm{E}[g(Y)|T; \eta_0],$$

which proves the result.

We now consider the details of the argument. Let h denote a bounded, real-valued function on the range of T. Then, since Z and $g(Y)$ are bounded,

$$\mathrm{E}[|Zh(T)|; \eta] < \infty \quad \text{and} \quad \mathrm{E}[|g(Y)h(T)|; \eta] < \infty;$$

by Lemma 5.2,

$$\mathrm{E}[Zh(T); \eta] = \exp\{k(\eta) - k(\eta_0)\} \mathrm{E}[Zh(T)\exp\{(\eta - \eta_0)^T T\}; \eta_0]$$

and

$$\mathrm{E}[g(Y)h(T); \eta] = \exp\{k(\eta) - k(\eta_0)\} \mathrm{E}[g(Y)h(T)\exp\{(\eta - \eta_0)^T T\}; \eta_0].$$

Let

$$h_0(T) = h(T)\exp\{(\eta - \eta_0)^T T\}.$$

Note that

$$\mathrm{E}[|h_0(T)|; \eta_0] = \exp\{k(\eta_0) - k(\eta)\} \mathrm{E}[|h(T)|; \eta] < \infty.$$

It follows that

$$\mathrm{E}[Zh_0(T); \eta_0] = \mathrm{E}[g(Y)h_0(T); \eta_0]$$

so that

$$E[Zh(T); \eta] = E[g(Y)h(T); \eta]$$

for all bounded h. Hence, by Theorem 2.6,

$$Z \equiv E[g(Y)|T; \eta_0] = E[g(Y)|T; \eta],$$

proving the result.   ■

Theorem 5.4. *Let Y denote a random variable with model function of the form*

$$\exp\{c(\theta)^T T(y) - A(\eta)\}h(y), \quad y \in \mathcal{Y},$$

where $\theta \in \Theta$ and $c : \Theta \to \mathbf{R}^m$. Let

$$\mathcal{H}_0 = \{\eta \in \mathcal{H}: \eta = c(\theta), \quad \theta \in \Theta\},$$

where $\mathcal{H}$ denotes the natural parameter space of the exponential family, and let Z denote a real-valued function on $\mathcal{Y}$.
  (i) If Z and $T(Y)$ are independent, then the distribution of Z does not depend on $\theta \in \Theta$.
  (ii) If $\mathcal{H}_0$ contains an open subset of $\mathbf{R}^m$ and the distribution of Z does not depend on $\theta \in \Theta$, then Z and $T(Y)$ are independent.

Proof.   We begin by reparameterizing the model in terms of the natural parameter $\eta = c(\theta)$ so that the model function can be written

$$\exp\{\eta^T T(y) - k(\eta)\}h(y), \quad y \in \mathcal{Y},$$

with parameter space $\mathcal{H}_0$.
  Suppose that Z and $T(Y)$ are independent. Define

$$\varphi(t; \eta) = E[\exp(itZ); \eta], \quad t \in \mathbf{R}, \quad \eta \in \mathcal{H}.$$

Then, by Lemma 5.2, for any $\eta_0 \in \mathcal{H}$

$$\varphi(t; \eta) = \exp\{k(\eta_0) - k(\eta)\}E[\exp(itZ)\exp\{(\eta - \eta_0)^T T(Y)\}; \eta_0], \quad t \in \mathbf{R}, \quad \eta \in \mathcal{H}.$$

Since Z and $T(Y)$ are independent,

$$\varphi(t; \eta) = \exp\{k(\eta_0) - k(\eta)\}E[\exp(itZ); \eta_0]E[\exp\{(\eta - \eta_0)^T T(Y)\}; \eta_0], \quad t \in \mathbf{R}, \quad \eta \in \mathcal{H}.$$

Since

$$E[\exp\{(\eta - \eta_0)^T T(Y)\}; \eta_0] = \exp\{k(\eta) - k(\eta_0)\}$$

and $E[\exp(itZ); \eta_0] = \varphi(t; \eta_0)$, it follows that, for all $\eta, \eta_0 \in \mathcal{H}$,

$$\varphi(t; \eta) = \varphi(t; \eta_0), \quad t \in \mathbf{R}$$

so that the distribution of Z does not depend on $\eta \in \mathcal{H}$ and, hence, it does not depend on $\eta \in \mathcal{H}_0$. This proves (i).
  Now suppose that the distribution of Z does not depend on $\eta \in \mathcal{H}_0$ and that there exists a subset of $\mathcal{H}_0$, $\mathcal{H}_1$, such that $\mathcal{H}_1$ is an open subset of $\mathbf{R}^m$. Fix $\eta_0 \in \mathcal{H}_1$ and let g denote

a bounded function on the range of Z; then there exists a $\delta_1 > 0$ such that $\exp[tg(Z)]$ is bounded for $|t| < \delta_1$. Hence, by Lemma 5.2, for any $\eta \in \mathcal{H}$,

$$\text{E}\{\exp[tg(Z)]; \eta\} = \exp\{k(\eta_0) - k(\eta)\}\text{E}[\exp[tg(Z)] \exp\{(\eta - \eta_0)^T T(Y)\}; \eta_0].$$

Using the fact that

$$\text{E}[\exp\{(\eta - \eta_0)^T T(Y)\}; \eta_0] = \exp\{k(\eta) - k(\eta_0)\},$$

it follows that, for all $\eta \in \mathcal{H}$ and all $|t| < \delta_1$,

$$\text{E}\{\exp[tg(Z) + (\eta - \eta_0)^T T(Y)]; \eta_0\} = \text{E}\{\exp[tg(Z)]; \eta\}\text{E}[\exp\{(\eta - \eta_0)^T T(Y)\}; \eta_0].$$

For $\delta > 0$, let

$$\mathcal{H}(\delta) = \{\eta \in \mathcal{H}: \|\eta - \eta_0\| < \delta\}$$

and let $\delta_2 > 0$ be such that $\mathcal{H}(\delta_2) \subset \mathcal{H}_1$; since $\mathcal{H}_1$ is an open subset of $\mathbf{R}^m$ and $\eta_0 \in \mathcal{H}_1$, such a δ_2 must exist. Then, since the distribution of $g(Z)$ does not depend on η for $\eta \in \mathcal{H}_0$, for $\eta \in \mathcal{H}(\delta_2)$ and $|t| < \delta_1$,

$$\text{E}\{\exp[tg(Z)]; \eta\} = \text{E}\{\exp[tg(Z)]; \eta_0\}.$$

It follows that, for all $\eta \in \mathcal{H}(\delta_2)$ and all $|t| < \delta_1$,

$$\text{E}\{\exp[tg(Z) + (\eta - \eta_0)^T T(Y)]; \eta_0\} = \text{E}\{\exp[tg(Z)]; \eta_0\}\text{E}[\exp\{(\eta - \eta_0)^T T(Y)\}; \eta_0].$$

That is, the joint moment-generating function of $g(Z)$ and $T(Y)$ can be factored into the product of the two marginal moment-generating functions. Hence, by Corollary 4.2 $g(Z)$ and $T(Y)$ are independent and by part (ii) of Theorem 2.1, Z and $T(Y)$ are independent, proving part (ii) of the theorem. ∎

Example **5.14** *(Bernoulli random variables).* Let $Y = (Y_1, \ldots, Y_n)$ where $Y_1, \ldots, Y_n$ are independent random variables such that

$$\text{Pr}(Y_j = 1; \theta) = 1 - \text{Pr}(Y_j = 0; \theta) = \theta, \quad j = 0, \ldots, n,$$

where $0 < \theta < 1$. Then, for all $y_1, \ldots, y_n$ in the set $\{0, 1\}$,

$$\text{Pr}\{Y = (y_1, \ldots, y_n); \theta\} = \theta^{\sum_{j=1}^{n} y_j} (1 - \theta)^{n - \sum_{j=1}^{n} y_j}.$$

It follows that the model function of Y can be written

$$\exp\left\{\log\left(\frac{\theta}{1 - \theta}\right) \sum_{j=1}^{n} y_j + n \log(1 - \theta)\right\},$$

and, hence, this is a one-parameter exponential family of distributions, with natural parameter $\eta = \log\theta - \log(1 - \theta)$ and $T(y) = \sum_{j=1}^{n} y_j$.

We have seen that the distribution of $T(Y)$ is a binomial distribution with parameters n and θ. Hence,

$$\text{Pr}\{Y = (y_1, \ldots, y_n) | T(Y) = t; \theta\} = \frac{\theta^{\sum_{j=1}^{n} y_j} (1 - \theta)^{n - \sum_{j=1}^{n} y_j}}{\binom{n}{t} \theta^t (1 - \theta)^{n - t}},$$

provided that $t = \sum_{j=1}^{n} y_j$. This probability simplifies to

$$\frac{1}{\binom{n}{t}}$$

for all $y_1, \ldots, y_n$ taking values in the set $\{0, 1\}$ such that $\sum_{j=1}^{n} y_j = t$.

That is, given that $\sum_{j=1}^{n} Y_j = t$, each possible arrangement of 1s and 0s summing to t is equally likely.   □

***Example* 5.15 (*Exponential random variables*).** Let $Y_1, \ldots, Y_n$ denote independent, identically distributed random variables each distributed according to the absolutely continuous distribution with density function

$$\theta \exp\{-\theta y\}, \quad y > 0$$

where $\theta > 0$. Then $Y = (Y_1, \ldots, Y_n)$ has model function

$$\theta^n \exp\left\{-\theta \sum_{j=1}^{n} y_j\right\}, \quad y = (y_1, \ldots, y_n) \in (0, \infty)^n;$$

hence, this is a one-parameter exponential family of distributions with natural parameter $\eta = -\theta$, $\mathcal{H} = (-\infty, 0)$, and $T(y) = \sum_{j=1}^{n} y_j$.

Let $a_1, \ldots, a_n$ denote real-valued, nonzero constants and let $Z = \sum_{j=1}^{n} a_j \log Y_j$. Then Z has moment-generating function

$$M_Z(t; \theta) = E\left\{\exp\left[t \sum_{j=1}^{n} a_j \log(Y_j)\right]; \theta\right\} = \prod_{j=1}^{n} E(Y_j^{a_j t}; \theta)$$

$$= \prod_{j=1}^{n} \frac{\Gamma(a_j t + 1)}{\theta^{a_j t + 1}} = \frac{\prod_{j=1}^{n} \Gamma(a_j t + 1)}{\theta^{n + t \sum_{j=1}^{n} a_j}}, \quad |t| < 1/\max(|a_1|, \ldots, |a_n|).$$

It follows that the distribution of Z does not depend on θ if and only if $\sum_{j=1}^{n} a_j = 0$.

Hence, since $\mathcal{H}_0 = \mathcal{H}$, by Theorem 5.4, $\sum_{j=1}^{n} a_j \log(Y_j)$ and $\sum_{j=1}^{n} Y_j$ are independent if and only if $\sum_{j=1}^{n} a_j = 0$.   □

In applying the second part of Theorem 5.4 it is important that $\mathcal{H}_1$ contains an open subset of $\mathbf{R}^m$. Otherwise, the condition that the distribution of Z does not depend on $\theta \in \Theta$ is not strong enough to ensure that Z and $T(Y)$ are independent. The following example illustrates this possibility.

***Example* 5.16.** Let Y_1 and Y_2 denote independent Poisson random variables such that Y_1 has mean θ and Y_2 has mean $1 - \theta$, where $0 < \theta < 1$. The model function for the distribution of $Y = (Y_1, Y_2)$ can then be written

$$\exp\{\log \theta y_1 + \log(1 - \theta) y_2\} \frac{1}{y_1! y_2!}, \quad y_1 = 0, 1, \ldots; y_2 = 0, 1, \ldots.$$

Hence, $c(\theta) = (\log \theta, \log(1 - \theta))$ and $T(y) = (y_1, y_2)$.

Let $Z = Y_1 + Y_2$. Then, by Example 4.15, Z has a Poisson distribution with mean $\theta + (1 - \theta) = 1$ so that the distribution of Z does not depend on θ. However, Z and (Y_1, Y_2) are clearly not independent; for instance, $\text{Cov}(Z, Y_1; \theta) = \theta$.

Note that

$$\mathcal{H}_0 = \{(\eta_1, \eta_2) \in \mathbf{R}^2 \colon \exp(\eta_1) + \exp(\eta_2) = 1\}$$
$$= \{(z_1, z_2) \in (\mathbf{R}^+)^2 \colon z_1 + z_2 = 1\}.$$

It follows that $\mathcal{H}_0$ is a one-dimensional subset of $\mathbf{R}^2$ and, hence, it does not contain an open subset of $\mathbf{R}^2$. $\square$

5.4 Hierarchical Models

Let X denote a random variable with a distribution depending on a parameter λ. Suppose that λ is not a constant, but is instead itself a random variable with a distribution depending on a parameter θ. This description yields a model for X with parameter θ.

More specifically, consider random variables X and λ, each of which may be a vector. The random variable λ is assumed to be unobserved and we are interested in the model for X. This model is specified by giving the conditional distribution of X given λ, along with the marginal distribution of λ. Both of these distributions may depend on the parameter θ, taking values in a set $\Theta \subset \mathbf{R}^m$, for some m. Hence, probabilities regarding X may be calculated by first conditioning on λ and then averaging with respect to the distribution of λ. For instance,

$$\Pr(X \le x; \theta) = \mathrm{E}[\Pr(X \le x | \lambda; \theta); \theta]$$

where, in this expression, the expectation is with respect to the distribution of λ. The result is a parametric model for X.

If the conditional distribution of X given λ is an absolutely continuous distribution, then the marginal distribution of X is also absolutely continuous. Similarly, if the conditional distribution of X given λ is discrete, the marginal distribution of X is discrete as well.

Theorem 5.5. *Let X and λ denote random variables such that the conditional distribution of X given λ is absolutely continuous with density function $p(x|\lambda; \theta)$ where $\theta \in \Theta$ is a parameter with parameter space Θ. Then the marginal distribution of X is absolutely continuous with density function*

$$p_X(x; \theta) = E[p(x|\lambda; \theta); \theta], \quad \theta \in \Theta.$$

Let X and λ denote random variables such that the conditional distribution of X given λ is discrete and that there exists a countable set $\mathcal{X}$, not depending on λ, and a conditional frequency function $p(x|\lambda; \theta)$ such that

$$\sum_{x \in \mathcal{X}} p(x|\lambda; \theta) = 1, \quad \theta \in \Theta$$

for all λ. Then the marginal distribution of X is discrete with frequency function

$$p_X(x; \theta) = E[p(x|\lambda; \theta); \theta], \quad x \in \mathcal{X}, \quad \theta \in \Theta.$$

Proof. First suppose that the conditonal distribution of X given λ is absolutely continuous. Let g denote a bounded continuous function on the range of X, which we take to be $\mathbf{R}^d$, and let Λ denote the range of λ. Then

$$E[g(X);\theta] = E\{E[g(x)|\lambda;\theta];\theta\} = \int_\Lambda \left\{ \int_{\mathbf{R}^d} g(x)p(x|\lambda;\theta)\,dx \right\} dF_\lambda(\lambda;\theta).$$

By Fubini's Theorem,

$$\int_\Lambda \left\{ \int_{\mathbf{R}^d} g(x)p(x|\lambda;\theta)\,dx \right\} dF_\lambda(\lambda;\theta) = \int_{\mathbf{R}^d} g(x) \left\{ \int_\Lambda p(x|\lambda;\theta)\,dF_\lambda(\lambda;\theta) \right\} dx$$

so that X has an absolutely continuous distribution with density p_X, as given above.

Now suppose that the conditional distribution of X given λ is discrete with frequency function $p(x|\lambda;\theta)$ and support $\mathcal{X}$. Then

$$\Pr(X = x;\theta) = \begin{cases} E\{p(x|\lambda;\theta);\theta\}, & \text{if } x \in \mathcal{X} \\ 0 & \text{if } x \notin \mathcal{X} \end{cases},$$

proving the result. ∎

Example 5.17 (Negative binomial distribution). Let X denote a random variable with a Poisson distribution with mean λ and suppose that λ has a gamma distribution with parameters α and β. Then the marginal distribution of X is discrete with frequency function

$$\begin{aligned} p(x;\alpha,\beta) &= \frac{\beta^\alpha}{\Gamma(\alpha)x!} \int_0^\infty \lambda^x \lambda^{\alpha-1} \exp\{-\lambda\}\,d\lambda \\ &= \frac{\beta^\alpha}{\Gamma(\alpha)x!} \frac{\Gamma(x+\alpha)}{(\beta+1)^{x+\alpha}} \\ &= \binom{x+\alpha-1}{\alpha-1} \frac{\beta^\alpha}{(\beta+1)^{x+\alpha}}, \quad x = 0, 1, \ldots; \end{aligned}$$

here $\alpha > 0$ and $\beta > 0$. This distribution is called the *negative binomial distribution* with parameters α and β.

This distribution has moment-generating function

$$M(t) = \beta^\alpha(1 + \beta - \exp\{t\})^{-\alpha}, \quad t < \log(1 + \beta).$$

It is straightforward to show that $E(X;\theta) = \alpha/\beta$ and

$$\text{Var}(X;\theta) = \frac{\alpha}{\beta} \frac{\beta+1}{\beta}.$$

The *geometric distribution* is a special case of the negative binomial distribution corresponding to $\alpha = 1$.

The Poisson distribution may be obtained as a limiting case of the negative binomial distribution. Suppose X has a negative binomial distribution with parameters α and β where $\alpha = \lambda\beta$, $\lambda > 0$. Then

$$\Pr(X = x; \lambda, \beta) = \binom{x + \lambda\beta - 1}{\lambda\beta - 1} \frac{\beta^{\lambda\beta}}{(\beta+1)^{x+\lambda\beta}}, \quad x = 0, 1, \ldots$$

and

$$\lim_{\beta\to\infty} \Pr(X = x; \lambda, \beta) = \frac{\lambda^x \exp\{-\lambda\}}{x!}, \quad x = 0, 1, \ldots. \qquad \square$$

Models for heterogeneity and dependence

The advantage of a hierarchical representation of a model is that in many applications it is natural to construct models in this way. For instance, hierarchical models are useful for incorporating additional heterogeniety into the model. Specifically, suppose that we are willing to tentatively assume that random variables $X_1, \ldots, X_n$ are independent and identically distributed, each with a distribution depending on a parameter λ. However, we may believe that there is more heterogeniety in the data than this model suggests. One possible explanation for this is that the assumption that the same value of λ applies to each X_j may be too strong. Hence, we might assume that the distribution on X_j depends on λ_j, $j = 1, \ldots, n$. However, taking $\lambda_1, \ldots, \lambda_n$ as arbitrary parameter values may potentially allow too much heterogeniety in the model. An alternative approach is to assume that $\lambda_1, \ldots, \lambda_n$ is a random sample from some distribution. The resulting distribution thus contains two sources of variation in the X_j: the variation inherent in the conditional distribution of X_j given λ, and the variation in the values in the sample $\lambda_1, \ldots, \lambda_n$.

***Example* 5.18** *(Negative binomial distribution).* Suppose that X has a negative binomial distribution, as described in Example 5.17. Then

$$\mathrm{E}(X; \theta) = \frac{\alpha}{\beta} \quad \text{and} \quad \mathrm{Var}(X; \theta) = \frac{\alpha}{\beta} \frac{\beta + 1}{\beta}$$

where $\theta = (\alpha, \beta)^T \in \mathbf{R}^2$. In Example 5.17 it was shown that the distribution of X may be viewed as a Poisson distribution with mean λ, where λ has a gamma distribution with parameters α and β.

If X has a Poisson distribution, then

$$\frac{\mathrm{Var}(X)}{\mathrm{E}(X)} = 1;$$

for the negative binomial distribution considered here,

$$\frac{\mathrm{Var}(X; \theta)}{\mathrm{E}(X; \theta)} = 1 + \frac{1}{\beta}.$$

Hence, β measures the overdispersion of X relative to that of the Poisson distribution. $\quad\square$

Hierarchical models are also useful for modeling dependence. For instance, as above, suppose that $X_1, \ldots, X_n$ are independent random variables, each with the same distribution, but with parameter values $\lambda_1, \ldots, \lambda_n$, respectively, where the λ_j are random variables. However, instead of assuming that $\lambda_1, \ldots, \lambda_n$ are independent, identically distributed random variables, we might assume that some of the λ_j are equal; this might be appropriate if there are certain conditions which affect more than one of $X_1, \ldots, X_n$. These relationships among $\lambda_1, \ldots, \lambda_n$ will induce dependence between $X_1, \ldots, X_n$. This idea is illustrated in the following example.

***Example* 5.19** *(Normal theory random effects model).* Let X_1 and X_2 denote real-valued random variables. Suppose that, given λ, X_1 and X_2 are independent, identically distributed random variables, each with a normal distribution with mean λ and standard deviation σ. Note that the same value of λ is assumed to hold for both X_1 and X_2.

Suppose that λ is distributed according to a normal distribution with mean μ and standard deviation τ. Note that X_1 has characteristic function

$$\mathrm{E}[\exp(itX_1)] = \mathrm{E}\{\mathrm{E}[\exp(itX_1|\lambda)]\} = \mathrm{E}\left[\exp\left(it\lambda - \frac{1}{2}\sigma^2 t^2\right)\right];$$
$$= \exp\left\{it\mu - \frac{1}{2}(\tau^2 + \sigma^2)t^2\right\}$$

it follows that X_1 is distributed according to a normal distribution with mean μ and variance $\tau^2 + \sigma^2$. Clearly, X_2 has the same marginal distribution as X_1.

However, X_1 and X_2 are no longer independent. Since

$$\mathrm{E}(X_1 X_2) = \mathrm{E}[\mathrm{E}(X_1 X_2|\lambda)] = \mathrm{E}(\lambda^2) = \tau^2 + \mu^2,$$

it follows that

$$\mathrm{Cov}(X_1, X_2) = \tau^2$$

and, hence, that the correlation of X_1 and X_2 is $\tau^2/(\tau^2 + \sigma^2)$. Hence, although, conditionally on λ, X_1 and X_2 are independent, marginally they are dependent random variables.

The distribution of (X_1, X_2) is called the *bivariate normal distribution*; its properties will be considered in detail in Chapter 8. $\square$

5.5 Regression Models

Consider a parametric model on $\mathcal{Y} \subset \mathbf{R}^d$, $\mathcal{P} = \{\mathrm{P}(\cdot; \lambda): \lambda \in \Lambda\}$. Suppose that $Y_1, \ldots, Y_n$ are independent random variables such that, for each $j = 1, \ldots, n$, the distribution of Y_j is the element of $\mathcal{P}$ corresponding to a parameter value λ_j. Hence, $Y_1, \ldots, Y_n$ are independent, but are not necessarily identically distributed.

Let $x_1, x_2, \ldots, x_n$ denote a known sequence of nonrandom vectors such that, for each $j = 1, \ldots, n$, there exists a function h such that

$$\lambda_j = h(x_j; \theta)$$

for some θ in a set Θ. Thus, the distribution of Y_j depends on the value of x_j, along with the value of θ and the function h. The vectors $x_1, \ldots, x_n$ are known as *covariates* or *explanatory variables*; the random variables $Y_1, \ldots, Y_n$ are called the *response variables*. The response variables and covariates are sometimes called *dependent* and *independent variables*, respectively; however, those terms will not be used here, in order to avoid confusion with the concept of independence, as discussed in Section 2.2.

In a regression model, the function h is known, while θ is an unknown parameter. Interest generally centers on the relationship between Y_j and x_j, $j = 1, \ldots, n$, as expressed through the function h. Regression models are very widely used in statistics; the goal of this section is to present a few examples illustrating some of the regression models commonly used.

Example 5.20 (Additive error models). Suppose that each distribution in $\mathcal{P}$ has a finite mean. Let

$$\mu(x_j, \theta) = \mathrm{E}(Y_j; \theta).$$

Then we may write

$$Y_j = \mu(x_j, \theta) + \epsilon_j, \quad j = 1, \ldots, n$$

where ϵ_j is simply $Y_j - \mu(x_j, \theta)$. If, for each j, the minimal support of the distribution of ϵ_j does not depend on the values of x_j and θ, we call the model an *additive error* model.

By construction, $\epsilon_1, \ldots, \epsilon_n$ are independent random variables each with mean 0 and with a distribution depending on θ. Often additional assumptions are made about the distribution of $\epsilon_1, \ldots, \epsilon_n$. For instance, it may be reasonable to assume that $\epsilon_1, \ldots, \epsilon_n$ are identically distributed with a specific distribution. $\square$

Example 5.21 (*Linear models*). Let $Y_1, \ldots, Y_n$ denote independent scalar random variables such that each Y_j follows an additive error model of the form

$$Y_j = \mu(x_j, \theta) + \epsilon_j.$$

Suppose further that θ may be written (β, σ), where β is a vector and $\sigma > 0$ is a scalar such that $\mu(x_j; \theta)$ is a linear function of β,

$$\mu(x_j, \theta) = x_j \beta$$

and the distribution of ϵ_j depends only on σ, the standard deviation of the distribution of Y_j. Thus, we may write

$$Y_j = x_j \beta + \sigma z_j, \quad j = 1, \ldots, n$$

where $z_1, \ldots, z_n$ have known distributions. A model of this type is called a *linear model*; when $z_1, \ldots, z_n$ are assumed to be standard normal random variables, it is called a *normal-theory* linear model. $\square$

Example 5.22 (*Linear exponential family regression models*). Let $Y_1, \ldots, Y_n$ denote independent scalar random variables such that Y_j has an exponential family distribution with model function of the form

$$p(y; \lambda_j) = \exp\left\{\lambda_j^T T(y) - k(\lambda_j)\right\} h(y),$$

as discussed in Section 5.3. Suppose that

$$\lambda_j = x_j \beta, \quad j = 1, \ldots, n$$

where, as above, $x_1, \ldots, x_n$ are fixed covariates and β is an unknown parameter. Hence, the density, or frequency function, of $Y_1, \ldots, Y_n$ is

$$\exp\left\{\beta^T \sum_{j=1}^n x_j^T T(y_j) - \sum_{j=1}^n k(x_j \beta)\right\} \prod_{j=1}^n h(y_j).$$

This is called a *linear exponential family regression model*; it is also a special case of a *generalized linear model*.

For instance, suppose that $Y_1, \ldots, Y_n$ are independent Poisson random variables such that Y_j has mean λ_j with $\log \lambda_j = x_j \beta$. Then $Y_1, \ldots, Y_n$ has frequency function

$$\exp\left\{\beta^T \sum_{j=1}^n x_j y_j - \sum_{j=1}^n \exp(x_j \beta)\right\} \prod_{j=1}^n \frac{1}{y_j!}, \quad y_j = 0, 1, 2, \ldots; \quad j = 1, \ldots, n. \quad \square$$

In some cases, the explanatory variables are random variables as well. Let (X_j, Y_j), $j = 1, \ldots, n$, denote independent random vectors such that the conditional distribution of Y_j given $X_j = x_j$ is the element of $\mathcal{P}$ corresponding to parameter value

$$\lambda_j = h(x_j; \theta)$$

for some known function h. Then the model based on the conditional distribution of $(Y_1, \ldots, Y_n)$ given $(X_1, \ldots, X_n) = (x_1, \ldots, x_n)$ is identical to the regression model considered earlier. This approach is appropriate provided that the distribution of the covariate vector $(X_1, \ldots, X_n)$ does not depend on θ, the parameter of interest. If the distribution of $(X_1, \ldots, X_n)$ is also of interest, or depends on parameters that are of interest, then a model for the distribution of $(X_1, Y_1), \ldots, (X_n, Y_n)$ would be appropriate.

***Example* 5.23.** Let $(X_1, Y_1), \ldots, (X_n, Y_n)$ denote independent, identically distributed pairs of real-valued random variables such that the conditional distribution of Y_j given $X_j = x$ is a binomial distribution with frequency function of the form

$$\binom{x}{y} \theta_1^y (1 - \theta_1)^{x-y}, \quad y = 0, 1, \ldots, x,$$

where $0 < \theta_1 < 1$, and the marginal distribution of X_j is a Poisson distribution with mean θ_2, $\theta_2 > 0$.

If only the parameter θ_1 is of interest, then a statistical analysis can be based on the conditional distribution of $(Y_1, \ldots, Y_n)$ given $(X_1, \ldots, X_n) = (x_1, \ldots, x_n)$, which has model function

$$\prod_{j=1}^{n} \binom{x_j}{y_j} \theta_1^{y_j} (1 - \theta_1)^{x_j - y_j}.$$

If both parameters θ_1 and θ_2 are of interest, a statistical analysis can be based on the distribution of $(X_1, Y_1), \ldots, (X_n, Y_n)$, which has model function

$$\prod_{j=1}^{n} \frac{\theta_2^{x_j}}{y_j!(x_j - y_j)!} \exp(-\theta_2) \theta_1^{y_j} (1 - \theta_1)^{x_j - y_j}. \qquad \square$$

5.6 Models with a Group Structure

For some models there is additional structure relating distributions with different parameter values and it is often possible to exploit this additional structure in order to simplify the distribution theory of the model. The following example illustrates this possibility.

***Example* 5.24 (*Normal distribution*).** Let X denote a random variable with a normal distribution with mean μ, $-\infty < \mu < \infty$ and standard deviation σ, $\sigma > 0$. Using characteristic functions it is straightforward to show that the distribution of X is identical to the distribution of $\mu + \sigma Z$, where Z has a standard normal distribution. Hence, we may write $X = \mu + \sigma Z$.

Let b and c denote constants with $c > 0$ and let $Y = b + cX$. Then we may write $Y = c\mu + b + (c\sigma)Z$ so that Y has the same distribution of X except that μ is modified to $c\mu + b$ and σ is modified to $c\sigma$. Hence, many properties of the distribution of Y may be obtained directly from the corresponding property of the distribution of X. For example, if we know that

$$E(X^4) = 3\sigma^4 + 6\mu^2\sigma^2 + \mu^4,$$

it follows immediately that

$$E(Y^4) = 3c^4\sigma^4 + 6(c\mu + b)^2 c^2\sigma^2 + (c\mu + b)^4.$$

Now consider n independent random variables $X_1, X_2, \ldots, X_n$ such that the marginal distribution of each X_j is the same as the distribution of X above. Then we may write $X_j = \mu + \sigma Z_j$, $j = 1, \ldots, n$, where $Z_1, \ldots, Z_n$ are independent standard normal random variables. Let $\bar{X} = \sum_{j=1}^n X_j/n$; since $\bar{X} = \mu + \sigma\bar{Z}$ where $\bar{Z} = \sum_{j=1}^n Z_j/n$, it follows that the relationship between $\bar{X}$ and $\bar{Z}$ is the same as the relationship between X_j and Z_j. For instance, if we know that $\bar{Z}$ has a normal distribution with mean 0 and standard deviation $1/\sqrt{n}$, it follows immediately that $\bar{X}$ has a normal distribution with mean μ and standard deviation $\sigma/\sqrt{n}$. The statistic $\bar{X}$ is an example of an *equivariant statistic*.

Consider the statistic

$$T = \frac{X_1 - \bar{X}}{X_2 - \bar{X}}.$$

It follows immediately from the facts that $X_j = \mu + \sigma Z_j$ and $\bar{X} = \mu + \sigma\bar{Z}$ that T has the same distribution as

$$\frac{Z_1 - \bar{Z}}{Z_2 - \bar{Z}}$$

so that the distribution of T does not depend on μ or σ. Hence, when studying the distributional properties of T we can assume that $\mu = 0$ and $\sigma = 1$. This is an example of an *invariant statistic*. $\square$

The goal of this section is to generalize the ideas presented in the previous example. Let X denote a random variable with range $\mathcal{X}$ and probability distribution P taking values in a set $\mathcal{P}$. The key idea in Example 5.24 is that there is an algebraic operation on the space $\mathcal{X}$ that corresponds to changes in the distribution of X; we will refer to these algebraic operations as *transformations*. Hence, we need to specify a set of transformations on $\mathcal{X}$ and relate them to the different distributions in $\mathcal{P}$. In order to do this, it is convenient to use the language of group theory.

Consider a transformation $g : \mathcal{X} \to \mathcal{X}$. We require that g is one-to-one, so that $g(x_1) = g(x_2)$ implies that $x_1 = x_2$, and onto, so that every $x_1 \in \mathcal{X}$ may be written $g(x_2)$ for some $x_2 \in \mathcal{X}$. We will be interested in a set of such transformations together with an operation that allows two transformations to be combined. There are a number of conditions such a combination must satisfy in order for the set of transformations to be useful. For instance, if two transformations are combined, they must form another transformation and every transformation must have an inverse transformation such that, if the transformation is combined with its inverse, an identity tranformation taking each $x \in \mathcal{X}$ back to x is formed.

More formally, let $\mathcal{G}$ denote a *group* of transformations $g : \mathcal{X} \to \mathcal{X}$. Recall that a group is a nonempty set $\mathcal{G}$ together with a binary operation $\circ$ such that the following conditions are satisfied

(G1) If $g_1, g_2 \in \mathcal{G}$ then $g_1 \circ g_2 \in \mathcal{G}$

(G2) If $g_1, g_2, g_3 \in \mathcal{G}$ then $(g_1 \circ g_2) \circ g_3 = g_1 \circ (g_2 \circ g_3)$

(G3) There exists an element $e \in \mathcal{G}$, called the identity element, such that for each $g \in \mathcal{G}$,
$$e \circ g = g \circ e = g$$

(G4) For each $g \in \mathcal{G}$, there exists an element $g^{-1} \in \mathcal{G}$ such that $g \circ g^{-1} = g^{-1} \circ g = e$.

We assume further that, for a suitable topology on $\mathcal{G}$, the operations

$$(g_1, g_2) \to g_1 \circ g_2$$

and $g \to g^{-1}$ are continuous. Often, the group $\mathcal{G}$ can be taken to be finite-dimensional Euclidean space with the topology on $\mathcal{G}$ taken to be the usual one. For simplicity, we will suppress the symbol $\circ$ when writing group operations so that, for example, $g_1 g_2 = g_1 \circ g_2$.

So far, we have only described how the transformations in $\mathcal{G}$ must relate to each other. However, it is also important to put some requirements on how the transformations in $\mathcal{G}$ act on $\mathcal{X}$.

Let $g_1, g_2 \in \mathcal{G}$. Since $g_2 x$ is an element of $\mathcal{X}$, we may calculate $g_1(g_2 x)$. We require that the result is the same as applying $g_1 g_2$ to x; that is,

(T1) $g_1(g_2 x) = (g_1 g_2)x$.

We also require that the identity element of the group, e, is also the identity transformation on $\mathcal{X}$:

(T2) $ex = x, \quad x \in \mathcal{X}$.

***Example* 5.25 (*Location and scale groups*).** In statistics, the most commonly used transformations are location and scale transformations. Here we take $\mathcal{X} = \mathbf{R}^n$.

First consider the group $\mathcal{G}_s$ of scale transformations. That is, $\mathcal{G}_s = \mathbf{R}^+$ such that for any $a \in \mathcal{G}_s$ and $x \in \mathcal{X}$, ax represents scalar multiplication. It is easy to see that this is a group with the group operation defined to be multiplication; that is, for a_1, a_2 in $\mathcal{G}_s$, $a_1 a_2$ denotes the multiplication of a_1 and a_2. The identity element of the group is 1 and $a^{-1} = 1/a$. For the topology on $\mathcal{G}_s$ we may take the usual topology on $\mathbf{R}$.

We may also consider the set $\mathcal{G}_l$ of location transformations. Then $\mathcal{G}_l = \mathbf{R}$ and for $b \in \mathcal{G}_l$ and $x \in \mathcal{X}$, $bx = x + b1_n$ where 1_n denotes the vector of length n consisting of all 1s. The group operation is simple addition, the identity element is 0, $b^{-1} = -b$, and the topology on $\mathcal{G}_l$ may be taken to be the $\mathbf{R}$-topology.

Now consider the group of location–scale transformations, $\mathcal{G}_{ls}$. The group $\mathcal{G}_{ls} = \mathbf{R}^+ \times \mathbf{R}$ such that for any $(a, b) \in \mathcal{G}_{ls}$ and any $x \in \mathcal{X}$,

$$(a, b)x = ax + b1_n.$$

Let (a_1, b_1) and (a_2, b_2) denote elements of $\mathcal{G}_{ls}$. Then

$$(a_1, b_1)[(a_2, b_2)x] = a_1(a_2 x + b_2 1_n) + b_1 1_n = a_1 a_2 x + (a_1 b_2 + b_1)1_n.$$

Hence, the group operation is given by

$$(a_1, b_1)(a_2, b_2) = (a_1 a_2, a_1 b_2 + b_1).$$

The identity element of the group is $(1, 0)$ and since

$$\left(\frac{1}{a}, -\frac{b}{a}\right)(a, b) = (a, b)\left(\frac{1}{a}, -\frac{b}{a}\right) = (1, 0),$$

$$(a, b)^{-1} = \left(\frac{1}{a}, -\frac{b}{a}\right).$$

The topology on $\mathcal{G}_{\mathrm{ls}}$ may be taken to be the usual topology on $\mathbf{R}^2$. $\square$

Transformation models

Recall that our goal is to use the group of transformations $\mathcal{G}$ to relate different distributions in $\mathcal{P}$. Consider a random variable X with range $\mathcal{X}$ and let $\mathcal{G}$ denote a group of transformations on $\mathcal{X}$. The set of probability distributions $\mathcal{P}$ is said to be *invariant with respect to* $\mathcal{G}$ if the following condition holds: if X has probability distribution $P \in \mathcal{P}$, then the probability distribution of gX, which can be denoted by P_1, is also an element of $\mathcal{P}$. That is, for every $P \in \mathcal{P}$ and every $g \in \mathcal{G}$ there exists a $P_1 \in \mathcal{P}$ such that, for all bounded continuous functions $h : \mathcal{X} \to \mathbf{R}$,

$$E_P[h(gX)] = E_{P_1}[h(X)]$$

where E_P denotes the expectation with respect to P and E_{P_1} denotes expectation with respect to P_1. In this case we may write $P_1 = gP$ so that we may view g as operating on $\mathcal{P}$ as well as on $\mathcal{X}$.

We have already considered one example of a class of distributions that is invariant with respect to a group of transformations when considering exchangeable random variables in Section 2.6.

Example 5.26 (*Exchangeable random variables*). Let $X_1, X_2, \ldots, X_n$ denote exchangeable real-valued random variables and let $\mathcal{G}$ denote the set of all permutations of $(1, 2, \ldots, n)$. Hence, if $X = (X_1, \ldots, X_n)$ and $g = (n, n-1, \ldots, 2, 1)$, for example, then

$$gX = (X_n, X_{n-1}, \ldots, X_1).$$

It is straightforward to show that $\mathcal{G}$ is a group and, by definition, the set of distributions of X is invariant with respect to $\mathcal{G}$. $\square$

Suppose $\mathcal{P}$ is a parametric family of distributions, $\mathcal{P} = \{P(\cdot; \theta): \theta \in \Theta\}$. If $\mathcal{P}$ is invariant with respect to $\mathcal{G}$ and if X is distributed according to the distribution with parameter θ, then gX is distributed according to the distribution with parameter $g\theta$, so that g may be viewed as acting on the parameter space Θ. In statistics, such a model is often called a *transformation model*.

Example 5.27 (*Exponential distribution*). Let $X_1, X_2, \ldots, X_n$ denote independent, identically distributed random variables, each distributed according to an exponential distribution with parameter $\theta > 0$. Hence, the vector $X = (X_1, \ldots, X_n)$ has density

$$p(x; \theta) = \theta^n \exp\left\{-\theta \sum_{j=1}^{n} x_j\right\}, \quad x_j > 0, \quad j = 1, \ldots, n.$$

For a function $h : \mathbf{R}^n \to \mathbf{R}$,

$$\mathrm{E}[h(X); \theta] = \int_0^\infty \cdots \int_0^\infty h(x) \theta^n \exp\left\{ -\theta \sum_{j=1}^n x_j \right\} dx_1 \cdots dx_n.$$

Consider the group of scale transformations $\mathcal{G}_s$. For $a \in \mathcal{G}_s$, using a change-of-variable for the integral,

$$\mathrm{E}[h(aX); \theta] = \int_0^\infty \cdots \int_0^\infty h(ax) \theta^n \exp\left\{ -\theta \sum_{j=1}^n x_j \right\} dx_1 \cdots dx_n$$

$$= \int_0^\infty \cdots \int_0^\infty h(x) \frac{\theta^n}{a^n} \exp\left\{ -(\theta/a) \sum_{j=1}^n x_j \right\} dx_1 \cdots dx_n.$$

It follows that if X has an exponential distribution with parameter θ, then aX has an exponential distribution with parameter value θ/a and, hence, this model is a transformation model with respect to $\mathcal{G}_s$. The elements of $\mathcal{G}_s$ are nonnegative constants; for $a \in \mathcal{G}_s$, the action of a on Θ is given by $a\theta = \theta/a$. $\square$

In many cases, the parameter space of a transformation model is isomorphic to the group of transformations $\mathcal{G}$. That is, there is a one-to-one mapping from $\mathcal{G}$ to Θ and, hence, the group of transformations may be identified with the parameter space of the model. In this case, the group $\mathcal{G}$ may be taken to be the parameter space Θ.

To see how such an isomorphism can be constructed, suppose the distribution of X is an element of $\mathcal{P}$ which is invariant with respect to a group of transformations $\mathcal{G}$. Fix some element θ_0 of Θ and suppose X is distributed according to the distribution with parameter θ_0. Then, for $g \in \mathcal{G}$, gX is distributed according to the distribution with parameter value $\theta_1 = g\theta_0$, for some $\theta_1 \in \Theta$. Hence, we can write θ_1 for g so that, if X is distributed according to the distribution with parameter θ_0, $\theta_1 X$ is distributed according to the distribution with parameter θ_1. If, for each $\theta \in \Theta$, there is a unique $g \in \mathcal{G}$ such that $\theta = g\theta_0$ and $g_1\theta_0 = g_2\theta_0$ implies that $g_1 = g_2$, then Θ and $\mathcal{G}$ are isomorphic and we can proceed as if $\Theta = \mathcal{G}$. The parameter value θ_0 may be identified with the identity element of $\mathcal{G}$.

Example 5.28 (Exponential distribution). Consider the exponential distribution model considered in Example 5.27; for simplicity, take $n = 1$. Let $g = a > 0$ denote a scale transformation. If X has a standard exponential distribution, then gX has an exponential distribution with parameter $\theta_1 = 1/a$.

The group $\mathcal{G}$ may be identified with Θ using the correspondence $a \to 1/\theta$. If X has a standard exponential distribution, then θX has an exponential distribution with parameter θ. Hence, $\theta X = X/\theta$; the identity element of the group is the parameter value corresponding to the standard exponential distribution, 1. The same approach may be used for a vector $(X_1, \ldots, X_n)$. $\square$

Example 5.29 (Location-scale models). Let X denote a real-valued random variable with an absolutely continuous distribution with density p_0 satisfying $p_0(x) > 0$, $x \in \mathbf{R}$. The location-scale model based on p_0 consists of the class of distributions of gX, $g \in \mathcal{G}_{ls}^{(1)}$ where $\mathcal{G}_{ls}^{(1)}$ denotes the group of location-scale transformations on $\mathbf{R}$.

Suppose X is distributed according to p_0 and let $g = (\sigma, \mu)$ denote an element of $\mathcal{G}_{\text{ls}}^{(1)}$. Then

$$E[h(gX)] = \int_{-\infty}^{\infty} h(gx)p_0(x)\,dx = \int_{-\infty}^{\infty} h(\sigma x + \mu)p_0(x)\,dx$$
$$= \int_{-\infty}^{\infty} h(x)p_0\left(\frac{x-\mu}{\sigma}\right)\frac{1}{\sigma}\,dx.$$

Hence, the distribution of gX is absolutely continuous with density function

$$p(x;\theta) = \frac{1}{\sigma}p_0\left(\frac{x-\mu}{\sigma}\right), \quad -\infty < x < \infty.$$

The model given by

$$\{p(\cdot;\theta): \theta = (\sigma, \mu), \ \sigma > 0, \ -\infty < \mu < \infty\}$$

is a transformation model with respect to $\mathcal{G}_{\text{ls}}^{(1)}$.

For instance, consider the case in which p_0 denotes the density of the standard normal distribution. Then the set of distributions of X is simply the set of normal distributions with mean μ and standard deviation σ.

Now consider independent, identically distributed random variables $X_1, \ldots, X_n$, each distributed according to an absolutely continuous distribution with density function of the form $p(\cdot;\theta)$, as given above. The density for $(X_1, \ldots, X_n)$ is of the form

$$\frac{1}{\sigma^n}\prod p_0\left(\frac{x_i - \mu}{\sigma}\right), \quad -\infty < x_i < \infty, \ i = 1, \ldots, n.$$

Clearly, the model for $(X_1, \ldots, X_n)$ is a transformation model with respect to $\mathcal{G}_{\text{ls}}$. $\quad \square$

Let $x \in \mathcal{X}$. The *orbit* of x is that subset of $\mathcal{X}$ that consists of all points that are obtainable from x using a transformation in $\mathcal{G}$; that is, the orbit of x is the set

$$O(x) = \{x_1 \in \mathcal{X}: x_1 = gx \text{ for some } g \in \mathcal{G}\}.$$

Example 5.30 (*Location-scale group*). Let $\mathcal{X} = \mathbf{R}^n$ and consider the group of location-scale transformations on $\mathcal{X}$. Then two elements of $\mathbf{R}^n$, x_1 and x_2, are on the same orbit if there exists $(a, b) \in \mathcal{G}_{\text{ls}}$ such that

$$x_1 = ax_2 + b1_n,$$

that is, if there exists a constant $a > 0$ such that the elements of $x_1 - ax_2$ are all equal.

For a given element $x \in \mathcal{X}$,

$$O(x) = \{x_1 \in \mathcal{X}: x_1 = ax + b1_n, \ a > 0, \ -\infty < b < \infty\}. \quad \square$$

Invariance

Now consider a function $T : \mathcal{X} \to \mathbf{R}^k$. We say that $T(X)$ is an *invariant statistic* with respect to $\mathcal{G}$ if, for all $g \in \mathcal{G}$ and all $x \in \mathcal{X}$, $T(gx) = T(x)$. That is, T is constant on the orbits of $\mathcal{X}$. Hence, if T is invariant, then the probability distribution of $T(gX)$ is the same for all $g \in \mathcal{G}$. In particular, if the family of probability distributions of X is invariant with respect to $\mathcal{G}$ then the distribution of $T(X)$ is the same for all $P \in \mathcal{P}$. Thus, in order to find the distribution of $T(X)$ we may choose a convenient element $P \in \mathcal{P}$ and determine the distribution of $T(X)$ under P.

Example 5.31 *(Location group).* Let $\mathcal{X} = \mathbf{R}^n$ and consider the group of location transformations on $\mathcal{X}$. Consider the function

$$T(x) = x - \left[\left(1_n^T x\right)/n\right] 1_n = \begin{pmatrix} x_1 - \bar{x} \\ \vdots \\ x_n - \bar{x} \end{pmatrix},$$

where $\bar{x} = \sum_{j=1}^n x_j/n$. Since $bx = x + b1_n$ for $b \in \mathcal{G}_1$,

$$\begin{aligned} T(bx) &= x + b1_n - \left[1_n^T(x + b1_n)/n\right] 1_n \\ &= x + b1_n - \left[\left(1_n^T x\right)/n\right] 1_n - b1_n \\ &= x - \left[\left(1_n^T x\right)/n\right] 1_n = T(x); \end{aligned}$$

hence, T is invariant with respect to $\mathcal{G}_1$.   □

Example 5.32 *(Cauchy distribution).* Let X_1, X_2, X_3, X_4 denote independent, identically distributed random variables, each with a normal distribution with mean μ and standard deviation σ. As noted earlier, this model for (X_1, X_2, X_3, X_4) is a transformation model with respect to $\mathcal{G}_{ls}$.

For $x \in \mathbf{R}^4$, $x = (x_1, x_2, x_3, x_4)$, let

$$T(x) = \frac{x_1 - x_2}{x_3 - x_4}.$$

This function is invariant with respect to $\mathcal{G}_{ls}^{(4)}$ since, for $g = (\sigma, \mu) \in \mathcal{G}_{ls}^{(4)}$,

$$T(gx) = \frac{(\sigma x_1 + \mu) - (\sigma x_2 + \mu)}{(\sigma x_3 + \mu) - (\sigma x_4 + \mu)} = \frac{x_1 - x_2}{x_3 - x_4} = T(x).$$

Hence, the distribution of $T(X)$, $X = (X_1, X_2, X_3, X_4)$, does not depend on the value of μ and σ under consideration.

For instance, take $\mu = 0$ and $\sigma^2 = 1/2$. Then X_1 and X_2 each have characteristic function

$$\varphi(t) = \exp(-t^2/4), \quad -\infty < t < \infty.$$

Hence, $X_1 - X_2$ has characteristic function

$$\varphi(t)\varphi(-t) = \exp(-t^2/2), \quad -\infty < t < \infty.$$

It follows that $X_1 - X_2$ has a standard normal distribution; similarly, $X_3 - X_4$ also has a standard normal distribution. Furthermore, $X_1 - X_2$ and $X_3 - X_4$ are independent. It follows from Example 3.13 that $T(X)$ has a standard Cauchy distribution.   □

The statistic T is said be a *maximal invariant* if it is invariant and any other invariant statistic is a function of T. That is, if T_1 is invariant, then there exists a function h such that, for each $x \in \mathcal{X}$,

$$T_1(x) = h(T(x)).$$

Theorem 5.6. *Let X denote a random variable with range $\mathcal{X}$ and suppose that the distribution of X is an element of*

$$\mathcal{P} = \{P(\cdot; \theta): \theta \in \Theta\}.$$

Let $\mathcal{G}$ denote a group of transformations from $\mathcal{X}$ to $\mathcal{X}$ and suppose that $\mathcal{P}$ is invariant with respect to $\mathcal{G}$. Let T denote a function from $\mathcal{X}$ to $\mathbf{R}^k$.

The following conditions are equivalent:

  (i) T is a maximal invariant
 (ii) Let x_1, x_2 denote elements of $\mathcal{X}$ such that $x_2 \notin O(x_1)$. Then T is constant on $O(x_i)$, $i = 1, 2$ and $T(x_1) \neq T(x_2)$.
(iii) Let $x_1, x_2 \in \mathcal{X}$. $T(x_1) = T(x_2)$ if and only if there exists $g \in \mathcal{G}$ such that $x_1 = gx_2$.

Proof.   We first show that conditions (ii) and (iii) are equivalent. Suppose that condition (ii) holds. If $T(x_1) = T(x_2)$ for some $x_1, x_2 \in \mathcal{X}$, then we must have $x_2 \in O(x_1)$ since otherwise condition (ii) implies that $T(x_1) \neq T(x_2)$. Since T is constant on $O(x_i)$, it follows that $T(x_1) = T(x_2)$ if and only if $x_2 \in O(x_1)$; that is, condition (iii) holds.

Now suppose condition (iii) holds. Clearly, T is constant on $O(x)$ for any $x \in \mathcal{X}$. Furthermore, if x_1, x_2 are elements of $\mathcal{X}$ such that $x_2 \notin O(x_1)$, there does not exist a g such that $x_1 = gx_2$ so that $T(x_1) \neq T(x_2)$. Hence, condition (ii) holds.

We now show that condition (iii) and condition (i) are equivalent. Suppose that condition (iii) holds and let T_1 denote an invariant statistic. T is maximal invariant provided that T_1 is a function of T. Define a function h as follows. If y is in the range of T so that $y = T(x)$ for some $x \in \mathcal{X}$, define $h(y) = T_1(x)$; otherwise, define $h(y)$ arbitrarily. Suppose x_1, x_2 are elements of $\mathcal{X}$ such that $T(x_1) = T(x_2)$. Under condition (iii), $x_1 = gx_2$ for some $g \in \mathcal{G}$ so that $T_1(x_1) = T_1(x_2)$; hence, h is well defined. Clearly, $h(T(x)) = T_1(x)$ so that T is a maximal invariant. It follows that (iii) implies (i).

Finally, assume that T is a maximal invariant, that is, that (i) holds. Clearly, $x_2 = gx_1$ implies that $T(x_1) = T(x_2)$. Suppose that there does not exist a $g \in \mathcal{G}$ satisfying $x_2 = gx_1$. Define a statistic T_1 as follows. Let y_1, y_2, y_3 denote distinct elements of $\mathbf{R}^k$. If $x \in O(x_1)$, $T_1(x) = y_1$, if $x \in O(x_2)$, $T_1(x) = y_2$, if x is not an element of either $O(x_1)$ or $O(x_2)$, then $T_1(x) = y_3$. Note that T_1 is invariant. It follows that there exists a function h such that $T_1(x) = h(T(x))$, $x \in \mathcal{X}$, so that $h(T(x_1)) \neq h(T(x_2))$; hence, $T(x_1) \neq T(x_2)$. Therefore condition (iii) holds.   ∎

Theorem 5.6 gives a useful description of a maximal invariant statistic. The range $\mathcal{X}$ of a random variable X can be divided into orbits. Two points x_1, x_2 lie on the same orbit if there exists a $g \in \mathcal{G}$ such that $x_2 = gx_1$. An invariant statistic is constant on orbits. An invariant statistic T is a maximal invariant if, in addition, it takes different values on different orbits. Hence, a maximal invariant statistic completely describes the differences between the orbits of $\mathcal{X}$; however, it does not give any information regarding the structure within each orbit.

Example 5.33 (Location group).   Let $\mathcal{X} = \mathbf{R}^n$ and consider the group of location transformations on $\mathcal{X}$. In Example 5.31 it was shown that the function

$$T(x) = x - \left[\left(1_n^T x \right) / n \right] 1_n$$

is invariant.

Suppose that x_1, x_2 are elements of $\mathbf{R}^n$. Then $T(x_1) = T(x_2)$ if and only if

$$x_1 - \left[\left(1_n^T x_1 \right) / n \right] 1_n = x_2 - \left[\left(1_n^T x_2 \right) / n \right] 1_n,$$

that is, if and only if

$$x_1 = x_2 + \left\{ \left[1_n^T(x_1 - x_2) \right] / n \right\} 1_n = x_2 + b^* 1_n$$

where $b^* = [1_n^T(x_1 - x_2)]/n$. It follows from part (iii) of Theorem 5.6 that T is a maximal invariant. □

Consider an invariant statistic T. According to part (ii) of Theorem 5.6, in order to show that T is not a maximal invariant, it is sufficient to find x_1, x_2 in $\mathcal{X}$ that lie on different orbits such that $T(x_1) = T(x_2)$.

***Example* 5.34 (*Cauchy distribution*).** Consider $\mathcal{X} = \mathbf{R}^4$ and for $x \in \mathbf{R}^4$, $x = (x_1, x_2, x_3, x_4)$, let

$$T(x) = \frac{x_1 - x_2}{x_3 - x_4}.$$

In Example 5.32, it is shown that T is invariant with respect to $\mathcal{G}_{ls}^{(4)}$.

Consider two elements of $\mathcal{X}$, $x = (1, 0, 1, 0)$ and $\tilde{x} = (1, 0, 2, 1)$. Note that since for $a > 0$,

$$(1, 0, 2, 1) - a(1, 0, 1, 0) = (1 - a, 0, 2 - a, 1)$$

there does not exist an a such that the elements of $x - a\tilde{x}$ are all equal. Hence, x and $\tilde{x}$ lie on different orbits; see Example 5.30. Since $T(x) = T(\tilde{x}) = 1$, it follows from part (ii) of Theorem 5.6 that T is not a maximal invariant. □

Equivariance

Consider a group $\mathcal{G}$ acting on a set $\mathcal{X}$ and let T denote a statistic, $T : \mathcal{X} \to \mathcal{Y}$ for some set $\mathcal{Y}$. Suppose that $\mathcal{G}$ also acts on $\mathcal{Y}$. The statistic T is said to be *equivariant* if for each $g \in \mathcal{G}$,

$$T(gx) = gT(x), \quad x \in \mathcal{X}.$$

Note that two different applications of the transformation g are being used in this expression: gx refers to the action of g on $\mathcal{X}$, while $gT(x)$ refers to the action of g on $\mathcal{Y}$.

Equivariance is an important concept in statistics. For instance, consider a transformation model for a random variable X, with respect to a group of transformations $\mathcal{G}$; let $\mathcal{X}$ denote the range of X and let Θ denote the parameter space. Let T denote an *estimator* of θ, a function $T : \mathcal{X} \to \Theta$. Hence, if $X = x$ is observed, we estimate the value of θ to be $T(x)$. The estimator is equivariant if the estimate corresponding to gx, $g \in \mathcal{G}$, is $gT(x)$.

***Example* 5.35 (*Estimation of a location parameter*).** Let X denote a real-valued random variable with an absolutely continuous distribution with density p_0 satisfying $p_0(x) > 0$, $x \in \mathbf{R}$. Consider the location model based on p_0 consisting of the class of distributions of gX, $g \in \mathcal{G}_1^{(1)}$ where $\mathcal{G}_1^{(1)}$ denotes the group of location transformations on $\mathbf{R}$.

Suppose X is distributed according to p_0 and let $g = \theta$ denote an element of $\mathcal{G}_1^{(1)}$. Then the distribution of gX is absolutely continuous with density function

$$p(x; \theta) = p_0(x - \theta), \quad -\infty < x < \infty.$$

The model given by

$$\{p(\cdot; \theta): \ -\infty < \theta < \infty\}$$

is a transformation model with respect to $\mathcal{G}_1^{(1)}$. For $g \in \mathcal{G}_1^{(1)}$ and $\theta \in \Theta = \mathbf{R}$, $g\theta = \theta + g$.

Now consider independent, identically distributed random variables $X_1, \ldots, X_n$, each distributed according to an absolutely continuous distribution with density function of the form $p(\cdot; \theta)$, as given above. The density for $(X_1, \ldots, X_n)$ is of the form

$$\prod_{j=1}^n p_0(x_j - \theta), \quad -\infty < x_j < \infty, \ i = j, \ldots, n.$$

and the model for $(X_1, \ldots, X_n)$ is a transformation model with respect to $\mathcal{G}_1$ and $g\theta = \theta + g$.

Now consider T, an estimator of the location parameter θ. The estimator is equivariant if, for any $b \in \mathbf{R}$,

$$T(x + b 1_n) = T(x) + b;$$

that is, T is equivariant if adding a constant b to each observation shifts the estimate of θ by b. □

Theorem 5.7. *Consider a space $\mathcal{X}$ and let $\mathcal{G}$ denote a group acting on $\mathcal{X}$. If a statistic T is equivariant then for each $x_1, x_2 \in \mathcal{X}$,*

$$T(x_1) = T(x_2) \ \textit{implies that} \ \ T(gx_1) = T(gx_2) \ \ \textit{for all} \ \ g \in \mathcal{G}.$$

Conversely, if

$$T(x_1) = T(x_2) \ \textit{implies that} \ \ T(gx_1) = T(gx_2) \ \ \textit{for all} \ \ g \in \mathcal{G},$$

then the action of $\mathcal{G}$ on $\mathcal{Y}$, the range of $T(X)$, may be defined so that T is equivariant.

Proof. Let T be an equivariant statistic and suppose $T(x_1) = T(x_2)$. Let $g \in \mathcal{G}$. Then, by the definition of equivariance,

$$T(gx_i) = gT(x_i), \quad i = 1, 2.$$

Hence,

$$T(gx_1) = gT(x_1) = gT(x_2) = T(gx_2).$$

Now suppose that $T(gx_1) = T(gx_2)$ for all $g \in \mathcal{G}$ whenever $T(x_1) = T(x_2)$. For $g \in \mathcal{G}$ and $x \in \mathcal{X}$, define $gT(x) = T(gx)$. Since $T(x_1) = T(x_2)$ implies $T(gx_1) = T(gx_2)$, $gT(x)$ is well defined. It remains to verify that (T1) and (T2) are satisfied.

Note

$$eT(x) = T(ex) = T(x),$$

verifying (T1) and that

$$g_1(g_2 T(x)) = g_1 T(g_2 x) = T(g_1 g_2 x) = T((g_1 g_2)x) = (g_1 g_2)T(x),$$

verifying (T2). The result follows. ■

Based on Theorem 5.7, it is tempting to conclude that T is equivariant if and only if $T(x_1) = T(x_2)$ implies that $T(gx_1) = T(gx_2)$. Note, however, that this statement does not require any specification of the action of $\mathcal{G}$ on $T(\mathcal{X})$. Theorem 5.7 states that there is a definition of the action of $\mathcal{G}$ such that T is equivariant. The following simple example illustrates this point.

Example 5.36. Let $\mathcal{X} = \mathbf{R}$ and let $\mathcal{G} = \mathbf{R}^+$. For $g \in \mathcal{G}$ define gx to be multiplication of g and x. Let $T(x) = |x|$. Clearly, $T(x_1) = T(x_2)$ implies that $T(gx_1) = T(gx_2)$. However, equivariance of T depends on how the action of $\mathcal{G}$ on $T(\mathcal{X})$ is defined. Suppose that $gT(x) = T(x)/g$. Then T is not equivariant since $T(gx) = |gx|$ while $gT(x) = |x|/g$. However, we may define the action of $\mathcal{G}$ on $T(\mathcal{X})$ in a way that makes T equivariant. In particular, we may use the definition $gT(x) = T(gx)$ so that, for $y \in T(\mathcal{X})$, gy denotes multiplication of g and y. $\square$

Consider a random variable X with range $\mathcal{X}$ and distribution in the set

$$\mathcal{P} = \{P(\cdot; \theta) : \theta \in \Theta\}.$$

Let $\mathcal{G}$ denote a group of transformations such that $\mathcal{P}$ is invariant with respect to $\mathcal{G}$ and that $\mathcal{G}$ and Θ are isomorphic.

Let T_1 denote a maximal invariant statistic. Then, as discussed above, $T_1(x)$ indicates the orbit on which x resides. However, T_1 does not completely describe the value of x because it provides no information regarding the location of x on its orbit. Let T_2 denote an equivariant statistic with range equal to Θ; hence, the action of a transformation on the range of T_2 is the same as the action on $\mathcal{G}$. Then $T_2(x)$ indicates the position of x on its orbit. The following theorem shows that $T_1(x)$ and $T_2(x)$ together are equivalant to x.

Theorem 5.8. *Let X denote a random variable with range $\mathcal{X}$ and suppose that the distribution of X is an element of*

$$\mathcal{P} = \{P(\cdot; \theta) : \theta \in \Theta\}.$$

Let $\mathcal{G}$ denote a group of transformations from $\mathcal{X}$ to $\mathcal{X}$ and suppose that $\mathcal{P}$ is invariant with respect to $\mathcal{G}$. Suppose that $\mathcal{G}$ and Θ are isomorphic.

Let $T : \mathcal{X} \to \mathbf{R}^m$, $T = (T_1, T_2)$, denote a statistic such that T_1 is a maximal invariant and T_2 is an equivariant statistic with range Θ. Then T is a one-to-one function on $\mathcal{X}$.

Proof.   The theorem holds provided that $T(x_1) = T(x_2)$ if and only if $x_1 = x_2$. Clearly, $x_1 = x_2$ implies that $T(x_1) = T(x_2)$; hence, assume that $T(x_1) = T(x_2)$.

Since $T_1(x_1) = T_1(x_2)$, x_1 and x_2 lie on the same orbit and, hence, there exists $g \in \mathcal{G}$ such that $x_2 = gx_1$. By the equivariance of T_2,

$$T_2(x_1) = T_2(x_2) = T_2(gx_1) = gT_2(x_1).$$

Note that $T(x_1)$ may be viewed as an element of $\mathcal{G}$. Let $\theta_1 = T_2(x_1)^{-1}$ and let θ_e denote the identity element of Θ. Then

$$\theta_e = T_2(x_1)\theta_1 = gT_2(x_1)\theta_1$$

so that

$$\theta_e = gT_2(x_1)\theta_1 = g.$$

Hence, $g = e$, the identity element of $\mathcal{G}$, and $x_1 = x_2$.   ∎

Therefore, under the conditions of Theorem 5.8, a random variable X may be written as $(T_1(X), T_2(X))$ where the distribution of $T_1(X)$ does not depend on θ and the distribution of $T_2(X)$ varies with θ in an equivariant manner.

Example* 5.37 *(Location model). As in Example 5.35, let $(X_1, \ldots, X_n)$ denote a random vector with density function of the form

$$\prod_{j=1}^{n} p_0(x_j - \theta), \quad x = (x_1, \ldots, x_n) \in \mathbf{R}^n$$

where p_0 is a density function on $\mathbf{R}$. Consider the model corresponding to $\theta \in \Theta = \mathbf{R}$. This model is invariant under the group of location transformations, as described in Example 5.25.

In Example 5.33 it is shown that the statistic

$$T_1(x) = (x_1 - \bar{x}, \ldots, x_n - \bar{x})^T, \quad \bar{x} = \frac{1}{n} \sum_{j=1}^{n} x_j,$$

is a maximal invariant; the statistic $T_2(x) = \bar{x}$ is equivariant, since $T_2(x + \theta 1_n) = T_2(x) + \theta$, and the range of T_2 is Θ. Hence, an observation $x \in \mathbf{R}^n$ may be described by $\bar{x}$, together with the residual vector $T_1(x)$.   □

Under the conditions of Theorem 5.8, the random variable X is equivalent to a maximal invariant statistic T_1 and an equivariant statistic T_2. It is important to note that these statistics are not unique, even taking into account the equivalence of statistics. Let $T_2(X)$ and $\tilde{T}_2(X)$ denote two equivariant statistics and let $h(x) = T_2(x)^{-1}\tilde{T}_2(x)$. Then, for any $g \in \mathcal{G}$,

$$h(gx) = T_2(gx)^{-1}\tilde{T}_2(gx) = g^{-1}T_2(x)^{-1}g\tilde{T}_2(x) = gg^{-1}T_2^{-1}\tilde{T}_2(x) = h(x).$$

It follows $T_2(X)^{-1}\tilde{T}_2(X)$ is an invariant statistic and, hence, a function of $T_1(x)$. That is, $\tilde{T}_2(X)$ is not a function of $T_2(X)$ alone.

These points are illustrated in the following example.

Example* 5.38 *(Exponential random variables). As in Example 5.28, let $X = (X_1, \ldots, X_n)$ where $X_1, \ldots, X_n$ are independent, identically distributed random variables, each with an exponential distribution with mean θ, $\theta \in \Theta = \mathbf{R}^+$. As shown in Example 5.28, this model is invariant under the group of scale transformations. Here $\mathcal{X} = (\mathbf{R}^+)^n$ and $x, \tilde{x} \in \mathcal{X}$ are in the same orbit if $x = a\tilde{x}$ for some $a > 0$.

For $x = (x_1, \ldots, x_n) \in \mathcal{X}$, let $T_1(x) = (x_2/x_1, x_3/x_1, \ldots, x_n/x_1)$. Clearly, T_1 is invariant and it is easy to see that if $x = a\tilde{x}$ for some $a > 0$, then $T_1(x) = T_1(\tilde{x})$; hence, $T_1(X)$ is a maximal invariant statistic.

Let

$$T_2(x) = \frac{1}{n} \sum_{j=1}^{n} x_j.$$

Note that $T_2(x)$ takes values in Θ and, for $a > 0$, $T_2(ax) = aT_2(x)$; hence, $T_2(X)$ is an equivariant statistic with range Θ. It now follows from Theorem 5.8 that X is equivalent to $(T_1(X), T_2(X))$. This can be verified directly by noting that

$$X_1 = \frac{nT_2(X)}{(1 + \sum_{j=2}^{n} X_j/X_1)}$$

and $X_j = X_1(X_j/X_1)$, $j = 2, \ldots, n$, so that X is a function of $(T_1(X), T_2(X))$.

As noted above, the statistics $T_1(X)$ and $T_2(X)$ used here are not unique; for instance, consider $\tilde{T}_2(X) = X_1$. Clearly X_1 is an equivariant statistic with range Θ so that X is equivalent to $(T_1(X), \tilde{T}_2(X))$. Also,

$$\tilde{T}_2(X)^{-1} T_2(X) = \frac{T_2(X)}{\tilde{T}_2(X)} = \frac{1}{n} \frac{\sum_{j=1}^{n} X_j}{X_1} = \frac{1}{n} \left(1 + \sum_{j=2}^{n} \frac{X_j}{X_1} \right)$$

is a function of $T_1(X)$. $\square$

5.7 Exercises

5.1 Suppose that n computer CPU cards are tested by applying power to the cards until failure. Let $Y_1, \ldots, Y_n$ denote the failure times of the cards. Suppose that, based on prior experience, it is believed that it is reasonable to assume that each Y_j has an exponential distribution with mean λ and that the failure times of different cards are independent. Give the model for $Y_1, \ldots, Y_n$ by specifying the model function, the parameter, and the parameter space. Is the parameter identifiable?

5.2 In the scenario considered in Exercise 5.1, suppose that testing is halted at time c so that only those failure times less than or equal to c are observed; here c is a known positive value. For card j, $j = 1, \ldots, n$, we record X_j, the time at which testing is stopped, and a variable D_j such that $D_j = 1$ if a failure is observed and $D_j = 0$ if testing is stopped because time c is reached. Hence, if $D_j = 0$ then $X_j = c$. Give the model for $(X_1, D_1), \ldots, (X_n, D_n)$, including the parameter and the parameter space. Is the parameter identifiable?

5.3 Let X and Y denote independent random variables. Suppose that

$$\Pr(X = 1) = 1 - \Pr(X = 0) = \lambda, \quad 0 < \lambda < 1;$$

if $X = 1$, then Y has a normal distribution with mean μ_1 and standard deviation σ, while if $X = 0$, Y has a normal distribution with mean μ_0 and standard deviation σ. Here μ_0 and μ_1 each take any real value, while $\sigma > 0$. Let $Y_1, \ldots, Y_n$ denote independent, identically distributed random variables such that Y_1 has the distribution of Y. Give the model for $Y_1, \ldots, Y_n$ by specifying the model function, the parameter, and the parameter space. Is the parameter identifiable?

5.4 As in Exercise 5.1, suppose that n CPU cards are tested. Suppose that for each card, there is a probability π that the card is defective so that it fails immediately. Assume that, if a card does not fail immediately, then its failure time follows an exponential distribution and that the failure times of different cards are independent. Let R denote the number of cards that are defective; let $Y_1, \ldots, Y_{n-R}$ denote the failure times of those cards that are not defective. Give the model for $R, Y_1, \ldots, Y_{n-R}$, along with the parameter and the parameter space. Is the parameter identifiable?

5.5 Consider a parametric model $\{P_\theta : \theta \in \Theta\}$ for a random variable X. Such a model is said to be *complete* if, for a real-valued function g,

$$E[g(X); \theta] = 0, \quad \theta \in \Theta,$$

implies

$$\Pr[g(X) = 0; \theta] = 1, \quad \theta \in \Theta.$$

For each of the models given below, determine if the model is complete.

(a) P_θ is the uniform distribution on $(0, \theta)$ and $\Theta = (0, \infty)$

(b) P_θ is the absolutely continuous distribution with density

$$\frac{2\theta^\theta}{\Gamma(\theta)} x^{2\theta-1} \exp(-\theta x^2), \quad x > 0,$$

and $\Theta = (0, \infty)$

(c) P_θ is the binomial distribution with frequency function of the form

$$p(x; \theta) = \binom{3}{x} \theta^x (1 - \theta)^{3-x}, \quad x = 0, 1, 2$$

and $\Theta = (0, 1)$.

5.6 Consider the family of absolutely continuous distributions with density functions

$$p(y; \theta) = \frac{\Gamma(\alpha + \beta)}{\Gamma(\alpha)\Gamma(\beta)} y^{\alpha-1}(1 - y)^{\beta-1}, \quad 0 < y < 1,$$

where $\theta = (\alpha, \beta)$ and $\Theta = (0, \infty) \times (0, \infty)$. This is known as the family of *beta distributions*. Show that this is a two-parameter exponential family by putting $p(y; \theta)$ in the form (5.1). Find the functions c, T, A, h and the set $\mathcal{Y}$.

5.7 Consider the family of gamma distributions described in Example 3.4 and Exercise 4.1. Show that this is a two-parameter exponential family. Find the functions c, T, A, h and the set $\mathcal{Y}$ in the representation (5.1).

5.8 Consider the family of absolutely continuous distributions with density function

$$p(y; \theta) = \frac{\theta}{y^{\theta+1}}, \quad y > 1$$

where $\theta > 0$. Show that this is a one-parameter exponential family of distributions and write p in the form (5.1), giving explicit forms for $c, T, A,$ and h.

5.9 Consider the family of discrete distributions with density function

$$p(y; \theta) = \binom{\theta_1 + y - 1}{y} \theta_2^{\theta_1}(1 - \theta_2)^y, \quad y = 0, 1, \ldots$$

where $\theta = (\theta_1, \theta_2) \in (0, \infty) \times (0, 1)$. Is this an exponential family of distributions? If so, write p in the form (5.1), giving explicit forms for $c, T, A,$ and h.

5.10 Let X denote a random variable with range $\mathcal{X}$ such that the set of possible distributions of X is a one-parameter exponential family. Let A denote a subset of $\mathcal{X}$ and suppose that X is only observed if $X \in A$; let Y denote the value of X given that it is observed. Is the family of distributions of Y a one-parameter exponential family?

5.11 Consider the family of absolutely continuous distributions with density functions

$$p(y; \theta) = \frac{\sqrt{\lambda}}{(2\pi y^3)^{\frac{1}{2}}} \exp\left\{ -\frac{\phi^2}{2\lambda} y + \phi - \frac{\lambda}{2y} \right\}, \quad y > 0$$

where $\theta = (\phi, \lambda) \in (\mathbf{R}^+)^2$. Is this an exponential family of distributions? If so, write p in the form (5.1), giving explicit forms for $c, T, A,$ and h.

5.12 Let Y denote a random variable with an absolutely continuous distribution with density $p(y; \theta)$, where p is given in Exercise 5.11. Find the cumulant-generating function of $Y + 1/Y - 2$.

5.13 Let Y denote a random variable with density or frequency function of the form

$$\exp\{\eta T(y) - k(\eta)\}h(y)$$

where $\eta \in H \subset \mathbf{R}$. Fix $\eta_0 \in H$ and let s be such that $E\{\exp[sT(Y)]; \eta_0\} < \infty$. Find an expression for $\text{Cov}(T(Y), \exp[sT(Y)]; \eta_0)$ in terms of the function k.

5.14 A family of distributions that is closely related to the exponential family is the family of exponential dispersion models. Suppose that a scalar random variable X has a density of the form

$$p(x; \eta, \sigma^2) = \exp\{[\eta x - k(\eta)]/\sigma^2\}h(x; \sigma^2), \quad \eta \in H$$

where for each fixed value of $\sigma^2 > 0$ the density p satisfies the conditions of a one-parameter exponential family distribution and H is an open set. The set of density functions $\{p(\cdot; \eta, \sigma^2): \eta \in H, \sigma^2 > 0\}$ is said to be an exponential dispersion model.

(a) Find the cumulant-generating function of X.

(b) Suppose that a random variable Y has the density function $p(\cdot; \eta, 1)$, that is, it has the distribution as X except that σ^2 is known to be 1. Find the cumulants of X in terms of the cumulants of Y.

5.15 Suppose Y is a real-valued random variable with an absolutely continuous distribution with density function

$$p_Y(y; \eta) = \exp\{\eta y - k(\eta)\}h(y), \quad y \in \mathcal{Y},$$

where $\eta \in H \subset \mathbf{R}$, and X is a real-valued random variable with an absolutely continuous distribution with density function

$$p_X(x; \eta) = \exp\{\eta x - \tilde{k}(\eta)\}\tilde{h}(x), \quad x \in \mathcal{X},$$

where $\eta \in \mathcal{H}$. Show that:

(a) if $k = \tilde{k}$, then Y and X have the same distribution, that is, for each $\eta \in H$, $F_Y(\cdot; \eta) = F_X(\cdot; \eta)$ where F_Y and F_X denote the distribution functions of Y and X, respectively

(b) if $E(Y; \eta) = E(X; \eta)$ for all $\eta \in H$ then Y and X have the same distribution

(c) if $\text{Var}(Y; \eta) = \text{Var}(X; \eta)$ for all $\eta \in H$ then Y and X do not necessarily have the same distribution.

5.16 Let X denote a real-valued random variable with an absolutely continuous distribution with density function p. Suppose that the moment-generating function of X exists and is given by $M(t)$, $|t| < t_0$.

Let Y denote a real-valued random variable with an absolutely continuous distribution with density function of the form

$$p(y; \theta) = c(\theta) \exp(\theta y)p(y),$$

where $c(\cdot)$ is a function of θ.

(a) Find requirements on θ so that $p(\cdot; \theta)$ denotes a valid probability distribution. Call this set Θ.

(b) Find an expression for $c(\cdot)$ in terms of M.

(c) Show that $\{p(\cdot; \theta): \theta \in \Theta\}$ is a one-parameter exponential family of distributions.

(d) Find the moment-generating function corresponding to $p(\cdot; \theta)$ in terms of M.

5.17 Let $Y_1, Y_2, \ldots, Y_n$ denote independent, identically distributed random variables such that Y_1 has density $p(y; \theta)$ where p is of the form

$$p(y; \theta) = \exp\{\eta T(y) - k(\eta)\}h(y), \quad y \in \mathcal{Y}$$

and η takes values in the natural parameter space H. Let $S = \sum_{j=1}^{n} T(Y_j)$.

Let $A \equiv A(Y_1, \ldots, Y_n)$ denote a statistic such that for each $\eta \in H$ the moment-generating function of A, $M_A(t; \eta)$, exists for t in a neighborhood of 0 and let $M_S(t; \eta)$ denote the moment-generating function of S.

(a) Find an expression for the joint moment-generating function of (A, S),

$$M(t_1, t_2; \eta) = E_\eta(\exp\{t_1 A + t_2 S\})$$

in terms of M_A and M_S.

(b) Suppose that for a given value of η, η_0,

$$\frac{\partial}{\partial \eta} E(A; \eta)\Big|_{\eta = \eta_0} = 0.$$

Find an expression for $\text{Cov}(S, A; \eta_0)$.

5.18 Let $Y_1, \ldots, Y_n$ denote independent binomial random variables each with index m and success probability θ. As an alternative to this model, suppose that Y_j is a binomial random variable with index m and success probability ϕ where ϕ has a beta distribution. The beta distribution is an absolutely continuous distribution with density function

$$\frac{\Gamma(\alpha + \beta)}{\Gamma(\alpha)\Gamma(\beta)}\phi^{\alpha-1}(1 - \phi)^{\beta-1}, \quad 0 < \phi < 1$$

where $\alpha > 0$ and $\beta > 0$. The distribution of $Y_1, \ldots, Y_n$ is sometimes called the beta-binomial distribution.

(a) Find the density function of Y_j.

(b) Find the mean and variance of Y_j.

(c) Find the values of the parameters of the distribution for which the distribution reduces to the binomial distribution.

5.19 Let (Y_{j1}, Y_{j2}), $j = 1, 2, \ldots, n$, denote independent pairs of independent random variables such that, for given values of $\lambda_1, \ldots, \lambda_n$, Y_{j1} has an exponential distribution with mean ψ/λ_j and that Y_{j2} has an exponential distribution with mean $1/\lambda_j$. Suppose further that $\lambda_1, \ldots, \lambda_n$ are independent random variables, each distributed according to an exponential distribution with mean $1/\phi$, $\phi > 0$. Show that the pairs (Y_{j1}, Y_{j2}), $j = 1, \ldots, n$, are identically distributed and find their common density function.

5.20 Let λ and $X_1, \ldots, X_n$ denote random variables such that, given λ, $X_1, \ldots, X_n$ are independent and identically distributed. Show that $X_1, \ldots, X_n$ are exchangeable.

Suppose that, instead of being independent and identically distributed, $X_1, \ldots, X_n$ are only exchangeable given λ. Are $X_1, \ldots, X_n$ exchangeable unconditionally?

5.21 Consider a linear exponential family regression model, as discussed in Example 5.22. Find an expression for the mean and variance of $T(Y_j)$.

5.22 Show that $\mathcal{G}_s$, $\mathcal{G}_l$, and $\mathcal{G}_{ls}$ each satisfy (G1)–(G4) and (T1) and (T2).

5.23 Let $\mathcal{P}$ denote the family of normal distributions with mean θ and standard deviation θ, where $\theta > 0$. Is this model invariant with respect to the group of location transformations? Is it invariant with respect to the group of scale transformations?

5.24 Let $Y_1, \ldots, Y_n$ denote independent, identically distributed random variables, each uniformly distributed on the interval (θ_1, θ_2), $\theta_1 < \theta_2$.

(a) Show that this is a transformation model and identify the group of transformations. Show the correspondence between the parameter space and the transformations.

(b) Find a maximal invariant statistic.

5.25 Let X denote an n-dimensional random vector and, for $g \in \mathbf{R}^n$, define the transformation

$$gX = X + g.$$

Let $\mathcal{G}$ denote the set of such transformations with g restricted to a set $A \subset \mathbf{R}^n$ and define $g_1 g_2$ to be vector addition, $g_1 + g_2$. Find conditions on A such that $\mathcal{G}$ is a group and that (T1) and (T2) are satisfied.

5.26 Consider the set of binomial distributions with frequency function of the form

$$\binom{n}{x} \theta^x (1 - \theta)^{n-x}, \quad x = 0, 1, \ldots, n$$

where n is fixed and $0 < \theta < 1$. For $x \in \mathcal{X} \equiv \{0, 1, \ldots, n\}$, define transformations g_0 and g_1 by $g_0 x = x$ and $g_1 x = n - x$.

(a) Define $g_0 g_1, g_1 g_0, g_0 g_0,$ and $g_1 g_1$ so that $\{g_0, g_1\}$, together with the binary operation defined by these values, is a group satisfying (T1) and (T2). Call this group $\mathcal{G}$.

(b) Show that the set of binomial distributions described above is invariant with respect to $\mathcal{G}$. Find $g_0 \theta$ and $g_1 \theta$ for $\theta \in (0, 1)$.

(c) Is the set of binomial distributions a transformation model with respect to $\mathcal{G}$?

(d) Let $x \in \mathcal{X}$. Find the orbit of x.

(e) Find a maximal invariant statistic.

5.27 Suppose that the random vector $(X_1, \ldots, X_n)$ has an absolutely continuous distribution with density function of the form

$$\prod_{j=1}^{n} p_0(x_j - \theta), \quad -\infty < x_j < \infty, \quad j = 1, \ldots, n,$$

where $\theta \in \mathbf{R}$ and p_0 is a density function on the real line. Recall that this family of distributions is invariant with respect to $\mathcal{G}_1$; see Example 5.35.

Let T denote an equivariant statistic. Show that the mean and variance of T do not depend on θ. Let S denote an invariant statistic. What can be said about the dependence of the mean and variance of S on θ?

5.28 Let $\mathcal{X} = \mathbf{R}^{n+m}$, let v denote the vector in $\mathcal{X}$ with the first n elements equal to 1 and the remaining elements equal to 0, and let u denote the vector in $\mathcal{X}$ with the first n elements equal to 0 and the remaining elements equal to 1. Let $\mathcal{G} = \mathbf{R}^2$ and for an element $g = (a, b) \in \mathcal{G}$, define the transformation

$$gx = x + av + bu, \quad x \in \mathcal{X}.$$

For each of the models given below, either show that the model is invariant with respect to $\mathcal{G}$ or show that it is not invariant with respect to $\mathcal{G}$. If a model is invariant with respect to $\mathcal{G}$, describe the action of $\mathcal{G}$ on the parameter space Θ; that is, for $g \in \mathcal{G}$ and $\theta \in \Theta$, give $g\theta$.

(a) $X_1, X_2, \ldots, X_{n+m}$ are independent, identically distributed random variables, each with a normal distribution with mean θ, $-\infty < \theta < \infty$.

(b) $X_1, X_2, \ldots, X_{n+m}$ are independent random variables, such that $X_1, \ldots, X_n$ each have a normal distribution with mean μ_1 and $X_{n+1}, \ldots, X_{n+m}$ each have a normal distribution with mean μ_2; here $\theta = (\mu_1, \mu_2) \in \mathbf{R}^2$.

(c) $X_1, X_2, \ldots, X_{n+m}$ are independent random variables such that, for each $j = 1, \ldots, n + m$, X_j has a normal distribution with mean μ_j; here $\theta = (\mu_1, \ldots, \mu_{n+m}) \in \mathbf{R}^{n+m}$.

(d) $X_1, X_2, \ldots, X_{n+m}$ are independent random variables such that $X_1, \ldots, X_n$ each have an exponential distribution with mean μ_1 and $X_{n+1}, \ldots, X_{n+m}$ each have an exponential distribution with mean μ_2; here $\theta = (\mu_1, \mu_2) \in (\mathbf{R}^+)^2$.

5.29 Consider the model considered in Exercise 5.28. For each of the statistics given below, either show that the statistic is invariant with respect to $\mathcal{G}$ or show that it is not invariant with respect to $\mathcal{G}$. If a statistic is invariant, determine if it is a maximal invariant.

(a) $T(x) = \sum_{j=1}^{n} x_j / n - \sum_{j=n+1}^{n+m} x_j / m$

(b) $T(x) = x_1 - \sum_{j=1}^{n} x_j / n$

(c) $T(x) = (x_2 - x_1, x_3 - x_1, \ldots, x_n - x_1, x_{n+2} - x_{n+1}, \ldots, x_{n+m} - x_{n+1})$

(d) $T(x) = (x_1 - \bar{x}, x_2 - \bar{x}, \ldots, x_{n+m} - \bar{x})$ where $\bar{x} = \sum_{j=1}^{n+m} x_j/(n+m)$.

5.30 Let $F_1, \ldots, F_n$ denote absolutely continuous distribution functions on the real line, where n is a fixed integer. Let Θ denote the set of all permutations of $(1, \ldots, n)$ and consider the model P for an n-dimensional random vector $X = (X_1, \ldots, X_n)$ consisting of distribution functions of the form

$$F(x_1, \ldots, x_n; \theta) = F_{\theta_1}(x_1) \cdots F_{\theta_n}(x_n),$$

$\theta = (\theta_1, \ldots, \theta_n) \in \Theta$. The sample space of X, $\mathcal{X}$, may be taken to be the subset of $\mathbf{R}^n$ in which, for $x = (x_1, \ldots, x_n) \in \mathcal{X}$, $x_1, \ldots, x_n$ are unique.

Let $\mathcal{G}$ denote the set of all permutations of $(1, \ldots, n)$ and, for $g \in \mathcal{G}$ and $x \in \mathcal{X}$, define

$$gx = (x_{g_1}, \ldots, x_{g_n}).$$

(a) Show that P is invariant with respect to $\mathcal{G}$.

(b) For $g \in \mathcal{G}$, describe the action of g on $\theta \in \Theta$. That is, by part (a), if X has parameter $\theta \in \Theta$, then gX has parameter $\theta_1 \in \Theta$; describe θ_1 in terms of g and θ.

(c) Let $x \in \mathcal{X}$. Describe the orbit of x. In particular, for the case $n = 3$, give the orbit of $(x_1, x_2, x_3) \in X$.

(d) Let $T(x) = (x_{(1)}, \ldots, x_{(n)})$ denote the vector of order statistics corresponding to a point $x \in \mathcal{X}$. Show that T is an invariant statistic with respect to $\mathcal{G}$.

(e) Is the statistic T defined in part (d) a maximal invariant statistic?

(f) For $x \in \mathcal{X}$, let $R(x)$ denote the vector of ranks of $x = (x_1, \ldots, x_n)$. Note that R takes values in Θ. Is R an equivariant statistic with respect to the action of g on Θ?

5.8 Suggestions for Further Reading

Statistical models and identifiability are discussed in many books on statistical theory; see, for example, Bickel and Doksum (2001, Chapter 1) and Gourieroux and Monfort (1989, Chapters 1 and 3). In the approach used here, the parameter is either identified or it is not; Manski (2003) considers the concept of *partial identification.*

Exponential families are discussed in Bickel and Doksum (2001, Section 1.6), Casella and Berger (2002, Section 3.4), and Pace and Salvan (1997, Chapter 5). Comprehensive treatments of exponential family models are given by Barndorff-Nielsen (1978) and Brown (1988). Exponential dispersion models, considered briefly in Exercise 5.14 are considered in Pace and Salvan (1997, Chapter 6). Schervish (1995, Chapter 8) contains a detailed treatment of hierarchical models and the statistical inference in these models; see also Casella and Berger (2002, Section 4.4).

Regression models play a central role in applied statistics. See, for example, Casella and Berger (2002, Chapters 11 and 12). Rao and Toutenburg (1999) is a comprehensive reference on statistical inference in a wide range of linear regression models. McCullagh and Nelder (1989) considers a general class of regression models that are very useful in applications. Transformation models and equivariance and invariance are discussed in Pace and Salvan (1997, Chapter 7) and Schervish (1995, Chapter 6); Eaton (1988) contains a detailed treatment of these topics.

6

Stochastic Processes

6.1 Introduction

Statistical methods are often applied to data exhibiting dependence; for instance, we may observe random variables $X_1, X_2, \ldots$ describing the properties of a system as it evolves in time. In these cases, a model is needed for the dependence structure of the observations. In the present chapter, we consider some probability models used for dependent data. The usual point of view is that the X_j are ordered in time, although this is not always the case.

In general, we are concerned with a collection of random variables

$$\{X_t \colon t \in \mathcal{T}\}.$$

For each t, X_t is a random variable; although the X_t may be vector-valued, here we only consider the case in which the X_t are real-valued. Such a collection of random variables is called a *stochastic process*. The index set $\mathcal{T}$ is often either a countable set, in which case we refer to $\{X_t \colon t \in \mathcal{T}\}$ as a *discrete time process*, or an interval, possibly infinite, in which case we refer to $\{X_t \colon t \in \mathcal{T}\}$ as a *continuous time process*.

In this chapter, we consider two cases, discrete time processes in which

$$\mathcal{T} = \mathbf{Z} \equiv \{0, 1, 2, \ldots\}$$

and continuous time processes with $\mathcal{T} = [0, \infty)$. Note that, in both cases, the starting point of $t = 0$ is arbitrary and other starting points, such as $t = 1$, could be used if convenient; this is sometimes done in the examples. Also, in some cases, it is mathematically convenient to assume that the processes have an infinite history so that

$$\mathcal{T} = \{\ldots, -1, 0, 1, \ldots\}$$

or $\mathcal{T} = (-\infty, \infty)$. We assume that there is a set $\mathcal{X} \subset \mathbf{R}$ such that, for each $t \in \mathcal{T}$, the random variable X_t takes values in $\mathcal{X}$; $\mathcal{X}$ is called the *state space* of the process.

Associated with each process $\{X_t \colon t \in \mathcal{T}\}$ is the set of *finite-dimensional distributions*. Fix a positive integer n and let $t_1, \ldots, t_n$ be elements of $\mathcal{T}$. Then the distribution of $(X_{t_1}, \ldots, X_{t_n})$ is finite-dimensional and may be handled by standard methods. The distributions of all such vectors for all choices of n and $t_1, \ldots, t_n$ are called the finite-dimensional distributions of the process. In general, the finite-dimensional distributions do not completely determine the distribution of the process. However, for discrete time processes it is true that finite-dimensional distributions determine the distribution of the process. That is, if for any n and any $t_1, \ldots, t_n$ in $\mathbf{Z}$,

$$(X_{t_1}, \ldots, X_{t_n}) \quad \text{and} \quad (Y_{t_1}, \ldots, Y_{t_n})$$

have the same distribution, then the two processes have the same distribution. The proofs of these results require rather sophisticated methods of advanced probability theory; see, for example, Port (1994, Chapter 16) and Billingsley (1995, Chapter 7).

6.2   Discrete Time Stationary Processes

Perhaps the simplest type of discrete time process is one in which $X_0, X_1, X_2, \ldots$ are independent, identically distributed random variables so that, for each i and j, X_i and X_j have the same distribution. A generalization of this idea is a *stationary process* in which, for any $n = 1, 2, \ldots$ and any integers $t_1, \ldots, t_n$,

$$(X_{t_1}, \ldots, X_{t_n}) \quad \text{and} \quad (X_{t_1+h}, \ldots, X_{t_n+h})$$

have the same distribution for any $h = 1, 2, \ldots.$

Example 6.1 (*Exchangeable random variables*). Let $X_0, X_1, \ldots$ denote a sequence of exchangeable random variables. Then, for any $n = 0, 1, \ldots, X_0, X_1, \ldots, X_n$ are exchangeable. It follows from Theorem 2.8 that any subset of $X_0, X_1, \ldots$ of size m has the same marginal distribution as any other subset of $X_0, X_1, \ldots$ of size m. Clearly, this implies that the condition for stationarity is satisfied so that the process $\{X_t \colon t \in \mathbf{Z}\}$ is stationary. $\square$

The following result gives a necessary and sufficient condition for stationarity that is often easier to use than the definition.

Theorem 6.1. *Let* $\{X_t \colon t \in \mathbf{Z}\}$ *denote a discrete time process and define*

$$Y_t = X_{t+1}, \quad t = 0, 1, 2, \ldots.$$

Then $\{X_t \colon t \in \mathbf{Z}\}$ *is stationary if and only if* $\{Y_t \colon t \in \mathbf{Z}\}$ *has the same distribution as* $\{X_t \colon t \in \mathbf{Z}\}.$

Proof.   Clearly if $\{X_t \colon t \in \mathbf{Z}\}$ is stationary then $\{X_t \colon t \in \mathbf{Z}\}$ and $\{Y_t \colon t \in \mathbf{Z}\}$ have the same distribution. Hence, assume that $\{X_t \colon t \in \mathbf{Z}\}$ and $\{Y_t \colon t \in \mathbf{Z}\}$ have the same distribution.
   Fix $n = 1, 2, \ldots$ and $t_1, \ldots t_n$ in $\mathbf{Z}$. Then, by assumption,

$$(X_{t_1}, \ldots, X_{t_n}) \overset{\mathcal{D}}{=} (X_{t_1+1}, \ldots, X_{t_n+1}).$$

Here the symbol $\overset{\mathcal{D}}{=}$ is used to indicate that two random variables have the same distribution. Hence,

$$\{X_t \colon t \in T\} \overset{\mathcal{D}}{=} \{X_{t+1}, t \in \mathbf{Z}\}.$$

It then follows that

$$\{X_{t+1} \colon t \in \mathbf{Z}\} \overset{\mathcal{D}}{=} \{Y_{t+1} \colon t \in \mathbf{Z}\}.$$

That is,

$$(X_{t_1}, \ldots, X_{t_n}) \overset{\mathcal{D}}{=} (X_{t_1+2}, \ldots, X_{t_n+2}).$$

Continuing in this way shows that, for any $h = 0, 1, \ldots,$

$$(X_{t_1}, \ldots, X_{t_n}) \overset{\mathcal{D}}{=} (X_{t_1+h}, \ldots, X_{t_n+h}),$$

proving the result. ∎

***Example* 6.2** (*First-order autoregressive process*). Let $Z_0, Z_1, Z_2, \ldots$ denote a sequence of independent, identically distributed random variables, each with a normal distribution with mean 0 and variance σ^2. Let $-1 < \rho < 1$ and define

$$X_t = \begin{cases} \frac{1}{\sqrt{(1-\rho^2)}} Z_0 & \text{if } t = 0 \\ \rho X_{t-1} + Z_t & \text{if } t = 1, 2, 3, \ldots \end{cases}.$$

The stochastic process $\{X_t : t \in \mathbf{Z}\}$ is called a *first-order autoregressive* process.

For each $t = 0, 1, 2, \ldots$, let $Y_t = X_{t+1}$. Define

$$\tilde{Z}_t = \begin{cases} \rho Z_0 + \sqrt{(1 - \rho^2)} Z_1 & \text{if } t = 0 \\ Z_{t+1} & \text{if } t = 1, 2, \ldots \end{cases};$$

then

$$Y_t = \begin{cases} \frac{1}{\sqrt{(1-\rho^2)}} \tilde{Z}_0 & \text{if } t = 0 \\ \rho Y_{t-1} + \tilde{Z}_t & \text{if } t = 1, 2, 3, \ldots \end{cases}.$$

It follows that the process $\{X_t : t \in \mathbf{Z}\}$ is stationary provided that $(\tilde{Z}_0, \tilde{Z}_1, \ldots)$ has the same distribution as $(Z_0, Z_1, \ldots)$. Clearly, $\tilde{Z}_0, \tilde{Z}_1, \ldots$ are independent and each of $\tilde{Z}_1, \tilde{Z}_2, \ldots$ is normally distributed with mean 0 and standard deviation σ. It follows that $\{X_t : t \in \mathbf{Z}\}$ has the same distribution as $\{Y_t : t \in \mathbf{Z}\}$ and, hence, $\{X_t : t \in \mathbf{Z}\}$ is stationary, provided that $\tilde{Z}_0$ has a normal distribution with mean 0 and standard deviation σ.

Since Z_0 and Z_1 are independent, identically distributed random variables each with characteristic function

$$\exp\left(-\frac{1}{2}\sigma^2 t^2\right), \quad -\infty < t < \infty,$$

it follows that the characteristic function of $\tilde{Z}_0$ is

$$\exp\left(-\frac{1}{2}\sigma^2\rho^2 t^2\right) \exp\left(-\frac{1}{2}\sigma^2(1 - \rho^2)t^2\right) = \exp\left(-\frac{1}{2}\sigma^2 t^2\right).$$

It follows that $\{X_t : t \in \mathbf{Z}\}$ is stationary.

Figure 6.1 contains plots of four randomly generated first-order autoregressive processes based on $\rho = 0, 1/2, -1/2, 9/10$ and $\sigma^2 = 1$. Note that, in these plots, the processes are presented as continuous functions, rather than as points at integer values of t. These functions are constructed by taking the value at an integer t to be X_t and then using line segments to connect X_t and X_{t+1}. □

The following theorem shows that certain functions of a stationary process yield another stationary process.

***Theorem* 6.2.** *Suppose $\{X_t : t \in \mathbf{Z}\}$ is a stationary process. For each $t \in \mathbf{Z}$ define*

$$Y_t = f(X_t, X_{t+1}, \ldots)$$

where f is a real-valued function on $\mathcal{X}^\infty$. Then $\{Y_t : t \in \mathbf{Z}\}$ is stationary.

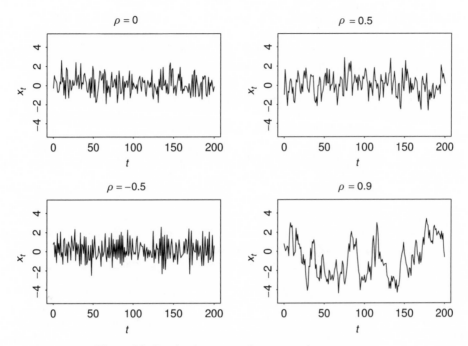

Figure 6.1. Randomly generated autoregressive processes.

Proof.   Let h denote a nonnegative integer and consider the event

$$A_h = \{Y_{1+h} \leq y_1, Y_{2+h} \leq y_2, \ldots\}.$$

We need to show that the probability of A_h does not depend on h. Note that

$$A_0 = \{f(X_1, X_2, \ldots,) \leq y_1, f(X_2, X_3, \ldots,) \leq y_2, \ldots\} = \{(X_1, X_2, \ldots) \in B\}$$

for some set B. Similarly, for $h = 1, 2, \ldots,$

$$A_h = \{(X_{1+h}, X_{2+h}, \ldots,) \in B\}.$$

Since the distribution of $(X_{1+h}, X_{2+h}, \ldots)$ is the same for all h, the result follows.   ∎

Example* 6.3 *(Moving differences).   Let $Z_0, Z_1, \ldots$ denote a sequence of independent, identically distributed random variables and define

$$X_t = Z_{t+1} - Z_t, \quad t = 0, 1, 2, \ldots.$$

It follows immediately from Theorem 6.2 that $\{X_t : t \in \mathbf{Z}\}$ is stationary. More generally, $\{X_t : t \in \mathbf{Z}\}$ is stationary provided only that $\{Z_t : t \in \mathbf{Z}\}$ is itself a stationary process.   □

Covariance-stationary processes

Consider a process $\{X_t : t \in \mathbf{Z}\}$ such that, for each $t \in \mathbf{Z}$, $\mathrm{E}(X_t^2) < \infty$. The *second-order properties* of this process are those that depend only on the *mean function*

$$\mu_t = \mathrm{E}(X_t), \quad t \in \mathbf{Z}$$

and the *covariance function*

$$K(s, t) = \mathrm{Cov}(X_t, X_s), \quad s, t \in \mathbf{Z}.$$

The process is said to be *second-order stationary*, or *covariance stationary*, if μ_{t+h} and $K(s+h, t+h)$ do not depend on h. Hence, $\{X_t : t \in \mathbf{Z}\}$ is covariance stationary provided that μ_t is constant, that is, does not depend on t, and $K(s, t)$ depends on s, t only through the difference $|s - t|$; in this case we write $\mu_t = \mu$ and $K(s, t) \equiv R(|s - t|)$ for some function R on $\mathbf{Z}$. The function R is known as the *autocovariance function* of the process; the *autocorrelation function* of the process is also useful and is given by

$$\rho(t) = R(t)/R(0), \quad t = 0, 1, \dots.$$

Example 6.4 (Moving differences). Let $\{Z_t : t \in \mathbf{Z}\}$ denote a covariance-stationary process and define

$$X_t = Z_{t+1} - Z_t, \quad t = 0, 1, \dots.$$

According to Example 6.3, $\{X_t : t \in \mathbf{Z}\}$ is stationary provided that $\{Z_t : t \in \mathbf{Z}\}$ is stationary; here we assume only that $\{Z_t : t \in \mathbf{Z}\}$ is covariance stationary.

Clearly, $E(X_t) = 0$ for all t. Let $\text{Cov}(Z_t, Z_s) = R_Z(|t - s|)$; then

$$\text{Cov}(X_t, X_{t+h}) = 2R_Z(|h|) - R_Z(|h - 1|) - R_Z(|h + 1|).$$

Since $|h - 1| = |-h + 1|$, it follows that $\text{Cov}(X_t, X_{t+h})$ depends on h only through $|h|$ so that $\{X_t : t \in \mathbf{Z}\}$ is covariance stationary. $\square$

Example 6.5 (Partial sums). Let $Z_0, Z_1, Z_2, \dots$ denote a sequence of independent, identically distributed random variables, each with mean 0 and standard deviation σ. Consider the process defined by

$$X_t = Z_0 + \cdots + Z_t, \quad t = 0, 1, \dots.$$

Then $E(X_t) = 0$ and $\text{Var}(X_t) = t\sigma^2$; hence, the process is not stationary.

The variance of the process can be stabilized by considering

$$Y_t = \frac{Z_0 + \cdots + Z_t}{\sqrt{(t + 1)}}, \quad t = 1, 2, \dots$$

which satisfies $E(Y_t) = 0$ and $\text{Var}(Y_t) = \sigma^2$ for all $t = 0, 1, 2, \dots$. However,

$$\text{Cov}(Y_t, Y_s) = \frac{\sigma^2}{\sqrt{[(s + 1)(t + 1)]}} \min(s + 1, t + 1)$$

so that $\{Y_t : t \in \mathbf{Z}\}$ is not covariance stationary. $\square$

6.3 Moving Average Processes

Let $\dots, \epsilon_{-1}, \epsilon_0, \epsilon_1, \dots$ denote a doubly infinite sequence of independent random variables such that, for each j, $E(\epsilon_j) = 0$ and $\text{Var}(\epsilon_j) = 1$ and let $\dots, \alpha_{-1}, \alpha_0, \alpha_1, \dots$ denote a doubly infinite sequence of constants. Consider the stochastic process $\{X_t : t \in \mathbf{Z}\}$ defined by

$$X_t = \sum_{j=-\infty}^{\infty} \alpha_j \epsilon_{t-j}, \quad t = 0, 1, \dots. \tag{6.1}$$

The process $\{X_t : t \in \mathbf{Z}\}$ is known as a *moving average process*. Two important special cases are the finite moving average process

$$X_t = \sum_{j=0}^{q} \alpha_j \epsilon_{t-j}, \quad t = 0, 1, \ldots,$$

where q is a fixed nonnegative integer, and the infinite moving average process,

$$X_t = \sum_{j=0}^{\infty} \alpha_j \epsilon_{t-j}, \quad t = 0, 1, \ldots.$$

In fact, it may be shown that a wide range of stationary processes have a representation of the form (6.1). We will not pursue this issue here; see, for example, Doob (1953).

Before proceeding we must clarify the exact meaning of (6.1). For each $n = 0, 1, \ldots$ define

$$X_{nt} = \sum_{j=-n}^{n} \alpha_j \epsilon_{t-j}, \quad t = 0, 1, \ldots.$$

Then, for each $t = 0, 1, \ldots, X_t$ is given by

$$X_t = \lim_{n \to \infty} X_{nt}. \tag{6.2}$$

Hence, we need a precise statement of the limiting operation in (6.2) and we must verify that the limit indeed exists.

The type of limit used in this context is a *limit in mean square*. Let $Y_0, Y_1, \ldots$ denote a sequence of real-valued random variables such that $E(Y_j^2) < \infty$, $j = 0, 1, \ldots$. We say that the sequence Y_n, $n = 0, 1, \ldots$, *converges in mean square* to a random variable Y if

$$\lim_{n \to \infty} E[(Y_n - Y)^2] = 0. \tag{6.3}$$

The following result gives some basic properties of this type of convergence; the proof is left as an exercise.

Theorem 6.3. *Let* $Y_0, Y_1, \ldots$ *denote a sequence of real-valued random variables such that* $E(Y_j^2) < \infty$ *for all* $j = 0, 1, \ldots$ *and let* Y *denote a real-valued random variable such that*

$$\lim_{n \to \infty} E[(Y_n - Y)^2] = 0.$$

(i) $E(Y^2) < \infty$.

(ii) *Suppose* Z *is real-valued random variable such that*

$$\lim_{n \to \infty} E[(Y_n - Z)^2] = 0.$$

   Then

$$\Pr(Z = Y) = 1.$$

(iii) $E(Y) = \lim_{n \to \infty} E(Y_n)$ *and* $E(Y^2) = \lim_{n \to \infty} E(Y_n^2)$.

Recall that, in establishing the convergence of a sequence of real numbers, it is often convenient to use the Cauchy criterion, which allows convergence to be established without knowing the limit by showing that, roughly speaking, the terms in the sequence become closer together as one moves down the sequence. See Section A3.2 in Appendix 3 for further details.

The same type of approach may be used to establish convergence in mean square of a sequence of random variables. Theorem 6.4 gives such a result; as in the case of convergence of a sequence of real numbers, the advantage of this result is that it may be shown that a limiting random variable exists without specifying the distribution of that random variable.

Theorem 6.4. *Let $Y_0, Y_1, \ldots$ denote a sequence of real-valued random variables such that $E(Y_j^2) < \infty$ for all $j = 0, 1, \ldots$. Suppose that for every $\epsilon > 0$ there exists an $N > 0$ such that*

$$E[(Y_m - Y_n)^2] < \epsilon$$

for all $m, n > N$; in this case we say that Y_n, $n = 0, 1, \ldots$, is Cauchy in mean square.

If Y_n, $n = 0, 1, \ldots$, is Cauchy in mean square, then there exists a random variable Y with $E(Y^2) < \infty$ such that

$$\lim_{n \to \infty} E[(Y_n - Y)^2] = 0.$$

Proof. Fix $j = 1, 2, \ldots$. Since $Y_0, Y_1, \ldots$ is Cauchy in mean square, we can find m, n, $m > n$, such that

$$E[(Y_m - Y_n)^2] \le \frac{1}{4^j}.$$

Thus, we can identify a subsequence $n_1, n_2, \ldots$ such that

$$E[(Y_{n_{j+1}} - Y_{n_j})^2] \le \frac{1}{4^j}, \quad j = 1, 2, \ldots.$$

Define

$$T_m = \sum_{j=1}^{m} |Y_{n_{j+1}} - Y_{n_j}|, \quad m = 1, 2, \ldots.$$

Let Ω denote the sample space of the underlying experiment so that, for each $m = 1, 2, \ldots, T_m \equiv T_m(\omega), \omega \in \Omega$. Note that the terms in the sum forming T_m are all nonnegative so that, for each $\omega \in \Omega$, either $\lim_{m \to \infty} T_m(\omega)$ exists or the sequence diverges. Define a random variable T by

$$T(\omega) = \lim_{m \to \infty} T_m(\omega), \quad \omega \in \Omega$$

if the limit exists; otherwise set $T(\omega) = \infty$.

Note that for any real-valued random variables Z_1, Z_2 such that $E(Z_j^2) < \infty$, $j = 1, 2$,

$$\left[E\left(Z_1^2\right)^{\frac{1}{2}} + E\left(Z_2^2\right)^{\frac{1}{2}} \right]^2 - E[(Z_1 + Z_2)^2] = 2E\left(Z_1^2\right)^{\frac{1}{2}} E\left(Z_2^2\right)^{\frac{1}{2}} - 2E(Z_1 Z_2) \ge 0,$$

by the Cauchy-Schwarz inequality. Hence,

$$E[(Z_1 + Z_2)^2]^{\frac{1}{2}} \le E\left(Z_1^2\right)^{\frac{1}{2}} + E\left(Z_2^2\right)^{\frac{1}{2}}.$$

It follows that

$$
\mathrm{E}\left(T_m^2\right)^{\frac{1}{2}} = \mathrm{E}\left[\left(\sum_{j=1}^{m} |Y_{n_j+1} - Y_{n_j}|\right)^2\right]^{\frac{1}{2}}
$$

$$
\leq \sum_{j=1}^{m} \mathrm{E}\left[|Y_{n_j+1} - Y_{n_j}|^2\right]^{\frac{1}{2}}
$$

$$
\leq \sum_{j=1}^{m} \frac{1}{2^j} \leq 1.
$$

Since, by Fatou's lemma,

$$
\mathrm{E}(T^2) \leq \liminf_{m \to \infty} \mathrm{E}(T_m^2),
$$

it follows that $\mathrm{E}(T^2) \leq 1$ and, hence, that $\Pr(T < \infty) = 1$. This implies that

$$
\sum_{j=1}^{m} \left(Y_{n_j+1} - Y_{n_j}\right)
$$

converges absolutely with probability 1.

Hence, the limit of

$$
Y_{n_1} + \sum_{j=1}^{m} \left(Y_{n_j+1} - Y_{n_j}\right)
$$

as $m \to \infty$ exists with probability 1 so that we may define a random variable

$$
Y = Y_{n_1} + \sum_{j=1}^{\infty} \left(Y_{n_{j+1}} - Y_{n_j}\right).
$$

Note that

$$
Y = \lim_{j \to \infty} Y_{n_j} \quad \text{with probability 1.}
$$

Consider $\mathrm{E}[(Y_n - Y)^2]$. Since $Y_{n_j} \to Y$ as $j \to \infty$,

$$
Y - Y_n = \lim_{j \to \infty} (Y_{n_j} - Y_n)
$$

and, by Fatou's lemma,

$$
\mathrm{E}[(Y_n - Y)^2] = \mathrm{E}\left[\lim_{j \to \infty} \left(Y_{n_j} - Y_n\right)^2\right] \leq \liminf_{j \to \infty} \mathrm{E}\left[\left(Y_{n_j} - Y_n\right)^2\right].
$$

Fix $\epsilon > 0$. Since $Y_0, Y_1, \ldots$ is Cauchy in mean square, for sufficiently large n, j,

$$
\mathrm{E}\left[\left(Y_n - Y_{n_j}\right)^2\right] \leq \epsilon.
$$

Hence, for sufficiently large n,

$$
\mathrm{E}[(Y_n - Y)^2] \leq \epsilon.
$$

Since $\epsilon > 0$ is arbitrary, it follows that

$$
\lim_{n \to \infty} \mathrm{E}[(Y_n - Y)^2] = 0.
$$

Furthermore, since $(a + b)^2 \leq 2a^2 + 2b^2$,

$$\mathrm{E}(Y^2) = \mathrm{E}\{[Y_n + (Y - Y_n)]^2\} < \infty,$$

proving the result. ∎

We now consider the existence of the stochastic process defined by (6.1).

Theorem 6.5. *Let* $\ldots, \epsilon_{-1}, \epsilon_0, \epsilon_0, \ldots$ *denote independent random variables such that, for each* j, $\mathrm{E}(\epsilon_j) = 0$ *and* $\mathit{Var}(\epsilon_j) = 1$ *and let* $\ldots, \alpha_{-1}, \alpha_0, \alpha_1, \ldots$ *denote constants.*
If

$$\sum_{j=-\infty}^{\infty} \alpha_j^2 < \infty$$

then, for each $t = 0, 1, \ldots$, *the limit*

$$\lim_{n \to \infty} \sum_{j=-n}^{n} \alpha_j \epsilon_{t-j}$$

exists in mean square.

Proof. Fix $t = 0, 1, \ldots$. For $n = 1, 2, \ldots$ define

$$X_{nt} = \sum_{j=-n}^{n} \alpha_j \epsilon_{t-j}.$$

For $m > n$,

$$X_{mt} - X_{nt} = \sum_{j=-m}^{-(n+1)} \alpha_j \epsilon_{t-j} + \sum_{j=n+1}^{m} \alpha_j \epsilon_{t-j}$$

and

$$\mathrm{E}[(X_{mt} - X_{nt})^2] = \sum_{j=-m}^{-(n+1)} \alpha_j^2 + \sum_{j=n+1}^{m} \alpha_j^2.$$

Let

$$A_n = \sum_{j=1}^{n} \alpha_j^2, \quad n = 1, 2, \ldots.$$

Since $\sum_{j=-\infty}^{\infty} \alpha_j^2 < \infty$, $\lim_{n \to \infty} A_n$ exists. It follows that A_n is a Cauchy sequence of real numbers: given $\epsilon > 0$ there exists an N_1 such that

$$|A_n - A_m| < \epsilon, \quad n, m > N_1.$$

That is, given $\epsilon > 0$, there exists an N_1 such that

$$\sum_{j=n+1}^{m} \alpha_j^2 < \epsilon/2, \quad n, m > N_1.$$

Similarly, there exists an N_2 such that

$$\sum_{j=-m}^{-(n+1)} \alpha_j^2 < \epsilon/2, \quad n, m > N_2.$$

Hence, given $\epsilon > 0$, there exists an N such that

$$E[(X_{nt} - X_{mt})^2] < \epsilon, \quad n, m > N.$$

That is, for each $t = 0, 1, \ldots,$ $X_{nt}, n = 1, 2, \ldots,$ is Cauchy in mean-square. The result now follows from Theorem 6.4.   ∎

The autocovariance function of a moving average process is given in the following theorem.

Theorem 6.6. *Let $\{X_t : t \in \mathbf{Z}\}$ be a moving average process of the form*

$$X_t = \sum_{j=-\infty}^{\infty} \alpha_j \epsilon_{t-j}, \quad t = 0, 1, \ldots$$

where $\ldots, \epsilon_{-1}, \epsilon_0, \epsilon_1, \ldots$ *is a sequence of independent random variables such that* $E(\epsilon_j) = 0$ *and* $\mathrm{Var}(\epsilon_j) = 1, j = \ldots, -1, 0, 1, \ldots,$ *and*

$$\sum_{j=-\infty}^{\infty} \alpha_j^2 < \infty.$$

Then $\{X_t : t \in \mathbf{Z}\}$ is a second-order stationary process with mean 0 and autocovariance function

$$R(h) = \sum_{j=-\infty}^{\infty} \alpha_j \alpha_{j+h}, \quad h = 0, \pm 1, \pm 2, \ldots.$$

If $\ldots, \epsilon_{-1}, \epsilon_0, \epsilon_0, \ldots$ *are identically distributed, then $\{X_t : t \in \mathbf{Z}\}$ is stationary.*

Proof.   Fix $h = 0, 1, \ldots$ and define $Y_t = X_{t+h}, \quad t = 0, 1, \ldots.$ Then

$$Y_t = \sum_{j=-\infty}^{\infty} \alpha_j \epsilon_{t+h-j} = \sum_{j=-\infty}^{\infty} \alpha_{j+h} \epsilon_{t-j}.$$

Fix t and define

$$X_{nt} = \sum_{j=-n}^{n} \alpha_j \epsilon_{t-j}$$

and

$$Y_{nt} = \sum_{j=-n}^{n} \alpha_{j+h} \epsilon_{t-j}.$$

Then

$$X_{nt} - Y_{nt} = \sum_{j=-n}^{n} (\alpha_j - \alpha_{j+h}) \epsilon_{t-j};$$

since

$$\sum_{j=-\infty}^{\infty} (\alpha_j - \alpha_{j+h})^2 \leq 4 \sum_{j=-\infty}^{\infty} \alpha_j^2 < \infty,$$

it follows that $X_{nt} - Y_{nt}$ converges to $X_t - Y_t$ in mean square. Hence,

$$\text{Var}(X_t - Y_t) = \lim_{n \to \infty} \text{Var}(X_{nt} - Y_{nt}) = \sum_{j=-\infty}^{\infty} (\alpha_j - \alpha_{j+h})^2.$$

Similarly,

$$\text{Var}(X_t) = \text{Var}(Y_t) = \sum_{j=-\infty}^{\infty} \alpha_j;$$

hence,

$$2\,\text{Cov}(X_t, Y_t) = 2\,\text{Cov}(X_t, X_{t+h}) = 2\sum_{j=-\infty}^{\infty} \alpha_j^2 - \sum_{j=-\infty}^{\infty} (\alpha_j - \alpha_{j+h})^2 = 2\sum_{j=-\infty}^{\infty} \alpha_j \alpha_{j+h},$$

proving the first result.

To prove the second result, let $Y_t = X_{t+1}$, $t = 0, 1, \ldots$. Then

$$Y_t = \sum_{j=-\infty}^{\infty} \alpha_j \epsilon_{t+1-j}, \quad t = 0, 1, \ldots.$$

Since $\ldots, \epsilon_{-1}, \epsilon_0, \epsilon_1, \ldots$ are identically distributed, we may write

$$Y_t = \sum_{j=-\infty}^{\infty} \alpha_j \epsilon_{t-j}, \quad t = 0, 1, \ldots.$$

It follows that the process $\{Y_t : t \in \mathbf{Z}\}$ has the same structure as $\{X_t : t \in \mathbf{Z}\}$. It now follows from Theorem 6.1 that $\{X_t : t \in \mathbf{Z}\}$ is stationary. ∎

Example 6.6 (Finite moving average process). Consider the qth-order finite moving average process,

$$X_t = \sum_{j=0}^{q} \alpha_j \epsilon_{t-j}, \quad t = 0, 1, \ldots,$$

where q is a fixed nonnegative integer; here $\epsilon_{-q}, \epsilon_{-q+1}, \ldots$ is a sequence of independent random variables such that, for each j, $\text{E}(\epsilon_j) = 0$ and $\text{Var}(\epsilon_j) = 1$ and $\alpha_0, \alpha_1, \ldots, \alpha_q$ are constants. This model is of the general form considered above with $\alpha_i = 0$, $i \leq -1$, $i \geq q+1$; hence, the condition of Theorem 6.6 is satisfied.

It follows that the covariance function of the process is given by

$$R(h) = \begin{cases} \sum_{j=0}^{q-h} \alpha_j \alpha_{j+h} & \text{if } h = 0, 1, \ldots, q \\ 0 & \text{if } h = q+1, q+2, \ldots \end{cases}.$$

Thus, observations sufficiently far apart in time are uncorrelated.

Figure 6.2 contains plots of four randomly generated moving average processes with $\alpha_j = 1$, $j = 1, 2, \ldots, q$, and with the ϵ_j taken to be standard normal random variables, for $q = 0, 1, 2, 5$. As in Figure 6.1, the processes are shown as continuous functions, rather than as points. □

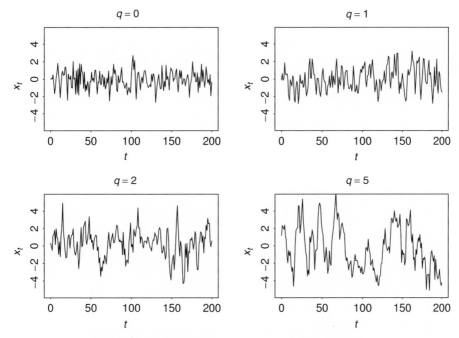

Figure 6.2. Randomly generated moving average processes.

Example 6.7   (Moving differences of a moving average process). Let $\{Z_t : t \in \mathbf{Z}\}$ denote a moving average process and let

$$X_t = Z_{t+1} - Z_t, \quad t = 1, 2, \ldots .$$

Write

$$Z_t = \sum_{j=-\infty}^{\infty} \alpha_j \epsilon_{t-j}$$

where $\sum_{j=-\infty}^{\infty} \alpha_j^2 < \infty$. Then we may write

$$X_t = \sum_{j=-\infty}^{\infty} \alpha_j \epsilon_{t+1-j} - \sum_{j=-\infty}^{\infty} \alpha_j \epsilon_{t-j} = \sum_{j=-\infty}^{\infty} (\alpha_{j+1} - \alpha_j)\epsilon_{t-j}.$$

Note that

$$\sum_{j=-\infty}^{\infty} (\alpha_{j+1} - \alpha_j)^2 \le 4 \sum_{j=-\infty}^{\infty} \alpha_j^2 < \infty;$$

it follows that $\{X_t : t \in \mathbf{Z}\}$ is also a moving average process.

By Theorem 6.6, the autocovariance function of the process is given by

$$R(h) = \sum_{j=-\infty}^{\infty} (\alpha_{j+1} - \alpha_j)(\alpha_{j+h+1} - \alpha_{j+h}).$$

Since

$$\sum_{j=-\infty}^{\infty} |\alpha_{j+1} - \alpha_j| \, |\alpha_{j+h+1} - \alpha_{j+h}| \leq \sum_{j=-\infty}^{\infty} (\alpha_{j+1} - \alpha_j)^2 < \infty,$$

we may write

$$R(h) = \sum_{j=-\infty}^{\infty} \alpha_{j+1}\alpha_{j+h+1} - \sum_{j=-\infty}^{\infty} \alpha_j \alpha_{j+h+1} - \sum_{j=-\infty}^{\infty} \alpha_{j+1}\alpha_{j+h} + \sum_{j=-\infty}^{\infty} \alpha_j \alpha_{j+h}$$

$$= 2 \sum_{j=-\infty}^{\infty} \alpha_j \alpha_{j+h} - \sum_{j=-\infty}^{\infty} \alpha_j \alpha_{j+h+1} - \sum_{j=-\infty}^{\infty} \alpha_{j+1}\alpha_{j+h}$$

$$= R_Z(|h|) - R_Z(|h+1|) - R_Z(|h-1|)$$

where R_Z denotes the autocovariance function of the process $\{Z_t : t \in \mathbf{Z}\}$. This is in agreement with Example 6.4. □

6.4 Markov Processes

If $X_0, X_1, \ldots$ are independent, real-valued random variables, then

$$\Pr(X_{n+1} \leq x | X_1, \ldots, X_n) = \Pr(X_{n+1} \leq x)$$

with probability 1. If $X_0, X_1, \ldots$ is a *Markov process*, this property is generalized to allow $\Pr(X_{n+1} \leq x | X_1, \ldots, X_n)$ to depend on X_n:

$$\Pr(X_{n+1} \leq x | X_1, \ldots, X_n) = \Pr(X_{n+1} \leq x | X_n).$$

That is, the conditional distribution of X_{n+1} given $X_1, \ldots, X_n$ does not depend on $X_1, \ldots, X_{n-1}$.

Example 6.8 (First-order autoregressive process). Consider the first-order autoregressive process considered in Example 6.2:

$$X_t = \begin{cases} \frac{1}{\sqrt{(1-\rho^2)}} Z_0 & \text{if } t = 0 \\ \rho X_{t-1} + Z_t & \text{if } t = 1, 2, \ldots \end{cases}$$

where $Z_0, Z_1, Z_2 \ldots$ is a sequence of independent, identically distributed random variables, each with a normal distribution with mean 0 and variance σ^2 and $-1 < \rho < 1$.

Note that we may write

$$X_t = \frac{\rho^t}{\sqrt{(1 - \rho^2)}} Z_0 + \rho^{t-1}Z_1 + \cdots + \rho Z_{t-1} + Z_t;$$

thus, for each $t = 0, 1, 2, \ldots, Z_{t+1}$ and $(X_0, \ldots, X_t)$ are independent. Using the fact that

$$X_{t+1} = \rho X_t + Z_{t+1}, \quad t = 1, 2, \ldots,$$

it follows that, for each $s \in \mathbf{R}$,

$$\mathrm{E}\{\exp(isX_{t+1})|X_0, \ldots, X_t\} = \mathrm{E}\{\exp(is\rho X_t) \exp(isZ_{t+1})|X_0, \ldots, X_t\}$$
$$= \exp(is\rho X_t)\mathrm{E}\{\exp(isZ_{t+1})\}.$$

Hence, the first-order autoregressive process is a Markov process. □

Markov chains

A *Markov chain* is simply a Markov process in which the state space of the process is a countable set; here we assume that the state space is finite and, without loss of generality, we take it to be $\{1, \ldots, J\}$ for some nonnegative integer J. A Markov chain may be either a discrete time or a continuous time process; here we consider only the discrete time case.

Since a Markov chain is a Markov process, the conditional distribution of X_{t+1} given $X_1, \ldots, X_t$ depends only on X_t. This conditional distribution is often represented by a matrix of *transition probabilities*

$$P_{ij}^{t,t+1} \equiv \Pr(X_{t+1} = j | X_t = i), \quad i, j = 1, \ldots, J.$$

If this matrix is the same for all t we say that the Markov chain has *stationary* transition probabilities; in the brief treatment here we consider only that case.

Hence, the properties of the process are completely determined by the transition probabilities P_{ij} along with the *initial distribution*, the distribution of X_0. Let P denote the $J \times J$ matrix with (i, j)th element P_{ij} and let p denote a $1 \times J$ vector with jth element

$$p_j = \Pr(X_0 = j), \quad j = 1, \ldots, J.$$

We will say that a process $\{X(t): t \in \mathbf{Z}\}$ has distribution $M(p, P)$ if it is a discrete time Markov chain with transition matrix P and initial distribution p.

Example 6.9 *(Two-state chain)*. Consider a Markov chain model with two states. Hence, the transition probability matrix is of the form

$$P = \begin{pmatrix} \alpha & 1 - \alpha \\ 1 - \beta & \beta \end{pmatrix}$$

where α and β take values in the interval $[0, 1]$; for simplicity, we assume that $0 < \alpha < 1$ and $0 < \beta < 1$. For instance,

$$\Pr(X_2 = 1 | X_1 = 1) = \alpha \quad \text{and} \quad \Pr(X_2 = 1 | X_1 = 2) = 1 - \beta.$$

The initial distribution is given by a vector of the form $(\theta, 1 - \theta)$ so that

$$\Pr(X_0 = 1) = 1 - \Pr(X_0 = 2) = \theta$$

where $0 < \theta < 1$. □

Example 6.10 *(Simple random walk with absorbing barrier)*. Suppose that, at time 0, a particle begins at position 0. At time 1, the particle remains at position 0 with probability $1/2$; otherwise the particle moves to position 1. Similarly, suppose that at time t the particle is at position m. At time $t + 1$ the particle remains at position m with probability $1/2$; otherwise the particle moves to position $m + 1$. When the particle reaches position J, where J is some fixed number, no further movement is possible. Hence, the transition probabilities have the form

$$P_{ij} = \begin{cases} 1/2 & \text{if } i < J \text{ and either } j = i \text{ or } j = i + 1 \\ 1 & \text{if } i = J \text{ and } j = J \\ 0 & \text{otherwise} \end{cases}.$$

For instance, for $J = 4$,

$$P = \begin{pmatrix} 1/2 & 1/2 & 0 & 0 \\ 0 & 1/2 & 1/2 & 0 \\ 0 & 0 & 1/2 & 1/2 \\ 0 & 0 & 0 & 1 \end{pmatrix}.$$

The initial distribution is given by a vector of the form $(1, 0, \ldots, 0)$ to reflect the fact that the particle begins at position 0. $\square$

The joint probability that $X_0 = i$ and $X_1 = j$ is given by $p_i P_{ij}$. The marginal probability that $X_1 = j$ may be written

$$\Pr(X_1 = j) = \sum_{i=1}^{J} \Pr(X_1 = j | X_0 = i) \Pr(X_0 = i) = \sum_{i=1}^{J} P_{ij} p_i.$$

Therefore, the vector of state probabilities for X_1 may be obtained from the vector of initial probabilities p and the transition matrix P by the matrix multiplication, pP. The vector of probabilities for X_2 may now be obtained from pP and P in a similar manner. These results are generalized in the following theorem.

Theorem 6.7. *Let $\{X_t : t \in \mathbf{Z}\}$ denote a discrete time process with distribution $M(p, P)$. Then*

    (i) $\Pr(X_0 = j_0, X_1 = j_1, \ldots, X_n = j_n) = p_{j_0} P_{j_0 j_1} P_{j_1 j_2} \cdots P_{j_{n-1} j_n}$
    (ii) The vector of state probabilities for X_n, $n = 1, 2, \ldots$, is given by pP^n.
    (iii) Let $r = 0, 1, 2, \ldots$. Then the distribution of $\{X_{r+t} : t \in \mathbf{Z}\}$ is $M(pP^r, P)$.

Proof. Part (i) follows directly from the calculation

$$\Pr(X_0 = j_0, X_1 = j_1, \ldots, X_n = j_n)$$
$$= \Pr(X_0 = j_0)\Pr(X_1 = j_1 | X_0 = j_0) \cdots \Pr(X_n = j_n | X_0 = j_0, \ldots, X_{n-1} = j_{n-1})$$
$$= \Pr(X_0 = j_0)\Pr(X_1 = j_1 | X_0 = j_0)\Pr(X_2 = j_2 | X_1 = j_1) \cdots \Pr(X_n = j_n | X_{n-1} = j_{n-1}).$$

Part (ii) may be established using induction. The result for $n = 1$ follows from the argument given before the theorem. Assume the result holds for $n = m$. Then

$$\Pr(X_{m+1} = j) = \sum_{i=1}^{J} \Pr(X_{m+1} = j | X_m = i)\Pr(X_m = i)$$

so that the vector of state probabilities is given by $(pP^m)P = pP^{m+1}$, proving the result.

To prove part (iii), it suffices to show that, for any $r = 1, 2, \ldots$, and any $n = 1, 2, \ldots$, the distributions of $(X_0, X_1, \ldots, X_n)$ and $(X_r, X_{r+1}, \ldots, X_{r+n})$ are identical. From part (i) of the theorem,

$$\Pr(X_0 = j_0, X_1 = j_1, \ldots, X_n = j_n) = p_{j_0} P_{j_0 j_1} P_{j_1 j_2} \cdots P_{j_{n-1} j_n} \tag{6.4}$$

and

$$\Pr(X_r = j_0, X_{r+1} = j_1, \ldots, X_{r+n} = j_n) = q_{j_0} P_{j_1 j_0} \cdots P_{j_{n-1} j_n} \tag{6.5}$$

where $q = (q_1, \ldots, q_J)$ denotes the vector of state probabilities for X_r. Note that (6.5) is of the same form as (6.4), except with the vector p replaced by q; from part (ii) of the theorem, $q = pP^r$, proving the result. ∎

Example* 6.11 *(Two-state chain). Consider the two-state Markov chain considered in Example 6.9. The vector of state probabilities for X_1 is given by

$$(\theta, \ 1-\theta) \begin{pmatrix} \alpha & 1-\alpha \\ 1-\beta & \beta \end{pmatrix} = (\theta\alpha + (1-\theta)(1-\beta), \ \ \theta(1-\alpha) + (1-\theta)\beta). \quad (6.6)$$

Hence,

$$\Pr(X_1 = 1) = 1 - \Pr(X_1 = 2) = \theta\alpha + (1-\theta)(1-\beta).$$

The position of the chain at time 2 follows the same model, except that the vector of initial probabilities $(\theta, 1 - \theta)$ is replaced by (6.6). Hence,

$$\Pr(X_{n+1} = 1) = \alpha\Pr(X_n = 1) + (1-\beta)\Pr(X_n = 2)$$
$$= (\alpha + \beta - 1)\Pr(X_n = 1) + (1-\beta).$$

Thus, writing $r_n = \Pr(X_n = 1)$, $c = \alpha + \beta - 1$, and $d = 1 - \beta$, we have the recursive relationship

$$r_{n+1} = cr_n + d, \quad n = 0, 1, 2, \ldots$$

with $r_0 = \theta$. It follows that

$$r_{n+1} = c(cr_{n-1} + d) + d = c^2 r_{n-1} + cd + d = c^2[cr_{n-2} + d] + cd + d$$
$$= c^3 r_{n-2} + (c^2 + c + 1)d,$$

and so on. Hence,

$$r_{n+1} = c^{n+1}r_0 + d\sum_{j=0}^{n} c^j = (\alpha + \beta - 1)^{n+1}\theta + (1-\beta)\frac{1 - (\alpha + \beta - 1)^{n+1}}{2 - (\alpha + \beta)}.$$

For the special case in which $\alpha + \beta = 1$,

$$\Pr(X_{n+1} = 1) = 1 - \beta, \quad n = 0, 1, 2, \ldots. \qquad \square$$

The matrix P gives the conditional distribution of X_{n+1} given $X_n = i$, for any $i = 1, \ldots, J$; hence, the probabilities in P are called the *one-step transition probabilities*. We may also be interested in the *m-step transition probabilities*

$$P_{ij}^{(m)} = \Pr(X_{n+m} = j | X_n = i).$$

The following theorem shows how the m-step transition probabilities can be obtained from P.

***Theorem* 6.8.** *Let $\{X_t : t \in \mathbf{Z}\}$ denote a discrete time process with distribution $M(p, P)$. Then, for any $r \leq m$, the m-step transition probabilities are given by*

$$P_{ij}^{(m)} = \sum_{k=1}^{J} P_{ik}^{(r)} P_{kj}^{(m-r)}.$$

Hence, the matrix $P^{(m)}$ with elements $P_{ij}^{(m)}$, $i = 1, \ldots, J$, $j = 1, \ldots, J$, satisfies

$$P^{(m)} = P^{(r)} P^{(m-r)}, \quad r = 0, \ldots, m$$

so that $P^{(m)} = P^m = P \times \cdots \times P$.

Proof. Since the process is assumed to have stationary transition probabilities,

$$P_{ij}^{(m)} = \Pr(X_m = j | X_0 = i) = \sum_{k=1}^{J} \Pr(X_m = j, X_r = k | X_0 = i)$$

$$= \sum_{k=1}^{J} \Pr(X_m = j | X_r = k, X_0 = i) \Pr(X_r = k | X_0 = i)$$

$$= \sum_{k=1}^{J} \Pr(X_m = j | X_r = k) \Pr(X_r = k | X_0 = i)$$

$$= \sum_{k=1}^{J} P_{kj}^{(m-r)} P_{ik}^{(r)},$$

proving the result. ∎

Example 6.12 *(Simple random walk with absorbing barrier).* Consider the simple random walk considered in Example 6.10. For the case $J = 4$, it is straightforward to show that the matrix of two-step transition probabilities is given by

$$\begin{pmatrix} 1/4 & 1/2 & 1/4 & 0 \\ 0 & 1/4 & 1/2 & 1/4 \\ 0 & 0 & 1/2 & 1/2 \\ 0 & 0 & 0 & 1 \end{pmatrix}.$$

The matrix of four-step transition probabilities is given by

$$\begin{pmatrix} 1/16 & 1/4 & 7/16 & 1/4 \\ 0 & 1/16 & 3/8 & 9/16 \\ 0 & 0 & 1/4 & 3/4 \\ 0 & 0 & 0 & 1 \end{pmatrix}. \qquad \square$$

Suppose that the distribution of X_1 is identical to that of X_0; that is, suppose that the vector of state probabilities for X_1 is equal to the vector of state probabilities for X_0. This occurs whenever

$$pP = p.$$

In this case p is said to be *stationary* with respect to P.

Theorem 6.9. *Let $\{X_t : t \in \mathbf{Z}\}$ denote a discrete time process with an $M(p, P)$ distribution. If p is stationary with respect to P, then $\{X_t : t \in \mathbf{Z}\}$ is a stationary process.*

Proof. Let $Y_t = X_{t+1}$, $t = 1, 2, \ldots$. According to Theorem 6.1, it suffices to show that $\{X_t : t \in \mathbf{Z}\}$ and $\{Y_t : t \in \mathbf{Z}\}$ have the same distribution. By Theorem 6.7, $\{Y_t : t \in \mathbf{Z}\}$ has distribution $M(pP, P)$. The result now follows from the fact that $pP = p$. ∎

***Example* 6.13 (*Two-stage chain*).** Consider the two-stage Markov chain considered in Examples 6.9 and 6.11. The initial distribution $p = (\theta, 1 - \theta)$ is stationary with respect to P whenever

$$\theta = \theta\alpha + (1 - \theta)(1 - \beta);$$

that is, whenever,

$$\frac{\theta}{1 - \theta} = \frac{1 - \beta}{1 - \alpha}$$

in which case

$$\theta = \frac{1 - \beta}{2 - (\alpha + \beta)}. \qquad \qquad \square$$

6.5   Counting Processes

A *counting process* is an integer-valued, continuous time process $\{X_t: t \geq 0\}$. Counting processes arise when certain events, often called "arrivals," occur randomly in time, with X_t denoting the number of arrivals occurring in the interval $[0, t]$. Counting processes are often denoted by $N(t)$ and we will use that notation here.

It is useful to describe a counting process in terms of the *interarrival times*. Let $T_1, T_2, \ldots$ be a sequence of nonnegative random variables and define

$$S_k = T_1 + \cdots + T_k.$$

Suppose that $N(t) = n$ if and only if

$$S_n \leq t \quad \text{and} \quad S_{n+1} > t;$$

Then $\{N(t): t \geq 0\}$ is a counting process. In the interpretation of the counting process in terms of random arrivals, T_1 is the time until the first arrival, T_2 is the time between the first and second arrivals, and so on. Then S_n is the time of the nth arrival.

If $T_1, T_2, \ldots$ are independent, identically distributed random variables, then the process is said to be a *renewal process*. If $T_1, T_2, \ldots$ has a stationary distribution then $\{N(t): t \geq 0\}$ is said to be a *stationary point process*.

***Example* 6.14 (*Failures with replacement*).** Consider a certain component that is subject to failure. Let T_1 denote the failure time of the original component. Upon failure, the original component is replaced by a component with failure time T_2. Assume that the process of failure and replacement continues indefinitely, leading to failure times $T_1, T_2, \ldots$; these failure times are modeled as nonnegative random variables. Let $\{N(t): t \geq 0\}$ denote the counting process corresponding to $T_1, T_2, \ldots$. Then $N(t)$ denotes the number of failures in the interval $[0, t]$. If $T_1, T_2, \ldots$ are independent, identically distributed random variables, then the counting process is a renewal process.

Figure 6.3 gives plots of four randomly generated counting processes of this type in which $T_1, T_2, \ldots$ are taken to be independent exponential random variables with $\lambda = 1/2, 1, 2, 5$, respectively. $\square$

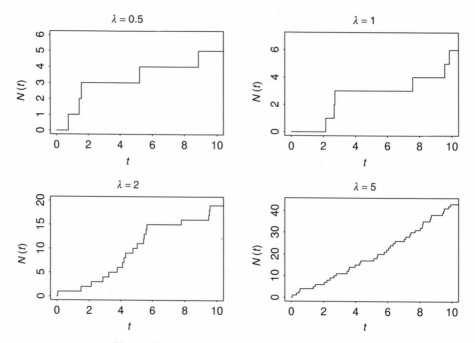

Figure 6.3. Randomly generated counting processes.

The mean value function of a counting process is given by

$$\mu(t) = E[N(t)], \quad t \geq 0.$$

Example* 6.15 *(A model for software reliability). Suppose that a particular piece of software has M errors, or "bugs." Let Z_j denote the testing time required to discover bug j, $j = 1, \ldots, M$. Assume that $Z_1, Z_2, \ldots, Z_M$ are independent, identically distributed random variables, each with distribution function F. Then S_1, the time until an error is detected, is the smallest value among $Z_1, Z_2, \ldots, Z_M$; S_2, the time needed to find the first two bugs, is the second smallest value among $Z_1, Z_2, \ldots, Z_M$, and so on.

Fix a time t. Then $N(t)$, the number of bugs discovered by time t, is a binomial random variable with parameter M and $F(t)$. Hence, the mean value function of the counting process $\{N(t): t \geq 0\}$ is given by

$$\mu(t) = MF(t), \quad t \geq 0. \qquad \square$$

Let $F_n(\cdot)$ denote the distribution function of $S_n = T_1 + \cdots + T_n$. The following result shows that the function $\mu(\cdot)$ can be calculated directly from $F_1, F_2, \ldots$.

***Theorem* 6.10.** *Let $\{N(t): t \geq 0\}$ denote a counting process and let S_n denote the time of the nth arrival. Then*

$$\mu(t) = \sum_{n=1}^{\infty} F_n(t)$$

where F_n denotes the distribution function of S_n.

If the number of possible arrivals is bounded by M, then

$$\mu(t) = \sum_{n=1}^{M} F_n(t).$$

Proof. Fix t. Let A_n denote the indicator of the event that the nth arrival occurs before t, that is, that $S_n \leq t$. Then $F_n(t) = \mathrm{E}(A_n)$ and

$$N(t) = \sum_{n=1}^{\infty} A_n.$$

Hence,

$$\mu(t) = \mathrm{E}[N(t)] = \sum_{n=1}^{\infty} \mathrm{E}(A_n) = \sum_{n=1}^{\infty} F_n(t),$$

proving the first result.

If there are at most M arrivals, then the indicator A_n is identically equal to 0 for $n = M + 1, M + 2, \ldots$. The second result follows. ∎

Poisson processes

The most important counting process is the *Poisson process*. A counting process $\{N(t): t \geq 0\}$ is said to be a Poisson process if the following conditions are satisfied:

(PP1) $N(0) = 0$

(PP2) $\{N(t): t \geq 0\}$ has independent increments: if

$$0 \leq t_0 \leq t_1 \leq \cdots \leq t_m,$$

then the random variables

$$N(t_1) - N(t_0), N(t_2) - N(t_1), \ldots, N(t_m) - N(t_{m-1})$$

are independent.

(PP3) There exists a constant $\lambda > 0$ such that, for any nonnegative s, t, $N(t+s) - N(s)$ has a Poisson distribution with mean λt.

The condition that differences of the form $N(t+s) - N(s)$ follow a Poisson distribution, condition (PP3), may be replaced by a condition on the behavior of $N(t)$ for small t, provided that the distribution of $N(t+s) - N(s)$ does not depend on s. Consider the following conditions.

(PP4) For any $t > 0$, the distribution of $N(t+s) - N(s)$ is the same for all $s \geq 0$.

(PP5) $\lim_{t \to 0} \Pr[N(t) \geq 2]/t = 0$ and for some positive constant λ,

$$\lim_{t \to 0} \frac{\Pr[N(t) = 1]}{t} = \lambda.$$

The equivalence of condition (PP3) and conditions (PP4) and (PP5) is established in the following theorem.

Theorem 6.11. *Suppose that a given counting process $\{N(t): t \geq 0\}$ satisfies conditions (PP1) and (PP2). The process satisfies condition (PP3) if and only if it satisfies conditions (PP4) and (PP5).*

Proof. First suppose that conditions (PP1), (PP2), and (PP3) are satisfied so that the process is a Poisson process. Then, for any $t \geq 0$, $N(t)$ has a Poisson process with mean λt. Hence,

$$\lim_{t \to 0} \Pr[N(t) = 1]/t = \lim_{t \to 0} \lambda \exp(-\lambda t) = \lambda$$

and

$$\Pr[N(t) \geq 2]/t = \sum_{j=2}^{\infty} \lambda^j t^{j-1} \exp(-\lambda t)/j!$$

$$\leq \lambda \sum_{j=1}^{\infty} (\lambda t)^j \exp(-\lambda t)/j!$$

$$\leq \lambda[1 - \exp(-\lambda t)].$$

It follows that

$$\limsup_{t \to 0} \Pr[N(t) \geq 2]/t \leq \limsup_{t \to 0} \lambda[1 - \exp(-\lambda t)] = 0.$$

Hence, (PP5) holds. Condition (PP4) follows directly from (PP3).

Now assume that (PP1), (PP2), (PP4), and (PP5) hold. Note that (PP5) implies that

$$\lim_{t \to 0}\{\Pr[N(t) = 0] - 1\}/t = -\lim_{t \to 0}\{\Pr[N(t) = 1] + \Pr[N(t) \geq 2]\}/t = -\lambda. \qquad (6.7)$$

Using (PP2) and (PP4),

$$\Pr[N(s + t) = 0] = \Pr[N(s) = 0, \quad N(s + t) - N(s) = 0] = \Pr[N(t) = 0]\Pr[N(s) = 0].$$

Hence,

$$\frac{\Pr[N(s + t) = 0] - \Pr[N(s) = 0]}{t} = \Pr[N(s) = 0]\frac{\Pr[N(t) = 0] - 1}{t}.$$

By (6.7),

$$\frac{d}{ds}\Pr[N(s) = 0] = -\lambda\Pr[N(s) = 0],$$

that is,

$$\frac{d}{ds} \log \Pr[N(s) = 0] = -\lambda.$$

Solving this differential equation yields

$$\Pr[N(s) = 0] = \exp\{-\lambda s\}, \quad s \geq 0.$$

Now consider $\Pr[N(s + t) = 1]$. Note that

$$\Pr[N(s + t) = 1] = \Pr[N(s) = 0]\Pr[N(t) = 1] + \Pr[N(s) = 1]\Pr[N(t) = 0].$$

Hence,

$$\frac{\Pr[N(s + t) = 1]}{\Pr[N(s + t) = 0]} = \frac{\Pr[N(t) = 1]}{\Pr[N(t) = 0]} + \frac{\Pr[N(s) = 1]}{\Pr[N(s) = 0]}.$$

Extending this argument shows that, for any $m = 1, 2, \ldots,$

$$\frac{\Pr[N(s) = 1]}{\Pr[N(s) = 0]} = m \frac{\Pr[N(s/m) = 1]}{\Pr[N(s/m) = 0]}.$$

Since this holds for any m,

$$\frac{\Pr[N(s) = 1]}{\Pr[N(s) = 0]} = \lim_{m \to \infty} m \frac{\Pr[N(s/m) = 1]}{\Pr[N(s/m) = 0]}.$$

Note that, from (6.7),

$$\lim_{m \to 0} \Pr[N(s/m) = 0] = 1$$

and, from (PP5), writing t for s/m,

$$\lim_{m \to \infty} m \Pr[N(s/m) = 1] = s \lim_{t \to 0} \Pr[N(t) = 1]/t = s\lambda.$$

It follows that

$$\Pr[N(s) = 1] = \lambda \Pr[N(s) = 0] = s\lambda \exp\{-\lambda s\}.$$

In a similar manner, we may write

$$\Pr[N(s) = 2] = \binom{m}{2} \Pr[N(s/m) = 1]^2 \Pr[N(s/m) = 0]^{m-2}$$
$$+ m \Pr[N(s/m) = 2] \Pr[N(s/m) = 0]^{m-1}$$

for any $m = 1, 2, \ldots.$ Using the expressions for $\Pr[N(s) = 0]$ and $\Pr[N(s) = 1]$ derived above, we have that, for any $m = 1, 2, \ldots,$

$$\Pr[N(s) = 2] = (1 - 1/m)(s\lambda)^2 \exp\{-s\lambda\}/2 + m\Pr[N(s/m) = 2] \exp\{-(1 - 1/m)s\lambda\}.$$

Letting $m \to \infty$ it follows that

$$\Pr[N(s) = 2] = \frac{(s\lambda)^2 \exp\{-s\lambda\}}{2!}.$$

The general case follows along similar lines:

$$\Pr[N(s) = n] = \binom{m}{n} \Pr[N(s/m) = 1]^n \Pr[N(s/m) = 0]^{m-n} + R$$
$$= \frac{m(m-1)\cdots(m-n+1)}{m^n}(\lambda s)^n \exp\{-\lambda s\} + R$$

where R is a finite sum, each term of which includes a factor of the form $\Pr[N(s/m) = j]$ for some $j \geq 2$. Letting $m \to \infty$, and using (PP5), yields the result. ∎

Distribution of the interarrival times

Consider a Poisson process $\{N(t): t \geq 0\}$ and let T_1 denote the time until the first arrival occurs. Since, for any $t > 0$,

$$X_1 > t \qquad \text{if and only if} \quad N(t) = 0,$$

it follows that

$$\Pr(X_1 \leq t) = 1 - \Pr(N(t) = 0) = 1 - \exp\{-\lambda t\}.$$

Hence, T_1 has an exponential distribution with rate parameter λ. More generally, the interarrival times $T_1, T_2, \ldots$ are independent, identically distributed exponential random variables. Conversely, if $T_1, T_2, \ldots$ are known to be independent exponential random variables then the counting process is a Poisson process. Thus, a counting process is a Poisson process if and only if the interarrival times are independent exponential random variables.

A formal proof of this result is surprisingly difficult and will not be attempted here; see, for example, Kingman (1993, Section 4.1). It is easy, however, to give an informal argument showing why we expect the result to hold.

We have seen that T_1 has an exponential distribution. Now consider

$$\Pr(T_2 > t | T_1 = t_1) = \Pr[N(t_1 + t) - N(t_1) = 0 | T_1 = t_1].$$

Since a Poisson process has independent increments, we expect that

$$\Pr[N(t_1 + t) - N(t_1) = 0 | T_1 = t_1] = \Pr[N(t_1 + t) - N(t_1) = 0] = \exp(-\lambda t). \quad (6.8)$$

Hence, T_1 and T_2 are independent and the marginal distribution of T_2 is an exponential distribution with parameter λ. This approach may be carried out indefinitely:

$$\Pr(T_m > t | T_1 = t_1, \ldots, T_{m-1} = t_{m-1})$$
$$= \Pr[N(t_1 + \cdots + t_{m-1} + t) - N(t_1 + \cdots + t_{m-1}) = 0]$$
$$= \exp\{-\lambda t\}.$$

The difficulty in carrying out this argument rigorously is in showing that (6.8) actually follows from (PP2). For instance, the event that $T_1 = t_1$ is the event that $N(t)$ jumps from 0 to 1 at $t = t_1$ and, hence, it is not a straightforward function of differences of the form $N(t_j) - N(t_{j-1})$ for some set of t_j.

6.6 Wiener Processes

A *Wiener process* or *Brownian motion* is a continuous time process $\{W_t : t \geq 0\}$ with the following properties:

(**W1**) $\Pr(W_0 = 0) = 1$

(**W2**) The process has *independent increments*: if

$$0 \leq t_0 \leq t_1 \leq \cdots \leq t_m,$$

then the random variables

$$W_{t_1} - W_{t_0}, W_{t_2} - W_{t_1}, \ldots, W_{t_m} - W_{t_{m-1}}$$

are independent.

(**W3**) For $t_2 > t_1 \geq 0$, $W_{t_2} - W_{t_1}$ has a normal distribution with mean 0 and variance $t_2 - t_1$.

Processes satisfying (W1)–(W3) can always be defined in such a way so as to be continuous; hence, we assume that (W4) is satisfied as well:

(**W4**) For every realization of the process, W_t is a continuous function of t.

Two basic properties of Wiener processes are given in the following theorem.

Theorem 6.12. *Let $\{W_t: t \geq 0\}$ denote a Wiener process.*

(i) For a given $h > 0$, let $Y_t = W_{h+t} - W_h$. Then $\{Y_t: t \geq 0\}$ is a Wiener process.

(ii) Let $K(t, s) = Cov(W_t, W_s)$. Then

$$K(t, s) = \min(t, s).$$

Proof. To verify part (i) it is enough to show that the process $\{Y_t: t \geq 0\}$ satisfies conditions (W1)–(W4). Clearly, $Y_0 = 0$ and since

$$Y_{t_1} - Y_{t_0} = W_{h+t_1} - W_{h+t_0},$$

$\{Y_t: t \geq 0\}$ has independent increments. Continuity of the sample paths of Y_t follows immediately from the continuity of Z_t. Hence, it suffices to show that (W3) holds.

For $0 \leq t_1 < t_2$,

$$Y_{t_2} - Y_{t_1} = W_{h+t_2} - W_{h+t_1}$$

has a normal distribution with mean 0 and variance $(h + t_2) - (h + t_1) = t_2 - t_1$, verifying (W3). Hence, $\{Y_t: t \geq 0\}$ is a Wiener process.

Suppose $t < s$. Then

$$K(t, s) = \text{Cov}(W_t, W_s) = \text{Cov}(W_t, W_t + W_s - W_t)$$
$$= \text{Cov}(W_t, W_t) + \text{Cov}(W_t, W_s - W_t).$$

By (W2), W_t and $W_s - W_t$ are independent. Hence, $K(t, s) = t$; the result follows. ∎

Example 6.16 (A transformation of a Wiener process). Let $\{W_t: t \geq 0\}$ denote a Wiener process and for some $c > 0$ define $Z_t = W_{c^2 t}/c, t \geq 0$. Note that $Z_0 = W_0$ so that $\text{Pr}(W_0 = 1)$. Let $0 \leq t_0 \leq t_1 \leq \cdots \leq t_m$; then

$$Z_{t_j} - Z_{t_{j-1}} = W_{c^2 t_j}/c - W_{c^2 t_{j-1}}/c, \quad j = 1, \ldots, m.$$

Hence, $\{Z_t: t \geq 0\}$ has independent increments. Furthermore, $Z_{t_j} - Z_{t_{j-1}}$ has a normal distribution with mean 0 and variance

$$\frac{c^2 t_j - c^2 t_{j-1}}{c^2} = t_j - t_{j-1}.$$

Finally, continuity of Z_t follows from continuity of W_t; hence, Z_t is a continuous function. It follows that $\{Z_t: t \geq 0\}$ is also a Wiener process. □

Rigorous analysis of Wiener process requires advanced results of probability theory and analysis that are beyond the scope of this book. Hence, in this section, we give an informal description of some of the properties of Wiener processes.

Irregularity of the sample paths of a Wiener process

By definition, the paths of a Wiener process are continuous; however, they are otherwise highly irregular. For instance, it may be shown that, with probability 1, a Wiener process $\{W_t: t \geq 0\}$ is nowhere differentiable. Although a formal proof of this fact is quite difficult (see, for example, Billingsley 1995, Section 37), it is not hard to see that it is unlikely that derivatives exist.

Consider

$$\frac{W_{t+h} - W_t}{h} \tag{6.9}$$

for small h; of course, the derivative of W_t is simply the limit of this ratio as $h \to 0$. Since $W_{t+h} - W_t$ has variance h, the difference $W_{t+h} - W_t$ tends to be of the same order as $\sqrt{h}$; for instance,

$$\mathrm{E}(|W_{t+h} - W_t|) = (2h/\pi)^{\frac{1}{2}}.$$

Thus, (6.9) tends to be of order $h^{-\frac{1}{2}}$, which diverges as $h \to 0$.

Let $f : [0, \infty) \to \mathbf{R}$ denote a continuous function of bounded variation and consider the quantity

$$Q_n(f) = \sum_{j=1}^{n}[f(j/n) - f((j-1)/n)]^2.$$

This is a measure of the variation of f over the interval $[0, 1]$. Note that

$$Q_n(f) \le \max_{1 \le j \le n} |f(j/n) - f((j-1)/n)| \sum_{j=1}^{n} |f(j/n) - f((j-1)/n)|.$$

Since f is continuous on $[0, \infty)$ it is uniformly continuous on $[0, 1]$ and, hence,

$$\lim_{n \to \infty} \max_{1 \le j \le n} |f(j/n) - f((j-1)/n)| = 0.$$

Since f is of bounded variation,

$$\sum_{j=1}^{n} |f(j/n) - f((j-1)/n)|$$

is bounded in n. Hence, $Q_n(f)$ approaches 0 as $n \to \infty$.

Now consider Q_n as applied to $\{W_t : t \ge 0\}$. By properties (W2) and (W3), $W_{j/n} - W_{(j-1)/n}$, $j = 1, \ldots, n$, are independent, identically distributed random variables, each with a normal distribution with mean 0 and variance $1/n$. Hence, for all $n = 1, 2, \ldots$,

$$\mathrm{E}\left\{ \sum_{j=1}^{n} (W_{j/n} - W_{(j-1)/n})^2 \right\} = 1;$$

that is, $\mathrm{E}[Q_n(W_t)] = 1$ for all $n = 1, 2, \ldots$. This suggests that the paths of a Wiener process are of unbounded variation, which is in fact true; see, for example, Billingsley (1968, Section 9).

The Wiener process as a martingale

Since, for any $s > t$, $W_s - W_t$ and W_t are independent, it follows that

$$\mathrm{E}[W_s | W_t] = \mathrm{E}[W_s - W_t | W_t] + \mathrm{E}[W_t | W_t] = W_t,$$

a property similar to that of martingales. In fact, a Wiener process is a (continuous time) martingale; see, for example, Freedman (1971). Although a treatment of continuous time martingales is beyond the scope of this book, it is not difficult to construct a discrete time martingale from a Wiener process. Let $\{W_t : t \ge 0\}$ denote a Wiener process and let $0 \le t_1 < t_2 < t_3 < \cdots$ denote an increasing sequence in $[0, \infty)$. For each $n = 1, 2, \ldots$, define

$Z_n = W_{t_n}$. Then, since there is a one-to-one correspondence between $\{Z_1, Z_2, \ldots, Z_n\}$ and $\{Z_1, Z_2 - Z_1, Z_3 - Z_2, \ldots, Z_n - Z_{n-1}\}$,

$$E[Z_{n+1}|Z_1, \ldots, Z_n] = E[Z_{n+1}|Z_1, Z_2 - Z_1, \ldots, Z_n - Z_{n-1}]$$
$$= Z_n + E[Z_{n+1} - Z_n|Z_1, Z_2 - Z_1, \ldots, Z_n - Z_{n-1}].$$

Since

$$E[Z_{n+1} - Z_n|Z_1, Z_2 - Z_1, \ldots, Z_n - Z_{n-1}]$$
$$= E[Z_{t_{n+1}} - Z_{t_n}|Z_{t_1}, Z_{t_2} - Z_{t_1}, \ldots, Z_{t_n} - Z_{t_{n-1}}]$$
$$= 0$$

by properties (W2) and (W3), it follows that

$$E[Z_{n+1}|Z_1, \ldots, Z_n] = Z_n;$$

that is, the process $\{Z_t: t = 1, 2, \ldots\}$ is a martingale.

6.7 Exercises

6.1 For some $q = 1, 2, \ldots$, let $Y_1, \ldots, Y_q$ and $Z_1, \ldots, Z_q$ denote real-valued random variables such that

$$E(Y_j) = E(Z_j) = 0, \quad j = 1, \ldots, q,$$

$$E\left(Y_j^2\right) = E\left(Z_j^2\right) = \sigma_j^2, \quad j = 1, \ldots, q,$$

for some positive constants $\sigma_1, \ldots, \sigma_q$,

$$E(Y_i Y_j) = E(Z_i Z_j) = 0 \quad \text{for} \quad i \neq j$$

and $E(Y_i Z_j) = 0$ for all i, j.

Let $\alpha_1, \ldots, \alpha_q$ denote constants and define a stochastic process $\{X_t: t \in \mathbf{Z}\}$ by

$$X_t = \sum_{j=1}^{q} [Y_j \cos(\alpha_j t) + Z_j \sin(\alpha_j t)], \quad t = 0, \ldots.$$

Find the mean and covariance functions of this process. Is the process covariance stationary?

6.2 Let $Z_{-1}, Z_0, Z_1, \ldots$ denote independent, identically distributed real-valued random variables, each with an absolutely continuous distribution. For each $t = 0, 1, \ldots$, define

$$X_t = \begin{cases} 1 & \text{if } Z_t > Z_{t-1} \\ -1 & \text{if } Z_t \leq Z_{t-1} \end{cases}.$$

Find the mean and covariance functions of the stochastic process $\{X_t: t \in \mathbf{Z}\}$. Is this process covariance stationary? Is the process stationary?

6.3 Let $\{X_t: t \in \mathbf{Z}\}$ and $\{Y_t: t \in \mathbf{Z}\}$ denote stationary stochastic processes. Is the process $\{X_t + Y_t: t \in \mathbf{Z}\}$ stationary?

6.4 Let $Z_0, Z_1, \ldots$ denote a sequence of independent, identically distributed random variables; let

$$X_t = \max\{Z_t, \ldots, Z_{t+s}\}, \quad t = 0, 1, \ldots$$

where s is a fixed positive integer and let

$$Y_t = \max\{Z_0, \ldots, Z_t\}, \quad t = 0, 1, \ldots.$$

Is $\{X_t: t \in \mathbf{Z}\}$ a stationary process? Is $\{Y_t: t \in \mathbf{Z}\}$ a stationary process?

6.5 Let $\{X_t: t \in \mathbf{Z}\}$ denote a stationary stochastic process and, for $s, t \in \mathbf{Z}$, let $K(s, t) = \mathrm{Cov}(X_s, X_t)$. Show that if

$$\lim_{s,t \to \infty} K(s, t) = K(0, 0)$$

then there exists a random variable X such that

$$\lim_{n \to \infty} \mathrm{E}[(X_n - X)^2] = 0.$$

6.6 Let $Y, Y_0, Y_1, \ldots$ denote real-valued random variables such that

$$\lim_{n \to \infty} \mathrm{E}[(Y_n - Y)^2] = 0.$$

Let $a, a_0, a_1, \ldots$ and $b, b_0, b_1, \ldots$ denote constants such that

$$\lim_{n \to \infty} a_n = a \quad \text{and} \quad \lim_{n \to \infty} b_n = b.$$

For $n = 0, 1, \ldots$, let $X_n = a_n Y_n + b_n$ and let $X = aY + b$. Does it follow that

$$\lim_{n \to \infty} \mathrm{E}[(X_n - X)^2] = 0?$$

6.7 Let $\{X_t: t \in \mathbf{Z}\}$ denote a moving average process. Define

$$Y_t = \sum_{j=0}^{m} c_j X_{t-j}, \quad t = 0, 1, \ldots$$

for some constants $c_0, c_1, \ldots, c_m$. Is $\{Y_t: t \in \mathbf{Z}\}$ a moving average process?

6.8 Let $\{X_t: t \in \mathbf{Z}\}$ denote a stationary stochastic process with autocovariance function $R(\cdot)$. The *autocovariance generating function* of the process is defined as

$$C(z) = \sum_{j=-\infty}^{\infty} R(j)z^j, \quad |z| \le 1.$$

Show that the autocorrelation function of the process can be obtained by differentiating $C(\cdot)$.

6.9 Let $\{X_t: t \in \mathbf{Z}\}$ denote a finite moving average process of the form

$$X_t = \sum_{j=0}^{q} \alpha_j \epsilon_{t-j}$$

where $\epsilon_0, \epsilon_1, \ldots$ are uncorrelated random variables each with mean 0 and finite variance σ^2. Let $C(\cdot)$ denote the autocovariance generating function of the process (see Exercise 6.7) and define

$$D(z) = \sum_{j=0}^{q} \alpha_j z^{t-j}, \quad |z| \le 1.$$

Show that

$$C(z) = \sigma^2 D(z)D(z^{-1}), \quad |z| \le 1.$$

6.10 Prove Theorem 6.3.

6.11 Let $R(\cdot)$ denote the autocovariance function of a stationary stochastic process. Show that $R(\cdot)$ is positive semi-definite in the sense that for all $t_1 < t_2 < \ldots < t_m$, where $m = 1, \ldots$, and all real numbers $z_1, z_2, \ldots, z_m$,

$$\sum_{i=1}^{m} \sum_{j=1}^{m} R(t_i - t_j)z_i z_j \ge 0.$$

6.12 Let $\{X_t: t \in \mathbf{Z}\}$ denote a finite moving average process of the form

$$X_t = \sum_{j=0}^{q} \alpha_j \epsilon_{t-j}$$

where $\epsilon_0, \epsilon_1, \ldots$ are uncorrelated random variables each with mean 0 and finite variance σ^2. Suppose that

$$\alpha_j = \frac{1}{q+1}, \quad j = 0, 1, \ldots, q.$$

Find the autocorrelation function of the process.

6.13 Let $\epsilon_{-1}, \epsilon_0, \epsilon_1, \ldots$ denote independent random variables, each with mean 0 and standard deviation 1. Define

$$X_t = \alpha_0 \epsilon_t + \alpha_1 \epsilon_{t-1}, \quad t = 0, 1, \ldots$$

where α_0 and α_1 are constants. Is the process $\{X_t: t \in \mathbf{Z}\}$ a Markov process?

6.14 Consider a Markov chain with state space $\{1, 2, \ldots, J\}$. A state i is said to *communicate* with a state j if

$$P_{ij}^{(n)} > 0 \qquad \text{for some } n = 0, 1, \ldots$$

and

$$P_{ji}^{(n)} > 0 \qquad \text{for some } n = 0, 1, \ldots.$$

Show that communication is an equivalence relation on the state space. That is, show that a state i communicates with itself, if i communicates with j then j communicates with i, and if i communicates with j and j communicates with k, then i communicates with k.

6.15 Let $\{X_t: t \in T\}$ denote a Markov chain with state space $\{1, 2, \ldots, J\}$. For each $t = 0, 1, \ldots$, let $Y_t = (X_t, X_{t+1})$ and consider the stochastic process $\{Y_t: t \in \mathbf{Z}\}$, which has state space

$$\{1, \ldots, J\} \times \{1, \ldots, J\}.$$

Is $\{Y_t: t \in T\}$ a Markov chain?

6.16 Let $Y_1, Y_2, \ldots$ denote independent, identically distributed random variables, such that

$$\Pr(Y_1 = j) = p_j, \quad j = 1, \ldots, J,$$

where $p_1 + \cdots + p_j = 1$. For each $t = 1, 2, \ldots$, let

$$X_t = \max\{Y_1, \ldots, Y_t\}.$$

Is $\{X_t: t \in \mathbf{Z}\}$ a Markov chain? If so, find the transition probability matrix.

6.17 Let P denote the transition probability matrix of a Markov chain and suppose that P is doubly stochastic; that is, suppose that the rows and columns of P both sum to 1. Find the stationary distribution of the Markov chain.

6.18 A counting process $\{N(t): t \geq 0\}$ is said to be a nonhomogeneous Poisson process with intensity function $\lambda(\cdot)$ if the process satisfies (PP1) and (PP2) and, instead of (PP3), for any nonnegative s, t, $N(t+s) - N(s)$ has a Poisson distribution with mean

$$\int_{s}^{t+s} \lambda(u) \, du.$$

Assume that $\lambda(\cdot)$ is a positive, continuous function defined on $[0, \infty)$.
Find an increasing, one-to-one function $h : [0, \infty) \mapsto [0, \infty)$ such that $\{\tilde{N}(t): t \geq 0\}$ is a Poisson process, where $\tilde{N}(t) = N(h(t))$.

6.19 Let $\{N(t): t \geq 0\}$ denote a nonhomogeneous Poisson process with intensity function $\lambda(\cdot)$. Let T_1 denote the time to the first arrival. Find $\Pr(T_1 \leq t)$ and $\Pr(T_1 \leq t | N(s) = n)$, $s > t$.

6.20 Let $\{N(t): t \geq 0\}$ denote a stationary point process with $N(0) = 0$. Show that there exists a constant $m \geq 0$ such that $E[N(t)] = mt$, $t > 0$.

6.21 Let $\{N(t)! : t \geq 0\}$ denote a Poisson process. Find the covariance function of the process,

$$K(t, s) = \text{Cov}(N(t), N(s)), \quad t \geq 0, \ s \geq 0.$$

6.22 Let $\{W_t: t \geq 0\}$ denote a Wiener process and define

$$X_t = Z_t - tZ_1, \quad 0 \leq t \leq 1.$$

The process $\{X_t: 0 \leq t \leq 1\}$ is known as a *Brownian bridge* process. Does the process $\{X_t: 0 \leq t \leq 1\}$ have independent increments?

6.23 Let $\{X_t: 0 \leq t \leq 1\}$ denote a Brownian bridge process, as described in Exercise 6.22. Find the covariance function of the process.

6.24 Let $\{W_t: t \geq 0\}$ denote a Wiener process and let

$$X_t = c(t)W_{f(t)}, \quad t \geq 0$$

where $c(\cdot)$ is a continuous function and $f(\cdot)$ is a continuous, strictly increasing function with $f(0) = 0$. Show that $\{X_t: t \geq 0\}$ satisfies (W1) and (W2) and find the distribution of $X_t - X_s$, $t > s$.

6.8   Suggestions for Further Reading

The topic of stochastic processes is a vast one and this chapter gives just a brief introduction to this field. General, mathematically rigorous, treatments of many topics in stochastic processes are given by Cramér and Leadbetter (1967) and Doob (1953); a more applications-oriented approach is taken by Parzen (1962) and Ross (1995). Karlin (1975) and Karlin and Taylor (1981) provide an in-depth treatment of a wide range of stochastic processes. Stationary and covariance-stationary processes are discussed in detail in Yaglom (1973); see also Ash and Gardner (1975), Cox and Miller (1965, Chapter 7), and Parzen (1962, Chapter 3).

Moving average processes are used extensively in statistical modeling; see, for example, Anderson (1975) and Fuller (1976). Markov processes are discussed in Cox and Miller (1965); Norris (1997) contains a detailed discussion of Markov chains. Stationary distributions of Markov chains play a central role in the limiting behavior of the process, a topic which is beyond the scope of this book; see, for example, Norris (1997).

Kingman (1993) gives a detailed discussion of Poisson processes; in particular, this reference considers in detail spatial Poisson processes. Wiener processes are discussed in Billingsley (1995, Chapter 37) and Freedman (1971).

7

Distribution Theory for Functions of Random Variables

7.1 Introduction

A common problem in statistics is the following. We are given random variables $X_1, \ldots, X_n$, such that the joint distribution of $(X_1 \ldots, X_n)$ is known, and we are interested in determining the distribution of $g(X_1, \ldots, X_n)$, where g is a known function. For instance, $X_1, \ldots, X_n$ might follow a parametric model with parameter θ and $g(X_1, \ldots, X_n)$ might be an estimator or a test statistic used for inference about θ. In order to develop procedures for inference about θ we may need certain characteristics of the distribution of $g(X_1, \ldots, X_n)$.

Example* 7.1 *(Estimator for a beta distribution). Let $X_1, \ldots, X_n$ denote independent, identically distributed random variables, each with an absolutely continuous distribution with density

$$\theta x^{\theta-1}, \quad 0 < x < 1$$

where $\theta > 0$. This is a special case of a *beta distribution*.

Consider the statistic

$$\frac{1}{n} \sum_{j=1}^{n} \log X_j;$$

this statistic arises as an estimator of the parameter θ. In carrying out a statistical analysis of this model, we may need to know certain characteristics of the distribution of this estimator. □

In the earlier chapters, problems of this type have been considered for specific examples; in this chapter we consider methods that can be applied more generally.

7.2 Functions of a Real-Valued Random Variable

First consider the case in which X is a real-valued random variable with a known distribution and we want to determine the distribution of $Y = g(X)$ where g is a known function. In principle, this is a straightforward problem. Let $\mathcal{X}$ denote the range of X and let $\mathcal{Y} = g(\mathcal{X})$ denote the range of Y so that $g : \mathcal{X} \to \mathcal{Y}$. For any subset A of $\mathcal{Y}$,

$$\Pr(Y \in A) = \Pr(X \in \{x \in \mathcal{X}: g(x) \in A\}) = \int_{\{x \in \mathcal{X}: g(x) \in A\}} dF_X(x), \qquad (7.1)$$

$$x \in g^{-1}(A)$$

yielding the distribution of Y. For instance, if X has a discrete distribution with frequency function p_X, then the distribution of Y is discrete with frequency function

$$p_Y(y) = \sum_{x \in \mathcal{X}:\, g(x)=y} p_X(x).$$

However, in more general cases, difficulties may arise when attempting to implement this approach. For instance, the set $\{x \in \mathcal{X} : g(x) \in A\}$ is often complicated, making computation of the integral in (7.1) difficult. The analysis is simplified if g is a one-to-one function.

Theorem 7.1. *Let X denote a real-valued random variable with range $\mathcal{X}$ and distribution function F_X. Suppose that $Y = g(X)$ where g is a <u>one-to-one function</u> on $\mathcal{X}$. Let $\mathcal{Y} = g(\mathcal{X})$ and let h denote the inverse of g.*

(i) Let F_Y denote the distribution function of Y. If g is an increasing function, then

$$F_Y(y) = F_X(h(y)), \quad y \in \mathcal{Y}$$

If g is a decreasing function, then

$$F_Y(y) = 1 - F_X(h(y)-), \quad y \in \mathcal{Y}.$$

(ii) If X has a discrete distribution with frequency function p_X, then Y has a discrete distribution with frequency function p_Y, where

$$p_Y(y) = p_X(h(y)), \quad y \in \mathcal{Y}.$$

(iii) Suppose that X has an absolutely continuous distribution with density function p_X and let g denote a continuously differentiable function. Assume that there exists an open subset $\mathcal{X}_0 \subset \mathcal{X}$ with $\Pr(X \in \mathcal{X}_0) = 1$ such that $|g'(x)| > 0$ for all $x \in \mathcal{X}_0$ and let $\mathcal{Y}_0 = g(\mathcal{X}_0)$. Then Y has an absolutely continuous distribution with density function p_Y, where

$$p_Y(y) = p_X(h(y))|h'(y)|, \quad y \in \mathcal{Y}_0.$$

Proof. If g is increasing on $\mathcal{X}$, then, for $y \in \mathcal{Y}$,

$$F_Y(y) = \Pr(Y \le y) = \Pr(X \le h(y)) = F_X(h(y)).$$

Similarly, if g is decreasing on $\mathcal{X}$, then

$$\Pr(Y \le y) = \Pr(X \ge h(y)) = 1 - \Pr(X < h(y)) = 1 - F(h(y)-).$$

Part (ii) follows from the fact that, for a one-to-one function g,

$$\Pr(Y = y) = \Pr(X = g(y)).$$

Consider a bounded function f defined on $\mathcal{Y}_0 = g(\mathcal{X}_0)$. Since $\Pr(X \in \mathcal{X}_0) = 1$,

$$E[f(Y)] = \int_{\mathcal{X}_0} f(g(x)) p_X(x)\, dx.$$

By the change-of-variable formula for integration,

$$E[f(Y)] = \int_{\mathcal{Y}_0} f(y) p_X(h(y))|h'(y)|\, dy.$$

It follows that the distribution of Y is absolutely continuous with density function

$$p_X(h(y))|h'(y)|, \quad y \in \mathcal{Y}_0;$$

part (iii) follows.   ■

Note that in applying part (iii) of Theorem 7.1, often the set $\mathcal{X}_0$ may be taken to be $\mathcal{X}$. Also note that parts (i) and (ii) of the theorem give $F_Y(y)$ only for $y \in \mathcal{Y}$; in typical cases, values of $F_Y(y)$ for $y \notin \mathcal{Y}$ can be obtained by inspection.

Example 7.2 (*Lognormal distribution*). Let X denote a random variable with a normal distribution with mean μ and standard deviation σ. Let $Y = \exp(X)$ so that $\log Y$ has a normal distribution; the distribution of Y is known as a *lognormal distribution*.
   Recall that the distribution of X is absolutely continuous with density function

$$p_X(x) = \frac{1}{\sigma\sqrt{(2\pi)}} \exp\left\{-\frac{1}{2\sigma^2}(x-\mu)^2\right\}, \quad -\infty < x < \infty.$$

We may write $Y = g(X)$ with $g(x) = \exp(x)$; then $g'(x) = \exp(x) > 0$ for all $x \in \mathbf{R}$ and g has inverse $h(y) = \log(y)$. Hence, $\mathcal{X}_0 = \mathcal{X} = \mathbf{R}$ and $\mathcal{Y}_0 = (0, \infty)$. It follows that the distribution of Y is absolutely continuous with density

$$p_Y(y) = \frac{1}{y}\frac{1}{\sigma\sqrt{(2\pi)}} \exp\left\{-\frac{1}{2\sigma^2}(\log(y)-\mu)^2\right\}, \quad y > 0.$$

Example 7.3 (*Empirical odds ratio*). Let X denote a random variable with a binomial distribution with parameters n and θ; then the distribution of X is discrete with frequency function

$$p_X(x) = \binom{n}{x}\theta^x(1-\theta)^{n-x}, \quad x = 0, \ldots, n.$$

Let

$$Y = \frac{X + 1/2}{n - X + 1/2};$$

hence, if X denotes the number of successes in n trials, Y denotes a form of the empirical odds ratio based on those trials. The function g is given by $g(x) = (x + 1/2)/(n - x + 1/2)$ with inverse

$$h(y) = \frac{(n + 1/2)y - 1/2}{1 + y}.$$

It follows that the distribution of Y is discrete with frequency function

$$p_Y(y) = \binom{n}{h(y)}\theta^{h(y)}(1-\theta)^{h(y)}$$

for values of y in the set

$$g(\mathcal{X}) = \left\{\frac{1}{2n+1}, \frac{3}{2n-1}, \frac{5}{2n-3}, \ldots, 2n+1\right\}. \qquad \square$$

Example 7.4 *(Weibull distribution).* Let X denote a random variable with a standard exponential distribution so that X has an absolutely continuous distribution with distribution function

$$F_X(x) = 1 - \exp(-x), \quad x > 0$$

and density

$$p_X(x) = \exp(-x), \quad x > 0.$$

Let $Y = X^{\frac{1}{\theta}}$ where $\theta > 0$. The function $g(x) = x^{\frac{1}{\theta}}$ has inverse $x = y^\theta$. It follows that Y has an absolutely continuous distribution with distribution function

$$F_Y(y) = 1 - \exp(-y^\theta), \quad y > 0$$

and density function

$$p_Y(y) = \theta y^{\theta-1} \exp(-y^\theta), \quad y > 0.$$

The distribution of Y is called a standard *Weibull distribution with index* θ. □

7.3   Functions of a Random Vector

In this section, we consider the extension of Theorem 7.1 to the case of a random vector. Let X denote a random vector with range $\mathcal{X}$; consider a function g on $\mathcal{X}$ and let $Y = g(X)$. Because of the possible complexity of the function g, even when it is one-to-one, an analogue of part (i) of Theorem 7.1 is not available. Part (ii) of Theorem 7.1, which does not use the dimension of X in any meaningful way, is simply generalized to the vector case. Part (iii) of the theorem, which is essentially the change-of-variable formula for integration, is also easily generalized by using the change-of-variable formula for integrals on a multidimensional space.

Recall that if g is a function from $\mathbf{R}^d$ to $\mathbf{R}^d$, then the *Jacobian matrix* of g is the $d \times d$ matrix with (i, j)th element given by $\partial g_i / \partial x_j$ where g_i denotes the ith component of the vector-valued function g; this matrix will be denoted by $\partial g / \partial x$. The *Jacobian* of g at x is the absolute value of the determinant of the Jacobian matrix at x, and is denoted by

$$\left| \frac{\partial g(x)}{\partial x} \right|.$$

determinant

Theorem 7.2. *Let X denote a d-dimensional random vector with an absolutely continuous distribution with density function p_X. Suppose that $Y = g(X)$ where $g : \mathcal{X} \to \mathbf{R}^d$ denotes a one-to-one continuously differentiable function. Let $\mathcal{X}_0$ denote an open subset of $\mathcal{X}$ such that $\Pr(X \in \mathcal{X}_0) = 1$ and such that the Jacobian of g is nonzero on $\mathcal{X}_0$. Then $Y = g(X)$ has an absolutely continuous distribution with density function p_Y, given by*

$$p_Y(y) = p_X(h(y)) \left| \frac{\partial h(y)}{\partial y} \right|, \quad y \in \mathcal{Y}_0,$$

dx → dy

where $\mathcal{Y}_0 = g(\mathcal{X}_0)$.

Proof. This result is essentially the change-of-variable formula for integrals. Let f denote a bounded real-valued function on $\mathcal{Y}_0$. Then, since $f(Y) = f(g(X))$,

$$E[f(Y)] = E[f(g(X))] = \int_{\mathcal{X}_0} f(g(x)) p_X(x) dx.$$

Using the change-of-variable $y = g(x)$, we have that

$$E[f(Y)] = \int_{\mathcal{Y}_0} f(y) p_X(h(y)) \left| \frac{\partial h(y)}{\partial y} \right| dy.$$

$x = h(y)$

$dx/dy = h'(y)$

$dx = h'(y)\, dy$.

The result follows. ∎

Note that the condition that the Jacobian of g is nonzero is identical to the condition that the Jacobian of h is finite.

Example 7.5 (Functions of standard exponential random variables). Let X_1, X_2 denote independent, standard exponential random variables so that $X = (X_1, X_2)$ has an absolutely continuous distribution with density function

$$p_X(x_1, x_2) = \exp\{-(x_1 + x_2)\}, \quad (x_1, x_2) \in (\mathbf{R}^+)^2.$$

Let $Y_1 = \sqrt{(X_1 X_2)}$ and $Y_2 = \sqrt{(X_1/X_2)}$. Hence,

$$Y = (Y_1, Y_2) = g(X) = (g_1(X), g_2(X))$$

where

$$g_1(x) = \sqrt{(x_1 x_2)} \quad \text{and} \quad g_2(x) = \sqrt{(x_1/x_2)}.$$

The inverse function is given by $h = (h_1, h_2)$ where

$$h_1(y) = y_1 y_2 \quad \text{and} \quad h_2(y) = y_1/y_2$$

which has Jacobian

$$\left| \frac{\partial h(y)}{\partial y} \right| = \frac{2y_1}{y_2}.$$

The set $\mathcal{X}_0$ may be taken to be $(\mathbf{R}^+)^2$ and $g(\mathcal{X}_0) = \mathcal{X}_0$.

It follows that the distribution of (Y_1, Y_2) is absolutely continuous with density function

$$p_Y(y_1, y_2) = \frac{2y_1}{y_2} \exp\{-y_1(1/y_2 + y_2)\}, \quad y_1 > 0, \ y_2 > 0. \qquad \square$$

Example 7.6 (Products of independent uniform random variables). Let $X_1, X_2, \ldots, X_n$ denote independent, identically distributed random variables, each with a uniform distribution on the interval $(0, 1)$. Let

$$Y_1 = X_1, \quad Y_2 = X_1 X_2, \quad \ldots, \quad Y_n = X_1 X_2 \cdots X_n.$$

Letting $X = (X_1, \ldots, X_n)$ and $Y = (Y_1, \ldots, Y_n)$ we have $Y = g(X)$ where $g = (g_1, \ldots, g_n)$ with

$$g_j(x) = \prod_{i=1}^{j} x_i, \quad x = (x_1, \ldots, x_n).$$

The inverse of this function is given by $h = (h_1, \ldots, h_n)$ with

$$h_j(y) = y_j/y_{j-1}, \quad j = 1, 2, \ldots, n;$$

here $y = (y_1, \ldots, y_n)$ and $y_0 = 1$.

The density function of X is given by

$$p_X(x) = 1, \quad x \in (0, 1)^n.$$

We may take $\mathcal{X}_0 = (0, 1)^n$; then

$$g(\mathcal{X}_0) = \{y = (y_1, \ldots, y_n) \in \mathbf{R}^n : 0 < y_n < y_{n-1} < \cdots < y_1 < 1\}.$$

It is easy to see that the matrix $\partial h(y)/\partial y$ is a lower triangular matrix with diagonal elements $1, 1/y_1, \ldots, 1/y_{n-1}$. Hence,

$$\left| \frac{\partial h(y)}{\partial y} \right| = \frac{1}{y_1 \cdots y_{n-1}}.$$

It follows that Y has an absolutely continuous distribution with density function

$$p_Y(y) = \frac{1}{y_1 \cdots y_{n-1}}, \quad 0 < y_n < y_{n-1} < \cdots < y_1 < 1. \qquad \square$$

Functions of lower dimension

Let X denote a random variable with range $\mathcal{X} \subset \mathbf{R}^d$, $d \geq 2$. Suppose we are interested in the distribution of $g_0(X)$ where $g_0 : \mathcal{X} \to \mathbf{R}^q$, $q < d$. Note that, since the dimension of $g_0(X)$ is less than the dimension of X, Theorem 7.2 cannot be applied directly. To use Theorem 7.2 in these cases, we can construct a function g_1 such that $g = (g_0, g_1)$ satisfies the conditions of Theorem 7.2. We may then use Theorem 7.2 to find the density of $g(X)$ and then marginalize to find the density of $g_0(X)$. This approach is illustrated on the following examples.

***Example* 7.7 (*Ratios of exponential random variables to their sum*).** Let $X_1, X_2, \ldots, X_n$ denote independent, identically distributed random variables, each with a standard exponential distribution. Let

$$Y_1 = \frac{X_1}{X_1 + \cdots + X_n}, \ Y_2 = \frac{X_2}{X_1 + \cdots + X_n}, \ \ldots, \ Y_{n-1} = \frac{X_{n-1}}{X_1 + \cdots + X_n}$$

and suppose we want to find the distribution of the random vector $(Y_1, \ldots, Y_{n-1})$.

Clearly, the function mapping $(X_1, \ldots, X_n)$ to $(Y_1, \ldots, Y_{n-1})$ is not one-to-one. Let $Y_n = X_1 + \cdots + X_n$. Then, writing $Y = (Y_1, \ldots, Y_n)$ and $X = (X_1, \ldots, X_n)$, we have $Y = g(X)$ where $g = (g_1, \ldots, g_n)$ is given by

$$g_n(x) = x_1 + \cdots + x_n, \quad g_j(x) = \frac{x_j}{g_n(x)}, \qquad j = 1, \ldots, n-1.$$

The function g is one-to-one, with inverse $h = (h_1, \ldots, h_n)$ where

$$h_j(y) = y_j y_n, \quad j = 1, \ldots, n-1, \quad \text{and} \quad h_n(y) = (1 - y_1 - \cdots - y_{n-1}) y_n.$$

Then

$$
\frac{\partial h(y)}{\partial y} = \begin{pmatrix} y_n & 0 & \cdots & 0 & y_1 \\ 0 & y_n & \cdots & 0 & y_2 \\ \vdots & \vdots & \vdots & \vdots & \vdots \\ 0 & 0 & \cdots & y_n & y_{n-1} \\ -y_n & -y_n & \cdots & -y_n & 1 - \sum_1^{n-1} y_j \end{pmatrix}.
$$

It follows that

$$
\left| \frac{\partial h(y)}{\partial y} \right| = y_n^{n-1}.
$$

The distribution of X is absolutely continuous with density function

$$
p_X(x) = \exp\left\{ -\sum_{j=1}^n x_j \right\}, \quad x = (x_1, \ldots, x_n) \in (\mathbf{R}^+)^n.
$$

Hence, we may take $\mathcal{X}_0 = (\mathbf{R}^+)^n$ and

$$
\mathcal{Y}_0 = g(\mathcal{X}_0) = \left\{ (y_1, \ldots, y_{n-1}) \in (0, 1)^{n-1} : \sum_{j=1}^{n-1} y_j \le 1 \right\} \times \mathbf{R}^+.
$$

It follows that the distribution of Y is absolutely continuous with density

$$
p_Y(y) = y_n^{n-1} \exp(-y_n), \quad y = (y_1, \ldots, y_n) \in \mathcal{Y}_0.
$$

To obtain the density of $(Y_1, \ldots, Y_{n-1})$, as desired, we need to marginalize, eliminating Y_n. This density is therefore given by

$$
\frac{\int_0^\infty y^{n-1} \exp(-y)\, dy = (n-1)!,}{(y_1, \ldots, y_{n-1}) \in \left\{ (y_1, \ldots, y_{n-1}) \in (0, 1)^{n-1} : \sum_{j=1}^{n-1} y_j \le 1 \right\}.}
$$

Hence, the density of $(Y_1, \ldots, Y_{n-1})$ is uniform on the simplex in $\mathbf{R}^{n-1}$. $\square$

***Example* 7.8 (*Estimator for a beta distribution*).** As in Example 7.1, let $X_1, \ldots, X_n$ denote independent, identically distributed random variables, each with an absolutely continuous distribution with density

$$
\theta x^{\theta-1}, \quad 0 < x < 1
$$

where $\theta > 0$ and consider the statistic

$$
Y_1 = -\frac{1}{n} \sum_{j=1}^n \log X_j.
$$

In order to use Theorem 7.2 we need to supplement Y_1 with functions $Y_2, \ldots, Y_n$ such that the transformation from $(X_1, \ldots, X_n)$ to $(Y_1, \ldots, Y_n)$ satisfies the conditions of Theorem 7.2.

Let $Y_j = -\log X_j$, $j = 2, \ldots, n$. Then $Y_j = g_j(X_1, \ldots, X_n)$, $j = 1, \ldots, n$, where

$$g_1(x_1, \ldots, x_n) = -\frac{1}{n} \sum_{j=1}^{n} \log x_j$$

and

$$g_j(x_1, \ldots, x_n) = -\log x_j, \quad j = 2, \ldots, n.$$

The function $g = (g_1, \ldots, g_n)$ has inverse $h = (h_1, \ldots, h_n)$ where

$$h_1(y_1, \ldots, y_n) = \exp\{-ny_1 + (y_2 + \cdots + y_n)\}$$

and

$$h_j(y_1, \ldots, y_n) = \exp(-y_j), \quad j = 2, \ldots, n.$$

It follows that the Jacobian of the transformation h is given by

$$\left| \frac{\partial h(y)}{\partial y} \right| = n \exp(-ny_1).$$

Here $\mathcal{X} = (0, 1)^n$ and since the Jacobian of h is finite for

$$y \in g(\mathcal{X}) = \{(y_1, \ldots, y_n) \in (0, \infty)^n : y_2 + \cdots + y_n < ny_1\},$$

we may take $\mathcal{X}_0 = \mathcal{X}$.

The density of $X = (X_1, \ldots, X_n)$ is given by

$$p_X(x_1, \ldots, x_n; \theta) = \theta^n (x_1 \cdots x_n)^{\theta - 1}, \quad 0 < x_j < 1, \quad j = 1, \ldots, n;$$

it follows that the density of $Y = (Y_1, \ldots, Y_n)$ is given by

$$\theta^n \exp\{-n(\theta - 1)y_1\} n \exp(-ny_1) = n\theta^n \exp(-n\theta y_1)$$

for $0 < y_j$, $j = 2, \ldots, n$, and $y_2 + \cdots + y_n < ny_1$.

In order to obtain the density of Y_1 we need to integrate out $y_2, \ldots, y_n$ from the joint density. Hence, the density of Y_1 is given by

$$n\theta^n \exp(n\theta y_1) \int_0^{ny_1} \cdots \int_0^{ny_1 - y_3 - \cdots - y_n} dy_2 \cdots dy_n = n\theta^n \exp(-n\theta y_1) \frac{(ny_1)^{n-1}}{(n-1)!}$$

where $0 < y_1 < \infty$. Note that this is the density of a gamma distribution. $\square$

Example 7.9 (Cauchy distribution). Let X_1, X_2 denote independent random variables each with a standard normal distribution and consider the distribution of $Y_1 = X_1/X_2$. In Example 3.13 the density of Y_1 was found using a method based on the characteristic function; here we determine the density function using a method based on Theorem 7.2.

In order to use the change-of-variable formula given in Theorem 7.2 we need to construct a one-to-one function. For instance, let $Y_2 = X_2$ and consider $Y = (Y_1, Y_2) = g(X) = (g_1(X), g_2(X))$ where

$$g_1(x) = x_1/x_2 \quad \text{and} \quad g_2(x) = x_2.$$

The inverse of this transformation is given by $h = (h_1, h_2)$ with

$$h_1(y) = y_1 y_2 \quad \text{and} \quad h_2(y) = y_2;$$

it follows that

$$\left|\frac{\partial h(y)}{\partial y}\right| = |y_2|.$$

Here we must take $\mathcal{X}_0 = \mathbf{R} \times [(-\infty, 0) \cup (0, \infty)]$ so that $\mathcal{Y}_0 = g(\mathcal{X}_0) = \mathcal{X}_0$.

The density function of (X_1, X_2) is given by

$$p_X(x) = \frac{1}{2\pi} \exp\left\{-\frac{1}{2}(x_1^2 + x_2^2)\right\}, \quad (x_1, x_2) \in \mathbf{R}^2.$$

Hence, the density function of (Y_1, Y_2) is given by

$$p_Y(y_1, y_2) = \frac{1}{2\pi} \exp\left\{-\frac{1}{2}(1 + y_1^2)y_2^2\right\} |y_2|, \quad (y_1, y_2) \in \mathbf{R}^2$$

and the marginal density of Y_1 is given by

$$\frac{1}{2\pi} \int_{-\infty}^{\infty} |y_2| \exp\left\{-\frac{1}{2}(1 + y_1^2)y_2^2\right\} dy_2 = \frac{1}{\pi} \int_0^{\infty} \exp\left\{-(1 + y_1^2)t\right\} dt$$

$$= \frac{1}{\pi (1 + y_1^2)}, \quad -\infty < y_1 < \infty.$$

This is the density of the standard Cauchy distribution. $\square$

The distributions considered in the following two examples occur frequently in the statistical analysis of normally distributed data.

***Example* 7.10 (*t*-distribution).** Let X_1 and X_2 denote independent random variables such that X_1 has a standard normal distribution and X_2 has a chi-squared distribution with ν degrees of freedom. The distribution of

$$Y_1 = \frac{X_1}{\sqrt{(X_2/\nu)}}$$

is called the *t*-distribution with ν degrees of freedom. The density of this distribution may be determined using Theorem 7.2. Let $Y_2 = X_2$. Writing $X = (X_1, X_2)$ and $Y = (Y_1, Y_2)$, $X = h(Y)$ where $h = (h_1, h_2)$,

$$h_1(y) = \frac{y_1 \sqrt{y_2}}{\sqrt{\nu}}, \qquad h_2(y) = y_2.$$

Hence,

$$\left|\frac{\partial h(y)}{\partial y}\right| = \sqrt{y_2}/\sqrt{\nu}.$$

The density of X is given by

$$p_X(x) = \frac{1}{\sqrt{(2\pi)}} \exp\left(-\frac{1}{2}x_1^2\right) \frac{1}{2^{\frac{\nu}{2}} \Gamma\left(\frac{\nu}{2}\right)} x_2^{\frac{\nu}{2}-1} \exp\left(-\frac{1}{2}x_2\right), \quad x \in \mathbf{R} \times (0, \infty).$$

Hence, by Theorem 7.2, Y has density

$$p_Y(y) = \frac{1}{\sqrt{(2\pi\nu)}} \frac{1}{2^{\frac{\nu}{2}} \Gamma\left(\frac{\nu}{2}\right)} y_2^{\frac{\nu-1}{2}} \exp\left\{-\frac{1}{2}\left(y_1^2/\nu + 1\right)y_2\right\}, \quad y \in \mathbf{R} \times (0, \infty).$$

It follows that the marginal density of Y_1 is given by

$$\frac{1}{\sqrt{(2\pi\nu)}2^{\frac{\nu}{2}}\Gamma\left(\frac{\nu}{2}\right)}\int_0^\infty y_2^{\frac{\nu+1}{2}-1}\exp\left\{-\frac{1}{2}(y_1^2/\nu+1)y_2\right\}dy_2$$

$$=\frac{1}{\sqrt{(\pi\nu)}}\frac{\Gamma\left(\frac{\nu+1}{2}\right)}{\Gamma\left(\frac{\nu}{2}\right)}(y_1^2/\nu+1)^{-(\nu+1)/2},\quad y_2\in\mathbf{R}.$$

This is the density of the t-distribution with ν degrees of freedom.   □

Example 7.11 (F-distribution). Let X_1 and X_2 denote independent random variables such that X_1 has a chi-squared distribution with ν_1 degrees of freedom and X_2 has a chi-squared distribution with ν_2 degrees of freedom. Let

$$Y_1=\frac{X_1/\nu_1}{X_2/\nu_2};$$

the distribution of Y_1 is called the *F-distribution with (ν_1,ν_2) degrees of freedom*. The density of this distribution may be determined using Theorem 7.2.

Let $Y_2=X_2$. Writing $X=(X_1,X_2)$ and $Y=(Y_1,Y_2)$, $X=h(Y)$ where $h=(h_1,h_2)$,

$$h_1(y)=\frac{\nu_1}{\nu_2}y_1y_2,\qquad h_2(y)=y_2.$$

Hence,

$$\left|\frac{\partial h(y)}{\partial y}\right|=\frac{\nu_1}{\nu_2}y_2.$$

The density of X is given by

$$p_X(x)=\frac{1}{2^{\frac{(\nu_1+\nu_2)}{2}}\Gamma(\frac{\nu_1}{2})\Gamma(\frac{\nu_2}{2})}x_1^{\frac{\nu_1}{2}-1}x_2^{\frac{\nu_2}{2}-1}\exp\left\{-\frac{1}{2}(x_1+x_2)\right\},\quad x\in(0,\infty)^2.$$

Hence, by Theorem 7.2, Y has density

$$p_Y(y)=\frac{(\nu_1/\nu_2)^{\frac{\nu_1}{2}}}{2^{\frac{(\nu_1+\nu_2)}{2}}\Gamma\left(\frac{\nu_1}{2}\right)\Gamma\left(\frac{\nu_2}{2}\right)}y_1^{\frac{\nu_1}{2}-1}y_2^{\frac{\nu_1+\nu_2}{2}-1}\exp\left\{-\frac{1}{2}\left(\frac{\nu_1}{\nu_2}y_1+1\right)y_2\right\},\quad y\in(0,\infty)^2.$$

It follows that the marginal density of Y_1 is given by

$$\frac{(\nu_1/\nu_2)^{\frac{\nu_1}{2}}}{2^{\frac{(\nu_1+\nu_2)}{2}}\Gamma\left(\frac{\nu_1}{2}\right)\Gamma\left(\frac{\nu_2}{2}\right)}y_1^{\frac{\nu_1}{2}-1}\int_0^\infty y_2^{\frac{\nu_1+\nu_2}{2}-1}\exp\left\{-\frac{1}{2}\left(\frac{\nu_1}{\nu_2}y_1+1\right)y_2\right\}dy_2$$

$$=\frac{\Gamma\left(\frac{\nu_1+\nu_2}{2}\right)}{\Gamma\left(\frac{\nu_1}{2}\right)\Gamma\left(\frac{\nu_2}{2}\right)}\left(\frac{\nu_1}{\nu_2}\right)^{\frac{\nu_1}{2}}\frac{y_1^{\frac{\nu_1}{2}-1}}{\left(\frac{\nu_1}{\nu_2}y_1+1\right)^{\frac{\nu_1+\nu_2}{2}}},\quad y_1\in(0,\infty).$$

This is the density of the F-distribution with (ν_1,ν_2) degrees of freedom.   □

Functions that are **not one-to-one**

Even in cases in which the dimension of Y is the same as the dimension of X, it is not possible to apply Theorem 7.2 if the function g is not one-to-one. However, if the set $\mathcal{X}$ may be partitioned into subsets such that g is one-to-one on each subset, then the change-of-variable formula may be applied on each subset. The results may then be combined to obtain the result for $g(X)$.

Theorem 7.3. *Let X denote a random vector with an absolutely continuous distribution with density function p_X. Let $\mathcal{X}_1, \cdots, \mathcal{X}_m$ denote disjoint open subsets of $\mathbf{R}^d$ such that*

$$\Pr\left(X \in \bigcup_{i=1}^{m} \mathcal{X}_i\right) = 1.$$

Let g denote a function on $\mathcal{X}$ and let $g^{(i)}$ denote the restriction of g to $\mathcal{X}_i$. Assume that, for each $i = 1, \ldots, m$, $g^{(i)}$ is one-to-one and continuously differentiable with inverse $h^{(i)}$ and the Jacobian $g^{(i)}$ is nonzero on $\mathcal{X}_i$. Then $Y = g(X)$ has an absolutely continuous distribution with density function p_Y, given by

$$p_Y(y) = \sum_{i=1}^{m} p_X\big(h^{(i)}(y)\big)\left|\frac{\partial h^{(i)}(y)}{\partial y}\right| I_{\{y \in \mathcal{Y}_i\}}, \quad y \in g\big(\cup_{i=1}^{m} \mathcal{X}_i\big)$$

where $\mathcal{Y}_i = g^{(i)}(\mathcal{X}_i)$.

Proof. Let f denote a bounded function on $\mathcal{Y}$. Then,

$$\mathrm{E}[f(Y)] = \int_{\mathcal{X}} f(g(x))p_X(x)\,dx = \sum_{i=1}^{m} \int_{\mathcal{X}_i} f\big(g^{(i)}(x)\big)p_X(x)\,dx.$$

On $\mathcal{X}_i$, $g^{(i)}$ is one-to-one and continuously differentiable so that the change-of-variable formula may be applied to the integral over $\mathcal{X}_i$. Hence,

$$\mathrm{E}[f(Y)] = \sum_{i=1}^{m} \int_{\mathcal{Y}_i} f(y)p_X\big(h^{(i)}(y)\big)\left|\frac{\partial h^{(i)}(y)}{\partial y}\right| dy.$$

The result follows by interchanging integration and summation, which is valid since the sum is finite. ■

Example 7.12 (Product and ratio of standard normal random variables). Let X_1, X_2 denote independent random variables, each with a standard normal distribution. Let

$$Y_1 = X_1 X_2 \quad \text{and} \quad Y_2 = \frac{X_1}{X_2}.$$

Writing $X = (X_1, X_2)$ and $Y = (Y_1, Y_2)$, it follows that $Y = g(X)$, $g = (g_1, g_2)$, where

$$g_1(x) = x_1 x_2 \quad \text{and} \quad g_2(x) = x_1/x_2.$$

Clearly this is not a one-to-one function on $\mathbf{R}^2$; for instance, (x_1, x_2) and $(-x_1, -x_2)$ yield the same value of $g(x)$, as do $(x_1, -x_2)$ and $(-x_1, x_2)$.

  The function is one-to-one on the four quadrants of $\mathbf{R}^2$. Hence, take $\mathcal{X}_1 = \mathbf{R}^+ \times \mathbf{R}^+$, $\mathcal{X}_2 = \mathbf{R}^+ \times \mathbf{R}^-$, $\mathcal{X}_3 = \mathbf{R}^- \times \mathbf{R}^+$, and $\mathcal{X}_4 = \mathbf{R}^- \times \mathbf{R}^-$. The restriction of g to each $\mathcal{X}_i$ is one-to-one, with inverses given by

$$h^{(1)}(y) = (\sqrt{(y_1 y_2)}, \sqrt{(y_1/y_2)}), \qquad h^{(2)}(y) = (\sqrt{(y_1 y_2)}, -\sqrt{(y_1/y_2)}),$$

$$h^{(3)}(y) = (-\sqrt{(y_1 y_2)}, \sqrt{(y_1/y_2)}), \qquad h^{(4)}(y) = (-\sqrt{(y_1 y_2)}, -\sqrt{(y_1/y_2)})$$

and Jacobians

$$\left|\frac{\partial h^{(i)}(y)}{\partial y}\right| = \frac{1}{2|y_2|}.$$

The set $g(\cup_{i=1}^4 \mathcal{X}_i)$ is $(\mathbf{R}^+)^2 \cup (\mathbf{R}^-)^2$. It is worth noting that the partition $(\mathcal{X}_1 \cup \mathcal{X}_2)$, $(\mathcal{X}_3 \cup \mathcal{X}_4)$ could also be used, although this choice introduces some minor complications.

The density of X is given by

$$p_X(x) = \frac{1}{2\pi} \exp\left\{-\frac{1}{2}\left(x_1^2 + x_2^2\right)\right\}, \quad x \in \mathbf{R}^2.$$

Consider the transformation on $\mathcal{X}_1$. Since $x_1^2 + x_2^2 = y_1 y_2 + y_1/y_2$, the contribution to the density of Y from $\mathcal{X}_1$ is

$$\frac{1}{4\pi |y_2|} \exp\left\{-\frac{1}{2}(y_1 y_2 + y_1/y_2)\right\}, \quad y \in (\mathbf{R}^+)^2.$$

It is easy to see that the same result holds for $\mathcal{X}_4$.

The contribution to the density from either $\mathcal{X}_2$ or $\mathcal{X}_3$ is the same, except that y is restricted to $(\mathbf{R}^-)^2$. Hence, the density function of Y is given by

$$\frac{1}{2\pi |y_2|} \exp\left\{-\frac{1}{2}(y_1 y_2 + y_1/y_2)\right\} I_{\{y \in (\mathbf{R}^+)^2\}} + \frac{1}{2\pi |y_2|} \exp\left\{-\frac{1}{2}(y_1 y_2 + y_1/y_2)\right\} I_{\{y \in (\mathbf{R}^-)^2\}}$$

$$= \frac{1}{2\pi |y_2|} \exp\left\{-\frac{1}{2}(y_1 y_2 + y_1/y_2)\right\}, \quad y \in (\mathbf{R}^+)^2 \cup (\mathbf{R}^-)^2. \qquad \square$$

Application of invariance and equivariance

When the distribution of X belongs to a parametric model, it is often convenient to take advantage of invariance or equivariance when determining the distribution of Y.

Let X denote a random variable with range $\mathcal{X}$ and suppose that the distribution of X is an element of

$$\mathcal{P} = \{P(\cdot; \theta): \theta \in \Theta\}$$

and that $\mathcal{P}$ is invariant with respect to some group of transformations. If Y is a function of X, and is an invariant statistic, then the distribution of Y does not depend on θ; hence, when determining the distribution of Y, we may assume that X is distributed according to $P_X(\cdot; \theta_0)$ where θ_0 is any convenient element of Θ. The resulting distribution for Y does not depend on the value chosen.

Example 7.13 (Ratios of exponential random variables to their sum). Let $X_1, X_2, \ldots, X_n$ denote independent, identically distributed random variables, each with an exponential distribution with parameter θ, $\theta > 0$. As in Example 7.8, let

$$Y_1 = \frac{X_1}{X_1 + \cdots + X_n}, \quad Y_2 = \frac{X_2}{X_1 + \cdots + X_n}, \quad \ldots, \quad Y_{n-1} = \frac{X_{n-1}}{X_1 + \cdots + X_n}.$$

Recall that the set of exponential distributions with parameter $\theta \in (0, \infty)$ forms a transformation model with respect to the group of scale transformations; see Example 5.27. Note that the statistic $(Y_1, \ldots, Y_{n-1})$ is invariant under scale transformations: multiplying each X_j by a constant does not change the value of $(Y_1, \ldots, Y_{n-1})$. Hence, to determine the distribution of $(Y_1, \ldots, Y_{n-1})$ we may assume that $X_1, \ldots, X_n$ are distributed according to a standard exponential distribution.

It follows that the distribution of $(Y_1, \ldots, Y_{n-1})$ is the same as that given in Example 7.7. $\square$

Equivariance may also be used to simplify the determination of the distribution of a statistic. Suppose that the distribution of a random variable X is an element of

$$\mathcal{P} = \{P(\cdot; \theta) \colon \theta \in \Theta\}$$

and that $\mathcal{P}$ is invariant with respect to some group of transformations. Suppose that we may identify the group of transformations with Θ so that if X is distributed according to the distribution with parameter value e, the identity element of the group, then θX is distributed according to $P(\cdot; \theta)$.

Suppose that the distributions in $\mathcal{P}$ are all absolutely continuous and that the density of $P(\cdot; \theta)$ is given by $p_X(\cdot; \theta)$. Then, by Theorem 7.2,

$$p_X(x; \theta) = p_X(\theta^{-1}x; e) \left| \frac{\partial \theta^{-1}x}{\partial x} \right|.$$

In computing the Jacobian here it is important to keep in mind that $\theta^{-1}x$ refers to the transformation θ^{-1} applied to x, not to the product of θ^{-1} and x.

Now let $Y = g(X)$ denote an equivariant statistic so that the set of distributions of Y also forms a transformation model with respect to Θ. Then we may determine the distribution of Y under parameter value θ using the following approach. First, we find the density function of Y under the identity element e, $p_Y(\cdot; e)$, using Theorem 7.2. Then the density function of Y under parameter value θ is given by

$$p_Y(y; \theta) = p_Y(\theta^{-1}y; e) \left| \frac{\partial \theta^{-1}y}{\partial y} \right|.$$

The advantage of this approach is that, in some cases, it is simpler to apply Theorem 7.2 to $p_X(\cdot; e)$ than to apply it to $p_X(\cdot; \theta)$ for an arbitrary value of $\theta \in \Theta$.

***Example* 7.14 (*Difference of uniform random variables*).** Let X_1, X_2 denote independent, identically distributed random variables, each distributed according to the uniform distribution on the interval (θ_1, θ_2), where $\theta_2 > \theta_1$. The uniform distribution on (θ_1, θ_2) is an absolutely continuous distribution with density function

$$p(x; \theta) = \frac{1}{\theta_2 - \theta_1}, \quad \theta_1 < x < \theta_2.$$

Suppose we are interested in the distribution of $X_2 - X_1$.

The family of uniform distributions on (θ_1, θ_2) with $-\infty < \theta_1 < \theta_2 < \infty$ is invariant under the group of location-scale transformations. Let Z_1, Z_2 denote independent random variables each uniformly distributed on the interval $(0, 1)$. Then the distribution of (X_1, X_2) is the same as the distribution of

$$(\theta_2 - \theta_1)(Z_1, Z_2) + \theta_1.$$

It follows that the distribution of $X_2 - X_1$ is the same as the distribution of

$$(\theta_2 - \theta_1)(Z_2 - Z_1);$$

hence, the distribution of $X_2 - X_1$ can be obtained by first obtaining the distribution of $Z_2 - Z_1$ and then using Theorem 7.1 to find the distribution of $X_2 - X_1$.

Let $Y_1 = Z_2 - Z_1$ and $Y_2 = Z_1$. Then the joint density of Y_1, Y_2 is

$$1, \quad 0 < y_1 < 1, \ \ 0 < y_1 + y_2 < 1.$$

Hence, the marginal density of Y_1 is

$$\begin{cases} \int_0^{1-y_1} dy_2 & \text{if } 0 \le y_1 < 1 \\ \int_{-y_1}^1 dy_2 & \text{if } -1 < y_1 < 0 \end{cases} = \begin{cases} 1 - y_1 & \text{if } 0 \le y_1 < 1 \\ 1 + y_1 & \text{if } -1 < y_1 < 0 \end{cases}$$

$$= 1 - |y_1|, \quad |y_1| < 1.$$

It follows that the distribution of $X_2 - X_1$ under parameter value (θ_1, θ_2) has density

$$\left(1 - \frac{|t|}{\theta_2 - \theta_1}\right) \frac{1}{\theta_2 - \theta_1}, \quad |t| < \theta_2 - \theta_1. \qquad \square$$

7.4   Sums of Random Variables

Let $X_1, \ldots, X_n$ denote a sequence of real-valued random variables. We are often interested in the distribution of $S = \sum_{j=1}^n X_j$. Whenever the distribution of X is absolutely continuous, the distribution of S may be determined using Theorem 7.2. However, the distribution of a sum arises so frequently that we consider it in detail here; in addition, there are some results that apply only to sums.

We begin by considering the characteristic function of S.

Theorem 7.4. *Let* $X = (X_1, \ldots, X_n)$ *where* $X_1, X_2, \ldots, X_n$ *denote real-valued random variables. Let* φ_X *denote the characteristic function of* X *and let* φ_S *denote the characteristic function of* $S = \sum_{j=1}^n X_j$. *Then*

$$\varphi_S(t) = \varphi_X(tv), \quad t \in \mathbf{R}$$

where $v = (1, \ldots, 1) \in \mathbf{R}^n$.

Proof.   The characteristic function of X is given by

$$\varphi_X(t) = \mathrm{E}[\exp\{it^T X\}], \quad t \in \mathbf{R}^n.$$

Since $S = v^T X$, the characteristic function of S is given by

$$\varphi_S(t) = \mathrm{E}[\exp\{itv^T X\}] = \varphi_X(tv), \quad t \in \mathbf{R},$$

verifying the theorem.   ■

Example 7.15 *(Sum of exponential random variables).* Let $X_1, \ldots, X_n$ denote independent, identically distributed random variables, each with density function

$$\lambda \exp\{-\lambda x\}, \quad x > 0$$

where $\lambda > 0$; this is the density function of the exponential distribution with parameter λ.

The characteristic function of this distribution is given by

$$\varphi(t) = \int_0^\infty \exp(itx) \lambda \exp(-\lambda x)\, dx = \frac{\lambda}{(\lambda - it)}, \quad -\infty < t < \infty.$$

It follows that the characteristic function of $X = (X_1, \ldots, X_n)$ is

$$\varphi_X(t_1, \ldots, t_n) = \prod_{j=1}^{n} \varphi(t_i).$$

Hence, the characteristic function of $S = \sum_{j=1}^{n} X_j$ is given by

$$\varphi_S(t) = \varphi(t)^n = \frac{\lambda^n}{(\lambda - it)^n}.$$

This is the characteristic function of the gamma distribution with parameters n and λ; see Example 3.4. It follows that S has a gamma distribution with parameters n and λ. $\quad\square$

The techniques described in the previous section can be used to find the density or frequency function of a sum.

Theorem 7.5. *Let $X = (X_1, \ldots, X_n)$ where $X_1, \ldots, X_n$ denotes a sequence of real-valued random variables, let $S = \sum_{j=1}^{n} X_j$ and let S denote the range of S.*

(i) Suppose X has a discrete distribution with frequency function p. Then S has a discrete distribution with frequency function p_S where

$$p_S(s) = \sum_{\{(x_1,\ldots,x_n):\sum_{j=1}^{n} x_j = s\}} p(x_1, \ldots, x_n), \ s \in S$$

(ii) Suppose X has an absolutely continuous distribution with density function p. Then S has an absolutely continuous distribution with density function p_S where

$$p_S(s) = \int_{-\infty}^{\infty} \cdots \int_{-\infty}^{\infty} p\left(s - \sum_{j=2}^{n} x_j, x_2, \ldots, x_n\right) dx_2 \cdots dx_n, s \in \mathbf{R}$$

Proof. Let f denote a bounded, real-valued function defined on S, the range of S. Then, if X has discrete distribution with frequency function f,

$$E[f(S)] = \sum_{(x_1,\ldots,x_n) \in \mathcal{X}} f(s) p(x_1, \ldots, x_n) = \sum_{s \in S} \sum_{\{(x_1,\ldots,x_n):\sum_{j=1}^{n} x_j = s\}} f(s) p(x_1, \ldots, x_n)$$

$$= \sum_{s \in S} f(s) \sum_{\{(x_1,\ldots,x_n):\sum_{j=1}^{n} x_j = s\}} p(x_1, \ldots, x_n);$$

part (i) of the theorem follows.

Now suppose that X has an absolutely continuous distribution with density function p. To prove part (ii), we can use Theorem 7.2 with the function

$$g(x) = (s, x_2, \ldots, x_n), \quad s = \sum_{j=1}^{n} x_j.$$

Then $Y = g(X)$ has density

$$p(y_1 - (y_2 + \cdots + y_n), y_2, \ldots, y_n);$$

note that the Jacobian here is equal to 1. Hence, the marginal density of $Y_1 = S$ is

$$\int_{-\infty}^{\infty} \cdots \int_{-\infty}^{\infty} p(y_1 - (y_2 + \cdots + y_n) y_2, \ldots, y_n) \, dy_2 \cdots dy_n;$$

rewriting this in terms of $s = y_1$ and $x_j = y_j$, $j = 2, \ldots, n$, proves the result. $\quad\blacksquare$

***Example* 7.16 (*One-parameter exponential family distribution*).** Consider a one-parameter exponential family of absolutely continuous distributions with density functions of the form

$$\exp\{c(\theta)y - d(\theta)\}h(y), \ y \in \mathcal{Y}$$

where $\theta \in \Theta, c : \Theta \to \mathbf{R}, d : \Theta \to \mathbf{R}$, and $h : \mathcal{Y} \to \mathbf{R}^+$.

Let $Y_1, Y_2, \ldots, Y_n$ denote independent, identically distributed random variables, each distributed according to this distribution. Then $(Y_1, \ldots, Y_n)$ has density

$$p(y; \theta) = \exp\left\{\sum_{j=1}^{n} c(\theta)y_j - n \, d(\theta)\right\} \prod_{j=1}^{n} h(y_j), \quad y = (y_1, \ldots, y_n) \in \mathcal{Y}^n.$$

It follows that $S = \sum_{j=1}^{n} Y_j$ has density

$$\int_{\mathcal{Y}} \cdots \int_{\mathcal{Y}} \exp\{c(\theta)(s - y_2 - \ldots - y_n) - d(\theta)\} h(s - y_2 - \cdots - y_n)$$

$$\times \exp\left\{\sum_{j=2}^{n} c(\theta)y_j - (n-1) \, d(\theta)\right\} \prod_{j=2}^{n} h(y_j) \, dy_2 \cdots dy_n$$

$$= \exp\{c(\theta)s - nd(\theta)\} \int_{\mathcal{Y}} \cdots \int_{\mathcal{Y}} h(s - y_2 - \cdots - y_n) \prod_{j=2}^{n} h(y_j) \, dy_2 \cdots dy_n.$$

Hence, the model for S is also a one-parameter exponential family model. $\square$

***Example* 7.17 (*Multinomial distribution*).** Let $X = (X_1, \ldots, X_m)$ denote a random vector with a discrete distribution with frequency function

$$p(x_1, \ldots, x_m; \theta_1, \ldots, \theta_m) = \binom{n}{x_1, x_2, \ldots x_m} \theta_1^{x_1} \theta_2^{x_2} \cdots \theta_m^{x_m},$$

for $x_j = 0, 1, \ldots, n, \ j = 1, \ldots, m, \ \sum_{j=1}^{m} x_j = n$; here $\theta_1, \ldots, \theta_m$ are nonnegative constants satisfying $\sum_{j=1}^{m} \theta_j = 1$. Recall that this is a *multinomial distribution* with parameters n and $(\theta_1, \ldots, \theta_m)$; see Example 2.2.

Let $S = X_1 + \cdots + X_{m-1}$. Then S has a discrete distribution with frequency function

$$p_S(s) = \sum_{\mathcal{X}_s} \binom{n}{x_1, \ldots, x_m} \theta_1^{x_1} \theta_2^{x_2} \cdots \theta_m^{x_m};$$

here

$$\mathcal{X}_s = \left\{(x_1, \ldots, x_{m-1}) \in \mathbf{Z}^m : \sum_{j=1}^{m-1} x_j = s\right\}.$$

Let $\eta = \sum_{j=1}^{m-1} \theta_j$ so that $\theta_m = 1 - \eta$. Then

$$p_S(s) = \eta^s (1 - \eta)^{n-s} \sum_{\mathcal{X}_s} \binom{n}{x_1, x_2, \ldots, x_m} \left(\frac{\theta_1}{\eta}\right)^{x_1} \cdots \left(\frac{\theta_{m-1}}{\eta}\right)^{x_{m-1}}$$

$$= \eta^s (1 - \eta)^{n-s} \sum_{\mathcal{X}_s} \frac{\binom{n}{x_1, \ldots, x_m}}{\binom{s}{x_1, \ldots, x_m}} \binom{s}{x_1, \ldots, x_m} \left(\frac{\theta_1}{\eta}\right)^{x_1} \cdots \left(\frac{\theta_{m-1}}{\eta}\right)^{x_{m-1}}.$$

Since

$$\frac{\binom{n}{x_1,\ldots,x_m}}{\binom{s}{x_1,\ldots,x_{m-1}}} = \frac{n!}{s!x_m!} = \binom{n}{s},$$

it follows that

$$p_S(s) = \binom{n}{s}\eta^s(1-\eta)^{n-s} \sum_{\mathcal{X}_s} \binom{s}{x_1,\ldots,x_{m-1}} \left(\frac{\theta_1}{\eta}\right)^{x_1} \left(\frac{\theta_2}{\eta}\right)^{x_2} \cdots \left(\frac{\theta_m}{\eta}\right)^{x_m}.$$

Note that

$$\binom{s}{x_1,\ldots,x_{m-1}} \left(\frac{\theta_1}{\eta}\right)^{x_1} \cdots \left(\frac{\theta_{m-1}}{\eta}\right)^{x_{m-1}}$$

is the frequency function of a multinomial distribution with parameters s and

$$\theta_1/\eta, \ldots, \theta_{m-1}/\eta.$$

Hence,

$$\sum_{\mathcal{X}_s} \binom{s}{x_1,\ldots,x_{m-1}} \left(\frac{\theta_1}{\eta}\right)^{x_1} \cdots \left(\frac{\theta_{m-1}}{\eta}\right)^{x_{m-1}} = 1$$

and, therefore,

$$p_S(s; \theta_1, \ldots, \theta_m) = \binom{n}{s}\eta^s(1-\eta)^{n-s}, \quad s = 0, \ldots, n$$

so that S has a binomial distribution with parameters n and $\sum_{j=1}^{m-1} \theta_j$. □

***Example* 7.18 (*Sum of uniform random variables*).** Let X_1, X_2 denote independent, identically distributed random variables, each with a uniform distribution on the interval $(0, 1)$; hence, (X_1, X_2) has an absolutely continuous distribution with density function

$$p(x_1, x_2) = 1, \quad (x_1, x_2) \in (0, 1)^2.$$

Let $S = X_1 + X_2$. Then S has an absolutely continuous distribution with density function

$$p_S(s) = \int_0^1 I_{\{0<s-x_2<1\}} I_{\{0<x_2<1\}} \, dx_2.$$

Note that $p_S(s)$ is nonzero only for $0 < s < 2$. Suppose $0 < s \le 1$; then

$$p_S(s) = \int_0^s dx_2 = s.$$

Suppose $1 < s < 2$, then

$$p_S(s) = \int_{s-1}^1 dx_2 = 2 - s.$$

It follows that S has density function

$$p_S(s) = \begin{cases} 0 & \text{if } s \le 0 \text{ or } s \ge 2 \\ s & \text{if } 0 < s \le 1 \\ 2 - s & 1 < s < 2 \end{cases}.$$

The distribution of S is called a *triangular distribution*. □

***Example* 7.19 (*Dirichlet distribution*).** Let (X, Y) denote a two-dimensional random vector with an absolutely continuous distribution with density function

$$p(x, y) = \frac{\Gamma(\alpha_1 + \alpha_2 + \alpha_3)}{\Gamma(\alpha_1)\Gamma(\alpha_2)\Gamma(\alpha_3)} x^{\alpha_1 - 1} y^{\alpha_2 - 1} (1 - x - y)^{\alpha_3 - 1},$$

where $x > 0$, $y > 0$, $x + y < 1$; here $\alpha_1, \alpha_2, \alpha_3$ are positive constants. Let $Z = 1 - X - Y$; then the distribution of (X, Y, Z) is an example of a *Dirichlet distribution*.

Consider the distribution of $X + Y$. According to Theorem 7.5, S has an absolutely continuous distribution with density function

$$\begin{aligned}
p_S(s) &= \frac{\Gamma(\alpha_1 + \alpha_2 + \alpha_3)}{\Gamma(\alpha_1)\Gamma(\alpha_2)\Gamma(\alpha_3)} \int_0^s (s - y)^{\alpha_1 - 1} y^{\alpha_2 - 1} (1 - s)^{\alpha_3 - 1} \, dy \\
&= \frac{\Gamma(\alpha_1 + \alpha_2 + \alpha_3)}{\Gamma(\alpha_1)\Gamma(\alpha_2)\Gamma(\alpha_3)} s^{\alpha_1 + \alpha_2 - 1} (1 - s)^{\alpha_3 - 1} \int_0^1 (1 - u)^{\alpha_1 - 1} u^{\alpha_2 - 1} \, du \\
&= \frac{\Gamma(\alpha_1 + \alpha_2 + \alpha_3)}{\Gamma(\alpha_1)\Gamma(\alpha_2)\Gamma(\alpha_3)} s^{\alpha_1 + \alpha_2 - 1} (1 - s)^{\alpha_3 - 1} \frac{\Gamma(\alpha_1)\Gamma(\alpha_2)}{\Gamma(\alpha_1 + \alpha_2)} \\
&= \frac{\Gamma(\alpha_1 + \alpha_2 + \alpha_3)}{\Gamma(\alpha_1 + \alpha_2)\Gamma(\alpha_3)} s^{\alpha_1 + \alpha_2 - 1} (1 - s)^{\alpha_3 - 1}, \quad 0 < s < 1.
\end{aligned}$$

This is the density function of a beta distribution; see Exercises 4.2 and 5.6. □

7.5 Order Statistics

Let $X_1, \ldots, X_n$ denote independent, identically distributed, real-valued random variables. The *order statistics* based on $X_1, X_2, \ldots, X_n$, denoted by $X_{(1)}, X_{(2)}, \ldots, X_{(n)}$, are simply the random variables $X_1, X_2, \ldots, X_n$ placed in ascending order. That is, let Ω denote the underlying sample space of the experiment; then, for each $\omega \in \Omega$,

$$X_{(1)}(\omega) = \min\{X_1(\omega), \ldots, X_n(\omega)\},$$

$X_{(2)}(\omega)$ is the second smallest value from $X_1(\omega), \ldots, X_n(\omega)$ and so on, up to $X_{(n)}(\omega)$, the maximum value from $X_1(\omega), \ldots, X_n(\omega)$. Hence, the random variables $X_{(1)}, \ldots, X_{(n)}$ satisfy the ordering

$$X_{(1)} \le X_{(2)} \le \cdots \le X_{(n)}.$$

There are at least two ways in which order statistics arise in statistics. One is that process generating the observed data might involve the order statistics of some of underlying, but unobserved, process. Another is that order statistics are often useful as summaries of a set of data.

***Example* 7.20 (*A model for software reliability*).** Consider the model for software reliability considered in Example 6.15. In that model, it is assumed that a piece of software has M errors or "bugs." Let Z_j denote the testing time required to discover bug j, $j = 1, \ldots, M$. Assume that $Z_1, Z_2, \ldots, Z_M$ are independent, identically distributed random variables.

Suppose that testing is continued until m bugs have been discovered. Then S_1, the time needed to find the first discovered bug, is the first order statistic $Z_{(1)}$; S_2, the time needed to find the first two bugs, is the second order statistic $Z_{(2)}$, and so on. Thus, a statistical analysis of this model would require the distribution of $(Z_{(1)}, \ldots, Z_{(m)})$. $\square$

Example 7.21 (The sample range). Let $X_1, \ldots, X_n$ denote independent, identically distributed random variables. One measure of the variability in the data $\{X_1, \ldots, X_n\}$ is the *sample range*, defined as the difference between the maximum and minimum values in the sample; in terms of the order statistics, the sample range is given by $X_{(n)} - X_{(1)}$. $\square$

The distribution theory of the order statistics is straightforward, at least in principle. Let F denote the distribution function of X_j, $j = 1, \ldots, n$. The event that $X_{(n)} \le t$ is equivalent to the event that $X_j \le t$, $j = 1, \ldots, n$. Hence, $X_{(n)}$ has distribution function $F_{(n)}$, given by

$$F_{(n)}(t) = F(t)^n.$$

Similarly, the event that $X_{(n-1)} \le t$ is equivalent to the event that at least $n - 1$ of the X_j are less than or equal to t. Hence, $X_{(n-1)}$ has distribution function $F_{(n-1)}$, given by

$$F_{(n-1)}(t) = F(t)^n + nF(t)^{n-1}(1 - F(t)).$$

This same approach can be used for any order statistic. The result is given in the following theorem; the proof is left as an exercise.

Theorem 7.6. *Let $X_1, X_2, \ldots, X_n$ denote independent, identically distributed real-valued random variables, each with distribution function F. Then the distribution function of $X_{(m)}$ is given by $F_{(m)}$ where*

$$F_{(m)}(t) = \sum_{i=m}^{n} \binom{n}{i} F(t)^i (1 - F(t))^{n-i}, \quad -\infty < t < \infty.$$

Example 7.22 (Pareto random variables). Let $X_1, X_2, \ldots, X_n$ denote independent, identically distributed random variables, each with an absolutely continuous distribution with density function

$$\theta x^{-(\theta+1)}, \quad x > 1,$$

where θ is a positive constant. Recall that this is a Pareto distribution; see Example 1.28.

The distribution function of this distribution is given by

$$F(t) = \int_1^t \theta x^{-(\theta+1)} \, dx = 1 - t^{-\theta}, \quad t > 1.$$

Hence, the distribution function of $X_{(m)}$, the mth order statistic, is given by

$$F_{(m)}(t) = \sum_{i=m}^{n} \binom{n}{i} (1 - t^{-\theta})^i (t^{-\theta})^{n-i}, \quad t > 1. \qquad \square$$

When the distribution given by F is either absolutely continuous or discrete, it is possible to derive the density function or frequency function, respectively, of the distribution.

Theorem 7.7. *Let $X_1, X_2, \ldots, X_n$ denote independent, identically distributed, real-valued random variables each with distribution function F and range $\mathcal{X}$.*

(i) *If X_1 has a discrete distribution with frequency function p, then $X_{(m)}$ has a discrete distribution with frequency function*

$$p_{(m)}(t) = \sum_{k=1}^{m} \binom{n}{m-k} F(t_0)^{m-k} \sum_{j=k}^{n+k-m} \binom{n-m+k}{j} p(t)^j [1 - F(t)]^{n-m+k-j},$$

for $t \in \mathcal{X}$; here t_0 is the largest element of $\mathcal{X}$ less than t.

(ii) *If X_1 has an absolutely continuous distribution with density function p, then $X_{(m)}$ has an absolutely continuous distribution with density function*

$$p_{(m)}(t) = n \binom{n-1}{m-1} F(t)^{m-1} [1 - F(t)]^{n-m} p(t), \quad -\infty < t < \infty.$$

Proof. First consider the case in which X_1 has a discrete distribution. Let t denote a fixed element of $\mathcal{X}$. Each observation $X_1, \ldots, X_n$ falls into one of three sets: $(-\infty, t_0]$, $\{t\}$, or $[t_1, \infty)$. Here t_0 denotes the largest element of $\mathcal{X}$ less than t, and t_1 denotes the smallest element of $\mathcal{X}$ greater than t. Let N_1, N_2, N_3 denote the number of observations falling into these three sets, respectively. Then

$$\Pr(X_{(m)} = t) = \sum_{k=1}^{m} \Pr(N_1 = m - k, \ N_2 \geq k) = \sum_{k=1}^{m} \sum_{j=k}^{n+k-m} \Pr(N_1 = m - k, \ N_2 = j).$$

Note that (N_1, N_2) has a multinomial distribution with

$$\Pr(N_1 = n_1, \ N_2 = n_2) = \binom{n}{n_1, n_2, n_3} F(t_0)^{n_1} p(t)^{n_2} (1 - F(t))^{n_3},$$

$n_1 + n_2 + n_3 = n$, where F and p denote the distribution function and frequency function, respectively, of the distribution of X_1. Hence,

$$\Pr(X_{(m)} = t)$$
$$= \sum_{k=1}^{m} \sum_{j=k}^{n+k-m} \binom{n}{m-k, j, n-m+k-j} F(t_0)^{m-k} p(t)^j (1 - F(t))^{n-m+k-j}$$
$$= \sum_{k=1}^{m} F(t_0)^{m-k} \binom{n}{m-k} \sum_{j=k}^{n+k-m} \binom{n-m+k}{j} p(t)^j (1 - F(t))^{n-m+k-j},$$

the result in part (i).

Now suppose that X_1 has an absolutely continuous distribution. Recall from Theorem 7.6 that $X_{(m)}$ has distribution function

$$F_{(m)}(t) = \sum_{i=m}^{n} \binom{n}{i} F(t)^i (1 - F(t))^{n-i},$$

and, since F is an absolutely continuous function, $F_{(m)}$ is absolutely continuous. Let t denote a continuity point of p; then, by Theorem 1.7, $F'(t)$ exists and

$$F'_{(m)}(t) = \sum_{i=m}^{n} i\binom{n}{i} F(t)^{i-1}(1 - F(t))^{n-i} p(t) - \sum_{i=m}^{n} (n-i)\binom{n}{i} F(t)^{i}(1 - F(t))^{n-1-i} p(t).$$

Using the identities

$$(i+1)\binom{n}{i+1} = n\binom{n-1}{i} = (n-i)\binom{n}{i},$$

it follows that

$$\sum_{i=m}^{m} i\binom{n}{i} F(t)^{i-1}(1 - F(t))^{n-i} = \sum_{j=m-1}^{n-1} (j+1)\binom{n}{j+1} F(t)^{j}(1 - F(t))^{n-1-j}$$

$$= n \sum_{j=m-1}^{n-1} \binom{n-1}{j} F(t)^{j}(1 - F(t))^{n-1-j}$$

and

$$\sum_{i=m}^{n} (n-i)\binom{n}{i} F(t)^{i}(1 - F(t))^{n-1-i} = n \sum_{i=m}^{n-1} \binom{n-1}{i} F(t)^{i}(1 - F(t))^{n-1-i}.$$

Hence,

$$F'_{(m)}(t) = n\binom{n-1}{j} F(t)^{j}(1 - F(t))^{n-1-j} p(t)\bigg|_{j=m-1}$$

$$= n\binom{n-1}{m-1} F(t)^{m-1}(1 - F(t))^{n-m} p(t),$$

proving part (ii).

Since p is continuous almost everywhere, it follows that $F'_{(m)}$ exists almost everywhere and, hence, part (iii) of Theorem 1.9 shows that $p_{(m)}(t) = F'_{(m)}(t)$ is a density function of $X_{(m)}$. ∎

Example 7.23 (Geometric random variables). Let $X_1, X_2, \ldots, X_n$ denote independent, identically distributed random variables, each distributed according to a discrete distribution with frequency function

$$\theta(1 - \theta)^{x}, \quad x = 0, 1, \ldots.$$

This is a geometric distribution; see Example 5.17. It is straightforward to show that the distribution function of this distribution is given by

$$F(x) = 1 - (1 - \theta)^{x+1}, \quad x = 0, 1, \ldots.$$

It follows that $X_{(m)}$, the mth order statistic, has a discrete distribution with frequency function

$p_{(m)}(t)$

$$= \sum_{k=1}^{m} \binom{n}{m-k} [1 - (1-\theta)^t]^{m-k} \sum_{j=k}^{n+k-m} \binom{n-m+k}{j} \theta^j (1-\theta)^{tj} (1-\theta)^{(t+1)(n-m+k-j)}$$

$$= \sum_{k=1}^{m} \binom{n}{m-k} [1 - (1-\theta)^t]^{m-k} (1-\theta)^{(t+1)(n-m+k)} \sum_{j=k}^{n+k-m} \binom{n-m+k}{j} \left(\frac{\theta}{1-\theta}\right)^j. \quad \square$$

***Example* 7.24 (*Uniform random variables*).** Let $X_1, X_2, \ldots, X_n$ denote independent, identically distributed random variables, each distributed according to the uniform distribution on $(0, 1)$. The density function of this distribution is $I_{\{0<t<1\}}$ and the distribution function is t, $0 < t < 1$. It follows that $X_{(m)}$, the mth order statistic, has an absolutely continuous distribution with density function

$$p_{(m)}(t) = n \binom{n-1}{m-1} t^{m-1} (1-t)^{n-m}, \quad 0 < t < 1.$$

This distribution is known as a *beta distribution* with parameters m and $n - m + 1$.

In general, a beta distribution with parameters α and β is an absolutely continuous distribution with density function

$$\frac{\Gamma(\alpha + \beta)}{\Gamma(\alpha)\Gamma(\beta)} x^{\alpha-1} (1-x)^{\beta-1}, \quad 0 < x < 1;$$

here $\alpha > 0$ and $\beta > 0$ are not restricted to be integers.

By Theorem 7.6, $X_{(m)}$ has distribution function

$$\sum_{i=m}^{n} \binom{n}{i} t^i (1-t)^{n-i}, \quad 0 < t < 1.$$

Hence, we obtain the useful result

$$n \binom{n-1}{m-1} \int_0^t u^{m-1} (1-u)^{n-m} \, du = \sum_{i=m}^{n} \binom{n}{i} t^i (1-t)^{n-i}, \quad 0 < t < 1. \qquad \square$$

Pairs of order statistics

An approach similiar to that used in Theorem 7.6 can be used to determine the distribution function of a pair of order statistics.

***Theorem* 7.8.** *Let $X_1, X_2, \ldots, X_n$ denote independent, identically distributed, real-valued random variables each with distribution function F. Let $X_{(1)}, X_{(2)}, \ldots, X_{(n)}$ denote the order statistics of $X_1, \ldots, X_n$ and let $m < r$.*
   Then

$\Pr(X_{(m)} \leq s, \; X_{(r)} \leq t)$

$$= \begin{cases} \sum_{i=m}^{n} \sum_{j=\max(0,r-i)}^{n-i} \binom{n}{i,j,n-i-j} F(s)^i [F(t) - F(s)]^{i-j} [1 - F(t)]^{n-i-j} & \textit{if } s < t \\ \Pr(X_{(r)} \leq t) & \textit{if } s \geq t \end{cases}.$$

Proof. Fix s, t; if $t \leq s$, then $X_{(r)} \leq t$ implies that $X_{(m)} \leq t \leq s, m < r$, so that

$$\Pr(X_{(m)} \leq s, \ X_{(r)} \leq t) = \Pr(X_{(r)} \leq t).$$

Now suppose that $s < t$; let N_1 denote the number of the observations $X_1, \ldots, X_n$ falling in the interval $(-\infty, s]$, let N_2 denote the number of observations falling in the interval $(s, t]$, and let $N_3 = n - N_1 - N_2$. Then, for $m < r$,

$$\Pr(X_{(m)} \leq s, \ X_{(r)} \leq t) = \Pr(N_1 \geq m, \ N_1 + N_2 \geq r)$$

$$= \sum_{i=m}^{n} \sum_{j=\max(0, r-i)}^{n-i} \Pr(N_1 = i, \ N_2 = j).$$

Since (N_1, N_2, N_3) has a multinomial distribution, with probabilities $F(s), F(t) - F(s), 1 - F(t)$, respectively, it follows that

$$\Pr(X_{(m)} \leq s, \ X_{(r)} \leq t)$$

$$= \sum_{i=m}^{n} \sum_{j=\max(0, r-i)}^{n-i} \binom{n}{i, \ j, \ n-i-j} F(s)^i [F(t) - F(s)]^j [1 - F(t)]^{n-i-j},$$

as stated in part (i). ∎

If the distribution function F in Theorem 7.8 is absolutely continuous, then the distribution of the order statistics $(X_{(m)}, X_{(r)})$ is absolutely continuous and the corresponding density function may be obtained by differentiation, as in Theorem 7.7. However, somewhat suprisingly, it turns out to be simpler to determine the density function of the entire set of order statistics and then marginalize to determine the density of the pair of order statistics under consideration.

Theorem 7.9. *Let $X_1, X_2, \ldots, X_n$ denote independent, identically distributed real-valued random variables each with distribution function F. Suppose that the distribution function F is absolutely continuous with density p. Then the distribution of $(X_{(1)}, \ldots, X_{(n)})$ is absolutely continuous with density function*

$$n! \, p(x_1) \cdots p(x_n), \quad x_1 < x_2 < \cdots < x_n. \quad \checkmark$$

Proof. Let τ denote a permutation of the integers $(1, \cdots, n)$ and let

$$\mathcal{X}(\tau) = \{x \in X^n : x_{\tau_1} < x_{\tau_2} < \cdots < x_{\tau_n}\} \cdot$$

where $\mathcal{X}$ denotes the range of X_1. Let

$$\mathcal{X}_0 = \cup_\tau \mathcal{X}(\tau)$$

where the union is over all possible permutations; note that

$$\Pr\{(X_1, \ldots, X_n) \in \mathcal{X}_0\} = 1$$

and, hence, we may proceed as if $\mathcal{X}_0$ is the range of $X = (X_1, \ldots, X_n)$.

Let $\tau X = (X_{\tau_1}, \ldots, X_{\tau_n})$, let $X_{(\cdot)} = (X_{(1)}, \ldots, X_{(n)})$ denote the vector of order statistics, and let h denote a bounded, real-valued function on the range of $X_{(\cdot)}$. Then

$$\mathrm{E}[h(X_{(\cdot)})] = \sum_\tau \mathrm{E}\{h(X_{(\cdot)}) \mathrm{I}_{\{X \in \mathcal{X}(\tau)\}}\}.$$

Note that, for $X \in \mathcal{X}(\tau)$, $X_{(\cdot)} = \tau X$. Hence,

$$\mathrm{E}\{h(X_{(\cdot)})\mathrm{I}_{\{X \in \mathcal{X}(\tau)\}}\} = \mathrm{E}\{h(\tau X)\mathrm{I}_{\{X \in \mathcal{X}(\tau)\}}\}.$$

Let τ_0 denote the identity permutation. Then the event that $X \in \mathcal{X}(\tau)$ is equivalent to the event that $\tau X \in \mathcal{X}(\tau_0)$. Hence,

$$\mathrm{E}[h(X_{(\cdot)})] = \sum_{\tau} \mathrm{E}\{h(\tau X)\mathrm{I}_{\{\tau X \in \mathcal{X}(\tau_0)\}}\}.$$

Since $X_1, \ldots, X_n$ are independent and identically distributed, for any permutation τ, τX has the same distribution as X. It follows that

$$\mathrm{E}[h(X_{(\cdot)})] = \sum_{\tau} \mathrm{E}\{h(X)\mathrm{I}_{\{X \in \mathcal{X}(\tau_0)\}}\} = n!\mathrm{E}[h(X)\mathrm{I}_{\{X \in \mathcal{X}(\tau_0)\}}];$$

the factor $n!$ is due to the fact that there are $n!$ possible permutations. The result follows.   ∎

As noted above, the density function of $(X_{(1)}, \ldots, X_{(n)})$ may be used to determine the density function of some smaller set of order statistics. The following lemma is useful in carrying out that approach.

Lemma 7.1. *Let p denote the density function of an absolutely continuous distribution on $\mathbf{R}$ and let F denote the corresponding distribution function. Then, for any $n = 1, 2, \ldots$, and any $a < b$,*

$$n! \int_{-\infty}^{\infty} \cdots \int_{-\infty}^{\infty} p(x_1) \cdots p(x_n)\mathrm{I}_{\{a < x_1 < x_2 < \cdots < x_n < b\}} \, dx_1 \cdots dx_n = [F(b) - F(a)]^n.$$

Proof.   Let $X_1, X_2, \ldots, X_n$ denote independent, identically distributed random variables, each distributed according to the distribution with distribution function F and density function p. Then, according to Theorem 7.9, the density function of $(X_{(1)}, \ldots, X_{(n)})$ is given by

$$n! p(x_1) \cdots p(x_n), \quad -\infty < x_1 < x_2 < \cdots < x_n < \infty.$$

It follows that

$$\Pr(a < X_{(1)} < \cdots < X_{(n)} < b)$$
$$= n! \int_{-\infty}^{\infty} \cdots \int_{-\infty}^{\infty} p(x_1) \cdots p(x_n)\mathrm{I}_{\{a < x_1 < x_2 < \cdots < x_n < b\}} \, dx_1 \cdots dx_n.$$

Note that the event $a < X_{(1)} < \cdots < X_{(n)} < b$ is simply the event that all observations fall in the interval (a, b). Hence,

$$\Pr(a < X_{(1)} < \cdots < X_{(n)} < b) = \Pr(a < X_1 < b, \ a < X_2 < b, \ldots, a < X_n < b)$$
$$= \Pr(a < X_1 < b) \cdots \Pr(a < X_n < b)$$
$$= [F(b) - F(a)]^n,$$

proving the result.   ∎

Using Lemma 7.1 together with Theorem 7.9 yields the density function of any pair of order statistics; note that the same approach may be used to determine the density function of any subset of the set of all order statistics.

Theorem 7.10. *Let $X_1, X_2, \ldots, X_n$ denote independent, identically distributed, real-valued random variables, each with an absolutely continuous distribution with density p and distribution function F. Let $X_{(1)}, X_{(2)}, \ldots, X_{(n)}$ denote the order statistics of $X_1, \ldots, X_n$ and let $m < r$.*

The distribution of $(X_{(m)}, X_{(r)})$ is absolutely continuous with density function

$$\frac{n!}{(m-1)!(r-m-1)!(n-r)!} F(x_m)^{m-1}[F(x_r) - F(x_m)]^{r-m-1}[1 - F(x_r)]^{n-r} p(x_m)p(x_r),$$

for $x_m < x_r$.

Proof. The density function of $(X_{(1)}, X_{(2)}, \ldots, X_{(n)})$ is given by

$$n! \, p(x_1) \cdots p(x_n), \quad -\infty < x_1 < x_2 < \cdots < x_n < \infty.$$

The marginal density of $(X_{(m)}, X_{(r)})$, $m < r$, is therefore given by

$$n! \int_{-\infty}^{\infty} \cdots \int_{-\infty}^{\infty} p(x_1) \cdots p(x_n) I_{\{x_1 < x_2 < \cdots < x_n\}} \, dx_1 \cdots dx_{m-1} \, dx_{m+1} \cdots dx_{r-1} \, dx_{r+1} \cdots dx_n.$$

Note that

$$I_{\{x_1 < x_2 < \cdots < x_n\}} = I_{\{x_1 < \cdots < x_m\}} I_{\{x_m < x_{m+1} < \cdots < x_r\}} I_{\{x_r < x_{r+1} < \cdots < x_n\}}.$$

By Lemma 7.1,

$$\int_{-\infty}^{\infty} \cdots \int_{-\infty}^{\infty} p(x_1) \cdots p(x_{m-1}) I_{\{x_1 < x_2 < \cdots < x_{m-1} < x_m\}} \, dx_1 \cdots dx_{m-1} = \frac{1}{(m-1)!} F(x_m)^{m-1},$$

$$\int_{-\infty}^{\infty} \cdots \int_{-\infty}^{\infty} p(x_{m+1}) \cdots p(x_{r-1}) I_{\{x_m < x_{m+1} < \cdots < x_{r-1} < x_r\}} \, dx_{m+1} \cdots dx_{r-1}$$

$$= \frac{1}{(r-m-1)!} [F(x_r) - F(x_m)]^{r-m-1},$$

and

$$\int_{-\infty}^{\infty} \cdots \int_{-\infty}^{\infty} p(x_{r+1}) \cdots p(x_n) I_{\{x_r < x_{r+1} < \cdots < x_{n-1} < x_n\}} \, dx_{r+1} \cdots dx_n$$

$$= \frac{1}{(n-r)!} [1 - F(x_r)]^{n-r}.$$

The result follows. ∎

Example 7.25 (Order statistics of exponential random variables). Let $X_1, X_2, \ldots, X_n$ denote independent, identically distributed random variables, each with an exponential distribution with parameter $\lambda > 0$; this distribution has density function $\lambda \exp(-\lambda x)$, $x > 0$, and distribution function $1 - \exp(-\lambda x)$, $x > 0$.

According to Theorem 7.9, $(X_{(1)}, \ldots, X_{(n)})$ has an absolutely continuous distribution with density function

$$n! \lambda^n \exp\left(-\lambda \sum_{j=1}^{n} x_j\right), \quad 0 < x_1 < x_2 < \cdots < x_n < \infty.$$

Let

$$Y_1 = X_{(1)}, \quad Y_2 = X_{(2)} - X_{(1)}, \ldots, Y_n = X_{(n)} - X_{(n-1)};$$

note that, if $X_1, \ldots, X_n$ denote event times of some random process, then $X_{(1)}, \ldots, X_{(n)}$ denote the ordered event times and $Y_1, \ldots, Y_n$ denote the inter-event times.

The density function of $Y = (Y_1, \ldots, Y_n)$ can be obtained from Theorem 7.2. We can write $(X_{(1)}, \ldots, X_{(n)}) = h(Y) \equiv (h_1(Y), \ldots, h_n(Y))$ where

$$h_j(y) = y_1 + \cdots + y_j, \quad j = 1, \ldots, n.$$

It follows that the Jacobian of h is 1 and the density of Y is given by

$$p_Y(y) = n! \lambda^n \exp\{-\lambda[y_1 + (y_1 + y_2) + \cdots + (y_1 + y_n)]\}$$

$$= n! \lambda^n \exp\left\{-\lambda \sum_{j=1}^{n} (n - j + 1) y_j\right\}, \quad y_j > 0, \quad j = 1, \ldots, n.$$

Since $n! = \prod_{j=1}^{n} (n - j + 1)$, it follows that $Y_1, \ldots, Y_n$ are independent exponential random variables such that Y_j has parameter $(n - j + 1)\lambda$. □

Example 7.26 (Range of uniform random variables). Let $X_1, X_2, \ldots, X_n$ denote independent, identically distributed random variables, each distributed according to a uniform distribution on $(0, 1)$. Consider the problem of determining the distribution of $X_{(n)} - X_{(1)}$, the difference between the maximum and minimum values in the sample.

The joint distribution of $(X_{(1)}, X_{(n)})$ is absolutely continuous with density function

$$n(n - 1)(x_n - x_1)^{n-2}, \quad 0 < x_1 < x_n < 1.$$

Let $T = X_{(n)} - X_{(1)}$ and $Y = X_{(1)}$. Then, using Theorem 7.2, the distribution of (T, Y) is absolutely continuous with density function

$$p(t, y) = n(n - 1)t^{n-2}, \quad 0 < y < t + y < 1.$$

Hence, the marginal density of T is

$$p_T(t) = \int_0^{1-t} n(n - 1)t^{n-2} \, dy = n(n - 1)t^{n-2}(1 - t), \quad 0 < t < 1.$$

Thus, the distribution of T is a beta distribution with parameters $n - 1$ and 2; see Example 7.24. □

7.6 Ranks

Let $X_1, \ldots, X_n$ denote independent, identically distributed, real-valued random variables and let $X_{(1)}, \ldots, X_{(n)}$ denote the corresponding order statistics. It is easy to see that $(X_1, \ldots, X_n)$ and $(X_{(1)}, \ldots, X_{(n)})$ are not equivalent statistics; in particular, it is not possible to reconstruct $X_1, \ldots, X_n$ given only $X_{(1)}, \ldots, X_{(n)}$. The missing information is the vector of *ranks* of $X_1, \ldots, X_n$.

The rank of X_i among $X_1, \ldots, X_n$ is its position in the order statistics and is defined to be the integer R_i, $1 \le R_i \le n$, satisfying $X_i = X_{(R_i)}$, provided that $X_1, \ldots, X_n$ are unique. Here we assume that common distribution of $X_1, X_2, \ldots, X_n$ is absolutely continuous so that $X_1, \ldots, X_n$ are unique with probability 1. Let $R = (R_1, \ldots, R_n)$ denote the vector of ranks.

The following theorem summarizes the properties of R.

Theorem 7.11. *Let $X_1, \ldots, X_n$ denote independent, identically distributed, real-valued random variables, each with an absolutely continuous distribution. Then*

(i) *The statistic $(R, X_{(\cdot)})$ is a one-to-one function of X.*

(ii) *$(R_1, \ldots, R_n)$ is uniformly distributed on the set of all permutations of $(1, 2, \ldots, n)$; that is, each possible value of $(R_1, \ldots, R_n)$ has the same probability.*

(iii) *$X_{(\cdot)}$ and R are independent*

(iv) *For any statistic $T \equiv T(X_1, \ldots, X_n)$ such that $\mathrm{E}(|T|) < \infty$,*

$$\mathrm{E}[T \mid R = r] = \mathrm{E}[T(X_{(r_1)}, X_{(r_2)}, \ldots, X_{(r_n)})]$$

where $r = (r_1, r_2, \ldots, r_n)$.

Proof. Clearly, $(R, X_{(\cdot)})$ is a function of $(X_1, \ldots, X_n)$. Part (i) of the theorem now follows from the fact that $X_j = X_{(R_j)}$, $j = 1, \ldots, n$.

Let τ denote a permutation of $(1, 2, \ldots, n)$ and let

$$\mathcal{X}(\tau) = \{x \in X^n : x_{\tau_1} < x_{\tau_2} < \cdots < x_{\tau_n}\}.$$

Let

$$\mathcal{X}_0 = \cup_\tau \mathcal{X}(\tau)$$

where the union is over all possible permutations of $(1, 2, \ldots, n)$. Note that

$$\mathrm{Pr}\{(X_1, \ldots, X_n) \in \mathcal{X}_0\} = 1$$

so that we may proceed as if the range of $(X_1, \ldots, X_n)$ is $\mathcal{X}_0$.

Let h denote a real-valued function of $R \equiv R(X)$ such that $\mathrm{E}[h(R)] < \infty$. Then

$$\mathrm{E}[h(R)] = \sum_\tau \mathrm{E}[h(R(X))\mathrm{I}_{\{X \in \mathcal{X}(\tau)\}}].$$

Note that, for $X \in \mathcal{X}(\tau)$, $R(X) = \tau$. Hence,

$$\mathrm{E}[h(R)] = \sum_\tau \mathrm{E}[h(\tau)\mathrm{I}_{\{X \in \mathcal{X}(\tau)\}}] = \sum_\tau h(\tau)\mathrm{Pr}(X \in \mathcal{X}(\tau)).$$

Let τ_0 denote the identity permutation. Then

$$X \in \mathcal{X}(\tau) \qquad \text{if and only if} \quad \tau X \in \mathcal{X}(\tau_0)$$

and, since the distribution of $(X_1, X_2, \ldots, X_n)$ is exchangeable, τX has the same distribution as X. Hence,

$$\mathrm{Pr}(X \in \mathcal{X}(\tau)) = \mathrm{Pr}(\tau X \in \mathcal{X}(\tau_0)) = \mathrm{Pr}(X \in \mathcal{X}(\tau_0)).$$

Since there are $n!$ possible permutations of $(1, 2, \ldots, n)$ and

$$\sum_\tau \mathrm{Pr}(X \in \mathcal{X}(\tau)) = 1,$$

it follows that $\Pr(X \in \mathcal{X}(\tau)) = 1/n!$ for each τ and, hence, that

$$E[h(R)] = \frac{1}{n!} \sum_{\tau} h(\tau)$$

proving part (ii).

Part (iii) follows along similar lines. Let g denote a bounded function of $X_{(\cdot)}$ and let h denote a bounded function of R.

Note that

$$
\begin{aligned}
E[g(X_{(\cdot)})h(R)] &= \sum_{\tau} E[g(X_{(\cdot)})h(R(X))\mathrm{I}_{\{X \in \mathcal{X}(\tau)\}}] \\
&= \sum_{\tau} E[g(\tau X)h(\tau)\mathrm{I}_{\{X \in \mathcal{X}(\tau)\}}] \\
&= \sum_{\tau} h(\tau)E[g(\tau X)\mathrm{I}_{\{\tau X \in \mathcal{X}(\tau_0)\}}] \\
&= \sum_{\tau} h(\tau)E[g(X)\mathrm{I}_{\{X \in \mathcal{X}(\tau_0)\}}] \\
&= E[g(X)\mathrm{I}_{\{X \in \mathcal{X}(\tau_0)\}}] \sum_{\tau} h(\tau) \\
&= n! E[g(X)\mathrm{I}_{\{X \in \mathcal{X}(\tau_0)\}}] \frac{1}{n!} \sum_{\tau} h(\tau) \\
&= E[g(X_{(\cdot)})]E[h(R)],
\end{aligned}
$$

proving part (iii).

Finally, part (iv) follows from the fact that any statistic $T \equiv T(X)$ may be written as $\bar{T}(R, X_{(\cdot)})$ and, by part (iii) of the theorem,

$$E[T|R = r] = E[\bar{T}(R, X_{(\cdot)})|R = r] = E[\bar{T}(r, X_{(\cdot)})] = E[T(X_{(r_1)}, \cdots, X_{(r_n)})]. \qquad \blacksquare$$

***Example* 7.27 (*Mean and variance of linear rank statistics*).** Let $R_1, R_2, \ldots, R_n$ denote the ranks of a sequence of independent and identically distributed real-valued random variables, each distributed according to an absolutely continuous distribution. Consider a statistic of the form

$$T = \sum_{j=1}^{n} a_j R_j$$

where $a_1, a_2, \ldots, a_n$ is a sequence of constants. Here we consider determination of the mean and variance of T.

Note that each R_j has the same marginal distribution. Since

$$\sum_{j=1}^{n} R_j = \frac{n(n+1)}{2}, \tag{7.2}$$

it follows that $E(R_j) = (n+1)/2$, $j = 1, \ldots, n$. Also, each pair (R_i, R_j) has the same marginal distribution so that $\mathrm{Cov}(R_i, R_j)$ does not depend on the pair (i, j).

Let $\sigma^2 = \mathrm{Var}(R_j)$ and $c = \mathrm{Cov}(R_i, R_j)$. By (7.2),

$$\mathrm{Var}\left(\sum_{j=1}^{n} R_j\right) = n\sigma^2 + n(n-1)c = 0$$

so that

$$c = -\frac{\sigma^2}{n-1}.$$

To find σ^2, note that $R_1 = j$ with probability $1/n$. Hence,

$$E\left(R_1^2\right) = \frac{1}{n}\sum_{j=1}^{n} j^2 = \frac{(n+1)(2n+1)}{6}.$$

Since $E(R_1) = (n+1)/2$, it follows that $\sigma^2 = (n^2-1)/12$ and $c = -(n+1)/12$.

Now consider the statistic T. The expected value of T is given by

$$E(T) = \frac{n+1}{2}\sum_{j=1}^{n} a_j;$$

the variance of T is given by

$$\text{Var}(T) = \sigma^2 \sum_{j=1}^{n} a_j^2 + 2c \sum_{i<j} a_i a_j = \frac{n^2-1}{12}\sum_{j=1}^{n} a_j^2 - \frac{n+1}{6}\sum_{i<j} a_i a_j.$$

For instance, consider $a_j = j$; when the data are collected in time order, the statistic $\sum_{j=1}^{n} j R_j$ may be used to test the hypothesis of a time trend in the data. If the data are in fact independent and identically distributed, this statistic has mean

$$\frac{n(n+1)^2}{4}$$

and variance

$$\frac{n^2+n}{12}\sum_{j=1}^{n} j^2 - \frac{n+1}{12}\sum_{j=1}^{n} j^3 = \frac{n^2(n+1)(n^2-1)}{144}. \qquad \Box$$

Example 7.28 (Conditional expectation of a sum of uniform random variables). Let $X_1, \ldots, X_n$ denote independent, identically distributed random variables, each uniformly distributed on $(0, 1)$ and let $a_1, \ldots, a_n$ denote a sequence of constants. Consider

$$E\left\{\sum_{j=1}^{n} a_j X_j | R_1, \ldots, R_n\right\}$$

where $(R_1, \ldots, R_n)$ denotes the vector of ranks.

Let $(r_1, \ldots, r_n)$ denote a permutation of $(1, \ldots, n)$. According to Theorem 7.11, part (iv),

$$E\left\{\sum_{j=1}^{n} a_j X_j | R_1 = r_1, \ldots, R_n = r_n\right\} = E\left\{\sum_{j=1}^{n} a_j X_{(r_j)}\right\} = \sum_{j=1}^{n} a_j E\{X_{(r_j)}\}.$$

From Example 7.19, we know that $X_{(m)}$ has a beta distribution with parameters m and $n - m + 1$; hence, it is straightforward to show that

$$E\{X_{(m)}\} = \frac{m}{n+1}.$$

It follows that

$$E\left\{\sum_{j=1}^{n} a_j X_j | R_1 = r_1, \ldots, R_n = r_n\right\} = \sum_{j=1}^{n} a_j \frac{r_j}{n+1}$$

so that

$$E\left\{\sum_{j=1}^{n} a_j X_j | R_1, \ldots, R_n\right\} = \frac{1}{n+1} \sum_{j=1}^{n} a_j R_j;$$

that is, $E\{\sum_{j=1}^{n} a_j X_j | R_1, \ldots, R_n\}$ is a linear rank statistic. $\square$

7.7   Monte Carlo Methods

Let X denote a random variable, possibly vector-valued, with distribution function F_X. Suppose we are interested in the probability $\Pr(g(X) \le y)$ where g is a real-valued function on the range of X and y is some specified value. For instance, this probability could represent a p-value or a coverage probability. In this chapter, we have discussed several methods of determining the distribution of $Y = g(X)$. However, these methods often require substantial mathematical analysis that, in some cases, is very difficult or nearly impossible.

Consider the following alternative approach. Suppose that we may construct a process that generates data with the same distribution as X; let $X_1, \ldots, X_N$ denote independent, identically distributed random variables, each with the same distribution as X and let

$$Y_j = g(X_j), \quad j = 1, \ldots, N.$$

Let

$$\hat{P}_N = \frac{1}{N} \sum_{j=1}^{N} I_{\{Y_j \le y\}}$$

denote the proportion of $Y_1, \ldots, Y_N$ that are less than or equal to y. Thus, if N is large enough, we expect that

$$\hat{P}_N \approx \Pr(Y \le y).$$

Hence, we use $\hat{P}_N$ as an estimate of $\Pr(Y \le y)$. In fact, any type of statistical method, such as a confidence interval, may be used to analyze the data generated in this manner.

This approach is known as the *Monte Carlo method*. The Monte Carlo method is a vast topic. In this section, we give only a brief overview of the method; for further details, see Section 7.9.

***Example* 7.29 (*Ratio of exponential random variables to their sum*).** Let $X = (X_1, X_2)$ denote a random vector such that X_1, X_2 are independent, identically distributed exponential random variables with mean λ and let $Y = X_1/(X_1 + X_2)$; see Example 7.7. Consider the probability $\Pr(Y \le 1/4)$ for $\lambda = 1$.

To estimate this probability, we can generate N pairs of independent standard exponential random variables, $(X_{11}, X_{21}), \ldots, (X_{1N}, X_{2N})$, and define

$$Y_j = \frac{X_{1j}}{X_{1j} + X_{2j}}, \quad j = 1, \ldots, N.$$

When this approach was used with $N = 1,000$, the estimate $\hat{P}_N = 0.244$ was obtained; when $N = 10,000$ was used, the estimate $\hat{P}_N = 0.2464$ was obtained. Recall that the distribution of Y is uniform on the interval $(0, 1)$; thus, the exact probability $\Pr(Y \leq 1/4)$ is $1/4$. $\quad \square$

Of course, use of the Monte Carlo method requires that we be able to generate random variables with a specified distribution. Standard statistical packages generally have procedures for generating data from commonly used distributions. In many cases if the density or distribution function of X is available, then it is possible to construct a method for generating data with the same distribution as X; this is particularly true if X is real-valued or is a random vector with independent components. These methods are generally based on a procedure to generate variables that are uniformly distributed on $(0, 1)$; such procedures are well-studied and widely available. However, depending on the exact distribution of X, the actual method required to generate the data may be quite sophisticated. The following two examples illustrate some simple methods that are often applicable.

Example 7.30 *(Generation of standard exponential random variables).* Consider generation of observations with a standard exponential distribution, required in Example 7.29. Let U denote a random variable that is uniformly distributed on $(0, 1)$ and let $X = -\log(1 - U)$. Then

$$\Pr(X \leq x) = \Pr(U \leq 1 - \exp(-x)) = 1 - \exp(-x), \quad x > 0.$$

Hence, a sequence of independent uniform random variables may be easily transformed to a sequence of independent standard exponential random variables.

This approach, sometimes called the *inversion method*, can be used whenever X has distribution function F on the real line and the quantile function corresponding to F is available; see the proof of Theorem 1.3. $\quad \square$

Example 7.31 *(Hierarchical models).* Suppose that the random variable X follows a hierarchical model, as discussed in Section 5.4. Specifically, suppose that the distribution of X can be described in two stages: the conditional distribution of X given a random variable λ and the marginal distribution of λ. If algorithms for generating data from the conditional distribution of X given λ and from the marginal distribution of λ are both available, then random variables from the distribution of X may be generated using a two-stage process. For each $j = 1, \ldots, N$, suppose that λ_j is drawn from the marginal distribution of λ; then we can draw X_j from the conditional distribution of X given $\lambda = \lambda_j$.

This method can also be used in cases in which the distribution of X is not originally described in terms of a hierarchical model, but it is possible to describe the distribution of X in terms of a hierarchical model. $\quad \square$

The primary advantage of the Monte Carlo method is that it may be used in (nearly) every problem; it is particularly useful in cases, such as the one in the following example, in which an exact analysis is very difficult.

Example 7.32 *(An implicitly defined statistic).* Let $X = (Z_1, \ldots, Z_n)$ where the Z_j are independent and identically distributed standard exponential random variables and consider the statistic $Y \equiv Y(Z_1, \ldots, Z_n)$ defined by the following equation:

$$\sum_{j=1}^{n} \left(Z_j^Y - 1 \right) \log Z_j = n.$$

This statistic arises in connection with estimation of the parameters of the Weibull distribution.

Since an explicit expression for Y in terms of $Z_1, \ldots, Z_n$ is not available, an exact expression for the distribution of Y is difficult, if not impossible, to determine. However, the Monte Carlo method is still easily applied; all that is needed is an algorithm for determining Y from a given value of $X = (Z_1, \ldots, Z_n)$. $\square$

There are a number of disadvantages to the Monte Carlo method. One is that, if the method is repeated, that is, if a new set of observations is generated, a new result for $\hat{P}_N$ is obtained. Although the variation in different values of $\hat{P}_N$ may be decreased by choosing a very large value of N, clearly it would be preferable if two sets of "identical" calculations would lead to identical results. This is particularly a problem in complex settings in which generation of each Y_j is time-consuming and, hence, N must be chosen to be relatively small. In view of this variation, it is standard practice to supplement each estimate of $\Pr(Y \leq y)$ with its standard error.

Example 7.33 (Ratio of exponential random variables to their sum). Consider the probability considered in Example 7.29. When the Monte Carlo analysis was repeated, the results $\hat{P}_N = 0.246$ for $N = 1,000$ and $\hat{P}_N = 0.2428$ for $N = 10,000$ were obtained. These may be compared to the results obtained previously. $\square$

A second drawback of the Monte Carlo method is that, because no formula for $\Pr(Y \leq y)$ is available, it may be difficult to see how the probability varies as different parameters in the problem vary.

Example 7.34 (Ratio of exponentials random variables to their sum). Recall Example 7.29. The probability of interest was $\Pr(Y \leq 1/4)$, calculated under the assumption that $\lambda = 1$. Suppose that we now want the same probability calculated under the assumption that $\lambda = 5$. Note that the distribution of the statistic

$$\frac{X_1}{X_1 + X_2}$$

does not depend on the value of λ. To see this, note that we may write $X_j = \lambda Z_j$, $j = 1, 2$, where Z_1, Z_2 are independent standard exponential random variables; this result also follows from the general results on invariance presented in Section 5.6.

When the Monte Carlo approach was used with $\lambda = 5$ and $N = 10,000$, the result was $\hat{P}_N = 0.2500$. Although this is close to the result calculated under $\lambda = 1$ (0.2464), it is not clear from these values that the two probabilities are exactly equal. $\square$

Despite these drawbacks, the Monte Carlo method is a very useful and powerful method. It is invaluable in cases in which an exact analysis is not available. Furthermore, the generality of the Monte Carlo method gives the statistical analyst more flexibility in choosing a statistical model since models do not have to be chosen on the basis of their analytical tractability. Also, even in cases in which an exact analysis is possible, results from a Monte

Carlo analysis are very useful as a check on the theoretical calculations. When approximations to $\Pr(Y \leq y)$ are used, as will be discussed in Chapters 11–14, the results of a Monte Carlo study give us a method of assessing the accuracy of the approximations.

The importance of the drawbacks discussed above can be minimized by more sophisticated Monte Carlo methods. For instance, many methods are available for reducing the variation in the Monte Carlo results. Also, before carrying out the Monte Carlo study, it is important to do a thorough theoretical analysis. For instance, in Example 7.34, even if the distribution of $X_1/(X_1 + X_2)$ is difficult to determine analytically, it is easy to show that the distribution of this ratio does not depend on the value of λ; thus, the results for $\lambda = 1$ can be assumed to hold for all $\lambda > 0$.

7.8 Exercises

7.1 Let X denote a random variable with a uniform distribution on the interval $(0, 1)$. Find the density function of

$$Y = \frac{X}{1 - X}.$$

7.2 Let X denote a random variable with a standard normal distribution. Find the density function of $Y = 1/X$.

7.3 Let X denote a random variable with a Poisson distribution with mean 1. Find the frequency function of $Y = X/(1 + X)$.

7.4 Let X denote a random variable with an F-distribution with ν_1 and ν_2 degrees of freedom. Find the density function of

$$Y = \frac{\nu_1}{\nu_2} \frac{X}{1 + (\nu_1/\nu_2)X}.$$

7.5 Let X_1 and X_2 denote independent, real-valued random variables with absolutely continuous distributions with density functions p_1 and p_2, respectively. Let $Y = X_1/X_2$. Show that Y has density function

$$p_Y(y) = \int_{-\infty}^{\infty} |z| p_1(zy) p_2(z) \, dz.$$

7.6 Let X_1, X_2 denote independent random variables such that X_j has an absolutely continuous distribution with density function

$$\lambda_j \exp(-\lambda_j x), \quad x > 0,$$

$j = 1, 2$, where $\lambda_1 > 0$ and $\lambda_2 > 0$. Find the density of $Y = X_1/X_2$.

7.7 Let X_1, X_2, X_3 denote independent random variables, each with an absolutely continuous distribution with density function

$$\lambda \exp\{-\lambda x\}, \quad x > 0$$

where $\lambda > 0$. Find the density function of $Y = X_1 + X_2 - X_3$.

7.8 Let X and Y denote independent random variables, each with an absolutely continuous distribution with density function

$$p(x) = \frac{1}{2} \exp\{-|x|\}, \quad -\infty < x < \infty.$$

Find the density function of $Z = X + Y$.

7.9 Let X_1 and X_2 denote independent random variables, each with a uniform distribution on $(0, 1)$. Find the density function of $Y = \log(X_1/X_2)$.

7.10 Let X and Y denote independent random variables such that X has a standard normal distribution and Y has a standard exponential distribution. Find the density function of $X + Y$.

7.11 Suppose $X = (X_1, X_2)$ has density function

$$p(x_1, x_2) = x_1^{-2} x_2^{-2}, \quad x_1 > 1, \quad x_2 > 1.$$

Find the density function of $X_1 X_2$.

7.12 Let $X_1, \ldots, X_n$ denote independent, identically distributed random variables, each of which is uniformly distributed on the interval $(0, 1)$. Find the density function of $T = \prod_{j=1}^{n} X_j$.

7.13 Let $X_1, X_2, \ldots, X_n$ denote independent, identically distributed random variables, each with an absolutely continuous distribution with density function $1/x^2$, $x > 1$, and assume that $n \geq 3$. Let

$$Y_j = X_j X_n, \quad j = 1, \ldots, n-1.$$

Find the density function of $(Y_1, \ldots, Y_{n-1})$.

7.14 Let X be a real-valued random variable with an absolutely continuous distribution with density function p. Find the density function of $Y = |X|$.

7.15 Let X denote a real-valued random variable with a t-distribution with ν degrees of freedom. Find the density function of $Y = X^2$.

7.16 Let X and Y denote independent discrete random variables, each with density function $p(\cdot; \theta)$ where $0 < \theta < 1$. For each of the choices of $p(\cdot; \theta)$ given below, find the conditional distribution of X given $S = s$ where $S = X + Y$.
(a) $p(j; \theta) = (1 - \theta)\theta^j, \quad j = 0, \ldots$
(b) $p(j; \theta) = (1 - \theta)[-\log(1 - \theta)]^j/j!, \quad j = 0, \ldots$
(c) $p(j; \theta) = \theta^{j+1}/[j(-\log(1 - \theta))], \quad j = 0, \ldots$
Suppose that $S = 3$ is observed. For each of the three distributions above, give the conditional probabilities of the pairs $(0, 3), (1, 2), (2, 1), (3, 0)$ for (X, Y).

7.17 Let X and Y denote independent random variables, each with an absolutely continuous distribution with density function

$$\frac{\alpha}{x^{\alpha+1}}, \quad x > 1$$

where $\alpha > 1$. Let $S = XY$ and $T = X/Y$. Find $\mathrm{E}(X|S)$ and $\mathrm{E}(T|S)$.

7.18 Let X denote a nonnegative random variable with an absolutely continuous distribution. Let F and p denote the distribution function and density function, respectively, of the distribution. The *hazard function* of the distribution is defined as

$$h(x) = \frac{p(x)}{1 - F(x)}, \quad x > 0.$$

Let $X_1, \ldots, X_n$ denote independent, identically distributed random variables, each with the same distribution as X, and let

$$Y = \min(X_1, \ldots, X_n).$$

Find the hazard function of Y.

7.19 Let X_1, X_2 denote independent random variables, each with a standard exponential distribution. Find $\mathrm{E}(X_1 + X_2|X_1 - X_2)$ and $\mathrm{E}(X_1 - X_2|X_1 + X_2)$.

7.20 Let $X_1, \ldots, X_n$ denote independent random variables such that X_j has a normal distribution with mean μ_j and standard deviation σ_j. Find the distribution of $\bar{X}$.

7.21 Let $X_1, \ldots, X_n$ denote independent random variables such that X_j has an absolutely continuous distribution with density function

$$p_j(x_j) = \Gamma(\alpha_j)^{-1} x_j^{\alpha_j - 1} \exp\{-x_j\}, \quad x_j > 0$$

where $\alpha_j > 0$, $j = 1, \ldots, n$. Find the density function of $Y = \sum_{j=1}^{n} X_j$.

7.22 Let $X_1, \ldots, X_n$ denote independent, identically distributed random variables, each with a standard exponential distribution. Find the density function of $R = X_{(n)} - X_{(1)}$.

7.23 Prove Theorem 7.6.

7.24 Let $X_1, X_2, \ldots, X_n$ be independent, identically distributed random variables, each with an absolutely continuous distribution with density function

$$\frac{1}{x^2}, \quad x > 1.$$

Let $X_{(j)}$ denote the jth order statistic of the sample. Find $E[X_{(j)}]$. Assume that $n \geq 2$.

7.25 Let X_1, X_2, X_3 denote independent, identically distributed random variables, each with an exponential distribution with mean λ. Find an expression for the density function of $X_{(3)}/X_{(1)}$.

7.26 Let $X_1, \ldots, X_n$ denote independent, identically distributed random variables, each with a uniform distribution on $(0, 1)$ and let $X_{(1)}, \ldots, X_{(n)}$ denote the order statistics. Find the correlation of $X_{(i)}$ and $X_{(j)}$, $i < j$.

7.27 Let $X_1, \ldots, X_n$ denote independent, identically distributed random variables, each with a standard exponential distribution. Find the distribution of

$$\sum_{j=1}^{n} (X_j - X_{(1)}).$$

7.28 Let $X = (X_1, \ldots, X_n)$ where $X_1, \ldots, X_n$ are independent, identically distributed random variables, each with an absolutely continuous distribution with range $\mathcal{X}$. Let $X_{(\cdot)} = (X_{(1)}, \ldots, X_{(n)})$ denote the vector of order statistics and $R = (R_1, \ldots, R_n)$ denote the vector of ranks corresponding to $(X_1, \ldots, X_n)$.

(a) Let h denote a real-valued function on $\mathcal{X}^n$. Show that if h is permutation invariant, then

$$h(X) = h(X_{(\cdot)}) \qquad \text{with probability 1}$$

and, hence, that $h(X)$ and R are independent.

(b) Does the converse hold? That is, suppose that $h(X)$ and R are independent. Does it follow that h is permutation invariant?

7.29 Let U_1, U_2 denote independent random variables, each with a uniform distribution on the interval $(0, 1)$. Let

$$X_1 = \sqrt{(-2 \log U_1)} \cos(2\pi U_2)$$

and

$$X_2 = \sqrt{(-2 \log U_1)} \sin(2\pi U_2).$$

Find the density function of (X_1, X_2).

7.30 Consider an absolutely continuous distribution with nonconstant, continuous density function p and distribution function F such that $F(1) = 1$ and $F(0) = 0$. Let $(X_1, Y_1), (X_2, Y_2), \ldots$ denote independent pairs of independent random variables such that each X_j is uniformly distributed on $(0, 1)$ and each Y_j is uniformly distributed on $(0, c)$, where

$$c = \sup_{0 \leq t \leq 1} p(t).$$

Define a random variable Z as follows. If $Y_1 \leq p(X_1)$, then $Z = X_1$. Otherwise, if $Y_2 \leq p(X_2)$, then $Z = X_2$. Otherwise, if $Y_3 \leq p(X_3)$, then $Z = X_3$, and so on. That is, $Z = X_j$ where

$$j = \min\{i\colon Y_i \leq p(X_i)\}.$$

(a) Show $c > 1$.

(b) Find the probability that the procedure has not terminated after n steps. That is, find the probability that $Z = X_j$ for some $j > n$. Based on this result, show that the procedure will eventually terminate.

(c) Find the distribution function of Z.

7.31 Let X denote a random variable with an absolutely continuous distribution with density function p and suppose that we want to estimate $E[h(X)]$ using Monte Carlo simulation, where $E[|h(X)|] < \infty$. Let $Y_1, Y_2, \ldots, Y_n$ denote independent, identically distributed random variables, each with an absolutely continuous distribution with density g. Assume that the distributions of X and Y_1 have the same support. Show that

$$E\left\{ \frac{1}{n} \sum_{j=1}^{n} \frac{p(Y_j)}{g(Y_j)} h(Y_j) \right\} = E[h(X)].$$

This approach to estimating $E[h(X)]$ is known as *importance sampling*; a well-chosen density g can lead to greatly improved estimates of $E[h(X)]$.

7.9 Suggestions for Further Reading

The problem of determining the distribution of a function of a random variable is discussed in many books on probability and statistics. See Casella and Berger (2002, Chapter 2) and Woodroofe (1975, Chapter 7) for elementary treatments and Hoffmann-Jorgenson (1994, Chapter 8) for a mathematically rigorous, comprehensive treatment of this problem.

Order statistics are discussed in Stuart and Ord (1994, Chapter 14) and Port (1994, Chapter 39). There are several books devoted to the distribution theory associated with order statistics and ranks; see, for example, Arnold, Balakrishnan, and Nagaraja (1992) and David (1981).

Monte Carlo methods are becoming increasingly important in statistical theory and methods. Robert and Casella (1999) gives a detailed account of the use of Monte Carlo methods in statistics; see also Hammersley and Handscomb (1964), Ripley (1987), and Rubinstein (1981).

8

Normal Distribution Theory

8.1 Introduction

The normal distribution plays a central role in statistical theory and practice, both as a model for observed data and as a large-sample approximation to the distribution of wide range of statistics, as will be discussed in Chapters 11–13. In this chapter, we consider in detail the distribution theory associated with the normal distribution.

8.2 Multivariate Normal Distribution

A d-dimensional random vector X has a *multivariate normal distribution* with mean vector $\mu \in \mathbf{R}^d$ and covariance matrix Σ if, for any $a \in \mathbf{R}^d$, $a^T X$ has a normal distribution with mean $a^T \mu$ and variance $a^T \Sigma a$. Here Σ is a $d \times d$ nonnegative-definite, symmetric matrix. Note that $a^T \Sigma a$ might be 0, in which case $a^T X = a^T \mu$ with probability 1. *degenerate*

The following result establishes several basic properties of the multivariate normal distribution.

Theorem 8.1. *Let X be a d-dimensional random vector with a multivariate normal distribution with mean vector μ and covariance matrix Σ.*

(i) *The characteristic function of X is given by*

$$\varphi(t) = \exp\left\{ it^T \mu - \frac{1}{2} t^T \Sigma t \right\}, \quad t \in \mathbf{R}^d.$$

(ii) *Let B denote a $p \times d$ matrix. Then BX has a p-dimensional multivariate normal distribution with mean vector $B\mu$ and covariance matrix $B\Sigma B^T$.*

(iii) *Suppose that the rank of Σ is $r < d$. Then there exists a $(d-r)$-dimensional subspace of $\mathbf{R}^d$, V, such that for any $v \in V$,*

$$\Pr\{v^T (X - \mu) = 0\} = 1.$$

There exists an $r \times d$ matrix C such that $Y = CX$ has a multivariate normal distribution with mean $C\mu$ and diagonal covariance matrix of full rank.

(iv) *Let $X = (X_1, X_2)$ where X_1 is p-dimensional and X_2 is $(d-p)$-dimensional. Write $\mu = (\mu_1, \mu_2)$ where $\mu_1 \in \mathbf{R}^p$ and $\mu_2 \in \mathbf{R}^{d-p}$, and write*

$$\Sigma = \begin{pmatrix} \Sigma_{11} & \Sigma_{12} \\ \Sigma_{21} & \Sigma_{22} \end{pmatrix}$$

where Σ_{11} is $p \times p$, $\Sigma_{12} = \Sigma_{21}^T$ is $p \times (d - p)$, and Σ_{22} is $(d - p) \times (d - p)$. Then X_1 has a multivariate normal distribution with mean vector μ_1 and covariance matrix Σ_{11}.

(v) Using the notation of part (iv), X_1 and X_2 are independent if and only if $\Sigma_{12} = 0$.

(vi) Let $Y_1 = M_1 X$ and $Y_2 = M_2 X$ where M_1 is an $r \times d$ matrix of constants and M_2 is an $s \times d$ matrix of constants. If $M_1 \Sigma M_2 = 0$ then Y_1 and Y_2 are independent.

(vii) Let $Z_1, \ldots, Z_d$ denote independent, identically distributed standard normal random variables and let $Z = (Z_1, \ldots, Z_d)$. Then Z has a multivariate normal distribution with mean vector 0 and covariance matrix given by the $d \times d$ identity matrix. A random vector X has a multivariate normal distribution with mean vector μ and covariance matrix Σ if and only if X has the same distribution as $\mu + \Sigma^{\frac{1}{2}} Z$.

Proof. Let $a \in \mathbf{R}^d$. Since $a^T X$ has a normal distribution with mean $a^T \mu$ and variance $a^T \Sigma a$, it follows that

$$E[\exp\{it(a^T X)\}] = \exp\left\{it(a^T \mu) - \frac{t^2}{2}(a^T \Sigma a)\right\}.$$

Hence, for any $t \in \mathbf{R}^d$,

$$E[\exp\{it^T X\}] = \exp\left\{it^T \mu - \frac{1}{2}t^T \Sigma t\right\},$$

proving part (i).

Let $a \in \mathbf{R}^p$. Then $B^T a \in \mathbf{R}^d$ so that $a^T B X$ has a normal distribution with mean $a^T B \mu$ and variance $a^T B \Sigma B^T a$; that is, for all $a \in \mathbf{R}^p$, $a^T (BX)$ has a normal distribution with mean $a^T (B\mu)$ and variance $a^T (B\Sigma B^T)a$. Part (ii) of the theorem now follows from the definition of the multivariate normal distribution.

Suppose that Σ has rank r; let $(\lambda_1, e_1), \ldots, (\lambda_r, e_r)$ denote the eigenvalue–eigenvector pairs of Σ, including multiplicities, corresponding to the nonzero eigenvalues so that

$$\Sigma = \lambda_1 e_1 e_1^T + \cdots + \lambda_r e_r e_r^T.$$

Consider the linear subspace of $\mathbf{R}^d$ spanned by $\{e_1, \ldots, e_r\}$ and let V denote the orthogonal complement of that space. Then, for any $v \in V$, $\Sigma v = 0$; hence, $v^T X$ has a normal distribution with mean $v^T \mu$ and variance 0, proving the first part of (iii). For the matrix C take the $r \times d$ matrix with jth row given by e_j^T. Then $C\Sigma C^T$ is the diagonal matrix with jth diagonal element λ_j. This proves the second part of (iii).

Part (iv) of the theorem is a special case of part (ii) with the matrix B taken to be of the form

$$B = (I_p \quad 0)$$

where I_p is the $p \times p$ identity matrix and 0 is a $p \times (d - p)$ matrix of zeros.

Let φ_1 denote the characteristic function of X_1, let φ_2 denote the characteristic function of X_2, and let φ denote the characteristic function of $X = (X_1, X_2)$. Then, from parts (i) and (iv) of the theorem,

$$\varphi_1(t) = \exp\left\{it^T \mu_1 - \frac{1}{2}t^T \Sigma_{11} t\right\}, \quad t \in \mathbf{R}^p,$$

$$\varphi_2(t) = \exp\left\{ it^T\mu_2 - \frac{1}{2}t^T\Sigma_{22}t \right\}, \quad t \in \mathbf{R}^{d-p},$$

and

$$\varphi(t) = \exp\left\{ it^T\mu - \frac{1}{2}t^T\Sigma t \right\}, \quad t \in \mathbf{R}^d;$$

here $\mu_j = \mathrm{E}(X_j)$, $j = 1, 2$.

Let $t_1 \in \mathbf{R}^p$, $t_2 \in \mathbf{R}^{d-p}$, and $t = (t_1, t_2)$. Then

$$t^T\mu = t_1^T\mu_1 + t_2^T\mu_2$$

$$(t_1, t_2)\begin{pmatrix}\Sigma_{11} & \Sigma_{12} \\ \Sigma_{12} & \Sigma_{22}\end{pmatrix}(t_1, t_2)$$

and

$$t^T\Sigma t = t_1^T\Sigma_{11}t_1 + t_2^T\Sigma_{22}t_2 + 2t_1^T\Sigma_{12}t_2.$$

$$= t_1\left(t_1\Sigma_{11} + t_2\Sigma_{12}\right) + \left(t_1\Sigma_{12} + t_2\Sigma_{22}\right)t_2$$

It follows that

$$\varphi(t) = \varphi_1(t_1)\varphi_2(t_2)\exp\left\{ -t_1^T\Sigma_{12}t_2 \right\}.$$

Part (v) of the theorem now follows from Corollary 3.3.

To prove part (vi), let

$$M = \begin{pmatrix} M_1 \\ M_2 \end{pmatrix}$$

and let $Y = MX$. Then, by part (ii) of the theorem, Y has a multivariate normal distribution with covariance matrix

$$\begin{pmatrix} M_1^T\Sigma M_1 & M_1^T\Sigma M_2 \\ M_2^T\Sigma M_1 & M_2^T\Sigma M_2 \end{pmatrix};$$

the result now follows from part (v) of the theorem.

Let $a = (a_1, \ldots, a_d) \in \mathbf{R}^d$. Then $a^T Z = \sum_{j=1}^d a_j Z_j$ has characteristic function

$$\mathrm{E}\left[\exp\left\{ it\sum_{j=1}^d a_j Z_j \right\} \right] = \prod_{j=1}^d \mathrm{E}[\exp\{ita_j Z_j\}] = \prod_{j=1}^d \exp\left(-\frac{1}{2}a_j^2 t^2 \right)$$

$$= \exp\left\{ -\frac{1}{2}\sum_{j=1}^d a_j^2 t^2 \right\}, \quad t \in \mathbf{R},$$

which is the characteristic function of a normal distribution with mean 0 and variance $\sum_{j=1}^d a_j^2$. Hence, Z has a multivariate normal distribution as stated in the theorem.

Suppose X has the same distribution as $\mu + \Sigma^{\frac{1}{2}}Z$. Then $a^T X$ has the same distribution as $a^T\mu + a^T\Sigma^{\frac{1}{2}}Z$, which is normal with mean $a^T\mu$ and variance

random

$$a^T\Sigma^{\frac{1}{2}}\Sigma^{\frac{1}{2}}a = a^T\Sigma a;$$

it follows that X has a multivariate normal distribution with mean vector μ and covariance matrix Σ.

Now suppose that X has a multivariate normal distribution with mean vector μ and covariance matrix Σ. Then, for any $a \in \mathbf{R}^d$, $a^T X$ has a normal distribution with mean $a^T\mu$ and variance $a^T\Sigma a$. Note that this is the same distribution as $a^T(\mu + \Sigma^{\frac{1}{2}}Z)$; it follows from

Corollary 3.2 that X has the same distribution as $\mu + \Sigma^{\frac{1}{2}} Z$. Hence, part (vii) of the theorem holds. ∎

Example 8.1 (Bivariate normal distribution). Suppose that $X = (X_1, X_2)$ has a two-dimensional multivariate normal distribution. Then Σ, the covariance matrix of the distribution, is of the form

$$\Sigma = \begin{pmatrix} \sigma_1^2 & \sigma_{12} \\ \sigma_{21} & \sigma_2^2 \end{pmatrix}$$

where σ_j^2 denotes the variance of X_j, $j = 1, 2$, and $\sigma_{12} = \sigma_{21}$ denotes the covariance of X_i, X_j.

We may write $\sigma_{12} = \rho \sigma_1 \sigma_2$ so that ρ denotes the correlation of X_1 and X_2. Then

$$\Sigma = \begin{pmatrix} \sigma_1 & 0 \\ 0 & \sigma_2 \end{pmatrix} \begin{pmatrix} 1 & \rho \\ \rho & 1 \end{pmatrix} \begin{pmatrix} \sigma_1 & 0 \\ 0 & \sigma_2 \end{pmatrix}.$$

It follows that Σ is nonnegative-definite provided that

$$\begin{pmatrix} 1 & \rho \\ \rho & 1 \end{pmatrix}$$

is nonnegative-definite. Since this matrix has eigenvalues $1 - \rho$, $1 + \rho$, Σ is nonnegative-definite for any $-1 \leq \rho \leq 1$. If $\rho = \pm 1$, then $X_1/\sigma_1 - \rho X_2/\sigma_2$ has variance 0. □

Example 8.2 (Exchangeable normal random variables). Consider a multivariate normal random vector $X = (X_1, X_2, \ldots, X_n)$ and suppose that $X_1, X_2, \ldots, X_n$ are exchangeable random variables. Let μ and Σ denote the mean vector and covariance matrix, respectively. Then, according to Theorem 2.8, each X_j has the same marginal distribution; hence, μ must be a constant vector and the diagonal elements of Σ must be equal. Also, each pair (X_i, X_j) must have the same distribution; it follows that the $\text{Cov}(X_i, X_j)$ is a constant, not depending on i, j. Hence, Σ must be of the form

$$\Sigma = \sigma^2 \begin{pmatrix} 1 & \rho & \rho & \cdots & \rho \\ \rho & 1 & \rho & \cdots & \rho \\ & & \vdots & & \\ \rho & \rho & \rho & \cdots & 1 \end{pmatrix}$$

for some constants $\sigma \geq 0$ and ρ. Of course, Σ is a valid covariance matrix only for certain values of ρ; see Exercise 8.8. □

Example 8.3 (Principal components). Let X denote a d-dimensional random vector with a multivariate normal distribution with mean vector μ and covariance matrix Σ. Consider the problem of finding the linear function $a^T X$ with the maximum variance; of course, the variance of $a^T X$ can be made large by choosing the elements of a to be large in magnitude. Hence, we require a to be a unit vector. Since

$$\text{Var}(a^T X) = a^T \Sigma a,$$

we want to find the unit vector a that maximizes $a^T \Sigma a$.

Let $(\lambda_1, e_1), \ldots, (\lambda_d, e_d)$ denote the eigenvalue–eigenvector pairs of Σ, $\lambda_1 \geq \lambda_2 \geq \cdots \geq \lambda_d$, so that

$$\Sigma = \lambda_1 e_1 e_1^T + \cdots + \lambda_d e_d e_d^T.$$

Note we may write $a = c_1 e_1 + \cdots c_d e_d$ for some scalar constants $c_1, c_2, \ldots, c_d$; since $a^T a = 1$, it follows that $c_1^2 + \cdots + c_d^2 = 1$.

Hence,

$$a^T \Sigma a = \lambda_1 c_1^2 + \cdots + \lambda_d c_d^2,$$

which is maximized, subject to the restriction $c_1^2 + \cdots + c_d^2 = 1$, by $c_1^2 = 1$, $c_2^2 = \cdots = c_d^2 = 0$. That is, the variance of $a^T X$ is maximized by taking a to be the eigenvector corresponding to the largest eigenvalue of Σ; $a^T X$ is called the *first principal component* of X. □

Example 8.4 (The multivariate normal distribution as a transformation model). Consider the class of multivariate normal distributions with mean vector $\mu \in \mathbf{R}^d$ and covariance matrix Σ, where Σ is an element of the set of all $d \times d$ positive-definite matrices; we will denote this set by $\mathcal{C}_d$.

For $A \in \mathcal{C}_d$ and $b \in \mathbf{R}^d$ let

$$(A, b)X = AX + b.$$

Consider the set of transformations $\mathcal{G}$ of the form (A, b) with $A \in \mathcal{C}_d$ and $b \in \mathbf{R}^d$. Since

$$(A_1, b_1)(A_0, b_0)X = A_1(A_0 X + b_0) + b_1 = A_1 A_0 X + A_1 b_0 + b_1,$$

define the operation

$$(A_1, b_1)(A_0, b_0) = (A_1 A_0, A_1 b_0 + b_1).$$

It is straightforward to show that $\mathcal{G}$ is a group with respect to this operation. The identity element of the group is $(I_d, 0)$ and the inverse operation is given by

$$(A, b)^{-1} = (A^{-1}, -A^{-1} b).$$

If X has a multivariate normal distribution with mean vector μ and covariance matrix Σ, then, by Theorem 8.1, $(A, b)X$ has a multivariate normal distribution with mean vector $A\mu + b$ and positive definite covariance matrix $A \Sigma A^T$. Clearly, the set of all multivariate normal distributions with mean vector $\mu \in \mathbf{R}^d$ and covariance matrix $\Sigma \in \mathcal{C}_d$ is invariant with respect to $\mathcal{G}$.

As discussed in Section 5.6, $\mathcal{G}$ may also be viewed as acting on the parameter space of the model $\mathcal{C}_d \times \mathbf{R}^d$; here

$$(A, b)(\Sigma, \mu) = (A \Sigma A^T, A\mu + b). \qquad \square$$

Density of the multivariate normal distribution

For the case in which the covariance matrix is positive-definite, it is straightforward to derive the density function of the multivariate normal distribution.

Theorem 8.2. *Let X be a d-dimensional random vector with a multivariate normal distribution with mean μ and covariance matrix Σ. If $|\Sigma| > 0$ then the distribution of X is* *absolutely continuous with density*

$$(2\pi)^{-\frac{d}{2}}|\Sigma|^{-\frac{1}{2}} \exp\left\{-\frac{1}{2}(x-\mu)^T\Sigma^{-1}(x-\mu)\right\}, \quad x \in \mathbf{R}^d.$$

Proof. Let Z be a d-dimensional vector of independent, identically distributed standard normal random variables. Then Z has density

$$\prod_{i=1}^{d} \frac{1}{(2\pi)^{\frac{1}{2}}} \exp\left\{-\frac{1}{2}z_i^2\right\} = (2\pi)^{-\frac{d}{2}} \exp\left\{-\frac{1}{2}z^Tz\right\}, \quad z \in \mathbf{R}^d.$$

By Theorem 8.1 part (vi), the density of X is given by the density of $\mu + \Sigma^{\frac{1}{2}}Z$. Let $W = \mu + \Sigma^{\frac{1}{2}}Z$; this is a one-to-one transformation since $|\Sigma| > 0$. Using the change-of-variable formula, the density of W is given by

$$(2\pi)^{-\frac{d}{2}} \exp\left\{-\frac{1}{2}(w-\mu)^T\Sigma^{-1}(w-\mu)\right\}\left|\frac{\partial z}{\partial w}\right|.$$

The result now follows from the fact that

$$\left|\frac{\partial z}{\partial w}\right| = |\Sigma|^{-\frac{1}{2}}. \quad \blacksquare$$

Example 8.5 (Bivariate normal distribution). Suppose X is a two-dimensional random vector with a bivariate normal distribution, as discussed in Example 8.1. The parameters of the distribution are the mean vector, (μ_1, μ_2), and the covariance matrix, which may be written

$$\Sigma = \begin{pmatrix} \sigma_1^2 & \rho\sigma_1\sigma_2 \\ \rho\sigma_1\sigma_2 & \sigma_2^2 \end{pmatrix}.$$

The density of the bivariate normal distribution may be written

$$\frac{1}{2\pi\sigma_1\sigma_2\sqrt{(1-\rho^2)}} \exp\left\{-\frac{1}{2(1-\rho^2)}\left[\left(\frac{x_1-\mu_1}{\sigma_1}\right)^2 \right.\right.$$
$$\left.\left. - 2\rho\frac{x_1-\mu_1}{\sigma_1}\frac{x_2-\mu_2}{\sigma_2} + \left(\frac{x_2-\mu_2}{\sigma_2}\right)^2\right]\right\},$$

for $(x_1, x_2) \in \mathbf{R}^2$. $\quad\square$

8.3 Conditional Distributions

An important property of the multivariate normal distribution is that the conditional distribution of one subvector of X given another subvector of X is also a multivariate normal distribution.

Theorem 8.3. *Let X be a d-dimensional random vector with a multivariate normal distribution with mean μ and covariance matrix Σ.*

Write $X = (X_1, X_2)$ where X_1 is p-dimensional and X_2 is $(d - p)$-dimensional, $\mu = (\mu_1, \mu_2)$ where $\mu_1 \in \mathbf{R}^p$ and $\mu_2 \in \mathbf{R}^{d-p}$, and

$$\Sigma = \begin{pmatrix} \Sigma_{11} & \Sigma_{12} \\ \Sigma_{21} & \Sigma_{22} \end{pmatrix}$$

where Σ_{11} is $p \times p$, $\Sigma_{12} = \Sigma_{21}^T$ is $p \times (d - p)$, and Σ_{22} is $(d - p) \times (d - p)$.

Suppose that $|\Sigma_{22}| > 0$. Then the conditional distribution of X_1 given $X_2 = x_2$ is a multivariate normal distribution with mean vector

$$\mu_1 + \Sigma_{12}\Sigma_{22}^{-1}(x_2 - \mu_2)$$

and covariance matrix

$$\Sigma_{11} - \Sigma_{12}\Sigma_{22}^{-1}\Sigma_{21}.$$

Proof.   Let

$$Z = \begin{pmatrix} I_p & -\Sigma_{12}\Sigma_{22}^{-1} \\ 0 & I_q \end{pmatrix} X; \quad \begin{pmatrix} X_1 \\ X_v \end{pmatrix}$$

then Z has a multivariate normal distribution with covariance matrix

$$\begin{pmatrix} \Sigma_{11} - \Sigma_{12}\Sigma_{22}^{-1}\Sigma_{21} & 0 \\ 0 & \Sigma_{22} \end{pmatrix}$$

where $q = d - p$. Write $Z = (Z_1, Z_2)$ where Z_1 has dimension p and Z_2 has dimension q. Note that $Z_2 = X_2$. Then, by part (v) of Theorem 8.1, Z_1 and X_2 are independent. It follows that the conditional distribution of Z_1 given X_2 is the same as the marginal distribution of Z_1, multivariate normal with mean $\mu_1 - \Sigma_{12}\Sigma_{22}^{-1}\mu_2$ and covariance matrix

$$\Sigma_{11} - \Sigma_{12}\Sigma_{22}^{-1}\Sigma_{21}.$$

Since

$$Z_1 = X_1 - \Sigma_{12}\Sigma_{22}^{-1}X_2,$$

$$X_1 = Z_1 + \Sigma_{12}\Sigma_{22}^{-1}X_2,$$

and the conditional distribution of X_1 given $X_2 = x_2$ is multivariate normal with mean given by

$$\mathrm{E}(Z_1|X_2 = x_2) + \Sigma_{12}\Sigma_{22}^{-1}x_2 = \mu_1 - \Sigma_{12}\Sigma_{22}^{-1}\mu_2 + \Sigma_{12}\Sigma_{22}^{-1}x_2 = \mu_1 + \Sigma_{12}\Sigma_{22}^{-1}(x_2 - \mu_2)$$

and covariance matrix

$$\Sigma_{11} - \Sigma_{12}\Sigma_{22}^{-1}\Sigma_{21},$$

proving the theorem.   ∎

Example 8.6   (Bivariate normal). Suppose that X has a bivariate normal distribution, as discussed in Example 8.5, and consider the conditional distribution of X_1 given $X_2 = x_2$. Then $\Sigma_{11} = \sigma_1^2$, $\Sigma_{12} = \rho\sigma_1\sigma_2$, and $\Sigma_{22} = \sigma_2^2$. It follows that this conditional distribution is normal, with mean

$$\mu_1 + \rho\frac{\sigma_1}{\sigma_2}(x_2 - \mu_2)$$

and variance

$$\sigma_1^2 - \rho^2 \sigma_1^2 = (1 - \rho^2)\sigma_1^2. \qquad \square$$

Example **8.7** *(Least squares).* Let X be a d-dimensional random vector with a multivariate normal distribution with mean μ and covariance matrix Σ. Write $X = (X_1, X_2)$, where X_1 is real-valued, and partition μ and Σ in a similar manner: $\mu = (\mu_1, \mu_2)$,

$$\Sigma = \begin{pmatrix} \Sigma_{11} & \Sigma_{12} \\ \Sigma_{21} & \Sigma_{22} \end{pmatrix}.$$

For a given $1 \times (d - 1)$ matrix A and a given scalar $a \in \mathbf{R}$, define

$$S(A, a) = \mathrm{E}[(X_1 - a - AX_2)^2] = (\mu_1 - a - A\mu_2)^2 + \Sigma_{11} + A\Sigma_{22}A^T - 2\Sigma_{12}A$$

and suppose we choose A and a to minimize $S(A, a)$.

First note that, given A, a must satisfy

$$a = \mu_1 - A\mu_2,$$

so that $(\mu_1 - a - A\mu_2)^2 = 0$. Hence, A may be chosen to minimize

$$A\Sigma_{22}A^T - 2\Sigma_{12}A^T. \tag{8.1}$$

Write $A = \Sigma_{12}\Sigma_{22}^{-1} + A_1$. Then

$$A\Sigma_{22}A^T - 2\Sigma_{12}A^T = A_1\Sigma_{22}A_1^T - \Sigma_{12}\Sigma_{22}^{-1}\Sigma_{21}. \tag{8.2}$$

Minimizing (8.1) with respect to A is equivalent to minimizing (8.2) with respect to A_1. Since Σ_{22} is nonnegative-definite, (8.2) is minimized by $A_1 = 0$; hence, (8.1) is minimized by $A = \Sigma_{12}\Sigma_{22}^{-1}$. That is, the affine function of X_2 that minimizes $\mathrm{E}[X_1 - (a + AX_2)]^2$ is given by

$$\mu_1 + \Sigma_{12}\Sigma_{22}^{-1}(X_2 - \mu_2),$$

which is simply $\mathrm{E}(X_1|X_2)$. This is to be expected given Corollary 2.2. $\quad\square$

Conditioning on a degenerate random variable

Theorem 8.3 may be extended to the case in which the conditioning random vector, X_2, has a singular covariance matrix.

Theorem **8.4.** *Let X be a d-dimensional random vector with a multivariate normal distribution with mean μ and covariance matrix Σ.*

Write $X = (X_1, X_2)$ where X_1 is p-dimensional and X_2 is $(d - p)$-dimensional, $\mu = (\mu_1, \mu_2)$ where $\mu_1 \in \mathbf{R}^p$ and $\mu_2 \in \mathbf{R}^{d-p}$, and

$$\Sigma = \begin{pmatrix} \Sigma_{11} & \Sigma_{12} \\ \Sigma_{21} & \Sigma_{22} \end{pmatrix}$$

where Σ_{11} is $p \times p$, $\Sigma_{12} = \Sigma_{21}$ is $p \times (d - p)$, and Σ_{22} is $(d - p) \times (d - p)$. Let $r = \mathrm{rank}(\Sigma_{22})$ and suppose that $r < d - p$. Then the conditional distribution of X_1 given $X_2 = x_2$ is a multivariate normal distribution with mean vector

$$\mu_1 + \Sigma_{12}\Sigma_{22}^{-}(x_2 - \mu_2)$$

and covariance matrix

$$\Sigma_{11} - \Sigma_{12}\Sigma_{22}^{-}\Sigma_{21},$$

provided that x_2 is such that for any vector a satisfying $a^T\Sigma_{22}a = 0$, $a^Tx_2 = a^T\mu_2$. Here Σ_{22}^{-} denotes the Moore–Penrose generalized inverse of Σ_{22}.

Proof.   By part (iii) of Theorem 8.1, there is a linear transformation of X_2 to (Y_1, Y_2) such that Y_1 is constant with probability 1 and Y_2 has a multivariate normal distribution with full-rank covariance matrix. Furthermore, $Y_2 = CX_2$ where C is an $r \times (d - p)$ matrix with rows taken to be the eigenvectors corresponding to nonzero eigenvalues of Σ_{22}; see the proof of Theorem 8.1. Hence, the conditional distribution of X_1 given X_2 is equivalent to the conditional distribution of X_1 given Y_2. Since the covariance matrix of Y_2 is of full-rank, it follows from Theorem 8.3 that this conditional distribution is multivariate normal with mean vector

$$\mu_1 + \Sigma_{13}\Sigma_{33}^{-1}(y_2 - \mu_3)$$

and covariance matrix

$$\Sigma_{11} - \Sigma_{13}\Sigma_{33}^{-1}\Sigma_{31}$$

where μ_3 denotes the mean of Y_2, Σ_{13} denotes the covariance of X_1 and Y_2, and Σ_{33} denotes the covariance matrix of Y_2.

By considering the transformation

$$\begin{pmatrix} X_1 \\ Y_2 \end{pmatrix} = \begin{pmatrix} I_p & 0 \\ 0 & C \end{pmatrix} \begin{pmatrix} X_1 \\ X_2 \end{pmatrix},$$

it follows from Theorem 8.1 that

$$\Sigma_{13} = \Sigma_{12}C^T,$$

$$\Sigma_{33} = C\Sigma_{22}C^T,$$

and $\mu_3 = C\mu_2$. Hence, the conditional distribution of X_1 given $X_2 = x_2$ is multivariate normal with mean

$$\mu_1 + \Sigma_{12}C^T[C\Sigma_{22}C^T]^{-1}C(x_2 - \mu_2)$$

and covariance matrix

$$\Sigma_{11} - \Sigma_{12}C^T[C\Sigma_{22}C^T]^{-1}C\Sigma_{21},$$

provided that x_2 is such that for any vector a such that a^TX_2 has variance 0, $a^Tx_2 = a^T\mu_2$; see Example 8.8 below for an illustration of this requirement.

Recall that $\Sigma_{22} = C^TDC$ where D is a diagonal matrix with diagonal elements taken to be the nonzero eigenvalues of Σ_{22}. Note that

$$\Sigma_{22}C^T[C\Sigma_{22}C^T]^{-1}C\Sigma_{22} = C^TD(CC^T)[(CC^T)D(CC^T)]^{-1}(CC^T)DC$$
$$= C^TDC = \Sigma_{22},$$

since (CC^T) and D are invertible. Hence,

$$\Sigma_{22}^{\dagger} \equiv C^T[C\Sigma_{22}C^T]^{-1}C$$

is a generalized inverse of Σ_{22}. Furthermore,

$$\Sigma_{22}^{\dagger} \Sigma_{22} \Sigma_{22}^{\dagger} = \Sigma_{22}^{\dagger}$$

and both $\Sigma_{22}^{\dagger} \Sigma_{22}$ and $\Sigma_{22} \Sigma_{22}^{\dagger}$ are symmetric. It follows that $\Sigma_{22}^{\dagger} = \Sigma_{22}^{-}$, the Moore–Penrose inverse of Σ_{22}. ∎

In fact, it has been shown by Rao (1973, Section 8a.2) that the result of Theorem 8.4 holds for any choice of generalized inverse.

Example 8.8 (Bivariate normal distribution). Suppose that X has a bivariate normal distribution, as discussed in Example 8.5, and consider the conditional distribution of X_1 given $X_2 = x_2$. Suppose that the covariance matrix of X_2 is singular; that is, suppose that $\sigma_2 = 0$. The Moore–Penrose generalized inverse of 0 is 0 so that the conditional distribution of X_1 given $X_2 = x_2$ is a normal distribution with mean μ_1 and variance σ_1^2. This holds provided that x_2 is such that for any vector a such that $a^T X_2$ has variance 0, $a^T x_2 = a^T \mu_2$; in this case, this means that we require that $x_2 = \mu_2$. Note that $X_2 = \mu_2$ with probability 1. □

8.4 Quadratic Forms

Much of this chapter has focused on the properties of linear functions of a multivariate normal random vector X; however, quadratic functions of X also often occur in statistical methodology. In this section, we consider functions of X of the form $X^T A X$ where A is a symmetric matrix of constants; such a function is called a *quadratic form*. We will focus on the case in which A is a nonnegative-definite matrix.

Example 8.9 (Sample variance). Let $X_1, \ldots, X_n$ denote independent, identically distributed, real-valued random variables such that X_j has a normal distribution with mean 0 and variance σ^2 and let

$$S^2 = \frac{1}{n-1} \sum_{j=1}^{n} (X_j - \bar{X})^2$$

where $\bar{X} = \sum_{j=1}^{n} X_j / n$.

Let $X = (X_1, \ldots, X_n)$. Then $\bar{X} = mX$ where m denotes a $1 \times n$ vector with each element taken to be $1/n$. Since, for any vector $c = (c_1, \ldots, c_n) \in \mathbf{R}^n$,

$$\sum_{j=1}^{n} c_j^2 = c^T c,$$

$$\sum_{j=1}^{n} (X_j - \bar{X})^2 = (X - nm^T mX)^T (X - nm^T mX) = X^T (I_n - nm^T m)^T (I_n - nm^T m)X.$$

Let

$$A = \frac{1}{n-1} (I_n - nm^T m)^T (I_n - nm^T m) = \frac{1}{n-1} (I_n - nm^T m);$$

then $S^2 = X^T A X$. □

Thus, the previous example shows that the sample variance is a quadratic form. Similarly, the sums of squares arising in regression analysis and the analysis of variance can also generally be expressed as quadratic forms. Quadratic forms also arise in the approximation of the distribution of certain test statistics.

Since the term in the exponent of the multivariate normal distribution, $-x^T \Sigma^{-1} x / 2$, is a quadratic form in x, it is a relatively simple matter to determine the moment-generating function and, hence, the cumulant-generating function, of a quadratic form.

Theorem 8.5. *Let X denote a d-dimensional random vector with a multivariate normal distribution with mean vector 0 and covariance matrix Σ, $|\Sigma| > 0$. Let A be a $d \times d$ nonnegative-definite, symmetric matrix and let $Q = X^T A X$. Then Q has cumulant-generating function*

$$K_Q(t) = -\frac{1}{2} \sum_{k=1}^{d} \log(1 - 2t\lambda_k), \quad |t| < \delta$$

where $\lambda_1, \ldots, \lambda_d$ are the eigenvalues of ΣA and $\delta > 0$. The jth cumulant of Q is given by

$$\sum_{k=1}^{d} 2^{j-1}(j-1)! \lambda_k^j.$$

Proof.   Consider $E[\exp\{t X^T A X\}]$, which is given by

$$\int_{\mathbf{R}^d} (2\pi)^{-\frac{d}{2}} |\Sigma|^{-\frac{1}{2}} \exp\left\{-\frac{1}{2} x^T \Sigma^{-1} x\right\} \exp\{t x^T A x\} \, dx$$

$$= \int_{\mathbf{R}^d} (2\pi)^{-\frac{d}{2}} |\Sigma|^{-\frac{1}{2}} \exp\left\{-\frac{1}{2} x^T [\Sigma^{-1} - 2t A] x\right\} \, dx.$$

Note that $|\Sigma^{-1} - 2t A|$ is a continuous function of t and is positive for $t = 0$; hence, there exists a $\delta > 0$ such that

$$|\Sigma^{-1} - 2t A| > 0 \quad \text{for} \quad |t| < \delta.$$

Thus, for $|t| < \delta$,

$$\int_{\mathbf{R}^d} (2\pi)^{-\frac{d}{2}} |\Sigma|^{-\frac{1}{2}} \exp\left\{-\frac{1}{2} x^T [\Sigma^{-1} - 2t A] x\right\} \, dx = |\Sigma|^{-\frac{1}{2}} |\Sigma^{-1} - 2t A|^{-\frac{1}{2}}$$

$$= |I_d - 2t \Sigma^{\frac{1}{2}} A \Sigma^{\frac{1}{2}}|^{-\frac{1}{2}}.$$

Let $B = \Sigma^{\frac{1}{2}} A \Sigma^{\frac{1}{2}}$. Note that B is a symmetric nonnegative-definite matrix so that we may write

$$B = P D P^T$$

where P is an orthogonal matrix and D is a diagonal matrix with diagonal elements given by the eigenvalues of B or, equivalently, the eigenvalues of ΣA. It follows that

$$|I_d - 2t P D P^T| = |P(I_d - 2t D) P^T| = |I_d - 2t D| = \prod_{j=1}^{n} (1 - 2t\lambda_j)$$

where $\lambda_1, \ldots, \lambda_d$ denote the eigenvalues of B. Hence, Q has cumulant-generating function

$$K_Q(t) = -\frac{1}{2}\sum_{k=1}^{d}\log(1 - 2t\lambda_k), \quad |t| < \delta;$$

the expression for the cumulants follows easily from differentiating K_Q. ∎

Recall that the sum of squared independent standard normal random variables has a chi-squared distribution (Example 3.7). That is, if Z is a d-dimensional random vector with a multivariate normal distribution with mean 0 and covariance matrix given by I_d, then the quadratic form $Z^T Z$ has a chi-squared distribution. A general quadratic form $X^T A X$ has a chi-squared distribution if it can be rewritten as the sum of squared independent standard normal random variables. A necessary and sufficient condition for this is given in the following theorem.

Theorem 8.6. *Let X denote a d-dimensional random vector with a multivariate normal distribution with mean 0 and covariance matrix Σ. Let A be a $d \times d$ nonnegative-definite, symmetric matrix and let $Q = X^T A X$. Q has a chi-squared distribution if and only if ΣA is idempotent. The degrees of freedom of the chi-squared distribution is the trace of ΣA.*

Proof.   From Theorem 8.5, the cumulant-generating function of Q is

$$K_Q(t) = -\frac{1}{2}\sum_{k=1}^{d}\log(1 - 2t\lambda_k), \quad |t| < \delta,$$

where $\lambda_1, \ldots, \lambda_d$ are the eigenvalues of ΣA and $\delta > 0$. If ΣA is idempotent, then each λ_k is either 0 or 1. Suppose that r eigenvalues are 1. Then

$$K_Q(t) = -\frac{1}{2}r\,\log(1 - 2t), \quad |t| < \delta,$$

which is the cumulant-generating function of a chi-squared random variable with r degrees of freedom. Hence, by Theorem 4.9, Q has a chi-squared distribution with r degrees of freedom.

Now suppose that

$$K_Q(t) = -\frac{1}{2}r\,\log(1 - 2t)$$

for some r; that is, suppose that

$$\sum_{k=1}^{d}\log(1 - 2t\lambda_k) = r\log(1 - 2t), \quad |t| < \delta,$$

for some $\delta > 0$. For any positive number λ,

$$\log(1 - 2t\lambda) = -\sum_{j=1}^{\infty}(2\lambda)^j t^j / j,$$

for sufficiently small $|t|$. Hence,

$$\sum_{k=1}^{d}\log(1 - 2t\lambda_k) = -\sum_{k=1}^{d}\sum_{j=1}^{\infty}(2\lambda_k)^j t^j / j$$

for all $|t| < \epsilon$, for some $\epsilon > 0$. It is straightforward to show that the double series converges absolutely and, hence, the order of summation can be changed:

$$\sum_{k=1}^{d} \log(1 - 2t\lambda_k) = -\sum_{k=1}^{d} \sum_{j=1}^{\infty} (2\lambda_k)^j t^j / j = -\sum_{j=1}^{\infty} \left(\sum_{k=1}^{d} \lambda_k^j \right) 2^j t^j / j$$

for all $|t| < \epsilon$. This implies that

$$\sum_{k=1}^{d} \lambda_k^j = r, \quad j = 1, 2, \ldots, \tag{8.3}$$

and, hence, that each λ_k is either 0 or 1.

To prove this last fact, consider a random variable W taking value λ_k, $k = 1, 2, \ldots, d$ with probability $1/d$; note that, by (8.3), all moments of W are equal to r/d. Since W is bounded, its moment-generating function exists; by Theorem 4.8 it is given by

$$1 + \frac{r}{d} \sum_{j=1}^{\infty} \frac{t^j}{j!} = 1 + \frac{r}{d} [\exp\{t\} - 1] = (1 - r/d) + (r/d) \exp\{t\},$$

which is the moment-generating function of a random variable taking the values 0 and 1 with probabilities $1 - r/d$ and r/d, respectively. The result follows. ∎

Example 8.10 *(Sum of squared independent normal random variables).* Let X_1, $X_2, \ldots, X_n$ denote independent, real-valued random variables, each with a normal distribution with mean 0. Let $\sigma_j^2 = \text{Var}(X_j)$, $j = 1, \ldots, n$, and suppose that $\sigma_j^2 > 0$, $j = 1, \ldots, n$. Consider a quadratic form of the form

$$Q = \sum_{j=1}^{n} a_j X_j^2$$

where $a_1, a_2, \ldots, a_n$ are given constants.

Let $X = (X_1, \ldots, X_n)$. Then X has a multivariate normal distribution with covariance matrix Σ, where Σ is a diagonal matrix with jth diagonal element given by σ_j^2. Let A denote the diagonal matrix with jth diagonal element a_j. Then $Q = X^T A X$.

It follows from Theorem 8.6 that Q has a chi-squared distribution if and only if ΣA is idempotent. Since ΣA is a diagonal matrix with jth diagonal element given by $a_j \sigma_j^2$, it follows that Q has a chi-squared distribution if and only if, for each $j = 1, \ldots, n$, either $a_j = 0$ or $a_j = 1/\sigma_j^2$. □

The same basic approach used in part (iii) of Theorem 8.1 can be used to study the joint distribution of two quadratic forms, or the joint distribution of a quadratic form and a linear function of a multivariate normal random vector.

Theorem 8.7. *Let X denote a d-dimensional random vector with a multivariate normal distribution with mean 0 and covariance matrix Σ. Let A_1 and A_2 be $d \times d$ nonnegative-definite, symmetric matrices and let $Q_j = X^T A_j X$, $j = 1, 2$.*

   (i) If $A_1 \Sigma A_2 = 0$ then Q_1 and Q_2 are independent.

   (ii) Let $Y = MX$ where M is an $r \times d$ matrix. If $A_1 \Sigma M^T = 0$ then Y and Q_1 are independent.

Proof. Let $\lambda_1, \ldots, \lambda_{r_1}$ denote the nonzero eigenvalues of A_1 and let $e_1, \ldots, e_{r_1}$ denote the corresponding eigenvectors; similarly, let $\gamma_1, \ldots, \gamma_{r_2}$ denote the nonzero eigenvalues of A_2 and let $v_1, \ldots, v_{r_2}$ denote the corresponding eigenvectors. Then

$$A_1 = \lambda_1 e_1 e_1^T + \cdots + \lambda_{r_1} e_{r_1} e_{r_1}^T$$

and

$$A_2 = \gamma_1 v_1 v_1^T + \cdots + \gamma_{r_2} v_{r_2} v_{r_2}^T.$$

Suppose $A_1 \Sigma A_2 = 0$. Then

$$e_k^T A_1 \Sigma A_2 v_j = \lambda_k \gamma_j e_k^T \Sigma v_j = 0$$

so that $e_k^T \Sigma v_j = 0$ for all $j = 1, \ldots, r_2$ and $k = 1, \ldots, r_1$. Let P_1 denote the matrix with columns $e_1, \ldots, e_{r_1}$ and let P_2 denote the matrix with columns $v_1, \ldots, v_{r_2}$. Then

$$P_1^T \Sigma P_2 = 0.$$

It follows that $P_1^T X$ and $P_2^T X$ are independent. Since Q_1 is a function of $P_1^T X$ and Q_2 is a function of $P_2^T X$, it follows that Q_1 and Q_2 are independent, proving part (i).

The proof of part (ii) is similar. As above, Q_1 is a function of $P_1 X$. Suppose that $A_1 \Sigma M^T = 0$. Since $A_1 = P_1 D P_1^T$ where D is a diagonal matrix with diagonal elements $\lambda_1, \ldots, \lambda_{r_1}$,

$$P_1 D P_1^T \Sigma M^T = 0.$$

It follows that $P_1^T \Sigma M^T = 0$; hence, by part (vi) of Theorem 8.1, $P_1^T X$ and MX are independent. The result follows. ∎

The following result gives a simple condition for showing that two quadratic forms are independent chi-squared random variables.

Theorem 8.8. *Let X denote a d-dimensional random vector with a multivariate normal distribution with mean 0 and covariance matrix I_d. Let A_1 and A_2 be $d \times d$ nonnegative-definite, symmetric matrices and let $Q_j = X^T A_j X$, $j = 1, 2$. Suppose that*

$$X^T X = Q_1 + Q_2.$$

Let r_j denote the rank of A_j, $j = 1, 2$. Q_1 and Q_2 are independent chi-squared random variables with r_1 and r_2 degrees of freedom, respectively, if and only if $r_1 + r_2 = d$.

Proof. Suppose Q_1 and Q_2 are independent chi-squared random variables with r_1 and r_2 degrees of freedom, respectively. Since, by Theorem 8.6, $X^T X$ has a chi-squared distribution with d degrees of freedom, clearly we must have $r_1 + r_2 = d$; for example, $E(X^T X) = d$, $E(Q_1) = r_1$, $E(Q_2) = r_2$, and $E(X^T X) = E(Q_1) + E(Q_2)$.

Suppose that $r_1 + r_2 = d$. Let $\lambda_1, \ldots, \lambda_{r_1}$ denote the nonzero eigenvalues of A_1 and let $e_1, \ldots, e_{r_1}$ denote the corresponding eigenvectors; similarly, let $\gamma_1, \ldots, \gamma_{r_2}$ denote the nonzero eigenvalues of A_2 and let $v_1, \ldots, v_{r_2}$ denote the corresponding eigenvectors. Then

$$A_1 = \lambda_1 e_1 e_1^T + \cdots + \lambda_{r_1} e_{r_1} e_{r_1}^T$$

and

$$A_2 = \gamma_1 v_1 v_1^T + \cdots + \gamma_{r_2} v_{r_2} v_{r_2}^T.$$

Note that

$$A_1 + A_2 = PDP^T$$

where D is a diagonal matrix with diagonal elements $\lambda_1, \ldots, \lambda_{r_1}, \gamma_1, \ldots, \gamma_{r_2}$ and P is a matrix with columns $e_1, \ldots, e_{r_1}, v_1, \ldots, v_{r_2}$; recall that $r_1 + r_2 = d$. Since

$$A_1 + A_2 = PDP^T = I_d,$$

and D has determinant

$$\lambda_1 \cdots \lambda_{r_1} \gamma_1 \cdots \gamma_{r_2} \neq 0,$$

it follows that $|P| \neq 0$.

Since $A_1 e_j = \lambda_j e_j$ and $(A_1 + A_2)e_j = e_j$, it follows that

$$A_2 e_j = (1 - \lambda_j)e_j, \quad j = 1, \ldots, r_1.$$

That is, either $\lambda_j = 1$ or e_j is an eigenvector of A_2. However, if e_j is an eigenvector of A_2, then two columns of P are identical, so that $|P| = 0$; hence, $\lambda_j = 1$, $j = 1, \ldots, r_1$; similarly, $\gamma_j = 1$, $j = 1, \ldots, r_2$. Furthermore, all the eigenvectors of A_1 are orthogonal to the eigenvectors of A_2 and, hence,

$$A_1 A_2 = 0 \quad \text{and} \quad A_2 A_1 = 0.$$

Also, since $(A_1 + A_2)A_1 = A_1$, it follows that A_1 is idempotent; similarly, A_2 is idempotent.

It now follows from Theorem 8.6 that Q_1 and Q_2 have chi-squared distributions. To prove independence of Q_1 and Q_2, note that

$$Q_1 = \lambda_1 Y_1^T Y_1 + \cdots + \lambda_{r_1} Y_{r_1}^T Y_{r_1}$$

where $Y_j = e_j^T X$, $j = 1, \ldots, r_1$. Similarly,

$$Q_2 = \gamma_1 Z_1^T Z_1 + \cdots + \gamma_{r_2} Z_{r_2}^T Z_{r_2}$$

where $Z_j = v_j^T X$. Since each e_j is orthogonal to each v_j, it follows from Theorem 8.1 that Y_i and Z_j are independent, $i = 1, \ldots, r_1$, $j = 1, \ldots, r_2$. Hence, Q_1 and Q_2 are independent. ∎

In the following corollary, the result in Theorem 8.8 is extended to the case of several quadratic forms; the proof is left as an exercise. This result is known as Cochran's Theorem.

Corollary 8.1. *Let X denote a d-dimensional random vector with a multivariate normal distribution with mean 0 and covariance matrix I_d. Let $A_1, A_2, \ldots, A_m$ be $d \times d$ nonnegative-definite, symmetric matrices and let $Q_j = X^T A_j X$, $j = 1, \ldots, m$, such that*

$$X^T X = Q_1 + Q_2 + \cdots + Q_m.$$

Let r_j denote the rank of A_j, $j = 1, \ldots, m$. $Q_1, Q_2, \ldots, Q_m$ are independent chi-squared random variables, such that the distribution of Q_j has r_j degrees of freedom, $j = 1, \ldots, m$, if and only if $r_1 + \cdots + r_m = d$.

Example* 8.11 *(Analysis of variance). Let X denote a d-dimensional random vector with a multivariate normal distribution with mean vector μ and covariance matrix given by I_d. Let $\mathcal{M}$ denote a p-dimensional linear subspace of $\mathbf{R}^d$ and let $P_{\mathcal{M}}$ be the matrix representing orthogonal projection onto $\mathcal{M}$; here orthogonality is with respect to the usual inner product on $\mathbf{R}^d$. Hence, for any $x \in \mathbf{R}^d$, $P_{\mathcal{M}}x \in \mathcal{M}$ and

$$(x - P_{\mathcal{M}}x)^T y = 0 \quad \text{for all} \quad y \in \mathcal{M}.$$

Note that $P_{\mathcal{M}}$ has rank r, the dimension of $\mathcal{M}$. Consider the linear transformation given by the matrix $I_d - P_{\mathcal{M}}$. It is easy to show that this matrix represents orthogonal projection onto the orthogonal complement of $\mathcal{M}$; hence, the rank of $I_d - P_{\mathcal{M}}$ is $d - r$. Since

$$X^T X = X^T (I_d - P_{\mathcal{M}})X + X^T P_{\mathcal{M}} X,$$

it follows that the quadratic forms $X^T (I_d - P_{\mathcal{M}})X$ and $X^T P_{\mathcal{M}} X$ are independent chi-squared random variables with $d - r$ and r degrees of freedom, respectively.

Now suppose that $\mathbf{R}^d$ may be written

$$\mathbf{R}^d = \mathcal{M}_1 \oplus \mathcal{M}_2 \oplus \cdots \oplus \mathcal{M}_J$$

where $\mathcal{M}_1, \mathcal{M}_2, \ldots, \mathcal{M}_J$ are orthogonal linear subspaces of $\mathbf{R}^d$ so that if $x_i \in \mathcal{M}_i$ and $x_j \in \mathcal{M}_j, i \neq j$,

$$x_i^T x_j = 0.$$

Let $P_{\mathcal{M}_j}$ denote orthogonal projection onto $\mathcal{M}_j$ and let r_j denote the dimension of $\mathcal{M}_j$, $j = 1, \ldots, J$; it follows that $r_1 + \cdots + r_J = d$. Let $Q_j = X^T P_{\mathcal{M}_j} X$, $j = 1, \ldots, J$. Then

$$X^T X = Q_1 + \cdots + Q_J$$

and $Q_1, \ldots, Q_J$ are independent chi-squared random variables such that Q_j has degrees of freedom r_j, $j = 1, \ldots, J$. $\square$

8.5 Sampling Distributions

In statistics, the results of this chapter are often applied to the case of independent real-valued, normally distributed random variables. In this section, we present some classic results in this area.

***Theorem* 8.9.** *Let* $X_1, \ldots, X_n$ *denote independent, identically distributed standard normal random variables. Let*

$$\bar{X} = \frac{1}{n} \sum_{j=1}^{n} X_j$$

and let

$$S^2 = \frac{1}{n-1} \sum_{j=1}^{n} (X_j - \bar{X})^2.$$

Then $\bar{X}$ has a normal distribution with mean 0 and variance $1/n$, $(n-1)S^2$ has a chi-squared distribution with $n-1$ degrees of freedom, and $\bar{X}$ and S^2 are independent.

Proof. Let m_0 denote a $1 \times n$ vector of all ones, let $m = m_0/n$ and let $X = (X_1, \ldots, X_n)$. Then $\bar{X} = mX$ and

$$(n-1)S^2 = (X - nm^T mX)^T (X - nm^T mX) = X^T(I_n - nm^T m)^T(I_n - nm^T m)X.$$

The marginal distribution of $\bar{X}$ follows from Theorem 8.1. Note that $mm^T = 1/n$. Hence,

$$(I - nm^T m)^T(I - nm^T m) = I - 2nm^T m + n^2 m^T(mm^T)m = I - nm^T m$$

so that, by Theorem 8.6, $(n-1)S^2$ has a chi-squared distribution with degrees of freedom equal to the trace of $I - nm^T m$. Note that each diagonal element of $I - nm^T m$ is $1 - 1/n = (n-1)/n$ so that the trace is $n-1$.

Finally, $(I - nm^T m)m^T = m^T - nm^T(mm^T) = 0$ so that, by Theorem 8.7, $\bar{X}$ and S are independent. ∎

The distribution of the ratio $\sqrt{n}\bar{X}/S$ now follows immediately from the definition of the t-distribution given in Example 7.10. The distribution of $n\bar{X}^2/S^2$ is also easily determined.

Corollary 8.2. *Let $X_1, \ldots, X_n$ denote independent, identically distributed standard normal random variables. Let*

$$\bar{X} = \frac{1}{n}\sum_{j=1}^{n} X_j$$

and let

$$S^2 = \sum_{j=1}^{n}(X_j - \bar{X})^2/(n-1).$$

Then

(i) $\sqrt{n}\bar{X}/S$ has a t-distribution with $n-1$ degrees of freedom.

(ii) $n\bar{X}^2/S^2$ has a F-distribution with $(1, n-1)$ degrees of freedom.

Proof. From Theorem 8.9, $\bar{X}$ and S are independent; $\sqrt{n}\bar{X}$ has a standard normal distribution, $n\bar{X}^2$ has a chi-squared distribution with 1 degree of freedom, and $(n-1)S^2$ has a chi-squared distribution with $n-1$ degrees of freedom. The results now follow easily from the definitions of the t- and F-distributions, given in Examples 7.10 and 7.11, respectively. ∎

The statistics $\bar{X}$ and S^2 considered in Theorem 8.9 may be interpreted as follows. Define the vector m_0 as in the proof of Theorem 8.9. Then the projection of a random vector X onto the space spanned by m_0 is $\bar{X}m_0$. The statistic

$$(n-1)S^2 = \sum_{j=1}^{n}(X_j - \bar{X})^2$$

may be viewed as the squared length of the residual vector, X minus its projection. Theorem 8.9 states that the projection $\bar{X}m_0$ and the length of the residual vector are independent random variables and that any linear function of the projection has a normal distribution.

This result holds much more generally and, in fact, this generalization follows almost immediately from the results given above.

Theorem 8.10. *Let X denote an n-dimensional random vector with a multivariate normal distribution with mean vector 0 and covariance matrix given by $\sigma^2 I_n$. Let $\mathcal{M}$ denote a p-dimensional linear subspace of $\mathbf{R}^n$ and let $P_\mathcal{M}$ be the matrix representing orthogonal projection onto $\mathcal{M}$.*

Let $a \in \mathbf{R}^n$ be such that $a^T P_\mathcal{M} a > 0$.

(i) $a^T P_\mathcal{M} X$ has a normal distribution with mean 0 and variance $(a^T P_\mathcal{M} a)\sigma^2$.

(ii) Let $S^2 = X^T(I_n - P_\mathcal{M})X/(n - p)$. Then $(n - p)S^2/\sigma^2$ has a chi-squared distribution with $n - p$ degrees of freedom.

(iii) S^2 and $P_\mathcal{M}X$ are independent.

(iv)

$$\frac{a^T P_\mathcal{M} X}{(a^T P_\mathcal{M} a)^{\frac{1}{2}}[X^T(I_d - P_\mathcal{M})X/(n - p)]^{\frac{1}{2}}}$$

has a t-distribution with $n - p$ degrees of freedom.

Proof. Let $Y = a^T P_\mathcal{M} X$ and $S^2 = X^T(I_n - P_\mathcal{M})X/(n - p)$. Since $P_\mathcal{M}^T(I_n - P_\mathcal{M}) = 0$, it follows from Theorem 8.7 that $P_\mathcal{M}X$ and S^2 are independent. From Theorem 8.1, $a^T P_\mathcal{M} X$ has a normal distribution with mean 0 and variance $(a^T P_\mathcal{M} a)\sigma^2$. From Theorem 8.6, $(n - p)S^2/\sigma^2$ has a chi-squared distribution with $n - p$ degrees of freedom. Part (iv) follows from the definition of the t-distribution. ∎

Example 8.12 (Simple linear regression). Let $Y_1, Y_2, \ldots, Y_n$ denote independent random variables such that, for each $j = 1, 2, \ldots, n$, Y_j has a normal distribution with mean $\beta_0 + \beta_1 z_j$ and variance σ^2. Here $z_1, z_2, \ldots, z_n$ are fixed scalar constants, not all equal, and β_0, β_1, and σ are parameters.

Let $Y = (Y_1, \ldots, Y_n)$ and let Z denote the $n \times 2$ matrix with jth row $(1 \quad z_j)$, $j = 1, \ldots, n$. Let $\mathcal{M}$ denote the linear subspace spanned by the columns of Z. Then

$$P_\mathcal{M} = Z(Z^T Z)^{-1} Z^T.$$

Let $\beta = (\beta_0 \quad \beta_1)$ and let

$$\hat{\beta} = (Z^T Z)^{-1} Z^T Y$$

so that $P_\mathcal{M} Y = Z\hat{\beta}$. Consider the distribution of

$$T = \frac{c^T(\hat{\beta} - \beta)}{[c^T Z(Z^T Z)^{-1} Z^T c]^{\frac{1}{2}} S}$$

where $S^2 = Y^T(I_d - P_\mathcal{M})Y/(n - 2)$ and $c \in \mathbf{R}^2$.

Let $X = Y - Z\beta$. Then X has a multivariate normal distribution with mean vector 0 and covariance matrix $\sigma^2 I_n$. Note that

$$\hat{\beta} - \beta = (Z^T Z)^{-1} Z^T X$$

and $c^T(\hat{\beta} - \beta) = a^T P_{\mathcal{M}} X$ where $a = Z(Z^T Z)^{-1} c$. It now follows from Theorem 8.10 that T has a t-distribution with $n - 2$ degrees of freedom. □

8.6 Exercises

8.1 Let X denote a d-dimensional random vector with a multivariate normal distribution with covariance matrix Σ satisfying $|\Sigma| > 0$. Write $X = (X_1, \ldots, X_d)$, where $X_1, \ldots, X_d$ are real-valued, and let ρ_{ij} denote the correlation of X_i and X_j for $i \neq j$.

Let R denote the $d \times d$ matrix with each diagonal element equal to 1 and (i, j)th element equal to ρ_{ij}, $i \neq j$. Find a $d \times d$ matrix V such that

$$\Sigma = V R V.$$

8.2 Let Y denote a d-dimensional random vector with mean vector μ. Suppose that there exists $m \in \mathbf{R}^d$ such that, for any $a \in \mathbf{R}^d$, $E(a^T Y) = a^T m$. Show that $m = \mu$.

8.3 Let $X = (X_1, \ldots, X_d)$ have a multivariate normal distribution with mean vector μ and covariance matrix Σ. For arbitrary nonnegative integers $i_1, \ldots, i_d$, find $\kappa_{i_1 \cdots i_d}$, the joint cumulant of $(X_1, \ldots, X_d)$ of order $(i_1, \ldots, i_d)$.

8.4 Let X denote a d-dimensional random vector with a multivariate normal distribution with mean μ and covariance matrix Σ. Let $(\lambda_1, e_1), \ldots, (\lambda_d, e_d)$ denote the eigenvalue–eigenvector pairs of Σ, $\lambda_1 \geq \lambda_2 \geq \cdots \geq \lambda_d$. For each $j = 1, \ldots, d$, let $Y_j = e_j^T X$ and let $Y = (Y_1, \ldots, Y_d)$. Find the covariance matrix of Y.

8.5 Let $X = (X_1, \ldots, X_d)$ where $X_1, \ldots, X_d$ are independent random variables, each normally distributed such that X_j has mean μ_j and standard deviation $\sigma > 0$. Let A denote a $d \times d$ matrix of constants. Show that

$$E(X^T A X) = \sigma^2 \operatorname{tr}(A) + \mu^T A \mu$$

where $\mu = (\mu_1, \ldots, \mu_d)$.

8.6 Let X_1 and X_2 denote independent, d-dimensional random vectors such that X_j has a multivariate normal distribution with mean vector μ_j and covariance matrix Σ_j, $j = 1, 2$. Let $X = X_1 + X_2$. Find the mean vector and covariance matrix of X. Does X have a multivariate normal distribution?

8.7 Consider a multivariate normal random vector $X = (X_1, X_2, \ldots, X_n)$ and suppose that $X_1, X_2, \ldots, X_n$ are exchangeable random variables, each with variance 1, and let Σ denote the covariance matrix of the distribution. Suppose that $\sum_{j=1}^{n} X_n = 1$ with probability 1; find Σ.

8.8 Consider a multivariate normal random vector $X = (X_1, X_2, \ldots, X_n)$ and suppose that $X_1, X_2, \ldots, X_n$ are exchangeable random variables, each with variance 1, and let Σ denote the covariance matrix of the distribution. Then

$$\Sigma = \begin{pmatrix} 1 & \rho & \rho & \cdots & \rho \\ \rho & 1 & \rho & \cdots & \rho \\ \rho & \rho & \rho & \cdots & 1 \end{pmatrix}$$

for some constant ρ; see Example 8.2. Find the eigenvalues of Σ and, using these eigenvalues, find restrictions on the value of ρ so that Σ is a valid covariance matrix.

8.9 Let $X = (X_1, X_2, \ldots, X_n)$ where $X_1, X_2, \ldots, X_n$ are independent, identically distributed random variables, each with a normal distribution with mean 0 and standard deviation σ. Let B denote an orthogonal $n \times n$ matrix and let $Y = (Y_1, \ldots, Y_n) = BX$. Find the distribution of $Y_1, \ldots, Y_n$.

8.10 Let X denote a d-dimensional random vector with a multivariate normal distribution with mean vector 0 and covariance matrix I_d. Let $\{v_1, \ldots, v_d\}$ be an orthonormal basis for $\mathbf{R}^d$ and let $Y_1, \ldots, Y_d$ denote real-valued random variables such that

$$X = Y_1 v_1 + \cdots + Y_d v_d;$$

then $Y_1, \ldots, Y_d$ are the coordinates of X with respect to $\{v_1, \ldots, v_d\}$.

(a) Find an expression for Y_j, $j = 1, \ldots, d$.

(b) Find the distribution of $(Y_1, \ldots, Y_d)$.

8.11 Let X denote a d-dimensional multivariate normal random vector with mean vector μ and covariance matrix $\sigma^2 I_d$ where $\sigma^2 > 0$. Let $\mathcal{M}$ denote a linear subspace of $\mathbf{R}^d$ such that $\mu \in \mathcal{M}$. Let $c \in \mathbf{R}^d$ be a given vector and consider $\mathrm{Var}(b^T X)$ where $b \in \mathbf{R}^d$ satisfies $b^T \mu = c^T \mu$. Show that $\mathrm{Var}(b^T X)$ is minimized by $b = P_{\mathcal{M}} c$.

8.12 Let X denote a d-dimensional random vector with a multivariate normal distribution with mean vector μ and covariance matrix given by I_d. Suppose that $\mathbf{R}^d$ may be written

$$\mathbf{R}^d = \mathcal{M}_1 \oplus \mathcal{M}_2 \oplus \cdots \oplus \mathcal{M}_J$$

where $\mathcal{M}_1, \mathcal{M}_2, \ldots, \mathcal{M}_J$ are orthogonal linear subspaces of $\mathbf{R}^d$. Let $P_{\mathcal{M}_j}$ denote orthogonal projection onto $\mathcal{M}_j$ and let $Y_j = P_{\mathcal{M}_j} X$, $j = 1, \ldots, d$. Show that $Y_1, \ldots, Y_J$ are independent.

8.13 Let X be a d-dimensional random vector with a multivariate normal distribution with mean vector 0 and covariance matrix Σ. Write $X = (X_1, X_2)$ where X_1 is p-dimensional and X_2 is $(d - p)$-dimensional, and

$$\Sigma = \begin{pmatrix} \Sigma_{11} & \Sigma_{12} \\ \Sigma_{21} & \Sigma_{22} \end{pmatrix}$$

where Σ_{11} is $p \times p$, $\Sigma_{12} = \Sigma_{21}^T$ is $p \times (d - p)$, and Σ_{22} is $(d - p) \times (d - p)$; assume that $|\Sigma_{22}| > 0$.

Find $\mathrm{E}[X_1^T X_1 | X_2 = x_2]$.

8.14 Let $X = (X_1, X_2, X_3)$ denote a three-dimensional random vector with a multivariate normal distribution with mean vector 0 and covariance matrix Σ. Assume that

$$\mathrm{Var}(X_1) = \mathrm{Var}(X_2) = \mathrm{Var}(X_3) = 1$$

and let

$$\rho_{ij} = \mathrm{Cov}(X_i, X_j), \quad i \neq j.$$

(a) Find the conditional distribution of (X_1, X_2) given $X_3 = x_3$.

(b) Find conditions on $\rho_{12}, \rho_{13}, \rho_{23}$ so that X_1 and X_2 are conditionally independent given $X_3 = x_3$.

(c) Suppose that any two of X_1, X_2, X_3 are conditionally independent given the other random variable. Find the set of possible values of $(\rho_{12}, \rho_{13}, \rho_{23})$.

8.15 Let X denote a d-dimensional random vector and let A denote a $d \times d$ matrix that is not symmetric. Show that there exists a symmetric matrix B such that $X^T A X = X^T B X$.

8.16 Let X denote a d-dimensional random vector with a multivariate normal distribution with mean 0 and covariance matrix I_d. Let A_1 and A_2 be $d \times d$ nonnegative-definite, symmetric matrices and let $Q_j = X^T A_j X$, $j = 1, 2$, and let r_j denote the rank of A_j, $j = 1, 2$. Show that Q_1 and Q_2 are independent chi-squared random variables if and only if one of the following equivalent conditions holds:

  (i) $A_1 A_1 = A_1$ and $A_2 A_2 = A_2$

  (ii) $r_1 + r_2 = r$

  (iii) $A_1 A_2 = A_2 A_1 = 0$.

8.17 Let $X_1, X_2, \ldots, X_n$ denote independent random variables such that X_j has a normal distribution with mean 0 and variance $\sigma_j^2 > 0$, $j = 1, \ldots, n$, and let

$$T = \sum_{j=1}^n (X_j - \bar{X})^2, \qquad \bar{X} = \frac{1}{n} \sum_{j=1}^n X_j.$$

Find conditions on $\sigma_1^2, \ldots, \sigma_n^2$ so that there exists a constant c such that cT has a chi-squared distribution with r degrees of freedom. Give expressions for c and r.

8.18 Let $Z_1, \ldots, Z_n$ denote independent random variables, each with a standard normal distribution, and let $\delta_1, \ldots, \delta_n$ denote real-valued constants.

Define a random variable X by

$$X = \sum_{j=1}^n (Z_j + \delta_j)^2.$$

The distribution of X is called a *noncentral chi-squared distribution* with n degrees of freedom.

(a) Show that the distribution of X depends on $\delta_1, \ldots, \delta_n$ only through

$$\delta^2 \equiv \sum_{j=1}^n \delta_j^2;$$

δ^2 is called the *noncentrality parameter* of the distribution.

(b) Find the mean and variance of X.

8.19 Suppose that X_1 and X_2 are independent random variables such that, for $j = 1, 2$, X_j has a noncentral chi-squared distribution with n_j degrees of freedom, $n_j = 1, 2, \ldots$, and noncentrality parameter $\gamma_j^2 \geq 0$. Does $X_1 + X_2$ have a noncentral chi-squared distribution? If so, find the degrees of freedom and the noncentrality parameter of the distribution.

8.20 Let X denote a d-dimensional random vector with a multivariate normal distribution with mean vector μ and covariance matrix Σ, which is assumed to be positive-definite. Let A denote a $d \times d$ symmetric matrix and consider the random variable

$$Q = X^T A X.$$

Find conditions on A so that Q has a noncentral chi-squared distribution. Find the degrees of freedom and the noncentrality parameter of the distribution.

8.21 Let $X = (X_1, \ldots, X_n)$ denote a random vector such that $X_1, \ldots, X_n$ are real-valued, exchangeable random variables and suppose that X_1 has a standard normal distribution. Let

$$S^2 = \sum_{j=1}^n (X_j - \bar{X})^2.$$

Let $\rho = \mathrm{Cov}(X_i, X_j)$. Find the values of ρ for which S^2 has a chi-squared distribution.

8.22 Let $X = (X_1, \ldots, X_n)$ denote a random vector such that $X_1, \ldots, X_n$ are real-valued, exchangeable random variables and suppose that X_1 has a standard normal distribution. Let

$$\bar{X} = \frac{1}{n} \sum_{j=1}^n X_j \quad \text{and} \quad S^2 = \frac{1}{n-1} \sum_{j=1}^n (X_j - \bar{X})^2.$$

Find the values of ρ for which $\bar{X}$ and S^2 are independent.

8.23 Let X denote a d-dimensional random vector with a multivariate normal distribution with mean vector 0 and covariance matrix given by the identity matrix. For a given linear subspace of $\mathbf{R}^d$, $\mathcal{M}$, define

$$D(\mathcal{M}) = \min_{m \in \mathcal{M}} \|X - m\|^2$$

where $|| \cdot ||$ denotes the Euclidean norm in $\mathbf{R}^d$, that is,

$$||x||^2 = x^T x.$$

(a) Find the distribution of $D(\mathcal{M})$.

(b) Let $\mathcal{M}_1$ and $\mathcal{M}_2$ denote linear subspaces of $\mathbf{R}^d$. Find conditions on $\mathcal{M}_1$ and $\mathcal{M}_2$ under which $D(\mathcal{M}_1)$ and $D(\mathcal{M}_2)$ are independent.

8.24 Let X denote an n-dimensional random vector with a multivariate normal distribution with mean vector 0 and covariance matrix given by $\sigma^2 I_n$. Let $\mathcal{M}_1$ and $\mathcal{M}_2$ denote orthogonal linear subspaces of $\mathbf{R}^n$ and let P_j denote the matrix representing orthogonal projection onto $\mathcal{M}_j$, $j = 1, 2$. Find the distribution of

$$\frac{X^T P_1 X}{X^T P_2 X}.$$

8.25 Prove Corollary 8.1.

8.7   Suggestions for Further Reading

An excellent reference for properties of the multivariate normal distribution and the associated sampling distributions is Rao (1973, Chapter 3). Stuart and Ord (1994, Chapters 15 and 16) contains a detailed discussion of the distribution theory of quadratic forms and distributions related to the normal distribution, such as the chi-squared distribution and the F-distribution. Many books on multivariate statistical analysis consider the multivariate normal distribution in detail; see, for example, Anderson (1984) and Johnson and Wichern (2002).

9

Approximation of Integrals

9.1 Introduction

Integrals play a fundamental role in distribution theory and, when exact calculation of an integral is either difficult or impossible, it is often useful to use an approximation. In this chapter, several methods of approximating integrals are considered. The goal here is not the determination of the numerical value of a given integral; instead, we are concerned with determining the properties of the integrals that commonly arise in distribution theory. These properties are useful for understanding the properties of the statistical procedures that are based on those integrals.

9.2 Some Useful Functions

There are a number of important functions that repeatedly appear in statistical calculations, such as the gamma function, the incomplete gamma function, and the standard normal distribution function. These functions are well-studied and their properties are well-understood; when an integral under consideration can be expressed in terms of one of these functions, the properties of the integral are, to a large extent, also well-understood. In this section, we consider the basic properties of these functions; further properties are presented in the remaining sections of this chapter.

Gamma function

The *gamma function* is defined by

$$\Gamma(x) = \int_0^\infty t^{x-1} \exp(-t)\, dt, \quad x > 0.$$

The most important property of the gamma function is its recursion property:

$$\Gamma(x + 1) = x\Gamma(x), \quad x > 0.$$

This, together with the fact that $\Gamma(1) = 1$, shows that

$$\Gamma(n + 1) = n!, \quad n = 0, 1, 2, \ldots$$

so that the gamma function represents a generalization of the factorial function to non-integer positive arguments. These properties are formally stated in the following theorem.

Theorem 9.1. $\Gamma(x + 1) = x\Gamma(x)$, $x > 0$, *and for* $n = 0, 1, \ldots, \Gamma(n + 1) = n!$.

Proof. Using integration-by-parts,

$$\Gamma(x + 1) = \int_0^\infty t^x \exp(-t)\, dt = -t^x \exp(-t)\Big|_0^\infty + x \int_0^\infty t^{x-1} \exp(-t)\, dt = x\Gamma(x).$$

Hence, for an integer n,

$$\Gamma(n + 1) = n\Gamma(n) = n(n - 1)\Gamma(n - 1) = \cdots = n!\Gamma(1);$$

the second result now follows from the easily verified fact that $\Gamma(1) = 1$. ∎

A function closely related to the gamma function is the *beta function*. Let r and s be nonnegative real numbers. Define

$$\beta(r, s) = \int_0^1 t^{r-1}(1 - t)^{s-1}\, dt;$$

this function appears in the normalizing constant of the density of the beta distribution. The following theorem gives an expression for the beta function in terms of the gamma function.

Theorem 9.2.

$$\beta(r, s) = \frac{\Gamma(r)\Gamma(s)}{\Gamma(r + s)}, \quad r > 0, \ s > 0.$$

Proof. Note that, for $r > 0$ and $s > 0$,

$$\Gamma(r)\Gamma(s) = \int_0^\infty t^{r-1} \exp(-t)\, dt \int_0^\infty t^{s-1} \exp(-t)\, dt$$

$$= \int_0^\infty \int_0^\infty t_1^{r-1} t_2^{s-1} \exp\{-(t_1 + t_2)\}\, dt_1\, dt_2.$$

Using the change-of-variable $x_1 = t_1 + t_2$, $x_2 = t_1/(t_1 + t_2)$, we may write

$$\int_0^\infty \int_0^\infty t_1^{r-1} t_2^{s-1} \exp\{-(t_1 + t_2)\}\, dt_1\, dt_2$$

$$= \int_0^1 \int_0^\infty x_1^{r+s-1} x_2^{r-1}(1 - x_2)^{s-1} \exp(-x_1)\, dx_1\, dx_2$$

$$= \int_0^1 x_2^{r-1}(1 - x_2)^{s-1}\, dx_2 \int_0^\infty x_1^{r+s-1} \exp(-x_1)\, dx_1$$

$$= \beta(r, s)\Gamma(r + s),$$

proving the result. ∎

The value of the gamma function increases very rapidly as the argument of the function increases. Hence, it is often more convenient to work with the log of the gamma function rather than with the gamma function itself; a plot of this function is given in Figure 9.1. Clearly, $\log \Gamma(x)$ satisfies the recursive relationship

$$\log \Gamma(x + 1) = \log x + \log \Gamma(x).$$

In addition, the following result shows that $\log \Gamma(x)$ is a convex function.

Theorem 9.3. *The function* $\log \Gamma(x)$, $x > 0$, *is convex.*

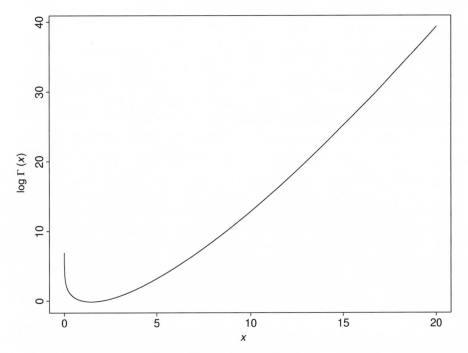

Figure 9.1. Log-gamma function.

Proof. Let $0 < \alpha < 1$. Then, for $x_1 > 0$, $x_2 > 0$,

$$\log \Gamma(\alpha x_1 + (1 - \alpha)x_2) = \log \int_0^\infty (t^{x_1})^\alpha (t^{x_2})^{1-\alpha} \frac{1}{t} \exp(-t)\,dt.$$

By the Hölder inequality,

$$\log \int_0^\infty (t^{x_1})^\alpha (t^{x_2})^{1-\alpha} \frac{1}{t} \exp(-t)\,dt \le \alpha \log \left(\int_0^\infty t^{x_1-1} \exp(-t)\,dt \right)$$
$$+ (1 - \alpha) \log \left(\int_0^\infty t^{x_2-1} \exp(-t)\,dt \right).$$

It follows that

$$\log \Gamma(\alpha x_1 + (1 - \alpha)x_2) \le \alpha \log \Gamma(x_1) + (1 - \alpha) \log \Gamma(x_2),$$

proving the result. ∎

Let

$$\psi(x) = \frac{d}{dx} \log \Gamma(x), \quad x > 0$$

denote the logarithmic derivative of the gamma function. The function ψ inherits a recursion property from the recursion property of the log-gamma function. This property is given in the following theorem; the proof is left as an exercise.

Theorem 9.4.

$$\psi(x + 1) = \psi(x) + \frac{1}{x}, \quad x > 0$$

For n = 1, 2, ...,

$$\psi(x + n) = \frac{1}{x} + \frac{1}{x+1} + \cdots + \frac{1}{x+n-1} + \psi(x), \quad x > 0.$$

Incomplete gamma function

The limits of the integral defining the gamma function are 0 and ∞. The *incomplete gamma function* is the function obtained by restricting the region of integration to $(0, y)$:

$$\gamma(x, y) = \int_0^y t^{x-1} \exp(-t)\, dt, \quad x > 0, \ y > 0.$$

Thus, considered as a function of y for fixed x, the incomplete gamma function is, aside from a normalization factor, the distribution function of the standard gamma distribution with index x.

The following theorem shows that, like the gamma function, the incomplete gamma function satisfies a recursive relationship; the proof is left as an exercise.

Theorem 9.5. *For $x > 0$ and $y > 0$,*

$$\gamma(x + 1, y) = x\gamma(x, y) - y^x \exp(-y).$$

It is sometimes convenient to consider the function

$$\Gamma(x, y) = \int_y^\infty t^{x-1} \exp(-t)\, dt = \Gamma(x) - \gamma(x, y).$$

When x is an integer, $\Gamma(x, y)/\Gamma(x)$, and, hence, $\gamma(x, y)/\Gamma(x)$, can be expressed as a finite sum.

Theorem 9.6. *For $n = 1, 2, \ldots,$*

$$\frac{\Gamma(n, y)}{\Gamma(n)} = \sum_{j=0}^{n-1} \frac{y^j}{j!} \exp(-y), \quad y > 0;$$

equivalently,

$$\frac{\gamma(n, y)}{\Gamma(n)} = 1 - \sum_{j=0}^{n-1} \frac{y^j}{j!} \exp(-y), \quad y > 0.$$

Proof. Note that, using the change-of-variable $s = t - y$,

$$\int_y^\infty t^{n-1} \exp(-t)\, dt = \exp(-y) \int_0^\infty (s + y)^{n-1} \exp(-s)\, ds$$

$$= \exp(-y) \int_0^\infty \sum_{j=0}^{n-1} \binom{n-1}{j} s^{n-1-j} y^j \exp(-s)\, ds$$

$$= \sum_{j=0}^{n-1} \exp(-y) y^j \binom{n-1}{j} \Gamma(n - j)$$

$$= \sum_{j=0}^{n-1} \exp(-y) \frac{y^j}{j!} (n - 1)!.$$

That is,

$$\Gamma(n, y) = \sum_{j=0}^{n-1} \frac{y^j}{j!} \exp(-y)\Gamma(n),$$

proving the result.   ∎

For general values of x, the following result gives two series expansions for $\gamma(x, y)$.

Theorem 9.7. *For $x > 0$ and $y > 0$,*

$$\gamma(x, y) = \sum_{j=0}^{\infty} \frac{(-1)^j}{j!} \frac{y^{x+j}}{x+j} \tag{9.1}$$

$$= \exp(-y) \sum_{j=0}^{\infty} \frac{\Gamma(x)}{\Gamma(x+j+1)} y^{x+j}. \tag{9.2}$$

Proof.   Recall that, for all x,

$$\exp(x) = \sum_{j=0}^{\infty} x^j/j!;$$

hence,

$$\gamma(x, y) = \int_0^y t^{x-1} \exp(-t) \, dt = \int_0^y t^{x-1} \sum_{j=0}^{\infty} (-1)^j t^j/j! \, dt$$

$$= \sum_{j=0}^{\infty} \frac{(-1)^j}{j!} \int_0^y t^{x+j-1} \, dt$$

$$= \sum_{j=0}^{\infty} \frac{(-1)^j}{j!} \frac{y^{x+j}}{x+j}.$$

Note that the interchanging of summation and integration is justified by the fact that

$$\sum_{j=0}^{\infty} \frac{1}{j!} \int_0^y t^{x+j-1} \, dt = \sum_{j=0}^{\infty} \frac{1}{j!} \frac{y^{x+j}}{x+j} \le y^x \exp(y) < \infty$$

for all y, x. This proves (9.1).

Now consider (9.2). Using the change-of-variable $u = 1 - t/y$, we may write

$$\gamma(x, y) = \int_0^y t^{x-1} \exp(-t) \, dt = \int_0^1 y^x(1-u)^{x-1} \exp\{-y(1-u)\} \, du$$

$$= y^x \exp(-y) \int_0^1 (1-u)^{x-1} \exp(yu) \, du$$

$$= y^x \exp(-y) \int_0^1 (1-u)^{x-1} \sum_{j=0}^{\infty} (yu)^j/j! \, du.$$

Since

$$\sum_{j=0}^{\infty} \int_0^1 (1-u)^{x-1}(yu)^j/j! \le \sum_{j=0}^{\infty} y^j/j! = \exp(y),$$

we may interchange summation and integration. Hence, by Theorem 9.2,

$$\gamma(x, y) = y^x \exp(-y) \sum_{j=0}^{\infty} \frac{y^j}{j!} \int_0^1 (1-u)^{x-1} u^j \, du$$

$$= y^x \exp(-y) \sum_{j=0}^{\infty} \frac{y^j}{j!} \frac{\Gamma(j+1)\Gamma(x)}{\Gamma(x+j+1)},$$

which, after simplification, is identical to (9.2). ∎

Standard normal distribution function
Let

$$\phi(z) = \frac{1}{\sqrt{(2\pi)}} \exp\left(-\frac{1}{2}z^2\right), \quad -\infty < z < \infty$$

denote the standard normal density function and let

$$\Phi(z) = \int_{-\infty}^z \phi(t) \, dt, \quad -\infty < z < \infty$$

denote the standard normal distribution function.

Although $\Phi(z)$ is defined by an integral over the region $(-\infty, z)$, calculation of $\Phi(z)$, for any value of z, only requires integration over a bounded region. Define

$$\Phi_0(z) = \int_0^z \phi(t) \, dt, \quad z \ge 0.$$

The following result shows that $\Phi(z)$ can be written in terms of $\Phi_0(z)$; also, by a change-of-variable in the integral defining Φ_0 it may be shown that Φ_0 is a special case of the incomplete gamma function $\gamma(\cdot, \cdot)$. The proof is left as an exercise.

Theorem 9.8.

$$\Phi(z) = \begin{cases} \frac{1}{2} - \Phi_0(-z) & \text{if } z < 0 \\[2mm] \frac{1}{2} + \Phi_0(z) & \text{if } z \ge 0 \end{cases}$$

and, for $z > 0$,

$$\Phi_0(z) = \frac{1}{2\sqrt{\pi}} \gamma\left(\frac{1}{2}, \frac{z^2}{2}\right).$$

Hence, using Theorem 9.8, together with the series expansions for $\gamma(\cdot, \cdot)$ given in Theorem 9.7, we obtain the following series expansions for $\Phi_0(\cdot)$; these, in turn, may be used to obtain a series expansion for $\Phi(\cdot)$. The result is given in the following theorem, whose proof is left as an exercise.

Theorem 9.9. *For z ≥ 0,*

$$\Phi_0(z) = \frac{1}{\sqrt{(2\pi)}} \sum_{j=0}^{\infty} \frac{(-1)^j}{2^j j!} \frac{z^{2j+1}}{2j+1} = \frac{1}{2\sqrt{2}} \exp\left(\frac{-z^2}{2}\right) \sum_{j=0}^{\infty} \frac{z^{2j+1}}{2^j \Gamma(j+3/2)}.$$

Since $\phi(z) > 0$ for all z, $\Phi(\cdot)$ is a strictly increasing function and, since it is a distribution function,

$$\lim_{z \to \infty} \Phi(z) = 1 \quad \text{and} \quad \lim_{z \to -\infty} \Phi(z) = 0.$$

The following result gives some information regarding the rate at which $\Phi(z)$ approaches these limiting values; the proof follows from L'Hospital's rule and is left as an exercise.

Theorem 9.10.

$$\lim_{z \to \infty} \frac{1 - \Phi(z)}{\phi(z)/z} = 1 \quad \text{and} \quad \lim_{z \to -\infty} \frac{\Phi(z)}{\phi(z)/z} = -1.$$

The following result gives precise bounds on

$$\frac{1 - \Phi(z)}{\phi(z)/z} \quad \text{and} \quad \frac{\Phi(z)}{\phi(z)/z}$$

that are sometimes useful.

Theorem 9.11. *For z > 0,*

$$\frac{z}{1+z^2}\phi(z) \leq 1 - \Phi(z) \leq \frac{1}{z}\phi(z).$$

For z < 0,

$$\frac{|z|}{1+z^2}\phi(z) < \Phi(z) < \frac{1}{|z|}\phi(z).$$

Proof.   Fix $z > 0$. Note that

$$\int_z^{\infty} \exp(-x^2/2)\,dx = \int_z^{\infty} \frac{1}{x} x \exp(-x^2/2)\,dx \leq \frac{1}{z} \int_z^{\infty} x \exp(-x^2/2)\,dx.$$

Since

$$\frac{d}{dx} \exp(-x^2/2) = -x \exp(-x^2/2),$$

$$\int_z^{\infty} x \exp(-x^2/2)\,dx = \exp(-z^2/2)$$

so that

$$\int_z^{\infty} \exp(-x^2/2)\,dx =\leq \frac{1}{z} \exp(-z^2/2)$$

or, equivalently,

$$1 - \Phi(z) \leq \frac{1}{z}\phi(z).$$

To verify the second inequality, note that

$$(1 + 1/z^2) \int_z^\infty \exp(-x^2/2)\,dx \geq \int_z^\infty (1 + 1/x^2)\,\exp(-x^2/2)\,dx.$$

Since

$$\frac{d}{dx}\frac{1}{x}\exp(-x^2/2) = (1 + 1/x^2)\,\exp(-x^2/2),$$

$$\int_z^\infty (1 + 1/x^2)\,\exp(-x^2/2)\,dx = \frac{1}{z}\exp(-z^2/2),$$

which proves the result for $z > 0$.

The result for $z < 0$ follows from the fact that, for $z < 0$, $\Phi(z) = 1 - \Phi(-z)$. ∎

9.3   Asymptotic Expansions

In many cases, exact calculation of integrals is not possible and approximations are needed. Here we consider aproximations as some parameter reaches a limiting value. In general discussions that parameter is denoted by n and we consider approximations as $n \to \infty$; different notation for the parameter and different limiting values are often used in specific examples. Note that, in this context, n is not necessarily an integer.

Let f denote a real-valued function of a real variable n. The series

$$\sum_{j=0}^\infty a_j n^{-j}$$

is said to be an *asymptotic expansion* of f if, for any $m = 1, 2, \ldots,$

$$f(n) = \sum_{j=0}^m a_j n^{-j} + \underline{R_{m+1}(n)}$$

where $R_{m+1}(x) = O(n^{-(m+1)})$ as $n \to \infty$; that is, where $n^{m+1} R_{m+1}(n)$ remains bounded as $n \to \infty$. This is often written

$$f(n) \sim \sum_{j=0}^\infty a_j n^{-j} \quad \text{as } n \to \infty.$$

It is important to note that this does not imply that

$$f(n) = \sum_{j=0}^\infty a_j n^{-j},$$

which would require additional conditions on the convergence of the series. An asymptotic expansion represents a sequence of approximations to $f(n)$ with the property that the order of the remainder term, as a power of n^{-1}, is higher than that of the terms included in the approximation. It is also worth noting that, for a given value of n, the approximation based on m terms in the expansion may be more accurate than the approximation based on $m + 1$ terms in the approximation.

In many cases, an entire asymptotic expansion is not needed; that is, we do not need to be able to compute an approximation to $f(n)$ with error of order $O(n^{-m})$ for any value of

m. It is often sufficient to be able to approximate $f(n)$ with error $O(n^{-m})$ for some given fixed value of m, often $m = 1$ or 2.

Example 9.1. Consider the function defined by

$$f(x) = \int_0^\infty \frac{1}{1 + t/x} \exp(-t) \, dt, \quad 0 < x < \infty.$$

Recall that, for $z \neq 1$,

$$1 + z + z^2 + \cdots + z^{m-1} = \frac{1 - z^m}{1 - z}.$$

Hence,

$$\frac{1}{1 + t/x} = 1 - \frac{t}{x} + \frac{t^2}{x^2} + \cdots + (-1)^{m-1} \frac{t^{m-1}}{x^{m-1}} + (-1)^m \frac{t^m}{x^m} \frac{1}{1 + t/x}$$

so that

$$f(x) = \sum_{j=0}^{m-1} (-1)^j \frac{j!}{x^j} + R_m(x)$$

where

$$|R_m(x)| = \int_0^\infty \frac{t^m}{x^m} \frac{1}{1 + t/x} \exp(-t) \, dt \le \frac{m!}{x^m}.$$

It follows that

$$\sum_{j=0}^\infty (-1)^j j!/x^j$$

is a valid asymptotic expansion of $f(x)$ as $x \to \infty$. Thus, we may write

$$\int_0^\infty \frac{1}{1 + t/x} \exp(-t) \, dt = 1 - \frac{1}{x} + O(x^{-2}),$$

$$\int_0^\infty \frac{1}{1 + t/x} \exp(-t) \, dt = 1 - \frac{1}{x} + \frac{2}{x^2} + O(x^{-3}),$$

and so on. For $x > 0$, let

$$\hat{f}_2(x) = 1 - \frac{1}{x} \quad \text{and} \quad \hat{f}_3(x) = 1 - \frac{1}{x} + \frac{2}{x^2}.$$

Table 9.1 gives the values of $\hat{f}_2(x)$ and $\hat{f}_3(x)$, together with $f(x)$, for several values of x. Although both approximations are inaccurate for small values of x, both are nearly exact for $x = 30$.

Note, however, that

$$\sum_{j=0}^\infty (-1)^j j!/x^j \neq \int_0^\infty \frac{1}{1 + t/x} \exp(-t) \, dt;$$

in fact, the series diverges for any value of x. □

Table 9.1. *Approximations in Example 9.1.*

x	$f(x)$	$\hat{f}_2(x)$	$\hat{f}_3(x)$
1	0.596	0	2.000
2	0.722	0.500	1.000
5	0.852	0.800	0.880
10	0.916	0.900	0.920
20	0.954	0.950	0.955
30	0.981	0.980	0.981

Integration-by-parts

One useful technique for obtaining asymptotic approximations to integrals is to repeatedly use integration-by-parts. This approach is illustrated in the following examples.

Example 9.2 (Incomplete beta function). Consider approximation of the integral

$$\int_0^x t^{\alpha-1}(1-t)^{\beta-1}\,dt,$$

where $\alpha > 0$ and $\beta > 0$, for small values of $x > 0$. This is the integral appearing in the distribution function corresponding to the beta distribution with density

$$\frac{\Gamma(\alpha+\beta)}{\Gamma(\alpha)\Gamma(\beta)}x^{\alpha-1}(1-x)^{\beta-1}, \quad 0 < x < 1$$

and it is known as the incomplete beta function.

Using integration-by-parts,

$$\int_0^x t^{\alpha-1}(1-t)^{\beta-1}\,dt = \frac{1}{\alpha}t^\alpha(1-t)^{\beta-1}\Big|_0^x + \frac{\beta-1}{\alpha}\int_0^x t^\alpha(1-t)^{\beta-2}\,dt$$

$$= \frac{1}{\alpha}x^\alpha(1-x)^{\beta-1} + \frac{\beta-1}{\alpha}\int_0^x t^\alpha(1-t)^{\beta-2}\,dt.$$

For $\beta \geq 2$,

$$\int_0^x t^\alpha(1-t)^{\beta-2}\,dt \leq \frac{1}{\alpha+1}x^{\alpha+1},$$

while for $0 < \beta < 2$,

$$\int_0^x t^\alpha(1-t)^{\beta-2}\,dt \leq (1-x)^{\beta-2}\frac{1}{\alpha+1}x^{\alpha+1};$$

hence, we may write

$$\int_0^x t^{\alpha-1}(1-t)^{\beta-1}\,dt = \frac{1}{\alpha}x^\alpha(1-x)^{\beta-1}[1+o(x)] \quad \text{as } x \to 0.$$

Alternatively, integration-by-parts may be used on the remainder term, leading to the expansion

$$\int_0^x t^{\alpha-1}(1-t)^{\beta-1}\,dt = \frac{1}{\alpha}x^\alpha(1-x)^{\beta-1}\left[1 + \frac{\beta-1}{\alpha+1}\frac{x}{1-x} + o(x^2)\right] \quad \text{as } x \to 0.$$

Further terms in the expansion may be generated in the same manner.

Table 9.2. *Approximations in Example 9.2.*

x	Exact	Approx.	Relative error (%)
0.50	0.102	0.103	1.4
0.20	0.0186	0.0186	0.12
0.10	0.00483	0.00483	0.024
0.05	0.00123	0.00123	0.0056

Table 9.2 contains values of the approximation

$$\frac{1}{\alpha}x^\alpha(1-x)^{\beta-1}\left[1+\frac{\beta-1}{\alpha+1}\frac{x}{1-x}\right]$$

for the case $\alpha=2$, $\beta=3/2$, along with the exact value of the integral and the relative error of the approximation. Note that for these choices of α and β, exact calculation of the integral is possible:

$$\int_0^x t(1-t)^{\frac{1}{2}}\,dt = \frac{4}{15}+\frac{2}{5}(1-x)^{\frac{5}{2}}-\frac{2}{3}(1-x)^{\frac{3}{2}}.$$

The results in Table 9.2 show that the approximation is extremely accurate even for relatively large values of x. $\quad\square$

Example 9.3 *(Normal tail probability).* Consider the function

$$\bar{\Phi}(z) \equiv \int_z^\infty \phi(t)\,dt$$

where ϕ denotes the density function of the standard normal distribution. Then, using integration-by-parts, we may write

$$\bar{\Phi}(z) = \int_z^\infty \phi(t)\,dt = \int_z^\infty \frac{1}{t}t\phi(t)\,dt$$

$$= -\frac{1}{t}\phi(t)\Big|_z^\infty - \int_z^\infty \frac{1}{t^2}\phi(t)\,dt$$

$$= \frac{1}{z}\phi(z) - \int_z^\infty \frac{1}{t^3}t\phi(t)\,dt$$

$$= \frac{1}{z}\phi(z) - \frac{1}{z^3}\phi(z) + \int_z^\infty \frac{1}{t^4}\phi(t)\,dt.$$

Hence,

$$\bar{\Phi}(z) = \frac{1}{z}\phi(z) - \frac{1}{z^3}\phi(z) + O\left(\frac{1}{z^4}\right)\bar{\Phi}(z)$$

as $z \to \infty$. That is,

$$\left[1+O\left(\frac{1}{z^4}\right)\right]\bar{\Phi}(z) = \phi(z)\left[\frac{1}{z}-\frac{1}{z^3}\right],$$

or,

$$\bar{\Phi}(z) = \phi(z)\frac{\frac{1}{z} - \frac{1}{z^3}}{1 + O\left(\frac{1}{z^4}\right)} = \phi(z)\left[\frac{1}{z} - \frac{1}{z^3} + O\left(\frac{1}{z^5}\right)\right]$$

as $z \to \infty$.

Note that this approach may be continued indefinitely, leading to the result that

$$\bar{\Phi}(z) = \phi(z)\left[\frac{1}{z} - \frac{1}{z^3} + \frac{3}{z^5} + \cdots + (-1)^k\frac{(2k)!}{2^k k!}\frac{1}{z^{2k+1}} + O\left(\frac{1}{z^{2k+2}}\right)\right]$$

as $z \to \infty$, for any $k = 0, 1, 2, \ldots$. □

9.4 Watson's Lemma

Note that an integral of the form

$$\int_0^\infty t^\alpha \exp(-nt)\,dt,$$

where $\alpha > -1$ and $n > 0$, may be integrated exactly, yielding

$$\int_0^\infty t^\alpha \exp(-nt)\,dt = \frac{\Gamma(\alpha + 1)}{n^{\alpha+1}}.$$

Now consider an integral of the form

$$\int_0^\infty h(t) \exp(-nt)\,dt$$

where h has an series representation of the form

$$h(t) = \sum_{j=0}^\infty a_j t^j.$$

Then, assuming that summation and integration may be interchanged,

$$\int_0^\infty h(t) \exp(-nt)\,dt = \sum_{j=0}^\infty a_j \int_0^\infty t^j \exp(-nt)\,dt = \sum_{j=0}^\infty a_j \frac{\Gamma(j+1)}{n^{j+1}}.$$

Note that the terms in this series have increasing powers of $1/n$; hence, if n is large, the value of the integral may be approximated by the first few terms in the series. *Watson's lemma* is a formal statement of this result.

Theorem 9.12 (Watson's lemma). *Let h denote a real-valued continuous function on $[0, \infty)$ satisfying the following conditions:*

(i) $h(t) = O(\exp(bt))$ as $t \to \infty$ for some constant b

(ii) there exist constants $c_0, c_1, \ldots, c_{m+1}, a_0, a_1, \ldots, a_{m+1}$,

$$-1 < a_0 < a_1 < \cdots < a_{m+1}$$

such that

$$\hat{h}_m(t) = \sum_{j=0}^m c_j t^{a_j}$$

satisfies

$$h(t) = \hat{h}_m(t) + O(t^{a_{m+1}}) \quad as \ t \to 0^+.$$

Consider the integral

$$I_n = \int_0^T h(t) \exp\{-nt\} \, dt, \quad 0 < T \leq \infty.$$

Then

$$I_n = \sum_{j=0}^m \frac{c_j \Gamma(a_j + 1)}{n^{a_j+1}} + O\left(\frac{1}{n^{a_{m+1}+1}}\right) \quad as \ n \to \infty.$$

Proof. Fix $0 < \epsilon < T$. There exists a constant K such that

$$|h(t)| \leq K \exp\{bt\}, \quad t \geq \epsilon.$$

Hence, for sufficiently large n,

$$\left| \int_\epsilon^T h(t) \exp\{-nt\} dt \right| \leq K \int_\epsilon^\infty \exp\{-(n-b)\} \, dt \leq \frac{K}{n-b} \exp\{-(n-b)\epsilon\}$$

$$= O\left(\frac{\exp\{-n\epsilon\}}{n}\right) \quad as \ n \to \infty.$$

Consider the integral

$$\int_0^\epsilon h(t) \exp\{-nt\} \, dt = \int_0^\epsilon \hat{h}_m(t) \exp\{-nt\} \, dt + \int_0^\epsilon R_m(t) \exp\{-nt\} \, dt$$

where $R_m(t) = O(t^{a_{m+1}})$ as $t \to 0^+$. For any $\alpha > -1$,

$$\int_0^\epsilon t^\alpha \exp\{-nt\} \, dt = \int_0^\infty t^\alpha \exp\{-nt\} \, dt - \int_\epsilon^\infty t^\alpha \exp\{-nt\} \, dt.$$

Note that

$$\int_0^\infty t^\alpha \exp\{-nt\} \, dt = \frac{\Gamma(\alpha + 1)}{n^{\alpha+1}}$$

and

$$\exp\{n\epsilon\} \int_\epsilon^\infty t^\alpha \exp\{-nt\} \, dt = \int_\epsilon^\infty t^\alpha \exp\{-n(t - \epsilon)\} \, dt$$

$$= \int_0^\infty (t + \epsilon)^\alpha \exp\{-nt\} \, dt = O(1)$$

as $n \to \infty$. It follows that

$$\int_0^\epsilon t^\alpha \exp\{-nt\} \, dt = \frac{\Gamma(\alpha + 1)}{n^{\alpha+1}} + O(\exp\{-n\epsilon\}).$$

Hence, for any $\epsilon > 0$,

$$\int_0^\epsilon \hat{h}_m(t) \exp\{-nt\} \, dt = \sum_{j=0}^m \frac{c_j \Gamma(a_j + 1)}{n^{a_j+1}} + O(\exp\{-n\epsilon\}).$$

Note that, since $R_m(t) = O(t^{a_{m+1}})$ as $t \to 0^+$, there exists a constant K_m such that for sufficiently small ϵ,

$$|R_m(t)| \leq K_m t^{a_{m+1}}.$$

Hence,

$$\int_0^\epsilon |R_m(t)| \exp\{-nt\} \, dt \leq K_m \int_0^\epsilon t^{a_{m+1}} \exp\{-nt\} \, dt \leq K_m \frac{\Gamma(a_{m+1} + 1)}{n^{a_{m+1}+1}}.$$

It follows that

$$
\begin{aligned}
I_n &= \int_0^T h(t) \exp\{-nt\} \, dt \\
&= \int_0^\epsilon \hat{h}_m(t) \exp\{-nt\} \, dt + \int_0^\epsilon R_m(t) \exp\{-nt\} \, dt + \int_\epsilon^T h(t) \exp\{-nt\} \, dt \\
&= \sum_{j=0}^m \frac{c_j \Gamma(a_j + 1)}{n^{a_j+1}} + O(\exp\{-n\epsilon\}) + O\left(\frac{1}{n^{a_{m+1}+1}}\right) + O\left(\frac{\exp\{-n\epsilon\}}{n}\right),
\end{aligned}
$$

proving the theorem. ∎

Example 9.4 (Exponential integral). Consider the *exponential integral* function defined by

$$E_1(z) = \int_z^\infty \frac{1}{u} \exp(-u) \, du;$$

this function arises in a number of different contexts. See, for example, Barndorff-Nielsen and Cox (1989, Example 3.3) for a discussion of a distribution with density function given by E_1.

Consider an asymptotic expansion for $E_1(z)$ for large z. In order to use Watson's lemma, the integral used to define E_1 must be rewritten. Note that, using the change-of-variable $t = u/z - 1$,

$$\int_z^\infty \frac{1}{u} \exp(-u) \, du = \exp(-z) \int_0^\infty \frac{1}{1+t} \exp(-zt) \, dt.$$

The function $h(t) = 1/(1+t)$ satisfies the conditions of Watson's lemma with $a_j = j$ and $c_j = (-1)^j$, $j = 0, 1, \ldots$. Hence, for any $m = 0, 1, \ldots$,

$$E_1(z) = \exp(-z) \left[\sum_{j=0}^m (-1)^j \frac{j!}{z^{j+1}} + O\left(\frac{1}{z^{m+1}}\right) \right].$$

Since this holds for all m we may write

$$E_1(z) \sim \exp(-z) \sum_{j=0}^\infty (-1)^j \frac{j!}{z^{j+1}} \quad \text{as } n \to \infty. \qquad \square$$

Example 9.5 (Ratio of gamma functions). Consider an asymptotic expansion of the ratio of gamma functions $\Gamma(z)/\Gamma(z+x)$ for large values of z and fixed x. By Theorem 9.2 we can write

$$\frac{\Gamma(z)}{\Gamma(z+x)} = \frac{1}{\Gamma(x)} \int_0^1 u^{z-1} (1-u)^{x-1} \, dx.$$

Hence, using the change-of-variable $t = -\log(u)$,

$$\frac{\Gamma(z)}{\Gamma(z+x)} = \frac{1}{\Gamma(x)} \int_0^\infty (1 - \exp(-t))^{x-1} \exp(-zt)\, dt$$

and an asymptotic expansion for this integral may be derived using Watson's lemma.

Note that we may write

$$[1 - \exp(-t)]^{x-1} = t^{x-1} \left(\frac{1 - \exp(-t)}{t}\right)^{x-1}$$

and, as $t \to 0$,

$$\left(\frac{1 - \exp(-t)}{t}\right)^{x-1} = 1 - \frac{1}{2}(x-1)t + O(t^2);$$

hence,

$$\left(1 - \exp(-t)\right)^{x-1} = t^{x-1} - \frac{1}{2}(x-1)t^x + O(t^{x+1}) \quad \text{as } t \to 0.$$

It now follows from Watson's lemma that

$$\int_0^\infty (1 - \exp(-t))^{x-1} \exp(-zt)\, dt = \frac{\Gamma(x)}{z^x} - \frac{1}{2}(x-1)\frac{\Gamma(x+1)}{z^{x+1}} + O\left(\frac{1}{z^{x+2}}\right)$$

and, hence, that

$$\frac{\Gamma(z)}{\Gamma(z+x)} = \frac{1}{z^x} - \frac{1}{2}\frac{x(x-1)}{z^{x+1}} + O\left(\frac{1}{z^{x+2}}\right) \quad \text{as } z \to \infty. \qquad \square$$

Watson's lemma may be generalized to allow the function h in

$$\int_0^T h(t) \exp\{-nt\}\, dt$$

to depend on n.

Theorem 9.13. *Let $h_1, h_2, \ldots$ denote a sequence of real-valued continuous functions on $[0, \infty)$ satisfying the following conditions:*

(i) $\sup_n h_n(t) = O(\exp(bt))$ as $t \to \infty$ for some constant b.

(ii) There exist constants $c_{n0}, c_{n1}, \ldots, c_{n(m+1)}$, $n = 1, 2, \ldots$, $a_0, a_1, \ldots, a_{m+1}$,

$$-1 < a_0 < a_1 < \cdots < a_{m+1}$$

such that

$$\hat{h}_{nm}(t) = \sum_{j=0}^m c_{nj} t^{a_j}$$

satisfies

$$h_n(t) = \hat{h}_{nm}(t) + R_{nm}(t)$$

where

$$\sup_n R_{nm}(t) = O(t^{a_{m+1}}) \quad \text{as } t \to 0^+.$$

Consider the integral

$$I_n = \int_0^{T_n} h_n(t) \exp\{-nt\} \, dt$$

where $T_1, T_2, \ldots$ are bounded away from 0. Then

$$I_n = \sum_{j=0}^m \frac{c_j \Gamma(a_j + 1)}{n^{a_j+1}} + O\left(\frac{1}{n^{a_{m+1}+1}}\right).$$

Proof. Fix $\epsilon > 0$ such that $\epsilon \le T_n$ for all n. There exists a constant K such that for all $n = 1, 2, \ldots,$

$$|h_n(t)| \le K \exp\{bt\}, \quad t \ge \epsilon.$$

Hence, for sufficiently large n,

$$\left| \int_\epsilon^{T_n} h_n(t) \exp\{-nt\} \, dt \right| \le K \int_\epsilon^\infty \exp\{-(n-b)t\} \, dt = \frac{K}{n-b} \exp\{-(n-b)\epsilon\}$$

$$= O\left(\frac{\exp\{-n\epsilon\}}{n}\right) \quad \text{as } n \to \infty.$$

Consider the integral

$$\int_0^\epsilon h_n(t) \exp\{-nt\} \, dt = \int_0^\epsilon \hat{h}_{nm}(t) \exp\{-nt\} \, dt + \int_0^\epsilon R_{nm}(t) \exp\{-nt\} \, dt.$$

For any $\epsilon > 0$,

$$\int_0^\epsilon \hat{h}_{nm}(t) \exp\{-nt\} \, dt = \sum_{j=0}^m \frac{c_{nj} \Gamma(a_j + 1)}{n^{a_j+1}} + O(\exp\{-n\epsilon\})$$

and

$$\int_0^\epsilon |R_{nm}(t)| \exp\{-nt\} \, dt \le K_m \frac{\Gamma(a_{m+1} + 1)}{n^{a_{m+1}+1}}.$$

The theorem follows from combining these results. ∎

Example 9.6 (*Tail probability of the t-distribution*). Consider the integral

$$\int_z^\infty \left(1 + \frac{y^2}{\nu}\right)^{-\frac{(\nu+1)}{2}} dy$$

which, aside from a normalizing constant, is the tail probability of the t-distribution with ν degrees of freedom. We will derive an asymptotic expansion for this integral as $\nu \to \infty$.

Using the change-of-variable

$$t = \frac{1}{2} \log\left(1 + \frac{y^2}{\nu}\right) - \frac{1}{2} \log\left(1 + \frac{z^2}{\nu}\right),$$

we may write

$$\int_z^\infty (1 + y^2/\nu)^{-\frac{(\nu+1)}{2}} \, dy = \frac{\sqrt{\nu}}{c_\nu^{\frac{\nu}{2}}} \int_0^\infty (1 - \exp(-2t)/c_\nu)^{-\frac{1}{2}} \exp(-\nu t) \, dt$$

Table 9.3. *Approximations in Example 9.6.*

z	ν	Exact	Approx.	Relative error (%)
1.5	2	0.136	0.174	27.5
1.5	5	0.0970	0.113	16.8
1.5	10	0.0823	0.0932	13.4
1.5	25	0.0731	0.0814	11.4
2.0	2	0.0918	0.0989	7.8
2.0	5	0.0510	0.0524	2.7
2.0	10	0.0367	0.0372	1.3
2.0	25	0.0282	0.0284	0.6
3.0	2	0.0477	0.0484	1.4
3.0	5	0.00150	0.00150	0.3
3.0	10	0.00667	0.00663	0.6
3.0	25	0.00302	0.00300	0.7

where $c_\nu = 1 + z^2/\nu$. To use Theorem 9.13, we may use the expansion

$$(1 - \exp(-2t)/c_\nu)^{-\frac{1}{2}} = \frac{\sqrt{c_\nu}}{\sqrt{(c_\nu - 1)}} \left[1 - \frac{1}{c_\nu - 1} t + \left(\frac{1}{c_\nu - 1} + \frac{3}{2} \frac{1}{(c_\nu - 1)^2} \right) t^2 + \cdots \right].$$

Note, however, that as $\nu \to \infty$, $c_\nu \to 1$, so that Theorem 9.13 can only be applied if, as $\nu \to \infty$, z^2/ν remains bounded away from 0. Hence, we assume that z is such that

$$\frac{z^2}{\nu} = b + o(1) \quad \text{as } \nu \to \infty,$$

for some $b > 0$.

Applying Theorem 9.13 yields the result

$$\int_z^\infty (1 + y^2/\nu)^{\frac{-(\nu+1)}{2}} \, dy$$

$$= \frac{1}{c_\nu^{(\frac{\nu-1}{2})}} \frac{1}{\sqrt{(c_\nu - 1)}} \left[\frac{1}{\nu} - \frac{1}{c_\nu - 1} \frac{1}{\nu^2} + \left(\frac{1}{c_\nu - 1} + \frac{3}{2} \frac{1}{(c_\nu - 1)^2} \right) \frac{1}{\nu^3} + O\left(\frac{1}{\nu^4} \right) \right]$$

as $\nu \to \infty$ and $z^2/\nu \to b > 0$.

When this expansion is used to approximate the integral, we will be interested in the value for some fixed values of ν and z. Hence, it is important to understand the relevance of the conditions that $\nu \to \infty$ and $z^2/\nu \to b > 0$ in this case. Since we are considering $\nu \to \infty$, we expect the accuracy of the approximation to improve with larger values of ν. However, for a fixed value of z, a larger value of ν yields a value of z^2/ν closer to 0. Hence, if ν is larger, we expect high accuracy only if z is large as well. That is, when approximating tail probabilities, we expect the approximation to have high accuracy only when approximating a small probability based on a moderate or large degrees of freedom.

Table 9.3 gives the approximations to the tail probability $\Pr(T \geq z)$, where T has a t-distribution with ν degrees of freedom, given by

$$\frac{\Gamma((\nu + 1)/2)}{\sqrt{(\nu\pi)}\Gamma(\nu/2)} \frac{1}{c_\nu^{(\frac{\nu-1}{2})}} \frac{1}{\sqrt{(c_\nu - 1)}} \left[\frac{1}{\nu} - \frac{1}{c_\nu - 1} \frac{1}{\nu^2} + \left(\frac{1}{c_\nu - 1} + \frac{3}{2} \frac{1}{(c_\nu - 1)^2} \right) \frac{1}{\nu^3} \right],$$

along with the exact value of this probability and the relative error of the approximation, for several choices of z and ν. Note that, for a fixed value of z, the relative error of the approximation decreases as ν increases. However, very high accuracy is achieved only if z is large as well. In fact, based on the results in this table, a large value of z appears to be at least as important as a large value of ν in achieving a very accurate approximation. □

Consider an integral of the form

$$\int_{-T_0}^{T_1} h(u) \exp\left\{-\frac{n}{2}u^2\right\} du;$$

clearly, this integral may be transformed into one that may be handled by Watson's lemma by using the change-of-variable $t = u^2/2$. This case occurs frequently enough that we give the result as a corollary below; it is worth noting that the case in which h depends on n can be handled in a similar manner.

Corollary 9.1. *Let h denote a real-valued continuous function on $[0, \infty)$ satisfying the following conditions:*

  (i) $h(t) = O(\exp(bt^2))$ *as* $|t| \to \infty$ *for some constant b*
  (ii) there exist constants $c_0, c_1, \ldots, c_{m+1}$ such that

$$h(t) = \sum_{j=0}^{m} c_j t^j + O(t^{m+1}) \quad \text{as } t \to 0.$$

Consider the integral

$$I_n = \int_{-T_0}^{T_1} h(t) \exp\left\{-\frac{n}{2}t^2\right\} dt$$

where $T_0 > 0$ and $T_1 > 0$. Then

$$I_n = \sum_{j=0}^{\lfloor \frac{m}{2} \rfloor} \frac{c_{2j} 2^{j+\frac{1}{2}} \Gamma(j + 1/2)}{n^{j+\frac{1}{2}}} + O\left(\frac{1}{n^{\frac{m}{2}+1}}\right) \quad \text{as } n \to \infty.$$

Proof.   First suppose that $T_1 = T_0 \equiv T$. Note that

$$\int_{-T}^{T} h(t) \exp\left\{-\frac{n}{2}t^2\right\} dt = \int_{0}^{T} [h(t) + h(-t)] \exp\left\{-\frac{n}{2}t^2\right\} dt$$

$$= \int_{0}^{\frac{T^2}{2}} \frac{h(\sqrt{(2u)}) + h(-\sqrt{(2u)})}{\sqrt{(2u)}} \exp\{-nu\} du$$

$$\equiv \int_{0}^{\frac{T^2}{2}} \bar{h}(u) \exp\{-nu\} du.$$

Since

$$h(t) = \sum_{j=0}^{m} c_j t^j + O(t^{m+1}) \quad \text{as } t \to 0,$$

$$\bar{h}(u) = \sum_{j=0}^{m} c_j 2^{\frac{i}{2}} u^{\frac{i}{2}} + \sum_{j=0}^{m} (-1)^j \frac{c_j 2^{\frac{i}{2}}}{\sqrt{(2u)}} + O\left(u^{\frac{m}{2}}\right)$$

$$= \sum_{j=0}^{\lfloor \frac{m}{2} \rfloor} 2^{j+\frac{1}{2}} c_{2j} u^{j-\frac{1}{2}} + O\left(u^{\frac{m}{2}}\right) \quad \text{as} \quad u \to 0^+$$

It follows from Watson's lemma that

$$\int_{-T}^{T} h(t) \exp\left\{-\frac{n}{2}t^2\right\} dt = \sum_{j=0}^{\lfloor \frac{m}{2} \rfloor} \frac{c_j 2^{j+\frac{1}{2}} \Gamma(j+1/2)}{n^{j+\frac{1}{2}}} + O\left(\frac{1}{n^{\frac{m}{2}+1}}\right) \quad \text{as } n \to \infty.$$

Now suppose that $T_1 > T_0$. Then

$$\int_{-T_0}^{T_1} h(t) \exp\left\{-\frac{n}{2}t^2\right\} dt = \int_{-T_0}^{T_0} h(t) \exp\left\{-\frac{n}{2}t^2\right\} dt + \int_{T_0}^{T_1} h(t) \exp\left\{-\frac{n}{2}t^2\right\} dt$$

$$= \int_{-T_0}^{T_0} h(t) \exp\left\{-\frac{n}{2}t^2\right\} dt + O\left(\exp\left\{-\frac{T_0^2}{2}n\right\}\right).$$

The result now follows as above.

Finally, suppose that $T_0 > T_1$. Since

$$\int_{-T_0}^{T_1} h(t) \exp\left\{-\frac{n}{2}t^2\right\} dt = \int_{-T_1}^{T_0} h(-t) \exp\left\{-\frac{n}{2}t^2\right\} dt,$$

the argument given above applies here as well, with h_1 in place of h, where $h_1(t) = h(-t)$. Clearly, h_1 has the same expansion as h, with coefficients $\bar{c}_j$, given by

$$\bar{c}_j = \begin{cases} c_j & \text{if } j \text{ is even} \\ -c_j & \text{if } j \text{ is odd} \end{cases}.$$

The result now follows from the fact that the expansion of the integral depends only on c_{2j}, $j = 0, 1, \ldots$.   ∎

Example 9.7 *(Expected value of a function of a normal random variable).* Let X denote a random variable with a normal distribution with mean 0 and variance n^{-1}; for instance, X may be a sample mean based on n independent identically distributed standard normal random variables. Let h denote a function satisfying the conditions of Corollary 9.1 for any $m = 1, 2, \ldots$ and consider $\mathrm{E}[h(X)]$. In particular, assume that

$$h(t) = \sum_{j=0}^{\infty} \frac{h^{(j)}(0)}{j!} t^j.$$

Note that

$$\mathrm{E}[h(X)] = \frac{\sqrt{n}}{\sqrt{(2\pi)}} \int_{-\infty}^{\infty} h(t) \exp\left(-\frac{n}{2}t^2\right) dt.$$

Hence, it follows immediately from Corollary 9.1 that, as $n \to \infty$,

$$\mathrm{E}[h(X)] \sim \frac{\sqrt{n}}{\sqrt{(2\pi)}} \sum_{j=0}^{\infty} \frac{h^{(2j)}(0)2^{j+\frac{1}{2}}\Gamma(j+1/2)}{(2j)! n^{j+\frac{1}{2}}} = \sum_{j=0}^{\infty} \frac{h^{(2j)}(0)2^j \Gamma(j+1/2)}{(2j)!\Gamma(1/2)} \frac{1}{n^j}. \quad \square$$

9.5   Laplace's Method

Laplace's method provides a futher generalization of the results of the previous section to the case in which the integral under consideration is not exactly of the form

$$\int_0^T h(t) \exp(-nt)\,dt$$

or

$$\int_{-T_0}^{T_1} h(t) \exp(-nt^2/2)\,dt.$$

Consider an integral of the form

$$\int_a^b h(y) \exp\{ng(y)\}\,dy.$$

Then, by changing the variable of integration, we can rewrite this integral in a form in which Corollary 9.1 may be applied. It is worth noting that Theorem 9.14 below may be generalized to the case in which g and h both depend on n; however, we will not consider such a generalization here.

Theorem 9.14. *Consider the integral*

$$I_n = \int_a^b h(y) \exp\{ng(y)\}\,dy, \quad -\infty \le a < b \le \infty$$

where

     (i) g is three-times differentiable on (a, b)
     (ii) h is twice-differentiable on (a, b) and $h(y) = O(\exp(dy^2))$ as $|y| \to \infty$, for some
         constant d
     (iii) g is maximized at $y = \hat{y}$, where $a < \hat{y} < b$
     (iv) $g'(\hat{y}) = 0$, $|g'(y)| > 0$ for all $y \ne \hat{y}$, and $g''(\hat{y}) < 0$.

If $|h(\hat{y})| > 0$, then

$$I_n = \exp\{ng(\hat{y})\} \frac{\sqrt{(2\pi)}h(\hat{y})}{[-ng''(\hat{y})]^{\frac{1}{2}}}[1 + O(n^{-1})] \quad as \; n \to \infty.$$

If $h(\hat{y}) = 0$, then

$$I_n = \exp\{ng(\hat{y})\} O\left(\frac{1}{n^{\frac{3}{2}}}\right) \quad as \; n \to \infty.$$

Proof. Note that

$$\int_a^b h(y) \exp\{ng(y)\}\,dy = \exp\{ng(\hat{y})\} \int_a^b h(y) \exp\{-n[g(\hat{y}) - g(y)]\}\,dy.$$

Consider the change-of-variable

$$u = \operatorname{sgn}(\hat{y} - y)\{2[g(\hat{y}) - g(y)]\}^{\frac{1}{2}}$$

so that u is a one-to-one function of y with

$$\frac{1}{2}u^2 = g(\hat{y}) - g(y).$$

Then

$$u\,du = -g'(y)\,dy.$$

Note that, since g is maximized at $\hat{y}$ and $|g'(y)| > 0$ for all $y \neq \hat{y}$, it follows that $g'(y) < 0$ for $y > \hat{y}$ and $g'(y) > 0$ for $y < \hat{y}$. Hence, u and $g'(y)$ have the same sign and

$$\left| \frac{dy}{du} \right| = \frac{u}{g'(y)}.$$

It follows that

$$\int_a^b h(y)\,\exp\{-n[g(\hat{y}) - g(y)]\}\,dy = \int_{u(a)}^{u(b)} h(y(u))\,\exp\left\{-\frac{n}{2}u^2\right\}\frac{u}{g'(y(u))}\,du$$

$$\equiv \int_{u(a)}^{u(b)} \bar{h}(u)\,\exp\left\{-\frac{n}{2}u^2\right\}\,du.$$

Under the conditions on h and g,

$$\bar{h}(u) = \bar{h}(0) + \bar{h}'(0)u + O(u^2) \quad \text{as } u \to 0$$

and, hence, by Corollary 9.1,

$$\int_{u(a)}^{u(b)} \bar{h}(u)\,\exp\left\{-\frac{n}{2}u^2\right\}\,du = \frac{\sqrt{2}\Gamma\left(\frac{1}{2}\right)}{\sqrt{n}}\bar{h}(0) + O\left(\frac{1}{n^{\frac{3}{2}}}\right) \quad \text{as } n \to \infty.$$

Hence, to complete the approximation we need to find

$$\bar{h}(0) \equiv \lim_{u \to 0} \frac{u h(y(u))}{g'(y(u))}.$$

Note that $u = 0$ if and only if $y = \hat{y}$; hence,

$$h(y(0)) = h(\hat{y}), \quad g'(y(0)) = 0$$

and

$$\bar{h}(0) = h(\hat{y}) \lim_{u \to 0} \frac{u}{g'(y(u))} \equiv h(\hat{y})L$$

where

$$L = \lim_{u \to 0} \frac{u}{g'(y(u))}.$$

By L'Hospital's rule,

$$L = \frac{1}{g''(\hat{y})y'(0)}.$$

We have seen that

$$y'(u) = \frac{u}{-g'(y(u))}$$

so that

$$y'(0) = \frac{1}{-g''(\hat{y})y'(0)} = -L$$

and, hence, that

$$L^2 = \frac{1}{-g''(\hat{y})}.$$

Finally, note that u and $g'(y(u))$ have the same sign so that $L \geq 0$. It follows that

$$L = \frac{1}{[-g''(\hat{y})]^{\frac{1}{2}}}$$

and

$$\bar{h}(0) = \frac{h(\hat{y})}{[-g''(\hat{y})]^{\frac{1}{2}}}.$$

Hence, if $|h(\hat{y})| > 0$, then

$$\int_{u(a)}^{u(b)} \bar{h}(u) \exp\left\{-\frac{n}{2}u^2\right\} du = \frac{\sqrt{2}\Gamma\left(\frac{1}{2}\right)}{\sqrt{n}} \frac{h(\hat{y})}{[-g''(\hat{y})]^{\frac{1}{2}}} + O\left(\frac{1}{n^{\frac{3}{2}}}\right)$$

so that

$$I_n = \exp\{ng(\hat{y})\} \frac{\sqrt{(2\pi)}h(\hat{y})}{[-ng''(\hat{y})]^{\frac{1}{2}}}[1 + O(n^{-1})] \quad \text{as} \quad n \to \infty.$$

If $h(\hat{y}) = 0$, then

$$\int_{u(a)}^{u(b)} \bar{h}(u) \exp\left\{-\frac{n}{2}u^2\right\} du = O\left(\frac{1}{n^{\frac{3}{2}}}\right)$$

so that

$$I_n = \exp\{ng(\hat{y})\} O\left(\frac{1}{n^{\frac{3}{2}}}\right) \quad \text{as} \quad n \to \infty. \quad \blacksquare$$

***Example* 9.8 (*Stirlings approximation*).** Consider approximation of the gamma function $\Gamma(z + 1)$ for large values of z. Note that, using the change-of-variable $y = t/z$, we may write

$$\Gamma(z+1) = \int_0^\infty t^z \exp(-t)\, dt = z^{z+1} \int_0^\infty \exp\{-z[y - \log(y)]\}\, dy.$$

Hence, we may apply Theorem 9.14 with $g(y) = \log(y) - y$, $n = z$, and $h(y) = 1$. It is straightforward to show that the conditions of Theorem 9.14 are satisfied with $\hat{y} = 1$, leading to the result that

$$\Gamma(z+1) = \sqrt{(2\pi)}z^{z+1/2} \exp(-z)[1 + O(z^{-1})] \quad \text{as} \quad z \to \infty.$$

Since $\Gamma(z) = \Gamma(z+1)/z$, Stirling's approximation to $\Gamma(z)$ is given by

$$\Gamma(z) = \sqrt{(2\pi)}z^{z-1/2} \exp(-z)[1 + O(z^{-1})] \quad \text{as} \quad z \to \infty. \qquad \square$$

Example 9.9 (*Ratio of two integrals*). Consider approximation of the ratio

$$\frac{\int_a^b h_1(y) \exp\{ng(y)\} \, dy}{\int_a^b h_2(y) \exp\{ng(y)\} \, dy}$$

where g and both h_1 and h_2 satisfy the conditions of Theorem 9.14, with $|h_1(\hat{y})| > 0$ and $|h_2(\hat{y})| > 0$. Here $\hat{y}$ is the maximizer of g and must satisfy $a < \hat{y} < b$.

Since, as $n \to \infty$,

$$\int_a^b h_j(y) \exp\{ng(y)\} \, dy = \exp\{ng(\hat{y})\} \frac{\sqrt{(2\pi)} h_j(\hat{y})}{[-ng''(\hat{y})]^{\frac{1}{2}}} [1 + O(n^{-1})], \quad j = 1, 2,$$

it follows that

$$\frac{\int_a^b h_1(y) \exp\{ng(y)\} \, dy}{\int_a^b h_2(y) \exp\{ng(y)\} \, dy} = \frac{h_1(\hat{y})}{h_2(\hat{y})} [1 + O(n^{-1})], \quad \text{as } n \to \infty. \qquad \square$$

Example 9.10. Let Y denote a real-valued random variable with an absolutely continuous distribution with density function

$$p(y; a) = c(a) \exp\{-2a \cosh(y)\}, \quad -\infty < y < \infty;$$

here a is a nonnegative constant and $c(a)$ is a constant chosen so that the density integrates to 1. This distribution arises in the following manner. Let X_1 and X_2 denote independent identically distributed gamma random variables, let $Y = \log(X_1/X_2)/2$ and let $A = (X_1 X_2)^{\frac{1}{2}}$. Then the conditional distribution of Y, given $A = a$, has the density $p(y; a)$.

Consider approximation of the constant $c(a)$ for large values of a; to do this, we need to approximate the integral

$$\int_{-\infty}^{\infty} \exp\{-2a \cosh(y)\} \, dy.$$

We may apply Theorem 9.14 with $g(y) = -2\cosh(y)$ and $h(y) = 1$. Hence, $\hat{y} = 0, g''(\hat{y}) = -2$, and

$$\int_{-\infty}^{\infty} \exp\{-2a \cosh(y)\} \, dy = \exp(-2a) \frac{\sqrt{(2\pi)}}{\sqrt{(2a)}} [1 + O(a^{-1})] \quad \text{as } a \to \infty.$$

It follows that

$$c(a) = \frac{\sqrt{a}}{\sqrt{\pi}} \exp(2a)[1 + O(a^{-1})] \quad \text{as } a \to \infty. \qquad \square$$

Laplace's method, as given in Theorem 9.14, applies to integrals of the form

$$\int_a^b h(y) \exp\{ng(y)\} \, dy$$

in which the maximum of g occurs at an interior point of (a, b); then, after changing the variable of integration, the integral can be approximated by Corollary 9.1. If the maximum of g occurs at either a or b, the same general approach may be used; however, in this case, the approximation is based on Theorem 9.12. The following result considers the case in which the maximum occurs at the lower endpoint a; a similar result may be derived for the case in which the maximum occurs at the upper endpoint.

Theorem 9.15. *Consider the integral*

$$I_n = \int_a^b h(y)\,\exp\{ng(y)\}\,dy, \quad -\infty < a < b \le \infty$$

where

 (i) *g is twice differentiable on $[a, b]$*
 (ii) *h is differentiable on $[a, b)$*
 (iii) *g is maximized at $y = a$*
 (iv) *$g'(y) < 0$ for all $a \le y < b$ and $h(a) \ne 0$.*

Then

$$I_n = \exp\{ng(a)\}\frac{h(a)}{-g'(a)}\frac{1}{n}[1 + O(n^{-1})] \quad as \quad n \to \infty.$$

Proof. Note that

$$\int_a^b h(y)\,\exp\{ng(y)\}\,dy = \exp\{ng(a)\}\int_a^b h(y)\,\exp\{-n[g(a) - g(y)]\}\,dy.$$

Consider the change-of-variable $u \equiv u(y) = g(a) - g(y)$; note that, since g is strictly decreasing, u is a one-to-one function of y. Also note that $u(a) = 0$.

It follows that

$$\int_a^b h(y)\,\exp\{-n[g(a) - g(y)]\}\,dy = \int_0^{u(b)} \frac{h(y(u))}{g'(y(u))}\exp\{-nu\}\,du$$

$$\equiv \int_0^{u(b)} \bar{h}(u)\,\exp\{-nu\}\,du.$$

Under the conditions on h and g,

$$\bar{h}(u) = \bar{h}(0) + O(u) \quad as \quad u \to 0$$

and, hence, by Watson's lemma,

$$\int_{u(a)}^{u(b)} \bar{h}(u)\,\exp\{-nu\}\,du = \bar{h}(0)\frac{1}{n} + O(n^{-2}) \quad as \quad n \to \infty.$$

The result now follows from the fact that

$$\bar{h}(0) = \frac{h(a)}{-g'(a)}. \quad \blacksquare$$

Example 9.11 (Pareto distribution). Let Y denote a real-valued random variable with an absolutely continuous distribution with density

$$p(y; \alpha) = \frac{\alpha}{y^{\alpha+1}}, \quad y \ge 1;$$

here $\alpha > 0$ is a parameter. This is a Pareto distribution. Consider an approximation to $E[\log(1 + Y^2); \alpha]$ for large values of α.

We may write

$$E[\log(1 + Y^2); \alpha] = \int_1^\infty \log(1 + y^2)\frac{\alpha}{y^{\alpha+1}}\,dy = \alpha\int_1^\infty \frac{\log(1 + y^2)}{y}\exp\{-\alpha\log(y)\}\,dy.$$

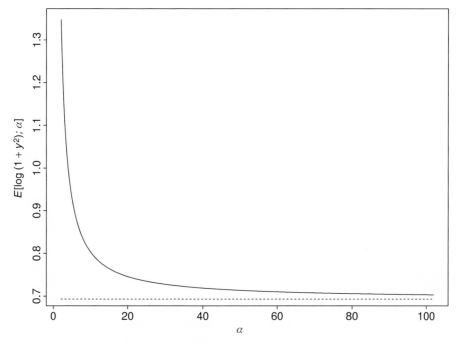

Figure 9.2. $E[\log(1 + Y^2); \alpha]$ in Example 9.11.

Note that, in the interval $[1, \infty)$, $-\log(y)$ is maximized at $y = 1$; hence, a Laplace approximation must be based on Theorem 9.15. Taking $g(y) = -\log(y)$ and $h(y) = \log(1 + y^2)/y$, the conditions of Theorem 9.15 are satisfied so that

$$\int_1^\infty \frac{\log(1 + y^2)}{y} \exp\{-\alpha \log(y)\}\, dy = \log(2)\frac{1}{\alpha}[1 + O(\alpha^{-1})]$$

so that

$$E[\log(1 + Y^2); \alpha] = \log(2)[1 + O(\alpha^{-1})] \quad \text{as } \alpha \to \infty.$$

Figure 9.2 contains a plot of $E[\log(1 + Y^2); \alpha]$ as a function of α, together with the approximation $\log(2)$, which is displayed as a dotted line. Note that, although the exact value of the expected value approaches $\log(2)$ as α increases, the convergence is relatively slow. For instance, the relative error of the approximation is still 1.4% when $\alpha = 100$. □

9.6 Uniform Asymptotic Approximations

Consider the problem of approximating an integral of the form

$$\int_z^\infty h(y) \exp(ng(y))\, dy,$$

as a function of z; for instance, we may be interested in approximating the tail probability function of a given distribution. Although Laplace's method may be used, in many cases, it has the undesirable feature that the form of the approximation depends on whether $\hat{y}$, the maximizer of g, is less than or greater than z. For instance, suppose that g is strictly decreasing on $(\hat{y}, \infty)$. If $z < \hat{y}$, then the approximation is based on Theorem 9.14; if $z \geq \hat{y}$, so that over the interval $[z, \infty)$ g is maximized at z, then the approximation is based on Theorem 9.15. Furthermore, in the case $z < \hat{y}$, the approximation does not depend on the value of z.

This is illustrated in the following example.

***Example* 9.12.** Consider approximation of the integral

$$\int_z^\infty \exp(x)\sqrt{n}\phi(\sqrt{n}x)\,dx$$

for large n, where $\phi(\cdot)$ denotes the standard normal density function. It is straightforward to show that the exact value of this integral is given by

$$\exp\{1/(2n)\}[1 - \Phi(\sqrt{n}z - 1/\sqrt{n})].$$

If $z < 0$ then, by Theorem 9.14,

$$\int_z^\infty \exp(x)\exp\left(-\frac{n}{2}x^2\right)dx = \frac{\sqrt{(2\pi)}}{\sqrt{n}}[1 + O(n^{-1})]$$

so that

$$\int_z^\infty \exp(x)\sqrt{n}\phi(\sqrt{n}x)\,dx = 1 + O(n^{-1}) \quad \text{as } n \to \infty.$$

If $z > 0$, then Theorem 9.15 must be used, leading to the approximation

$$\int_z^\infty \exp(x)\exp\left(-\frac{n}{2}x^2\right)dx = \frac{\exp\left(-\frac{n}{2}z^2\right)}{z}\frac{\exp(z)}{n}[1 + O(n^{-1})]$$

so that

$$\int_z^\infty \exp(x)\sqrt{n}\phi(\sqrt{n}x)\,dx = \phi(\sqrt{n}z)\frac{\exp(z)}{\sqrt{n}z}[1 + O(n^{-1})] \quad \text{as } n \to \infty.$$

Hence,

$$\int_z^\infty \exp(x)\sqrt{n}\phi(\sqrt{n}x)\,dx = \begin{cases} 1 + O(n^{-1}) & \text{if } z < 0 \\ [\phi(\sqrt{n}z)\exp(z)/(\sqrt{n}z)][1 + O(n^{-1})] & \text{if } z > 0 \end{cases}. \quad (9.3)$$

Since z^2 has derivative 0 at $z = 0$, neither Theorem 9.14 nor Theorem 9.15 can be used when $z = 0$.

In addition to the fact that the form of the approximation depends on the sign of z, the approach based on Laplace's method also has disadvantage that the approximations are not valid uniformly in z. That is, the $O(n^{-1})$ terms in (9.3) refer to asymptotic properties for each fixed z, not to the maximum error over a range of z values.

For instance, suppose that $z_n = z_0/\sqrt{n}$, where $z_0 > 0$ is a fixed constant. If the approximation in (9.3) is valid uniformly for all z in a neighborhood of 0 then

$$\sup_{0 \leq z \leq \epsilon} \frac{\int_z^\infty \exp(x)\sqrt{n}\phi(\sqrt{n}x)\,dx}{\phi(\sqrt{n}z)\exp(z)/(\sqrt{n}z)} = 1 + O(n^{-1})$$

for some $\epsilon > 0$ and, hence,

$$\frac{\int_{z_n}^{\infty} \exp(x)\sqrt{n}\phi(\sqrt{n}x)\,dx}{\phi(\sqrt{n}z_n)\exp(z_n)/(\sqrt{n}z_n)} = 1 + O(n^{-1}) \quad \text{as } n \to \infty.$$

However,

$$\phi(\sqrt{n}z_n)\exp(z_n)/(\sqrt{n}z_n) = \phi(z_0)\frac{\exp(z_0/\sqrt{n})}{z_0} = \phi(z_0)\frac{1}{z_0}\left[1 + O\left(\frac{1}{\sqrt{n}}\right)\right].$$

The exact value of the integral in this case is

$$\exp\{1/(2n)\}[1 - \Phi(z_0 - 1/\sqrt{n})] = 1 - \Phi(z_0) + \phi(z_0)\frac{1}{\sqrt{n}} + O\left(\frac{1}{n}\right).$$

Hence, if the approximation is valid to the order stated, then

$$1 - \Phi(z_0) = \phi(z_0)\frac{1}{z_0}\left[1 + O\left(n^{-\frac{1}{2}}\right)\right],$$

that is,

$$1 - \Phi(z_0) = \phi(z_0)\frac{1}{z_0}.$$

It follows that (9.3) is guaranteed to hold only valid for fixed values of z. $\quad\square$

In this section, we present a asymptotic approximation to integrals of the form

$$\int_{z}^{\infty} h(x)\sqrt{n}\phi(\sqrt{n}x)\,dx \tag{9.4}$$

that overcomes both of the drawbacks illustrated in the previous example. Specifically, this approximation has the properties that the form of the approximation does not depend on the value of z and that the approximation is valid uniformly in z. The approximation takes advantage of the fact that the properties of the integral (9.4) when h is a constant are well-known; in that case,

$$\int_{z}^{\infty} h(x)\sqrt{n}\phi(\sqrt{n}x)\,dx = h(z)[1 - \Phi(\sqrt{n}z)].$$

Note that it is generally necessary to do a preliminary transformation to put a given integral into the form (9.4).

Theorem 9.16. *Consider an integral of the form*

$$\int_{z}^{\infty} h_n(x)\sqrt{n}\phi(\sqrt{n}x)\,dx$$

where $h_1, h_2, \ldots$ is a sequence of functions such that

$$\sup_{n} |h_n^{(j)}(x)| \le c_j \exp\left(d_j x^2\right), \quad j = 0, 1, 2$$

for some constants $c_0, c_1, c_2, d_0, d_1, d_2$.
Then, for all $M < \infty$,

$$\int_{z}^{\infty} h_n(x)\sqrt{n}\phi(\sqrt{n}x)\,dx = [1 - \Phi(\sqrt{n}z)]\left[h_n(0) + O\left(\frac{1}{n}\right)\right] + \frac{h_n(z) - h_n(0)}{\sqrt{n}z}\phi(\sqrt{n}z),$$

as $n \to \infty$, uniformly in $z \le M$.

Proof.   We may write

$$\int_z^\infty h_n(x)\sqrt{n}\phi(\sqrt{n}x)\,dx = \int_z^\infty h_n(0)\sqrt{n}\phi(\sqrt{n}x)\,dx$$
$$+ \int_z^\infty x\frac{h_n(x) - h_n(0)}{x}\sqrt{n}\phi(\sqrt{n}x)\,dx$$
$$= h_n(0)[1 - \Phi(\sqrt{n}z)] + \int_z^\infty x\frac{h_n(x) - h_n(0)}{x}\sqrt{n}\phi(\sqrt{n}x)\,dx.$$

Let

$$g_n(x) = \frac{h_n(x) - h_n(0)}{x}.$$

Note that

$$\sqrt{n}x\phi(\sqrt{n}x) = -\frac{d}{dx}\phi(\sqrt{n}x);$$

hence, using integration-by-parts,

$$\int_z^\infty g_n(x)\sqrt{n}x\phi(\sqrt{n}x)\,dx = -\frac{1}{n}g_n(x)\sqrt{n}\phi(\sqrt{n}x)\Big|_z^\infty + \frac{1}{n}\int_z^\infty g_n'(x)\sqrt{n}\phi(\sqrt{n}x)\,dx$$
$$= g_n(z)\frac{\phi(\sqrt{n}z)}{\sqrt{n}} + \frac{1}{n}\int_z^\infty g_n'(x)\sqrt{n}\phi(\sqrt{n}x)\,dx.$$

It follows that

$$\int_z^\infty h_n(x)\sqrt{n}\phi(\sqrt{n}x)\,dx = h_n(0)[1 - \Phi(\sqrt{n}z)] + g_n(z)\frac{\phi(\sqrt{n}z)}{\sqrt{n}}$$
$$+ \frac{1}{n}\int_z^\infty g_n'(x)\sqrt{n}\phi(\sqrt{n}x)\,dx.$$

Hence, the theorem holds provided that

$$\frac{1}{n}\int_z^\infty g_n'(x)\sqrt{n}\phi(\sqrt{n}x)\,dx = [1 - \Phi(\sqrt{n}z)]O\left(\frac{1}{n}\right),$$

where the $O(1/n)$ term holds uniformly for $z \le M$, for any $M < \infty$.

Note that

$$g_n'(x) = \frac{h_n'(x)}{x} - \frac{h_n(x) - h_n(0)}{x^2} = -\frac{h_n(x) - h_n(0) - xh_n'(x)}{x^2}.$$

Using Taylor's series approximations,

$$h_n(x) = h_n(0) + h_n'(0) + \frac{1}{2}h_n''(x_1)x^2$$

and

$$h_n'(x) = h_n'(0) + h_n''(x_2)x$$

where $|x_j| \le |x|$, $j = 1, 2$. Hence,

$$g_n'(x) = \frac{1}{2}h_n''(x_1) - h_n''(x_2)$$

and

$$\sup_n |g_n'(x)| \le \frac{3}{2} \sup_n |h_n''(x)| \le \frac{3}{2} c_2 \exp\left(d_2 x^2\right)$$

and

$$\sup_n |g_n'(x) - 1| \le c_3 \exp\left(d_3 x^2\right)$$

for some constants c_2, d_2, c_3, d_3. It follows that

$$\left| \int_z^\infty g_n'(x)\sqrt{n}\phi(\sqrt{n}x)\,dx - \int_z^\infty \sqrt{n}\phi(\sqrt{n}x)\,dx \right| \le c_3 \int_z^\infty \sqrt{n}\phi(\sqrt{(n - 2d_3)}x)\,dx.$$

Note that

$$\int_z^\infty \sqrt{n}\phi(\sqrt{(n - 2d_3)}x)\,dx = \frac{1}{\sqrt{(1 - 2d_3/n)}}[1 - \Phi(\sqrt{(n - 2d_3)}z)].$$

Hence,

$$\frac{1}{n}\int_z^\infty g_n'(x)\sqrt{n}\phi(\sqrt{n}x)\,dx = [1 - \Phi(\sqrt{n}z)]\frac{1}{n}[1 + R_n(z)]$$

where

$$|R_n(z)| \le c_3 \frac{1}{\sqrt{(1 - 2d_3/n)}} \frac{1 - \Phi(\sqrt{(n - 2d_3)}z)}{1 - \Phi(\sqrt{n}z)}.$$

The result follows provided that, for any $M < \infty$,

$$\sup_{z \le M} |R_n(z)| = O(1) \quad \text{as} \quad n \to \infty.$$

First note that, for $z < 1/\sqrt{n}$

$$|R_n(z)| \le c_3 \frac{1}{\sqrt{(1 - 2d_3/n)}} \frac{1}{1 - \Phi(1)}.$$

By Theorem 9.11,

$$1 - \Phi((\sqrt{(n - 2d_3)}z) \le \frac{1}{\sqrt{(n - 2d_3)}} \frac{1}{z} \phi(\sqrt{(n - 2d_3)}z)$$

and

$$1 - \Phi(\sqrt{n}z) \ge \frac{\sqrt{n}z^2}{1 + nz^2} \frac{1}{z} \phi(\sqrt{n}z).$$

Hence, for $z \ge 1/\sqrt{n}$,

$$|R_n(z)| \le \frac{2c_3}{1 - 2d_3/n} \exp\left(d_3 z^2\right).$$

It follows that

$$\sup_{z \le M} |R_n(z)| \le \max\left\{ c_3 \frac{1}{\sqrt{(1 - 2d_3/n)}} \frac{1}{1 - \Phi(1)}, \frac{2c_3}{1 - 2d_3/n} \exp\left(d_3 M^2\right) \right\},$$

proving the result. ∎

***Example* 9.13.** Consider approximation of the integral

$$Q_n(z) \equiv \int_z^\infty \exp(x)\sqrt{n}\phi(\sqrt{n}x)\, dx, \quad z \in \mathbf{R},$$

for large n, as discussed in Example 9.12. We may use Theorem 9.16, taking $h(x) = \exp(x)$. Then

$$\int_z^\infty \exp(x)\sqrt{n}\phi(\sqrt{n}x)\, dx = [1 - \Phi(\sqrt{n}z)]\left[1 + O\left(\frac{1}{n}\right)\right] + \frac{\exp(z) - 1}{\sqrt{n}z}\phi(\sqrt{n}z)$$

and an approximation to $Q_n(z)$ is given by

$$\bar{Q}_n(z) = 1 - \Phi(\sqrt{n}z) + \frac{\exp(z) - 1}{\sqrt{n}z}\phi(\sqrt{n}z).$$

Suppose $z = z_0/\sqrt{n}$. Then

$$\bar{Q}_n(z) = 1 - \Phi(z_0) + \frac{\exp(z_0/\sqrt{n}) - 1}{z_0}\phi(z_0) = 1 - \Phi(z_0) + \phi(z_0)\frac{1}{\sqrt{n}} + O\left(\frac{1}{n}\right).$$

Using the expansion for $Q_n(z)$ given in Example 9.12, it follows that, for $z = z_0/\sqrt{n}$, $\bar{Q}_n(z) = Q_n(z) + O(1/n)$ as $n \to \infty$, as expected from Theorem 9.16. □

***Example* 9.14 (*Probability that a gamma random variable exceeds its mean*).** Let X denote a random variable with a standard gamma distribution with index z. Then

$$\Pr(X \geq c\mathrm{E}(X)) = \frac{1}{\Gamma(z)}\int_{cz}^\infty t^{z-1}\exp(-t)\, dt$$

where c is a positive constant. We consider approximation of the integral in this expression for large values of z.

Using the change-of-variable $y = t/z$, we may write

$$\int_{cz}^\infty t^{z-1}\exp(-t)\, dt = z^z\int_c^\infty y^{z-1}\exp(-zy)\, dy$$

$$= z^z\int_c^\infty \frac{1}{y}\exp\{-z(y - \log(y))\}\, dy.$$

Hence, consider approximation of the integral

$$\int_c^\infty \frac{1}{y}\exp\{-z(y - \log(y))\}\, dy.$$

The first step in applying Theorem 9.16 is to write the integral in the form (9.4). Note that the function $y - \log(y)$ is decreasing for $y < 1$ and increasing for $y > 1$ with minimum value 1 at $y = 1$. Hence, consider the transformation

$$x = \mathrm{sgn}(y - 1)\{2[y - \log(y) - 1]\}^{\frac{1}{2}}.$$

This is a one-to-one function of y with

$$\frac{1}{2}x^2 = y - \log(y) - 1$$

and

$$dy = \frac{y}{y - 1}x\, dx.$$

It follows that

$$\int_c^\infty \frac{1}{y} \exp\{-z(y - \log(y))\}\, dy = \exp(-z) \int_{x(c)}^\infty \frac{x}{y(x) - 1} \exp\left(-\frac{z}{2}x^2\right) dx$$

$$= \frac{\sqrt{(2\pi)}}{\sqrt{z}} \exp(-z) \int_{x(c)}^\infty \frac{x}{y(x) - 1} \sqrt{z}\phi(\sqrt{z}x)\, dx$$

where

$$x(c) = \text{sgn}(c - 1)\{2[c - \log(c) - 1]\}^{\frac{1}{2}}$$

and $y(x)$ solves

$$y - \log(y) = \frac{1}{2}x^2 + 1.$$

By Theorem 9.16,

$$\int_{x(c)}^\infty \frac{x}{y(x) - 1} \sqrt{z}\phi(\sqrt{z}x)\, dx = [1 - \Phi(\sqrt{z}x(c))]\left[h(0) + O\left(\frac{1}{z}\right)\right]$$

$$+ \frac{x(c)/(c - 1) - h(0)}{\sqrt{z}x(c)}\phi(\sqrt{z}x(c))$$

where

$$h(0) = \lim_{x \to 0} \frac{x}{y(x) - 1} = \lim_{y \to 1} \frac{[2(y - \log(y) - 1)]^{\frac{1}{2}}}{|y - 1|} = 1.$$

Hence,

$$\int_{x(c)}^\infty \frac{x}{y(x) - 1} \sqrt{z}\phi(\sqrt{z}x)\, dx = [1 - \Phi(\sqrt{z}x(c))]\left[1 + O\left(\frac{1}{z}\right)\right]$$

$$+ \frac{x(c)/(c - 1) - 1}{\sqrt{z}x(c)}\phi(\sqrt{z}x(c))$$

and

$$\Pr(X \geq cE(X)) = \frac{1}{\Gamma(z)}\frac{\sqrt{(2\pi)}}{\sqrt{z}} \exp(-z)z^z[1 - \Phi(\sqrt{z}x(c))]\left[1 + O\left(\frac{1}{z}\right)\right]$$

$$+ \frac{x(c)/(c - 1) - 1}{\sqrt{z}x(c)}\phi(\sqrt{z}x(c)).$$

Using Stirling's approximation for $\Gamma(z)$ in this expression yields

$$\Pr(X \geq cE(X)) = [1 - \Phi(\sqrt{z}x(c))]\left[1 + O\left(\frac{1}{z}\right)\right]$$

$$+ \frac{x(c)/(c - 1) - 1}{\sqrt{z}x(c)}\phi(\sqrt{z}x(c)) \quad \text{as } z \to \infty.$$

For comparison, we may consider approximations based on Laplace's method. The function $g(y) = \log(y) - y$ is maximized at $y = 1$ and is strictly decreasing on the interval $(1, \infty)$. Hence, if $c < 1$, we may use the approximation given in Theorem 9.14, leading to the approximation

$$\Pr(X \geq cE(X)) = \frac{z^{\frac{z-1}{2}}}{\Gamma(z)}\exp(-z)\sqrt{(2\pi)}[1 + O(z^{-1})]. \tag{9.6}$$

Table 9.4. *Approximations in Example 9.18.*

z	c	Exact	Uniform	Laplace
1	1/2	0.607	0.604	0.922
1	3/4	0.472	0.471	0.922
1	4/3	0.264	0.264	1.054
1	2	0.135	0.135	0.271
2	1/2	0.736	0.735	0.960
2	3/4	0.558	0.558	0.960
2	4/3	0.255	0.255	0.741
2	2	0.0916	0.0917	0.147
5	1/2	0.891	0.891	0.983
5	3/4	0.678	0.678	0.983
5	4/3	0.206	0.206	0.419
5	2	0.0293	0.0293	0.0378

Using Stirling's approximation to the gamma function shows that

$$\Pr(X \geq c\mathrm{E}(X)) = 1 + O(z^{-1}) \quad \text{as } z \to \infty.$$

For the case $c > 1$, $g(y)$ is maximized at $y = c$ and we may use the approximation given in Theorem 9.15; this yields the approximation

$$\Pr(X \geq c\mathrm{E}(X)) = \frac{\exp(-cz)c^z}{\Gamma(z)(c-1)} \frac{1}{z}[1 + O(z^{-1})]. \tag{9.7}$$

An approximation for the case $c = 1$ is not available using Theorem 9.15 since $\log(y) - y$ has derivative 0 at $y = 1$.

Table 9.4 contains the uniform approximation given by (9.5), with the $O(1/z)$ term omitted, together with the Laplace approximation given by (9.6) and (9.7), again with the $O(1/z)$ terms omitted and the exact value of $\Pr(X \geq c\mathrm{E}(X))$ for several values of c and z. These results show that the uniform approximation is nearly exact for a wide range of c and z values, while the Laplace approximation in nearly useless for the values of c and z considered. □

9.7 Approximation of Sums

The methods discussed thus far in this chapter may be applied to integrals of the form

$$\int_{-\infty}^{\infty} g(x)\,dF(x)$$

whenever F is absolutely continuous. In this section, we consider the approximation of sums; these methods may be applicable when the distribution function F is a step function so that an integral with respect to F reduces to a sum.

Consider a sum of the form

$$\sum_{j=m}^{r} f(j)$$

where m and r are nonnegative integers $m \leq r$. Here any or all of m, r, and f may depend on a parameter n and we consider an approximation to this sum as $n \to \infty$.

One commonly used approach to approximating this type of sum is to first approximate the sum by an integral and then approximate the integral using one of the methods discussed in this chapter. The basic result relating sums and integrals is known as the *Euler-Maclaurin summation formula*. The following theorem gives a simple form of this result. The general result incorporates higher derivatives of f; see, for example, Andrews, Askey, and Roy (1999, Appendix D) or Whittaker and Watson (1997, Chapter 13). Thus, the approximations derived in this section tend to be rather crude; however, they illustrate the basic approach to approximating sums.

Theorem 9.17. *Let f denote a continuously differentiable function on $[m, r]$ where m and r are integers, $m \leq r$. Then*

$$\sum_{j=m}^{r} f(j) = \int_{m}^{r} f(x)\, dx + \frac{1}{2}[f(m) + f(r)] + \int_{m}^{r} P(x) f'(x)\, dx$$

where

$$P(x) = x - \lfloor x - \frac{1}{2} \rfloor.$$

Proof. The result clearly holds whenever $m = r$ so assume that $m < r$. Let j denote an integer in $[m, r)$. Consider the integral

$$\int_{j}^{j+1} P(x) f'(x)\, dx.$$

Note that, on the interval $(j, j + 1)$, $P(x) = x - j - 1/2$. By integration-by-parts,

$$\int_{j}^{j+1} P(x) f'(x)\, dx = (x - j - 1/2) f(x) \Big|_{j}^{j+1} - \int_{j}^{j+1} f(x)\, dx$$

$$= \frac{1}{2}[f(j + 1) + f(j)] - \int_{j}^{j+1} f(x)\, dx.$$

Hence,

$$\int_{m}^{r} P(x) f'(x)\, dx = \sum_{j=m}^{r-1} \int_{j}^{j+1} P(x) f'(x)\, dx$$

$$= \frac{1}{2}\left[\sum_{j=m+1}^{r} f(j) + \sum_{j=m}^{r-1} f(j) \right] - \int_{m}^{r} f(x)\, dx$$

$$= \sum_{j=m}^{r} f(j) - \frac{1}{2}[f(m) + f(r)] - \int_{m}^{r} f(x)\, dx,$$

proving the result. ∎

The same approach may be used with infinite sums, provided that the sum and the terms in the approximation converge appropriately.

Corollary 9.2. *Let f denote a continuously differentiable function on $[m, \infty]$ where m is an integer. Assume that*

$$\sum_{j=m}^{\infty} |f(j)| < \infty, \quad \int_{m}^{\infty} |f(x)| \, dx < \infty$$

and

$$\int_{m}^{\infty} |f'(x)| \, dx < \infty.$$

Then

$$\sum_{j=m}^{\infty} f(j) = \int_{m}^{\infty} f(x) \, dx + \frac{1}{2} f(m) + \int_{m}^{\infty} P(x) f'(x) \, dx$$

where

$$P(x) = x - \lfloor x - \frac{1}{2}.$$

Proof. By Theorem 9.17, for any $r = m, m+1, \ldots$,

$$\sum_{j=m}^{r} f(j) = \int_{m}^{r} f(x) \, dx + \frac{1}{2} [f(m) + f(r)] + \int_{m}^{r} P(x) f'(x) \, dx. \tag{9.8}$$

Note that, under the conditions of the corollary,

$$\int_{m}^{r} |P(x)| \, |f'(x)| \, dx < \infty$$

and

$$\lim_{r \to \infty} f(r) = 0.$$

The result now follows from taking limits in (9.8). ∎

Using Theorem 9.17, the sum

$$\sum_{j=m}^{r} f(j)$$

may be approximated by the integral

$$\int_{m}^{r} f(x) \, dx$$

with error

$$\frac{1}{2} [f(m) + f(r)] + \int_{m}^{r} P(x) f'(x) \, dx$$

or it may be approximated by

$$\int_{m}^{r} f(x) \, dx + \frac{1}{2} [f(m) + f(r)]$$

with error

$$\int_m^r P(x) f'(x) \, dx. \tag{9.9}$$

***Example* 9.15 (*Discrete uniform distribution*).** Let X denote a discrete random variable with a uniform distribution on the set $\{0, 1, \ldots, m\}$ for some positive integer m; hence,

$$\Pr(X = j) = \frac{1}{m+1}, \quad j = 0, \ldots, m.$$

Let f denote a bounded, real-valued function on $[0, 1]$ and consider the expected value $\mathrm{E}[f(X/m)]$. Let U denote an absolutely continuous random variable with a uniform distribution on $(0, 1)$. Here we consider the approximation of $\mathrm{E}[f(X/m)]$ by $\mathrm{E}[f(U)]$ for large m. We assume that f is differentiable and that f' satisfies the Lipschitz condition

$$|f'(s) - f'(t)| \leq K|s - t|, \quad s, t \in [0, 1]$$

for some constant K.

Using Theorem 9.17,

$$
\begin{aligned}
\mathrm{E}[f(X/m)] &= \frac{1}{m+1} \sum_{j=0}^{m} f(j/m) \\
&= \frac{1}{m+1} \int_0^m f(x/m) \, dx + \frac{1}{2(m+1)} [f(0) + f(1)] \\
&\quad + \frac{1}{m(m+1)} \int_0^m (x - \lfloor x - 1/2 \rfloor) f'(x/m) \, dx.
\end{aligned}
$$

Changing the variable of integration,

$$\frac{1}{m+1} \int_0^m f(x/m) \, dx = \frac{m}{m+1} \int_0^1 f(u) du = \frac{m}{m+1} \mathrm{E}[f(U)].$$

Note that, for $j \leq x < j + 1$,

$$x - \lfloor x - 1/2 \rfloor = x - j - 1/2;$$

hence,

$$
\begin{aligned}
\int_0^m (x - \lfloor x - 1/2 \rfloor) f'(x/m) \, dx &= \sum_{j=0}^{m-1} \int_j^{j+1} (x - j - 1/2) f'(x/m) \, dx \\
&= \sum_{j=0}^{m-1} \int_{-\frac{1}{2}}^{\frac{1}{2}} u f'\left(\frac{u + j + 1/2}{m}\right) du.
\end{aligned}
$$

Since

$$\int_{-\frac{1}{2}}^{\frac{1}{2}} c u \, du = 0$$

for any constant c,

$$\int_{-\frac{1}{2}}^{\frac{1}{2}} u f'\left(\frac{u+j+1/2}{m}\right) du = \int_{-\frac{1}{2}}^{\frac{1}{2}} u \left[f'\left(\frac{u+j+1/2}{m}\right) \right.$$
$$\left. - f'\left(\frac{j+1/2}{m}\right) \right] du, \quad j = 0, \ldots, m-1.$$

Using the Lipschitz condition on f'

$$\left| \int_{-\frac{1}{2}}^{\frac{1}{2}} u f'\left(\frac{u+j+1/2}{m}\right) du \right| \leq \frac{K}{m} \int_{-\frac{1}{2}}^{\frac{1}{2}} |u|^2 du = \frac{K}{12m}, \quad j = 0, \ldots, m-1$$

so that

$$\left| \int_0^m (x - \lfloor x - 1/2 \rfloor) f'(x/m) \, dx \right| \leq \frac{K}{12}$$

and, hence, that

$$\frac{1}{m(m+1)} \int_0^m (x - \lfloor x - 1/2 \rfloor) f'(x/m) \, dx = O\left(\frac{1}{m^2}\right) \quad \text{as } m \to \infty.$$

It follows that

$$\mathrm{E}\left[f\left(\frac{X}{m}\right) \right] = \frac{m}{m+1} \mathrm{E}[f(U)]$$
$$+ \frac{1}{2(m+1)} [f(0) + f(1)] + O\left(\frac{1}{m^2}\right) \quad \text{as } m \to \infty. \qquad \square$$

In order for the result in Theorem 9.17 to be useful, we need some information regarding the magnitude of the integral (9.9), as in the previous example. A particularly simple bound is available for the case in which f is a monotone function.

Corollary 9.3. *Let f denote a continuously differentiable monotone function on $[m, r]$ where m and r are integers, $m \leq r$. Then*

$$\left| \sum_{j=m}^{r} f(j) - \int_m^r f(x) \, dx - \frac{1}{2} [f(m) + f(r)] \right| \leq \frac{1}{2} |f(r) - f(m)| \qquad (9.10)$$

and

$$\left| \sum_{j=m}^{r} f(j) - \int_m^r f(x) \, dx \right| \leq \max\{|f(m)|, |f(r)|\}. \qquad (9.11)$$

Suppose that f is a decreasing function and

$$\lim_{r \to \infty} f(r) = 0.$$

Then

$$\left| \sum_{j=m}^{\infty} f(j) - \int_m^{\infty} f(x) \, dx - \frac{1}{2} f(m) \right| \leq \frac{1}{2} |f(m)|. \qquad (9.12)$$

Proof. Equation (9.10) follows from Theorem 9.17 provided that

$$\int_m^r P(x)f'(x)\,dx \le \frac{1}{2}|f(r) - f(m)|.$$

Consider the case in which f is a decreasing function; the case in which f is increasing follows from a similar argument.

Since f is decreasing, $f'(x) \le 0$ on $[m, r]$ and, since $|P(x)| \le 1/2$ for all x,

$$\left| \int_m^r P(x)f'(x)\,dx \right| \le -\frac{1}{2}\int_m^r f'(x)\,dx = \frac{1}{2}[f(m) - f(r)],$$

proving (9.10).

Equation (9.11) follows from (9.10) using the fact that

$$\left| \sum_{j=m}^r f(j) - \int_m^r f(x)\,dx \right| \le \left| \sum_{j=m}^r f(j) - \int_m^r f(x)\,dx - \frac{1}{2}[f(m) + f(r)] \right|$$
$$+ \frac{1}{2}|f(r) - f(m)|$$

and (9.12) follows from (9.10) by taking limits as $r \to \infty$. ∎

Example 9.16 (Stirling's approximation). Consider the function $\log(n!)$, $n = 0, 1, \ldots$. Note that

$$\log(n!) = \sum_{j=1}^n \log(j).$$

Since $\log(x)$ is a strictly increasing function, by Corollary 9.3,

$$\log(n!) = \int_1^n \log(x)\,dx + R_n = n \log(n) - n + 1 + R_n$$

where

$$|R_n| \le \log(n)/2.$$

Hence,

$$\log(n!) = n \log(n) - n + O(\log(n)) \quad \text{as } n \to \infty,$$

which is a crude form of Stirling's approximation. By Example 9.8, we know that the $O(\log(n))$ term can be expanded as

$$\frac{1}{2}\log(2\pi) + \frac{1}{2}\log(n) + O(n^{-1}). \qquad \square$$

Example 9.17 (Tail probability of the logarithmic series distribution). Let X denote a discrete random variable such that

$$\Pr(X = j) = c(\theta)\theta^j/j, \quad j = 1, 2, \ldots,$$

where $0 < \theta < 1$ and

$$c(\theta) = -\frac{1}{\log(1 - \theta)}.$$

This is known as a *logarithmic series distribution*. Consider approximation of the tail probability

$$c(\theta) \sum_{j=n}^{\infty} \theta^j / j$$

for large n.

Note that θ^x / x is a decreasing function so that by Corollary 9.3,

$$\sum_{j=n}^{\infty} \theta^j / j = \int_n^{\infty} \frac{1}{x} \theta^x \, dx + R_n$$

where

$$|R_n| \leq \frac{1}{2} \theta^n / n.$$

We may write

$$\int_n^{\infty} x^{-1} \theta^x \, dx = \frac{\theta^n}{\log(\theta)} \int_0^{\infty} \frac{1}{1 + u / \log(\theta)} \exp\{nu\} \, du;$$

hence, by Watson's lemma,

$$\int_n^{\infty} x^{-1} \theta^x \, dx = \frac{\theta^n}{\log(\theta)} \left[\frac{1}{n} + O\left(\frac{1}{n^2}\right) \right].$$

Therefore, based on this approach, the magnitude of the remainder term R_n is the same as that of the integral itself, θ^n / n. Hence, all we can conclude is that

$$\sum_{j=n}^{\infty} \theta^j / j = \theta^n O\left(\frac{1}{n}\right) \quad \text{as} \quad n \to \infty. \qquad \square$$

The previous example shows that the approximations given in Corollary 9.3 are not very useful whenever the magnitude of $|f(r) - f(m)|$ is of the same order as the magnitude of the sum $f(m) + \cdots + f(r)$. This can occur whenever the terms in the sum increase or decrease very rapidly; then f' tends to be large and, hence, the remainder terms in Corollary 9.3 tend to be large. In some of these cases, *summation-by-parts* may be used to create an equivalent sum whose terms vary more slowly. This result is given in the following theorem; the proof is left as an exercise.

Theorem 9.18. *Consider sequences $x_1, x_2, \ldots$ and $y_1, y_2, \ldots$. Let m and r denote integers such that $m \leq r$. Define*

$$S_j = x_m + \cdots + x_j, \quad j = m, \ldots, r.$$

Then

$$\sum_{j=m}^{r} x_j y_j = \sum_{j=m}^{r-1} S_j (y_j - y_{j+1}) + S_r y_r.$$

If

$$\sum_{j=m}^{\infty} |x_j y_j| < \infty, \quad \sum_{j=m}^{\infty} |x_j| < \infty$$

and $\lim_{r\to\infty} y_r = 0$, *then*

$$\sum_{j=m}^{\infty} x_j y_j = \sum_{j=m}^{\infty} S_j(y_j - y_{j+1}).$$

Example 9.18 *(Tail probability of the logarithmic series distribution).* Consider approximation of the sum

$$\sum_{j=n}^{\infty} \theta^j / j,$$

where $0 < \theta < 1$, as discussed in Example 9.17. We can apply Theorem 9.18, taking $y_j = 1/j$ and $x_j = \theta^j$. Hence,

$$S_j = \theta^n \frac{1 - \theta^{j-n+1}}{1 - \theta}, \quad j = n, \dots .$$

It follows that

$$\sum_{j=n}^{\infty} \frac{\theta^j}{j} = \sum_{j=n}^{\infty} \theta^n \frac{1 - \theta^{j-n+1}}{1 - \theta} \frac{1}{j(j+1)}$$

$$= \frac{\theta^n}{1 - \theta} \left[\sum_{j=n}^{\infty} \frac{1}{j(j+1)} - \sum_{j=n}^{\infty} \frac{\theta^{j-n+1}}{j(j+1)} \right].$$

Using Corollary 9.3 it is straightforward to show that

$$\sum_{j=n}^{\infty} \frac{1}{j(j+1)} = \int_n^{\infty} \frac{1}{x(x+1)} dx + O\left(\frac{1}{n^2}\right)$$

$$= \log(1 + 1/n) + O\left(\frac{1}{n^2}\right) = \frac{1}{n} + O\left(\frac{1}{n^2}\right).$$

Since

$$\sum_{j=n}^{\infty} \frac{\theta^{j-n+1}}{j(j+1)} = \sum_{j=1}^{\infty} \frac{\theta^j}{(n+j-1)(n+j)} = O\left(\frac{1}{n^2}\right),$$

it follows that

$$\sum_{j=n}^{\infty} \frac{\theta^j}{j} = \frac{\theta^n}{1 - \theta} \frac{1}{n} \left[1 + O\left(\frac{1}{n}\right)\right] \quad \text{as } n \to \infty. \qquad \Box$$

9.8 Exercises

9.1 Prove Theorem 9.4.

9.2 Show that the beta function $\beta(\cdot, \cdot)$ satisfies

$$\beta(r, s) = \int_0^{\infty} \frac{t^{r-1}}{(1+t)^{s+r}} dt, \quad r > 0, s > 0.$$

9.3 Prove Theorem 9.5.

Exercises 9.4 and 9.5 use the following definition.

The *incomplete beta function* is defined as

$$I(r, s, x) = \int_0^x t^{r-1}(1 - t)^{s-1}\, dt,$$

where $r > 0$, $s > 0$, and $0 \le x \le 1$.

9.4 Show that, for all $r > 0$, $s > 0$, and $0 \le x \le 1$,

$$1 - I(r, s, x) = I(s, r, 1 - x).$$

9.5 Show that, for all $r > 1$, $s > 1$, and $0 \le x \le 1$,

$$I(r, s, x) = I(r, s - 1, x) - I(r + 1, s - 1, x).$$

9.6 Prove Theorem 9.8.

9.7 Prove Theorem 9.9.

9.8 Prove Theorem 9.10.

9.9 Suppose that, as $n \to \infty$,

$$f(n) = a_0 + \frac{a_1}{n} + \frac{a_2}{n^2} + O\left(\frac{1}{n^3}\right)$$

and

$$g(n) = b_0 + \frac{b_1}{n} + \frac{b_2}{n^2} + O\left(\frac{1}{n^3}\right)$$

for some constants $a_0, a_1, a_2, b_0, b_1, b_2$ such that $b_0 \ne 0$.
Find constants c_0, c_1, c_2 such that

$$\frac{f(n)}{g(n)} = c_0 + \frac{c_1}{n} + \frac{c_2}{n^2} + O\left(\frac{1}{n^2}\right).$$

9.10 Show that

$$\Gamma(x + 1) = \lim_{z \to \infty} z^x \beta(x, z), \quad x > 0.$$

9.11 Let X denote a random variable with an absolutely continuous distribution with density function

$$\frac{\beta^\alpha}{\Gamma(\alpha)} x^{\alpha-1} \exp(-\beta x), \quad x > 0.$$

Let h denote a function such that

$$h(t) = \sum_{j=0}^\infty h^{(j)}(0) t^j$$

and such that $h(t) = O(\exp(at^2))$ as $|t| \to \infty$ for some constant a. Find an asymptotic expansion for $\mathrm{E}[h(X)]$ as $\beta \to \infty$, with α remaining fixed.

9.12 Let $\Gamma(\cdot, \cdot)$ denote the incomplete gamma function. Show that, for fixed x,

$$\Gamma(x, y) = y^{x-1} \exp(-y)\left[1 + \frac{x - 1}{y} + \frac{(x - 1)(x - 2)}{y^2} + O\left(\frac{1}{y^3}\right)\right]$$

as $y \to \infty$.

9.13 Let Y denote a real-valued random variable with an absolutely continuous distribution with density function

$$p(y; a) = c(a) \exp\{-2a \cosh(y)\}, \quad -\infty < y < \infty;$$

see Example 9.10. Find an approximation to $\mathrm{E}[\cosh(Y)]$ that is valid for large a.

9.14 Let $\psi(\cdot)$ denote the logarithmic derivative of the gamma function. Show that

$$\psi(z) = \log z + O\left(\frac{1}{z}\right) \quad \text{as} \quad z \to \infty.$$

9.15 Let Y denote a real-valued random variable with an absolutely continuous distribution with density function

$$c(\alpha) \exp\{\alpha \cos(y)\}, \quad -\pi < y < \pi,$$

where $\alpha > 0$ and $c(\alpha)$ is a normalizing constant; this is a *von Mises distribution*. Find an approximation to $c(\alpha)$ that is valid for large values of α.

9.16 Let X denote a real-valued random variable with an absolutely continuous distribution with density function

$$\frac{1}{\alpha}(1 - x)^{\alpha - 1}, \quad 0 < x < 1.$$

Find an approximation to $E[\exp(X)]$ that is valid for large α.

9.17 Let Y denote a real-valued random variable with an absolutely continuous distribution with density function

$$p(y) = \frac{\sqrt{n}}{\sqrt{(2\pi)}} y^{-\frac{3}{2}} \exp\left\{-\frac{n}{2}(y + 1/y - 2)\right\}, \quad y > 0;$$

this is an inverse gaussian distribution. Find an approximation to $\Pr(Y \geq y)$, $y > 0$, that is valid for large n.

9.18 Let Y denote a random variable with an absolutely continuous distribution with density function

$$\frac{1}{\beta(\alpha, \alpha)} y^{\alpha - 1}(1 - y)^{\alpha - 1}, \quad 0 < y < 1,$$

where $\alpha > 0$; this is a beta distribution that is symmetric about $1/2$. Find an approximation to $\Pr(Y \geq y)$ that is valid for large α.

9.19 Euler's constant, generally denoted by γ, is defined by

$$\gamma = \lim_{n \to \infty} \left\{ \sum_{j=1}^{n} \frac{1}{j} - \log(n) \right\}.$$

Give an expression for γ in terms of the function P defined in Theorem 9.17.

9.20 Prove Theorem 9.18.

9.21 Consider the sum

$$\sum_{j=1}^{\infty} \frac{1}{j^{\alpha}} \quad \text{where} \quad \alpha > 1;$$

this is the *zeta function*, evaluated at α. Show that

$$\sum_{j=1}^{\infty} \frac{1}{j^{\alpha}} = \frac{1}{\alpha - 1} + O(1) \quad \text{as} \quad \alpha \to 1^{+}.$$

9.9 Suggestions for Further Reading

The functions described in Section 9.2 are often called *special functions* and they play an important role in many fields of science. See Andrews et al. (1999) and Temme (1996) for detailed discussions of special functions; in particular, Temme (1996, Chapter 11) discusses many special functions that are useful in statistics. Many imporant properties of special functions have been catalogued in Erdélyi (1953a,b).

The basic theory of asymptotic expansions is outlined in de Bruijn (1956) and Erdelyi (1956). Asymptotic expansions of integrals, including Watson's lemma and related results, are discussed in Andrews et al. (1999, Appendix C), Bleistein and Handelsman (1986), Temme (1996, Chapter 2), and Wong (1989). The form of Watson's lemma and Laplace's method presented here is based on Breitung (1994, Chapter 4). Evans and Swartz (2000) discuss the problem of approximating integrals that are useful in statistics; see also Barndorff-Nielsen and Cox (1989).

The uniform asymptotic approximation presented in Section 9.6 is known as Temme's method (Temme 1982); see Jensen (1995, Chapter 3) for further discussion of this result and some generalizations. The approximation of sums and the Euler-Maclaurin summation formula is discussed in de Bruijn (1956, Chapter 3); see also Andrews et al. (1999, Appendix D). Exercise 9.15 is discussed further in Barndorff-Nielsen and Cox (1989).

10

Orthogonal Polynomials

10.1 Introduction

Let F denote a distribution function on the real line such that

$$\int_{-\infty}^{\infty} |x|^r \, dF(x) < \infty$$

for all $r = 0, 1, \ldots$. A set of functions $\{f_0, f_1, \ldots\}$ is said to be *orthogonal with respect to F* if

$$\int_{-\infty}^{\infty} f_j(x) f_k(x) \, dF(x) = 0 \quad \text{for } j \neq k.$$

Suppose that, for each $n = 0, 1, \ldots$, f_n is a polynomial of degree n which we will denote by p_n; assume that the coefficient of x^n in $p_n(x)$ is nonzero. Then $\{p_0, p_1, \ldots\}$ are said to be *orthogonal polynomials with respect to F*.

Orthogonal polynomials are useful in a number of different contexts in distribution theory. For instance, they may be used to approximate functions or they may be used in the exact or approximate calculation of certain integrals; they also play a central role in asymptotic expansions for distribution functions, as will be discussed in Chapter 14. In this chapter, we give the basic properties of orthgonal polynomials with respect to a distribution function, along with some applications of these ideas.

10.2 General Systems of Orthogonal Polynomials

Let $\{p_0, p_1, \ldots\}$ denote orthogonal polynomials with respect to a distribution function F. Then any finite subset of $\{p_0, p_1, \ldots\}$ is linearly independent. A formal statement of this result is given in the following lemma; the proof is left as an exercise.

***Lemma* 10.1.** *Let $\{p_0, p_1, \ldots\}$ denote orthogonal polynomials with respect to a distribution function F. Then, for any integers $n_1 < n_2 < \cdots < n_m$ and any real numbers $\alpha_1, \alpha_2, \ldots, \alpha_m$,*

$$\alpha_1 p_{n_1}(x) + \cdots + \alpha_m p_{n_m}(x) = 0 \quad a.e. \ (F)$$

if and only if $\alpha_1 = \cdots = \alpha_m = 0$.

An important consequence of Lemma 10.1 is that, for any $j = 0, 1, \ldots$, the function x^j has a unique representation in terms of $p_0, p_1, \ldots, p_j$; that is,

$$x^j = \alpha_0 p_0(x) + \alpha_1 p_1(x) + \cdots + \alpha_j p_j(x)$$

for some unique set of constants $\alpha_0, \ldots, \alpha_j$. Hence, for each $n = 0, 1, \ldots,$

$$\int_{-\infty}^{\infty} x^j p_n(x) \, dF(x) = 0, \quad j = 0, 1, \ldots, n - 1. \tag{10.1}$$

Furthermore, as the following theorem shows, this property characterizes p_n.

Theorem 10.1. *Let $\{p_0, p_1, \ldots\}$ denote a set of orthogonal polynomials with respect to a distribution function F. Let f denote a polynomial of degree n. Then*

$$\int_{-\infty}^{\infty} x^j f(x) \, dF(x) = 0, \quad j = 0, 1, \ldots, n - 1 \tag{10.2}$$

if and only if for some $\alpha \neq 0$

$$f(x) = \alpha p_n(x) \quad a.e.(F).$$

Proof. Suppose that $f = \alpha p_n$ for some $\alpha \neq 0$. Fix $j = 0, \ldots, n - 1$. Since x^j may be written as a linear function of $p_0, \ldots, p_{n-1}$, it follows that (10.2) holds.

Now suppose that (10.2) holds for some polynomial f of degree n. Let c_n denote the coefficient of x^n in $f(x)$ and let d_n denote the coefficient of x^n in $p_n(x)$. Then

$$f(x) - \frac{c_n}{d_n} p_n(x) = g(x)$$

where g is a polynomial of degree at most $n - 1$. Let $\alpha = c_n/d_n$. Then

$$\int_{-\infty}^{\infty} (f(x) - \alpha p_n(x))^2 \, dF(x) = \int_{-\infty}^{\infty} g(x) f(x) \, dF(x) - \alpha \int_{-\infty}^{\infty} g(x) p_n(x) \, dF(x).$$

By (10.2),

$$\int_{-\infty}^{\infty} g(x) f(x) \, dF(x) = 0$$

and, by (10.1),

$$\int_{-\infty}^{\infty} g(x) p_n(x) \, dF(x) = 0.$$

It follows that

$$\int_{-\infty}^{\infty} (f(x) - \alpha p_n(x))^2 \, dF(x) = 0$$

and, hence, that

$$f(x) - \alpha p_n(x) = 0 \quad a.e. \, (F),$$

proving the result. ■

Construction of orthogonal polynomials

In many cases, orthogonal polynomials with respect to a given distribution may be easily constructed from the moments of that distribution.

Theorem 10.2. *Let*

$$m_n = \int_{-\infty}^{\infty} x^n \, dF(x), \quad n = 0, 1, \dots.$$

For each $n = 0, 1, \dots$, let

$$p_n(x) = \det \begin{pmatrix} x^n & x^{n-1} & \cdots & 1 \\ m_n & m_{n-1} & \cdots & m_0 \\ m_{n+1} & m_n & \cdots & m_1 \\ \vdots & \vdots & \cdots & \vdots \\ m_{2n-1} & m_{2n-2} & \cdots & m_{n-1} \end{pmatrix}.$$

If, for some $N = 0, 1, 2, \dots$,

$$\alpha_n \equiv \det \begin{pmatrix} m_{n-1} & \cdots & m_0 \\ m_n & \cdots & m_1 \\ \vdots & \cdots & \vdots \\ m_{2n-2} & \cdots & m_{n-1} \end{pmatrix} \neq 0,$$

for $n = 0, 1, \dots, N$, then $\{p_0, p_1, \dots, p_N\}$ are orthogonal polynomials with respect to F.

Proof. Clearly, p_n is a polynomial of degree n with coefficient of x^n given by $\alpha_n \neq 0$. Hence, by Theorem 10.1, it suffices to show that

$$\int_{-\infty}^{\infty} p_n(x) x^j \, dF(x) \, dx = 0, \quad j = 0, 1, \dots, n - 1.$$

Note that, for $j = 0, 1, \dots, n - 1$,

$$x^j p_n(x) = \det \begin{pmatrix} x^{n+j} & x^{n+j-1} & \cdots & x^j \\ m_n & m_{n-1} & \cdots & m_0 \\ m_{n+1} & m_n & \cdots & m_1 \\ \vdots & \vdots & \cdots & \vdots \\ m_{2n-1} & m_{2n-2} & \cdots & m_{n-1} \end{pmatrix}$$

and, since this determinant is a linear function of $x^{n+j}, x^{n+j-1}, \dots, x^j$,

$$\int_{-\infty}^{\infty} p_n(x) x^j \, dF(x) \, dx = \det \begin{pmatrix} m_{n+j} & m_{n+j-1} & \cdots & m_j \\ m_n & m_{n-1} & \cdots & m_0 \\ m_{n+1} & m_n & \cdots & m_1 \\ \vdots & \vdots & \cdots & \vdots \\ m_{2n-1} & m_{2n-2} & \cdots & m_{n-1} \end{pmatrix}.$$

Since the first row of this matrix is identical to one of the subsequent rows, it follows that the determinant is 0; the result follows. ∎

In this section, the ideas will be illustrated using the Legendre polynomials; in the following section, other families of orthogonal polynomials will be discussed.

***Example* 10.1 (*Legendre polynomials*).** Let F denote the distribution function of the uniform distribution on $(-1, 1)$; the orthogonal polynomials with respect to this distribution are known as the Legendre polynomials. They are traditionally denoted by $P_0, P_1, \ldots$ and we will use that notation here.

Note that

$$\int_{-\infty}^{\infty} x^n \, dF(x) = \frac{1}{2} \int_{-1}^{1} x^n \, dx = \begin{cases} \frac{1}{n+1} & \text{if } n \text{ is even} \\ 0 & \text{if } n \text{ is odd} \end{cases}.$$

Hence, using the procedure described in Theorem 10.2, we have that $P_0(x) = 1$,

$$P_1(x) = \det \begin{pmatrix} x & 1 \\ 0 & 1 \end{pmatrix} = x$$

and

$$P_2(x) = \det \begin{pmatrix} x^2 & x & 0 \\ 1/3 & 0 & 1 \\ 0 & 1/3 & 0 \end{pmatrix} = -\frac{1}{3}x^2 + \frac{1}{9}.$$

It is easy to verify directly that these polynomials are orthogonal with respect to F. □

If $p_0, p_1, \ldots$ are orthogonal polynomials with respect to some distribution function F, then so are $\alpha_0 p_0, \alpha_1 p_1, \ldots$ for any nonzero constants $\alpha_0, \alpha_1, \ldots$. Hence, orthogonal polynomials are generally standardized in some way. Typically, one of the following standardizations is used: the coefficient of x^n in $p_n(x)$ is required to be 1, it is required that $p_n(1) = 1$, or p_n must satisfy

$$\int_{-\infty}^{\infty} p_n(x)^2 \, dF(x) = 1. \tag{10.3}$$

***Example* 10.2 (*Legendre polynomials*).** Consider the Legendre polynomials described in Example 10.1. If we require that the polynomials have lead coefficient 1, then

$$P_0(x) = 1, \quad P_1(x) = x, \quad \text{and} \quad P_2(x) = x^2 - \frac{1}{3}.$$

If we require that $p_n(1) = 1$,

$$P_0(x) = 1, \quad P_1(x) = x, \quad \text{and} \quad P_2(x) = \frac{3}{2}x^2 - \frac{1}{2}.$$

If we require that (10.3) holds, then

$$P_0(x) = 1, \quad P_1(x) = \sqrt{3}x, \quad \text{and} \quad P_2(x) = \sqrt{5}\frac{3}{2}x^2 - \frac{\sqrt{5}}{2}.$$

For the Legendre polynomials, the second of these standardizations is commonly used and that is the one we will use here. □

The following result gives another approach to finding orthogonal polynomials.

***Theorem* 10.3.** *Let F denote the distribution function of an absolutely continuous distribution with support $[a, b]$, $-\infty \le a < b \le \infty$, and let p denote the corresponding density* function. Let $\{p_0, p_1, \ldots\}$ *denote a system of orthogonal polynomials with respect to* F. *Suppose there exists a polynomial g satisfying the following conditions:*

(i) *For each* $n = 0, 1, \ldots$,

$$f_n(x) \equiv \frac{1}{p(x)} \frac{d^n}{dx^n} [g(x)^n p(x)]$$

is a polynomial of degree n.

(ii) *For each* $n = 1, 2, \ldots$ *and each* $j = 0, 1, \ldots, n-1$,

$$\lim_{x \to b} x^j \frac{d^m}{dx^m} [g(x)^n p(x)] = \lim_{x \to a} x^j \frac{d^m}{dx^m} [g(x)^n p(x)] = 0, \quad m = 1, 2, \ldots, n-1.$$

Then there exist constants $c_0, c_1, \ldots$ *such that* $f_n = c_n p_n, n = 0, 1, \ldots$.

Proof. Fix n. By Theorem 10.1, it suffices to show that, for each $j = 0, 1, \ldots, n-1$,

$$\int_a^b x^j f_n(x) \, dF(x) = 0.$$

Using integration-by-parts,

$$\int_a^b x^j f_n(x) \, dF(x) \equiv \int_a^b x^j \frac{d^n}{dx^n} [g(x)^n p(x)] \, dx$$

$$= x^j \frac{d^{n-1}}{dx^{n-1}} [g(x)^n p(x)] \Big|_a^b - j \int_a^b x^{j-1} \frac{d^{n-1}}{dx^{n-1}} [g(x)^n p(x)] \, dx$$

$$= -j \int_a^b x^{j-1} \frac{d^{n-1}}{dx^{n-1}} [g(x)^n p(x)] \, dx.$$

Continuing in this way,

$$\int_a^b x^j \frac{d^n}{dx^n} [g(x)^n p(x)] \, dx = (-1)^j j! \frac{d^{n-j}}{dx^{n-j}} [g(x)^n p(x)] \Big|_a^b = 0.$$

Since $0 \leq n - j \leq n - 1$, the result follows. ∎

Hence, when the conditions of Theorem 10.3 hold, we can take the orthogonal polynomials to be

$$p_n(x) = \frac{1}{c_n p(x)} \frac{d^n}{dx^n} [g(x)^n p(x)] \tag{10.4}$$

for some constants $c_0, c_1, \ldots$. This is known as *Rodrigue's formula*.

Example 10.3 (*Legendre polynomials*). In order to determine Rodrigue's formula for the Legendre polynomials, it suffices to find a polynomial g such that

$$\frac{d^n}{dx^n} g(x)^n$$

is a polynomial of degree n and

$$\lim_{x \to 1} x^j \frac{d^m}{dx^m} g(x)^n = \lim_{x \to -1} x^j \frac{d^m}{dx^m} g(x)^n = 0$$

for $j = 0, 1, \ldots, n-1$ and $m = 1, 2, \ldots, n-1$.

Suppose g is a polynomial of degree r. Then $g(x)^n$ is a polynomial of degree nr and

$$\frac{d^n}{dx^n} g(x)^n$$

is a polynomial of degree $n(r-1)$; hence, g must be a polynomial of degree 2. Since all polynomials are bounded on $[-1, 1]$, in order to satisfy the second condition, it suffices that

$$\lim_{x \to 1} g(x) = \lim_{x \to -1} g(x) = 0.$$

Writing $g(x) = ax^2 + bx + c$, we need $a + c = 0$ and $b = 0$; that is, g is of the form $c(x^2 - 1)$. It follows that orthogonal polynomials with respect to the uniform distribution on $(-1, 1)$ are given by

$$\frac{d^n}{dx^n}(x^2 - 1)^n.$$

Note that

$$\frac{d^n}{dx^n}(x^2 - 1)^n = n!(2x)^n + Q(x)$$

where $Q(x)$ is a sum in which each term contains a factor $x^2 - 1$. Hence, the standardized polynomials that equal 1 at $x = 1$ are given by

$$\frac{1}{n!2^n} \frac{d^n}{dx^n}(x^2 - 1)^n. \qquad \square$$

Zeros of orthogonal polynomials and integration

Consider a function $g : \mathbf{R} \to \mathbf{R}$. A *zero* of g is a number r, possibly complex, such that $g(r) = 0$. If g is a polynomial of degree n, then g can have at most n zeros. A zero r is said to have *multiplicity* α if

$$g(r) = g'(r) = \cdots = g^{(\alpha-1)}(r) = 0$$

and $g^{(\alpha)}(r) \neq 0$. A zero is said to be *simple* if its multiplicity is 1.

Let g denote an nth degree polynomial and let $r_1, \ldots, r_m$ denote the zeros of g such that r_j has multiplicity α_j, $j = 1, \ldots, m$. Then $\sum_{j=1}^m \alpha_j = n$ and g can be written

$$g(x) = a(x - r_1)^{\alpha_1} \cdots (x - r_m)^{\alpha_m}$$

for some constant a.

The zeros of orthogonal polynomials have some useful properties.

Theorem 10.4. *Let $\{p_0, p_1, \ldots\}$ denote orthogonal polynomials with respect to F and let $[a, b]$, $-\infty \leq a < b \leq \infty$, denote the support of F. Then, for each $n = 0, 1, \ldots$, p_n has n simple real zeros, each of which takes values in (a, b).*

Proof. Fix n. Let k denote the number of zeros in (a, b) at which p_n changes sign; hence, $0 \leq k \leq n$.

Assume that $k < n$ and and let $x_1 < x_2 < \cdots < x_k$ denote the zeros in (a, b) at which p_n changes sign. Consider the polynomial

$$f(x) = (x - x_1)(x - x_2) \cdots (x - x_k).$$

Since this is a polynomial of degree k, it follows from Theorem 10.1 that

$$\int_a^b f(x)p_n(x)\,dF(x) = 0. \tag{10.5}$$

Note that $f(x)$ changes sign at each x_j, $j = 1, \ldots, k$, so that $w(x) = f(x)p_n(x)$ is always of the same sign. Without loss of generality, we assume that $w(x) > 0$ for all x. Hence,

$$\int_a^b w(x)\,dF(x) = \int_a^b f(x)p_n(x)\,dF(x)\,dx > 0.$$

This contradicts (10.5) so that we must have $k = n$. Thus, p_n has n simple zeros in (a, b); however, p_n has only n zeros so that all zeros of p_n lie in (a, b) and are simple. ■

***Example* 10.4 (*Legendre polynomials*).** The Legendre polynomial $P_1(x) = x$ has one zero, at $x = 0$. The second Legendre polynomial,

$$P_2(x) = \frac{3}{2}x^2 - \frac{1}{2}$$

has zeros at $x = \pm 1/\sqrt{3}$. It may be shown that the third Legendre polynomial is given by

$$P_3(x) = \frac{5}{2}x^3 - \frac{3}{2}x,$$

which has zeros at $x = \pm\sqrt{(.6)}$ and $x = 0$. □

Let $p_0, p_1, \ldots$ denote orthogonal polynomials with respect to a distribution function F and let x_1 denote the zero of p_1. Let $f(x) = ax + b$, where a and b are constants. Then $f(x) = cp_1(x) + d$ for some constants c and d; since $p_1(x_1) = 0$ we must have $d = f(x_1)$. It follows that

$$\int_{-\infty}^{\infty} f(x)\,dF(x) = c\int_{-\infty}^{\infty} p_1(x)\,dF(x) + f(x_1).$$

Since p_1 is orthogonal to all constant functions,

$$\int_{-\infty}^{\infty} p_1(x)\,dF(x) = 0;$$

hence, for any linear function f,

$$\int_{-\infty}^{\infty} f(x)\,dF(x) = f(x_1).$$

That is, the integral with respect to F of any linear function can be obtained by simply evaluating that function at x_1. The following result generalizes this method to an orthogonal polynomial of arbitrary order.

***Theorem* 10.5.** *Let* $\{p_0, p_1, \ldots\}$ *denote orthogonal polynomials with respect to F. For a given value of $n = 1, 2, \ldots$, let $x_1 < x_2 < \ldots < x_n$ denote the zeros of p_n. Then there exist constants $\lambda_1, \lambda_2, \ldots, \lambda_n$ such that, for any polynomial f of degree $2n - 1$ or less,*

$$\int_{-\infty}^{\infty} f(x)\,dF(x) = \sum_{j=1}^{n} \lambda_j f(x_j).$$

For each $k = 1, \ldots, n$, λ_k *is given by*

$$\lambda_k = \int_{-\infty}^{\infty} \frac{p_n(x)}{(x - x_k)p_n'(x_k)} \, dF(x).$$

Proof. Fix n and consider a polynomial f of degree less than or equal to $2n - 1$. Note that, by Theorem 10.4, the zeros of p_n are simple and, hence, $|p_n'(x_j)| > 0$, $j = 1, \ldots, n$. Define a function h by

$$h(x) = \sum_{j=1}^{n} f(x_j) \frac{p_n(x)}{(x - x_j)p_n'(x_j)}.$$

Note that

$$p_n(x) = \alpha(x - x_1) \cdots (x - x_n)$$

for some constant $\alpha \neq 0$; this may be seen by noting that the function on the right is a polynomial of degree n with the same zeros as p_n. Hence, for each $j = 1, 2, \ldots, n$,

$$\frac{p_n(x)}{(x - x_j)}$$

is a polynomial of degree $n - 1$ so that h is also a polynomial of degree $n - 1$. It follows that $h - f$ is a polynomial of degree less than or equal to $2n - 1$. Note that, for each $j = 1, 2, \ldots, n$,

$$\lim_{x \to x_k} h(x) = \sum_{j \neq k} f(x_j) \frac{p_n(x_k)}{(x_k - x_j)p_n'(x_j)} + f(x_k) = f(x_k);$$

hence, $h - f$ has zeros at $x_1, \ldots, x_n$. It follows that

$$h(x) - f(x) = (x - x_1) \cdots (x - x_n)q(x) \equiv p_n(x)r(x) \tag{10.6}$$

where q and r are polynomials each of degree at most $n - 1$.

Writing

$$f(x) = h(x) - p_n(x)r(x),$$

and using the fact that r is a polynomial of degree at most $n - 1$,

$$\int_{-\infty}^{\infty} f(x) \, dF(x) = \int_{-\infty}^{\infty} h(x) \, dF(x) - \int_{-\infty}^{\infty} p_n(x)r(x) \, dF(x) = \int_{-\infty}^{\infty} h(x) \, dF(x).$$

The result now follows from the fact that

$$\int_{-\infty}^{\infty} h(x) \, dF(x) = \sum_{j=1}^{n} f(x_j) \int_{-\infty}^{\infty} \frac{p_n(x)}{(x - x_j)p_n'(x_j)} \, dF(x) = \sum_{j=1}^{n} \lambda_j f(x_j)$$

where $\lambda_1, \ldots, \lambda_n$ are given in the statement of the theorem. ∎

Example 10.5 (*Legendre polynomials*). Consider the Legendre polynomial

$$P_2(x) = \frac{3}{2}x^2 - \frac{1}{2},$$

which has zeros at $\pm 1/\sqrt{3}$. Hence, we may write

$$P_2(x) = \frac{3}{2}(x - 1/\sqrt{3})(x + 1/\sqrt{3}).$$

Then

$$\lambda_1 = \frac{1}{2} \int_{-1}^{1} \frac{P_2(x)}{(x - 1/\sqrt{3})P_2'(-1/\sqrt{3})} \, dx = \frac{1}{2}$$

and

$$\lambda_2 = \frac{1}{2} \int_{-1}^{1} \frac{P_2(x)}{(x + 1/\sqrt{3})P_2'(1/\sqrt{3})} \, dx = \frac{1}{2}.$$

Hence, any polynomial f of degree 3 or less can be calculated by

$$\int_{-1}^{1} f(x) \, dF(x) = \frac{f(1/\sqrt{3}) + f(-1/\sqrt{3})}{2}.$$

This may be verified directly by integrating $1, x, x^2$, and x^3.   □

The method described in Theorem 10.5 can also be used as the basis for a method of numerical integration; this will be discussed in Section 10.4.

Completeness and approximation

Let $\{p_0, p_1, \ldots\}$ denote orthogonal polynomials with respect to F. We say that $\{p_0, p_1, \ldots\}$ is *complete* if the following condition holds: suppose f is a function such that

$$\int_{-\infty}^{\infty} f(x)^2 \, dF(x) < \infty$$

and

$$\int_{-\infty}^{\infty} f(x)p_n(x) \, dF(x) = 0, \quad n = 0, 1, \ldots;$$

then $f = 0$ almost everywhere (F).

***Example* 10.6 (*Completeness of the Legendre polynomials*).** Let F denote the distribution function of the uniform distribution on $(-1, 1)$ and let f denote a function such that

$$\int_{-1}^{1} f(x)^2 \, dF(x) < \infty$$

and

$$\int_{-1}^{1} f(x)P_n(x) \, dF(x) = 0, \quad n = 0, 1, \ldots$$

where P_n denotes the nth Legendre polynomial.

Let q_n be an arbritrary polynomial of degree n. Then

$$\int_{-1}^{1} (f(x) - q_n(x))^2 \, dF(x) = \int_{-1}^{1} f(x)^2 \, dF(x) + \int_{-1}^{1} q_n(x)^2 \, dF(x).$$

Fix $\epsilon > 0$. By the Weierstrass approximation theorem, there exists a polynomial q_n such that

$$\sup_{|x| \le 1} |f(x) - q_n(x)| < \epsilon.$$

Hence,

$$\int_{-1}^{1} f(x)^2 \, dF(x) \le \epsilon^2 - \int_{-1}^{1} q_n(x)^2 \, dF(x) \le \epsilon^2.$$

Since ϵ is arbitrary,

$$\int_{-1}^{1} f(x)^2 \, dF(x) = 0,$$

establishing completeness.

Note that this argument shows that any set of orthogonal polynomials on a bounded interval is complete. $\square$

Completeness plays an important role in the approximation of functions by series of orthogonal polynomials, as shown by the following theorem.

Theorem 10.6. *Let* $\{p_0, p_1, \ldots\}$ *denote orthogonal polynomials with respect to* F *and define*

$$\bar{p}_n(x) = \frac{p_n(x)}{[\int_{-\infty}^{\infty} p_n(x)^2 \, dF(x)]^{\frac{1}{2}}}, \quad n = 0, 1, \ldots.$$

Let f *denote a function satisfying*

$$\int_{-\infty}^{\infty} f(x)^2 \, dF(x) < \infty$$

and let

$$\alpha_n = \int_{-\infty}^{\infty} f(x) \bar{p}_n(x) \, dF(x), \quad n = 0, 1, \ldots.$$

For $n = 0, 1, 2, \ldots$ *define*

$$\hat{f}_n(x) = \sum_{j=0}^{n} \alpha_j \bar{p}_j(x).$$

If $\{p_0, p_1 \ldots\}$ *is complete, then*
  (i) $\lim_{n \to \infty} \int_{-\infty}^{\infty} [\hat{f}_n(x) - f(x)]^2 \, dF(x) = 0$
  (ii)

$$\int_{-\infty}^{\infty} [\hat{f}_n(x) - f(x)]^2 \, dF(x) \le \int_{-\infty}^{\infty} f(x)^2 \, dF(x) - \sum_{j=0}^{n} \alpha_j^2, \quad n = 1, 2, \ldots$$

(iii) $\sum_{j=0}^{\infty} \alpha_j^2 = \int_{-\infty}^{\infty} f(x)^2 \, dF(x)$

(iv) For any constants $\beta_0, \beta_1, \ldots, \beta_n$,

$$\int_{-\infty}^{\infty} [\hat{f}_n(x) - f(x)]^2 \, dF(x) \le \int_{-\infty}^{\infty} \left[\sum_{j=0}^{n} \beta_j \bar{p}_j(x) - f(x) \right]^2 \, dF(x),$$

that is, $\hat{f}_n$ is the best approximation to f among all polynomials of degree n, using the criterion

$$\int_{-\infty}^{\infty} [g(x) - f(x)]^2 \, dF(x).$$

Proof.   We first show that the sequence $\hat{f}_1, \hat{f}_2, \ldots$ converges to some function f_0 in the sense that

$$\lim_{n \to \infty} \int_{-\infty}^{\infty} (\hat{f}_n(x) - f_0(x))^2 \, dF(x) = 0.$$

Note that, for $m > n$,

$$\int_{-\infty}^{\infty} [\hat{f}_m(x) - \hat{f}_n(x)]^2 \, dF(x) = \sum_{j=n+1}^{m} \alpha_j^2;$$

hence, under the conditions of the theorem, for any $\epsilon > 0$ there exists an N such that

$$\int_{-\infty}^{\infty} [\hat{f}_m(x) - \hat{f}_n(x)]^2 \, dF(x) \le \epsilon$$

for $n, m > N$.

The construction of the function f_0 now follows as in the proof of Theorem 6.4; hence, only a brief sketch of the argument is given here.

There exists a subsequence $n_1, n_2, \ldots$ such that

$$\int_{-\infty}^{\infty} [\hat{f}_{n_{j+1}}(x) - \hat{f}_{n_j}(x)]^2 \, dF(x) \le \frac{1}{4^j}, \quad j = 1, 2, \ldots.$$

For each $m = 1, 2, \ldots$, define a function T_m by

$$T_m(x) = \sum_{j=1}^{m} |\hat{f}_{n_{j+1}}(x) - \hat{f}_{n_j}(x)|.$$

Then, for each x, either $T_1(x), T_2(x), \ldots$ has a limit or the sequence diverges to ∞. Define

$$T(x) = \lim_{m \to \infty} T_m(x)$$

if the limit exists; otherwise set $T(x) = \infty$. As in the proof of Theorem 6.4, it may be shown that the set of x for which $T(x) < \infty$ has probability 1 under F; for simplicity, assume that $T(x) < \infty$ for all x. It follows that

$$\sum_{j=1}^{\infty} [\hat{f}_{n_{j+1}}(x) - \hat{f}_{n_j}(x)]$$

converges absolutely and, hence, we may define a function

$$f_0(x) = \hat{f}_{n_1}(x) + \sum_{j=1}^{\infty} [\hat{f}_{n_{j+1}}(x) - \hat{f}_{n_j}(x)] = \lim_{j \to \infty} \hat{f}_{n_j}(x).$$

It now follows, as in the proof of Theorem 6.4, that

$$\lim_{n\to\infty} \int_{-\infty}^{\infty} (\hat{f}_n(x) - f_0(x))^2\, dF(x) = 0.$$

Now return to the proof of the theorem. Write $f(x) = f_0(x) + d(x)$. Then, for each $j = 1, 2, \ldots,$

$$\int_{-\infty}^{\infty} f(x)\bar{p}_j(x)\, dF(x) = \int_{-\infty}^{\infty} f_0(x)\bar{p}_j(x)\, dF(x) + \int_{-\infty}^{\infty} d(x)\bar{p}_j(x)\, dF(x).$$

Note that, by the Cauchy-Schwarz inequality,

$$\left| \int_{-\infty}^{\infty} (\hat{f}_n(x) - f_0(x))\bar{p}_j(x)\, dF(x) \right|$$

$$\leq \left[\int_{-\infty}^{\infty} (\hat{f}_n(x) - f_0(x))^2\, dF(x) \right]^{\frac{1}{2}} \left[\int_{-\infty}^{\infty} \bar{p}_j(x)^2\, dF(x) \right]^{\frac{1}{2}}.$$

Since

$$\int_{-\infty}^{\infty} \bar{p}_j(x)^2\, dF(x) = 1, \quad j = 1, \ldots,$$

and

$$\int_{-\infty}^{\infty} (\hat{f}_n(x) - f_0(x))^2\, dF(x) \to 0 \quad \text{as} \quad n \to \infty,$$

it follows that

$$\int_{-\infty}^{\infty} f_0(x)\bar{p}_j(x)\, dF(x) = \lim_{n\to\infty} \hat{f}_n(x)\bar{p}_j(x)\, dF(x) = \alpha_j. \tag{10.7}$$

Since

$$\int_{-\infty}^{\infty} f(x)\bar{p}_j(x)\, dF(x) = \alpha_j,$$

it follows that

$$\int_{-\infty}^{\infty} d(x)\bar{p}_j(x)\, dF(x) = 0, \quad j = 0, 1, \ldots$$

so that

$$\int_{-\infty}^{\infty} d(x)p_j(x)\, dF(x) = 0, \quad j = 0, 1, \ldots.$$

By completeness of $\{p_0, p_1, \ldots\}$, $d = 0$ and, hence,

$$\lim_{n\to\infty} \int_{-\infty}^{\infty} (\hat{f}_n(x) - f(x))^2\, dF(x) = 0. \tag{10.8}$$

This verifies part (i) of the theorem.

To show parts (ii) and (iii), note that

$$\int_{-\infty}^{\infty} (\hat{f}_n(x) - f(x))^2 \, dF(x) = \int_{-\infty}^{\infty} f(x)^2 \, dF(x) - \int_{-\infty}^{\infty} \hat{f}_n(x)^2 \, dF(x)$$

$$= \int_{-\infty}^{\infty} f(x)^2 \, dF(x) - \sum_{j=0}^{n} \alpha_j^2.$$

This proves that part (ii) and part (iii) now follows from (10.8).

Finally, consider part (iv). Note that

$$\int_{-\infty}^{\infty} \left[\sum_{j=0}^{n} \beta_j \bar{p}_j(x) - f(x) \right]^2 \, dF(x) = \int_{-\infty}^{\infty} f(x)^2 \, dF(x) - 2 \sum_{j=0}^{n} \beta_j \alpha_j + \sum_{j=0}^{n} \beta_j^2.$$

Hence,

$$\int_{-\infty}^{\infty} \left[\sum_{j=0}^{n} \beta_j \bar{p}_j(x) - f(x) \right]^2 \, dF(x) - \int_{-\infty}^{\infty} [\hat{f}_n(x) - f(x)]^2 \, dF(x)$$

$$= \sum_{j=0}^{n} \beta_j^2 - 2 \sum_{j=0}^{n} \beta_j \alpha_j + \sum_{j=0}^{n} \alpha_j^2$$

$$= \sum_{j=0}^{n} (\alpha_j - \beta_j)^2,$$

proving the result.   ∎

Hence, according to Theorem 10.6, if $\{p_0, p_1, \ldots\}$ is complete and

$$\int_{-\infty}^{\infty} f(x)^2 \, dF(x) < \infty,$$

the function f may be written

$$f(x) = \sum_{j=0}^{\infty} \alpha_j \frac{p_j(x)}{\int_{-\infty}^{\infty} p_j(x)^2 \, dF(x)}$$

for constants $\alpha_0, \alpha_1, \ldots$ given by

$$\alpha_n = \frac{\int_{-\infty}^{\infty} f(x) p_n(x) \, dF(x)}{\int_{-\infty}^{\infty} p_n(x)^2 \, dF(x)}.$$

In interpreting the infinite series in this expression, it is important to keep in mind that it means that

$$\lim_{n \to \infty} \int_{-\infty}^{\infty} \left[f(x) - \sum_{j=0}^{n} \alpha_j \frac{p_j(x)}{\int_{-\infty}^{\infty} p_j(x)^2 \, dF(x)} \right]^2 \, dF(x) = 0.$$

It is not necessarily true that for a given value of x the numerical series

$$\sum_{j=0}^{n} \alpha_j \frac{p_j(x)}{\int_{-\infty}^{\infty} p_j(x)^2 \, dF(x)}$$

converges to $f(x)$ as $n \to \infty$.

***Example* 10.7 (*Series expansion for a density*).** Consider a absolutely continuous distribution on $[-1, 1]$. Suppose the moments of the distribution are available, but the form of the density is not. We may approximate the density function by a lower-order polynomial by using an expansion in terms of the Legendre polynomials.

Consider a quadratic approximation of the form

$$a + bx + cx^2;$$

in terms of the Legendre polynomials, this approximation is

$$\left(a + \frac{1}{3}c\right) P_0(x) + bP_1(x) + \frac{2}{3}c P_2(x).$$

Using the approach of Theorem 10.6, the approximation to the density based on P_0, P_1, and P_2 is of the form

$$P_0(x) + \mu P_1(x) + \left[\frac{3}{2}(\mu^2 + \sigma^2) - \frac{1}{2}\right] P_2(x)$$

where μ and σ are the mean and standard deviation, respectively, of the distribution.

Hence, an approximation to the density is given by

$$\frac{5}{4} - \frac{1}{3}(\mu^2 + \sigma^2) + \mu x + \left(\mu^2 + \sigma^2 - \frac{3}{4}\right) x^2.$$

Although this function must integrate to 1, it is not guaranteed to be nonnegative. If $\mu = 0$, it is straightforward to show that it is nonnegative.   □

10.3   Classical Orthogonal Polynomials

Although orthogonal polynomials may be constructed for any distribution for which all the moments exist, there are only a few distributions that are often used in this context. One is the uniform distribution on $(-1, 1)$, leading to the Legendre polynomials discussed in the previous section. Others include the standard normal distribution, which leads to the *Hermite polynomials*, and the standard exponential distribution, which leads to the *Laguerre polynomials*. In this section, we consider the properties of the Hermite and Laguerre polynomials.

Hermite polynomials

Orthogonal polynomials with respect to the standard normal distribution function are known as the *Hermite polynomials*. These polynomials are often normalized by taking the coefficient of x^n in $H_n(x)$ to be 1 and that is the standardization that we will use here.

The following result shows that the Hermite polynomials may be generated using the procedure described in Theorem 10.3 using the function $g(x) = 1$. That is, we may take

$$H_n(x) = \frac{(-1)^n}{\phi(x)} \frac{d^n}{dx^n} \phi(x), \quad n = 0, 1, \ldots \tag{10.9}$$

where

$$\phi(x) = \frac{1}{\sqrt{(2\pi)}} \exp\left(-\frac{1}{2}x^2\right), \quad -\infty < x < \infty$$

denotes the standard normal density function.

Theorem 10.7. *For each $n = 0, 1, \ldots$, the nth Hermite polynomial is given by (10.9).*

Proof.   According to Theorem 10.3, the polynomials given by (10.9) are orthogonal with respect to the standard normal distribution provided that

$$\frac{1}{\phi(x)} \frac{d^n}{dx^n} \phi(x)$$

is a polynomial of degree n and that

$$\lim_{|x| \to \infty} x^j \frac{d^n}{dx^n} \phi(x) = 0 \qquad (10.10)$$

for all n and j. Once these are established, the result follows provided that the coefficient of x^n in (10.9) is 1.

For each $n = 0, 1, \ldots$, define a function q_n by

$$\frac{d^n}{dx^n} \phi(x) = q_n(x)\phi(x).$$

Note that $q_0(x) = 1$ and, hence, is a polynomial of degree 0. Assume that q_m is a polynomial of degree m; we will show that this implies that q_{m+1} is a polynomial of degree $m + 1$. It will then follow by induction that q_n is a polynomial of degree n for $n = 0, 1, \ldots$.

Note that

$$\frac{d^{m+1}}{dx^{m+1}} \phi(x) = q_{m+1}(x)\phi(x) = \frac{d}{dx} q_m(x)\phi(x)$$

so that

$$q_{m+1}(x)\phi(x) = q_m'(x)\phi(x) - xq_m(x)\phi(x);$$

hence,

$$q_{m+1}(x) = q_m'(x) - xq_m(x).$$

Under the assumption that q_m is a polynomial of degree m, it follows that q_{m+1} is a polynomial of degree $m + 1$. Hence, for all $n = 0, 1, \ldots$, q_n is a polynomial of degree n.

Since

$$\lim_{|x| \to \infty} x^j \phi(x) = 0$$

for all $j = 0, 1, \ldots$, (10.9) follows from the fact that

$$\frac{d^n}{dx^n} \phi(x) = q_n(x)\phi(x)$$

where q_n is a polynomial of degree n.

Finally, note that, if the coefficient of x^m in $q_m(x)$ is 1, then, since

$$q_{m+1}(x) = q_m'(x) - xq_m(x),$$

the coefficient of x^{m+1} in $q_{m+1}(x)$ is -1. The result now follows from the facts that $q_0(x) = 1$ and $H_n(x) = (-1)^n q_n(x)$.   ■

The following corollary gives another approach to constructing the Hermite polynomials; the proof follows from Theorem 10.7 and is left as an exercise.

Corollary 10.1. *For each $n = 0, 1, \ldots,$ let H_n denote the function defined by (10.9). Then*

$$H_{n+1}(x) = x H_n(x) - H_n'(x), \quad n = 0, 1, 2, \ldots.$$

Starting with $H_0(x) = 1$, it is straightforward to use Corollary 10.1 to determine the first several Hermite polynomials. The results are given in the following corollary; the proof is left as an exercise.

Corollary 10.2. *Let H_n denote the nth Hermite polynomial, $n = 1, 2, 3, 4$. Then*

$$H_1(x) = x, \quad H_2(x) = x^2 - 1,$$
$$H_3(x) = x^3 - 3x, \quad H_4(x) = x^4 - 6x^2 + 3.$$

Equation (10.9) can be used to find

$$\int_{-\infty}^{\infty} H_n(x)^2 \phi(x) \, dx.$$

The result is given in the following corollary; the proof is left as an exercise.

Corollary 10.3. *For each $n = 1, 2, \ldots,$ let H_n denote the nth Hermite polynomial. Then*

$$\int_{-\infty}^{\infty} H_n(x)^2 \phi(x) \, dx = n!.$$

Using the expression (10.9) for H_n, it is straightforward to calculate integrals of the form

$$\int_{-\infty}^{x} H_n(t) \phi(t) \, dt.$$

Theorem 10.8. *Let H_n denote the nth Hermite polynomial. Then*

$$\int_{-\infty}^{x} H_n(t) \phi(t) \, dt = -H_{n-1}(x) \phi(x).$$

Proof.   Note that

$$\int_{-\infty}^{x} H_n(t) \phi(t) \, dt = \int_{-\infty}^{x} (-1)^n \frac{d^n}{dt^n} \phi(t) \, dt$$
$$= (-1)^n \frac{d^{n-1}}{dt^{n-1}} \phi(t) \Big|_{-\infty}^{x} = -H_{n-1}(t) \phi(t) \Big|_{-\infty}^{x} = -H_{n-1}(x) \phi(x),$$

proving the result.   ∎

Hence, any integral of the form

$$\int_{-\infty}^{x} f(t) \phi(t) \, dt,$$

where f is a polynomial, can be integrated exactly in terms of $H_0, H_1, \ldots, \phi$, and the standard normal distribution function Φ.

***Example* 10.8   (*Chi-squared distribution function*).** For $x > 0$, consider calculation of the integral

$$\frac{1}{\sqrt{(2\pi)}} \int_0^x t^{\frac{1}{2}} \exp(-t/2) \, dt,$$

which is the distribution function of the chi-squared distribution with 3 degrees of freedom.

Using the change-of-variable $u = \sqrt{t}$,

$$\frac{1}{\sqrt{(2\pi)}} \int_0^x t^{\frac{1}{2}} \exp(-t/2) \, dt = 2 \int_0^{\sqrt{x}} u^2 \phi(u) \, du$$

$$= 2 \left[\int_{-\infty}^{\sqrt{x}} u^2 \phi(u) \, du - \frac{1}{2} \right].$$

Since $u^2 = H_2(u) - 1$, using Theorem 10.8 we have that

$$\frac{1}{\sqrt{(2\pi)}} \int_0^x t^{\frac{1}{2}} \exp(-t/2) \, dt = 2 \left[\Phi(\sqrt{x}) - \frac{1}{2} - \sqrt{x} \phi(\sqrt{x}) \right]. \qquad \square$$

It may be shown that the Hermite polynomials are complete; see, for example, Andrews, Askey, and Roy (1999, Chapter 6). Hence, the approximation properties outlined in Theorem 10.6 are valid for the Hermite polynomials.

***Example* 10.9   (*Gram-Charlier expansion*).** Let p denote a density function on the real line and let Φ and ϕ denote the distribution function and density function, respectively, of the standard normal distribution. Assume that

$$\int_{-\infty}^{\infty} \frac{p(x)^2}{\phi(x)} \, dx < \infty.$$

Under this assumption

$$\int_{-\infty}^{\infty} \frac{p(x)^2}{\phi(x)^2} \, d\Phi(x) < \infty$$

so that the function p/ϕ has an expansion of the form

$$\frac{p(x)}{\phi(x)} = \sum_{j=0}^{\infty} \alpha_j H_j(x)$$

where the constants $\alpha_0, \alpha_1, \ldots$ are given by

$$\alpha_j = \int_{-\infty}^{\infty} H_j(x) p(x) \phi(x) \, dx / \sqrt{(j!)};$$

note that $\alpha_0 = 1$.

Hence, the function p has an expansion of the form

$$p(x) = \phi(x) \left[1 + \sum_{j=1}^{\infty} \alpha_j H_j(x) \right].$$

This is known as a *Gram-Charlier expansion* of the density p. In interpreting this result it is important to keep in mind that the limiting operation refers to mean-square, not pointwise, convergence. $\square$

Laguerre polynomials

The Laguerre polynomials, which will be denoted by $L_0, L_1, \ldots$, are orthogonal polynomials with respect to the standard exponential distribution. Here we use the standardization that

$$\int_0^\infty L_n(x)^2 \exp(-x)\,dx = 1, \quad n = 0, 1, \ldots.$$

The Laguerre polynomials may be generated using the procedure described in Theorem 10.3 using the function $g(x) = x$. That is, we may take

$$L_n(x) = \frac{1}{n!} \exp(x)\frac{d^n}{dx^n}x^n \exp(-x), \quad n = 0, 1, \ldots. \tag{10.11}$$

Theorem 10.9. *The Laguerre polynomials are given by* (10.11).

Proof. Using Leibnitz's rule with (10.11) shows that

$$L_n(x) = \sum_{j=0}^n (-1)^j \binom{n}{j}\frac{x^j}{j!}$$

and, hence, L_n is a polynomial of degree n. Furthermore, it may be shown that

$$\int_0^\infty L_n(x)^2 \exp(-x)\,dx = 1;$$

this result is left as an exercise.

The result now follows from Theorem 10.3 provided that, for each $n = 1, 2, \ldots$,

$$\lim_{x \to 0} Q_{nmj}(x) = \lim_{x \to \infty} Q_{nmj}(x) = 0$$

for $j = 0, 1, \ldots, n-1$ and $m = 1, 2, \ldots, n-1$, where

$$Q_{nmj}(x) = x^j \frac{d^m}{dx^m}[x^n \exp(-x)].$$

Note that, for $m = 1, 2, \ldots, n-1$, $Q_{nmj}(x)$ is always of the form

$$x^{(n-m+j)}R(x) \exp(-x)$$

where R is a polynomial of degree m. The result follows from the fact that $n - m + j \geq 1$. ∎

The following corollary simply restates the expression for L_n derived in the proof of Theorem 10.9.

Corollary 10.4. *For each* $n = 0, 1, \ldots$, *let* L_n *denote the nth Laguerre polynomial. Then*

$$L_n(x) = \sum_{j=0}^n (-1)^j \binom{n}{j}\frac{x^j}{j!}.$$

Hence,

$$L_1(x) = x - 1, \quad L_2(x) = \frac{1}{2}x^2 - 2x + 1, \quad L_3(x) = -\frac{1}{6}x^3 + \frac{3}{2}x^2 - 3x + 1.$$

The Laguerre polynomials, like the Hermite polynomials, are complete (Andrews et al. 1999, Chapter 6).

***Example* 10.10** (*Series expansion of a density*). In Example 10.9, a series expansion of a density function based on the Hermite polynomials was considered. The same approach may be used with the Laguerre polynomials.

Let p denote a density function on $(0, \infty)$ and let F denote the distribution function of the standard exponential distribution. Assume that

$$\int_0^\infty p(x)^2 \exp(x) \, dx < \infty.$$

Under this assumption

$$\int_0^\infty \frac{p(x)^2}{\exp(-x)^2} \, dF(x) < \infty$$

so that the function $p(x)/\exp(-x)$ has an expansion of the form

$$\frac{p(x)}{\exp(-x)} = \sum_{j=0}^\infty \alpha_j L_j(x)$$

where the constants $\alpha_0, \alpha_1, \ldots$ are given by

$$\alpha_j = \int_0^\infty L_j(x) p(x) \exp(-x) \, dx;$$

note that

$$\int_0^\infty L_j(x)^2 \exp(-x) \, dx = 1$$

and that $\alpha_0 = 1$.

Hence, the function p has an expansion of the form

$$p(x) = \exp(-x) \left[1 + \sum_{j=1}^\infty \alpha_j L_j(x) \right]. \qquad \square$$

10.4 Gaussian Quadrature

One important application of orthogonal polynomials is in the development of methods of numerical integration. Here we give only a brief description of this area; see Section 10.6 for references to more detailed discussions.

Consider the problem of calculating the integral

$$\int_a^b f(x) \, dF(x)$$

where F is a distribution function on $[a, b]$ and $-\infty \le a < b \le \infty$.

Let $\{p_0, p_1, \ldots\}$ denote orthogonal polynomials with respect to F. Fix n and let $a < x_1 < x_2 < \cdots < x_n < b$ denote the zeros of p_n. Then, by Theorem 10.5, there exist constants

$\lambda_1, \lambda_2, \ldots, \lambda_n$ such that if f is a polynomial of degree $2n - 1$ or less,

$$\int_a^b f(x)\,dF(x) = \sum_{j=1}^n \lambda_j f(x_j).$$

Now suppose that f is not a polynomial, but may be approximated by a polynomial. Write

$$f(x) = \sum_{j=0}^{2n-1} \beta_j x^j + R_{2n}(x)$$

where $\beta_0, \ldots, \beta_{2n-1}$ are given constants and R_{2n} is a remainder term. Then

$$\int_a^b f(x)\,dF(x) = \int_a^b \hat{f}_{2n-1}(x)\,dF(x) + \int_a^b R_{2n}(x)\,dF(x)$$

where

$$\hat{f}_{2n-1}(x) = \sum_{j=0}^{2n-1} \beta_j x^j.$$

Since $\hat{f}_{2n-1}$ is a polynomial of degree $2n - 1$, it may be integrated exactly:

$$\int_a^b \hat{f}_{2n-1}(x)\,dF(x) = \sum_{j=1}^n \lambda_j \hat{f}_{2n-1}(x_j).$$

The function $\hat{f}_{2n-1}$ may be approximated by f so that

$$\int_a^b \hat{f}_{2n-1}(x)\,dF(x) \doteq \sum_{j=1}^n \lambda_j f(x_j)$$

and, provided that

$$\int_a^b R_{2n}(x)\,dF(x)$$

is small,

$$\int_a^b f(x)\,dF(x) \doteq \sum_{j=1}^n \lambda_j f(x_j).$$

This approach to numerical integration is known as *Gaussian quadrature*. Clearly, this method will work well whenever the function f being integrated can be approximated accurately by a polynomial over the range of integration.

In order to use Gaussian quadrature, we need to know the zeros of an orthogonal polynomial p_n, along with the corresponding weights $\lambda_1, \ldots, \lambda_n$. There are many published sources containing this information, as well as computer programs for this purpose; see Section 10.6 for further details.

Example 10.11. Let X denote a random variable with a standard exponential distribution and consider computation of $E[g(X)]$ for various functions g. Since

$$E[g(X)] = \int_0^\infty g(x) \exp(-x)\,dx,$$

we may use Gaussian quadrature based on the Laguerre polynomials. For illustration, consider the case $n = 5$. The zeros of L_5 and the corresponding weights $\lambda_0, \ldots, \lambda_5$ are available, for example, in Abramowitz and Stegun (1964, Table 25.9). Using these values, an approximation to $E[g(X)]$ is, roughly,

$$.5218g(.26) + .3987g(1.41) + .0759g(3.60) + .0036g(7.09) + .00002g(12.64);$$

for the numerical calculations described below, 10 significant figures were used.

Recall that this approximation is exact whenever g is a polynomial of degree 9 or less. In general, the accuracy of the approximation will depend on how well g may be approximated by a 9-degree polynomial over the interval $(0, \infty)$. For instance, suppose $g(x) = \sin(x)$; then $E[g(X)] = 1/2$ while the approximation described above is 0.49890, a relative error of roughly 0.2%. However, if $g(x) = x^{-\frac{1}{2}}$, then $E[g(X] = \sqrt{\pi}/2$ and the approximation is 1.39305, an error of roughly 21%. $\square$

10.5 Exercises

10.1 Prove Lemma 10.1.

10.2 Let F denote the distribution function of the absolutely continuous distribution density function

$$2x \exp\{-x^2\}, \quad x > 0.$$

Find the first three orthogonal polynomials with respect to F.

10.3 Let F denote the distribution function of the discrete distribution with frequency function

$$\frac{1}{2^{x+1}}, \quad x = 0, 1, \ldots.$$

Find the first three orthogonal polynomials with respect to F.

10.4 Let F denote a distribution function on **R** such that all moments of the distribution exist and the distribution is symmetric about 0, in the sense that

$$F(x) + F(-x) = 1, \quad -\infty < x < \infty.$$

Let $p_0, p_1, \ldots$ denote orthogonal polynomials with respect to F. Show that the orthogonal polynomials of even order include only even powers of x and that the orthogonal polynomials of odd order include only odd powers of x.

10.5 Let $\{p_0, p_1, \ldots\}$ denote orthogonal polynomials with respect to a distribution function F and suppose that the polynomials are standardized so that

$$\int_{-\infty}^{\infty} p_j(x)^2 \, dF(x) = 1, \quad j = 0, 1, \ldots.$$

Fix $n = 0, 1, \ldots$ and define a function $K_n : \mathbf{R} \times \mathbf{R} \mapsto \mathbf{R}$ by

$$K_n(x, y) = \sum_{j=0}^{n} p_j(x)p_j(y).$$

Show that, for any polynomial q of degree n or less,

$$q(x) = \int_{-\infty}^{\infty} K_n(x, y)q(y) \, dF(y).$$

10.6   Let $\{p_0, p_1, \ldots\}$ denote orthogonal polynomials with respect to a distribution function F and suppose that the polynomials are standardized so that

$$\int_{-\infty}^{\infty} p_j(x)^2 \, dF(x) = 1, \quad j = 0, 1, \ldots.$$

Fix $n = 0, 1, \ldots$ and let K_n denote the function defined in Exercise 10.5.
Show that for any polynomial g of degree n or less and any $z \in \mathbf{R}$,

$$g(z)^2 \leq K_n(z, z) \int_{-\infty}^{\infty} g(x)^2 \, dF(x).$$

10.7   Let $\{p_0, p_1, \ldots\}$ denote orthogonal polynomials with respect to a distribution function F and suppose that the polynomials are standardized so that the coefficient of x^n in $p_n(x)$ is 1. Let $\mathcal{G}_n$ denote the set of all polynomials of degree n with coefficient of x^n equal to 1. Find the function $g \in \mathcal{G}_n$ that minimizes

$$\int_{-\infty}^{\infty} g(x)^2 \, dF(x).$$

10.8   Let $\{p_0, p_1, \ldots\}$ denote orthogonal polynomials with respect to a distribution function F and suppose that the polynomials are standardized so that the coefficient of x^n in $p_n(x)$ is 1. For each $n = 1, 2, \ldots$ let β_n denote the coefficient of x^{n-1} in $p_n(x)$ and let

$$h_n = \int_{-\infty}^{\infty} p_n(x)^2 \, dF(x).$$

Show that $p_0, p_1, \ldots$ satisfy the three-term recurrence relationship

$$p_{n+1}(x) = (x + \beta_{n+1} - \beta_n) p_n(x) - \frac{h_n}{h_{n-1}} p_{n-1}(x).$$

10.9   Let $\{p_0, p_1, \ldots\}$ denote orthogonal polynomials with respect to a distribution function F. Show that, for all $m, n = 0, 1, \ldots,$

$$p_m(x) p_n(x) = \sum_{j=0}^{n+m} a(j, m, n) p_j(x)$$

where the constants $a(0, m, n), a(1, m, n), \ldots$ are given by

$$a(j, m, n) = \frac{\int_{-\infty}^{\infty} p_m(x) p_n(x) p_j(x) \, dF(x)}{\int_{-\infty}^{\infty} p_j(x)^2 \, dF(x)}.$$

10.10   Prove Corollary 10.1.

10.11   Prove Corollary 10.2.

10.12   Prove Corollary 10.3.

10.13   Find the three-term recurrence relationship (see Exercise 10.8) for the Hermite polynomials.

10.14   Some authors define the Hermite polynomials to be polynomials orthogonal with respect to the absolutely continuous distribution with density function

$$c \, \exp\left(-\frac{1}{2} x^2\right), \quad -\infty < x < \infty,$$

where c is a constant. Let $\bar{H}_0, \bar{H}_1, \ldots$ denote orthogonal polynomials with respect to this distribution, standardized so that the coefficient of x^n in $\bar{H}_n(x)$ is 2^n. Show that

$$\bar{H}_n(x) = 2^{\frac{n}{2}} H_n(x\sqrt{2}), \quad n = 0, 1, \ldots.$$

10.15 Show that the Hermite polynomials satisfy

$$\sum_{n=0}^{\infty} \frac{H_n(x)}{n!} z^n = \exp(xz - z^2/2), \quad x \in \mathbf{R}, \ z \in \mathbf{R}.$$

Using this result, give a relationship between $H_n(0)$, $n = 0, 1, \ldots$, and the moments of the standard normal distribution.

10.16 Let L_n denote the nth Laguerre polynomial. Show that

$$\int_0^{\infty} L_n(x)^2 \exp(-x) \, dx = 1.$$

10.17 Let p_0, p_1, p_2 denote the orthogonal polynomials found in Exercise 10.3. Find the zeros x_1, x_2 of p_2 and the constants λ_1, λ_2 such that

$$\sum_{j=0}^{\infty} f(j) \frac{1}{2^{j+1}} = \lambda_1 f(x_1) + \lambda_2 f(x_2)$$

for all polynomials f of degree 3 or less.

10.6 Suggestions for Further Reading

Orthogonal polynomials are a classical topic in mathematics. Standard references include Freud (1971), Jackson (1941), and Szegö (1975); see also Andrews et al. (1999, Chapters 5–7) and Temme (1996, Chapter 6). Many useful properties of the classical orthogonal polynomials are given in Erdélyi (1953b, Chapter X).

Gaussian quadrature is discussed in many books covering numerical integration; see, for example, Davis and Rabinowitz (1984). Evans and Swartz (2000, Chapter 5) and Thisted (1988, Chapter 5) discuss Gaussian quadrature with particular emphasis on statistical applications. Tables listing the constants needed to implement these methods are available in Abramowitz and Stegun (1964) and Stroud and Secrest (1966).

11

Approximation of Probability Distributions

11.1 Introduction

Consider a random vector Y, with a known distribution, and suppose that the distribution of $f(Y)$ is needed, for some given real-valued function $f(\cdot)$. In Chapter 7, general approaches to determining the distribution of $f(Y)$ were discussed. However, in many cases, carrying out the methods described in Chapter 7 is impractical or impossible. In these cases, an alternative approach is to use an asymptotic approximation to the distribution of the statistic under consideration. This approach allows us to approximate distributional quantities, such as probabilities or moments, in cases in which exact computation is not possible. In addition, the approximations, in contrast to exact results, take a few basic forms and, hence, they give insight into the structure of distribution theory. Asymptotic approximations also play a fundamental role in statistics.

Such approximations are based on the concept of *convergence in distribution*. Let $X_1, X_2, \ldots$ denote a sequence of real-valued random variables and let F_n denote the distribution function of X_n, $n = 1, 2, \ldots$. Let X denote a real-valued random variable with distribution function F. If

$$\lim_{n \to \infty} F_n(x) = F(x)$$

for each x at which F is continuous, then the sequence $X_1, X_2, \ldots$ is said to *converge in distribution* to X as $n \to \infty$, written

$$X_n \xrightarrow{\mathcal{D}} X \quad \text{as} \quad n \to \infty.$$

In this case, probabilities regarding X_n may be approximated using probabilities based on X; that is, the limiting distribution function F may then be used as an approximation to the distribution functions in the sequence $F_1, F_2, \ldots$. The property that $X_n \xrightarrow{\mathcal{D}} X$ as $n \to \infty$ is simply the property that the approximation error decreases to 0 as n increases to ∞. The distribution of X is sometimes called the *asymptotic distribution* of the sequence $X_1, X_2, \ldots$.

Example 11.1 *(Sequence of Bernoulli random variables).* For each $n = 1, 2, \ldots$, let X_n denote a random variable such that

$$\Pr(X_n = 1) = 1 - \Pr(X_n = 0) = \theta_n;$$

322

here $\theta_1, \theta_2, \ldots$ is a sequence of real numbers in the interval $(0, 1)$. Let F_n denote the distribution function of X_n. Then

$$F_n(x) = \begin{cases} 0 & \text{if } x < 0 \\ \theta_n & \text{if } 0 \le x < 1. \\ 1 & \text{if } x \ge 1 \end{cases}$$

Suppose that the sequence $\theta_n, n = 1, 2, \ldots,$ converges and $\boxed{\text{let } \theta = \lim_{n \to \infty} \theta_n.}$ Then

$$\lim_{n \to \infty} F_n(x) = F(x) \equiv \begin{cases} 0 & \text{if } x < 0 \\ \theta & \text{if } 0 \le x < 1. \\ 1 & \text{if } x \ge 1 \end{cases}$$

Let X denote a random variable such that

$$\Pr(X = 1) = 1 - \Pr(X = 0) = \theta.$$

Then $X_n \overset{D}{\to} X$ as $n \to \infty$.

If the sequence $\theta_n, n = 1, 2, \ldots,$ does not have a limit, then $X_n, n = 1, 2, \ldots$ does not converge in distribution. $\square$

An important property of the definition of convergence in distribution is that we require that the sequence $F_n(x), n = 1, 2, \ldots,$ converges to $F(x)$ only for those x that are continuity points of F. Hence, if F is not continuous at x_0, then the behavior of the sequence $F_n(x_0)$, $n = 1, 2, \ldots,$ plays no role in convergence in distribution. The reason for this is that requiring that $F_n(x), n = 1, 2, \ldots,$ converges to $F(x)$ at points at which F is discontinuous is too strong of a requirement; this is illustrated in the following example.

Example 11.2 (Convergence of a sequence of degenerate random variables). Let $X_1, X_2, \ldots$ denote a sequence of random variables such that

$$\Pr(X_n = 1/n) = 1, n = 1, 2, \ldots.$$

Hence, when viewed as a deterministic sequence, $X_1, X_2, \ldots$ has limit 0.

Let F_n denote the distribution function of X_n. Then

$$F_n(x) = \begin{cases} 0 & \text{if } x < 1/n \\ 1 & \text{if } x \ge 1/n \end{cases}.$$

Not continuous

Fix x. Clearly,

$$\lim_{n \to \infty} F_n(x) = \begin{cases} 0 & \text{if } x < 0 \\ 1 & \text{if } x > 0 \end{cases}.$$

Consider the behavior of the sequence $F_n(0), n = 1, 2, \ldots.$ Since $0 < 1/n$ for every $n = 1, 2, \ldots,$ it follows that

$$\lim_{n \to \infty} F_n(0) = 0$$

so that

Not Right continuous Left Limit (RCLL).

$$\lim_{n \to \infty} F_n(x) = G(x) \equiv \begin{cases} 0 & \text{if } x \le 0 \\ 1 & \text{if } x > 0 \end{cases}.$$

Note that, since G is not right-continuous, it is not the distribution function of any random variable.

Of course, we expect that $X_1, X_2, \ldots$ converges in distribution to a random variable identically equal to 0; such a random variable has distribution function

$$F(x) = \begin{cases} 0 & \text{if } x < 0 \\ 1 & \text{if } x \geq 0 \end{cases}.$$

Thus, $\lim_{n \to \infty} F_n(x) = F(x)$ at all $x \neq 0$, that is, at all x at which F is continuous. Hence, $X_n \overset{D}{\to} 0$ as $n \to \infty$, where 0 may be viewed as the random variable equal to 0 with probability 1. However, if we require convergence of $F_n(x)$ for all x, X_n would not have a limiting distribution. $\square$

Often the random variables under consideration will require some type of standardization in order to obtain a useful convergence result.

Example 11.3 (*Minimum of uniform random variables*). Let $Y_1, Y_2, \ldots$ denote a sequence of independent, identically distributed random variables, each with a uniform distribution on $(0, 1)$. Suppose we are interested in approximating the distribution of $T_n = \min(Y_1, \ldots, Y_n)$. For each $n = 1, 2, \ldots$, T_n has distribution function

$$H_n(t) = \Pr(T_n \leq t) = \Pr\{\min(Y_1, \ldots, Y_n) \leq t\}$$
$$= \begin{cases} 0 & \text{if } t < 0 \\ 1 - (1 - t)^n & \text{if } 0 \leq t < 1. \\ 1 & \text{if } t \geq 1 \end{cases}$$

Fix t. Then

$$\lim_{n \to \infty} H_n(t) = \begin{cases} 0 & \text{if } t \leq 0 \\ 1 & \text{if } t > 0 \end{cases}$$

so that, as $n \to \infty$, T_n converges in distribution to the random variable identically equal to 0. Hence, for any $t > 0$, $\Pr\{\min(Y_1, \ldots, Y_n) \leq t\}$ can be approximated by 1. Clearly, this approximation will not be very useful, or very accurate.

Now consider standardization of T_n. For each $n = 1, 2, \ldots$, let $X_n = n^\alpha T_n \equiv n \min(Y_1, \ldots, Y_n)$ where α is a given constant. Then, for each $n = 1, 2, \ldots$, X_n has distribution function

$$F_n(x; \alpha) = \Pr(X_n \leq x) = \Pr\{\min(Y_1, \ldots, Y_n) \leq x/n^\alpha\}$$
$$= \begin{cases} 0 & \text{if } x < 0 \\ 1 - (1 - x/n^\alpha)^n & \text{if } 0 \leq x < n^\alpha. \\ 1 & \text{if } x \geq n^\alpha \end{cases}$$

Fix $x > 0$. Then

$$\lim_{n \to \infty} F_n(x; \alpha) = \begin{cases} 1 & \text{if } \alpha < 1 \\ 1 - \exp(-x) & \text{if } \alpha = 1. \\ 0 & \text{if } \alpha > 1 \end{cases}$$

Thus, if $\alpha < 1$, X_n converges in distribution to the degenerate random variable 0, while if $\alpha > 1$, X_n does not converge in distribution. However, if $\alpha = 1$,

$$\lim_{n \to \infty} F_n(x; \alpha) = \begin{cases} 0 & \text{if } x < 0 \\ 1 - \exp(-x) & \text{if } 0 \leq x < \infty, \end{cases}$$

which is the distribution function of a standard exponential distribution. Hence, if X is a random variable with a standard exponential distribution, then $n \min(Y_1, \ldots, Y_n) \overset{D}{\to} X$ as $n \to \infty$.

For instance, an approximation to

$$\Pr\{\min(Y_1, \ldots, Y_{10}) \le 1/10\} = \Pr\{10 \min(Y_1, \ldots, Y_{10}) \le 1\}$$

is given by $1 - \exp(-1) \doteq 0.632$; the exact probability is 0.651.   □

The examples given above all have an important feature; in each case the distribution functions $F_1, F_2, \ldots$ are available. In this sense, the examples are not typical of those in which convergence in distribution plays an important role. The usefulness of convergence in distribution lies in the fact that the limiting distribution function may be determined in cases in which the $F_n, n = 1, 2, \ldots$, are not available. Many examples of this type are given in Chapters 12 and 13.

11.2   Basic Properties of Convergence in Distribution

Recall that there are several different ways in order to characterize the distribution of a random variable. For instance, if two random variables X and Y have the same characteristic function, or if $E[f(X)] = E[f(Y)]$ for all bounded, continuous, real-valued functions f, then X and Y have the same distribution; see Corollary 3.1 and Theorem 1.11, respectively, for formal statements of these results.

The results below show that these characterizations of a distribution can also be used to characterize convergence in distribution. That is, convergence in distribution is equivalent to convergence of expected values of bounded, continuous functions and is also equivalent to convergence of characterstic functions. We first consider expectation.

Theorem 11.1. *Let $X_1, X_2, \ldots$ denote a sequence of real-valued random variables and let X denote a real-valued random variable. Let $\mathcal{X}$ denote a set such that $\Pr(X_n \in \mathcal{X}) = 1, n = 1, 2, \ldots$ and $\Pr(X \in \mathcal{X}) = 1$.*

$$X_n \overset{D}{\to} X \quad as \quad n \to \infty$$

if and only if

$$E[f(X_n)] \to E[f(X)] \quad as \quad n \to \infty$$

for all bounded, continuous, real-valued functions f on $\mathcal{X}$.

Proof.   Suppose that $X_n \overset{D}{\to} X$ as $n \to \infty$ and let F denote the distribution function of X. In order to show that

$$E[f(X_n)] \to E[f(X)] \quad as \quad n \to \infty$$

for all bounded, continuous, real-valued functions f, we consider two cases. In case 1, the random variables $X, X_1, X_2, \ldots$ are bounded; case 2 removes this restriction.

Case 1: Suppose that there exists a constant M such that, with probability 1,

$$|X_n| \leq M, \quad n = 1, 2, \ldots$$

and $|X| \leq M$. We may assume, without loss of generality, that M is a continuity point of F.

Consider a bounded, continuous, function $f : \mathbf{R} \to \mathbf{R}$ and let $\epsilon > 0$. Let $x_1, x_2, \ldots, x_m$ denote continuity points of F such that

$$-M = x_0 < x_1 < \cdots < x_{m-1} < x_m < x_{m+1} = M$$

and

$$\max_{1 \leq i \leq m} \sup_{x_i \leq x \leq x_{i+1}} |f(x) - f(x_i)| \leq \epsilon.$$

Define

$$f_m(x) = f(x_i) \quad \text{for} \quad x_i \leq x < x_{i+1}.$$

Then, as $n \to \infty$,

$$\int_{-M}^{M} f_m(x) \, dF_n(x) = \sum_{i=1}^{m} f(x_i)[F_n(x_{i+1}) - F_n(x_i)]$$

$$\to \sum_{i=1}^{m} f(x_i)[F(x_{i+1}) - F(x_i)]$$

$$= \int_{-M}^{M} f_m(x) \, dF(x).$$

Hence, for sufficiently large n,

$$\left| \int_{-M}^{M} f_m(x)[dF_n(x) - dF(x)] \right| \leq \epsilon.$$

It follows that

$$\left| \int_{-M}^{M} f(x)[dF_n(x) - dF(x)] \right|$$

$$\leq \left| \int_{-M}^{M} [f(x) - f_m(x)][dF_n(x) - dF(x)] \right| + \left| \int_{-M}^{M} f_m(x)[dF_n(x) - dF(x)] \right|$$

$$\leq 3\epsilon.$$

Since ϵ is arbitrary, it follows that

$$\lim_{n \to \infty} E[f(X_n)] = E[f(X)].$$

Case 2: For the general case in which the $X_1, X_2, \ldots$ and X are not necessarily bounded, let $0 < \epsilon < 1$ be arbitrary and let M and $-M$ denote continuity points of F such that

$$\int_{-M}^{M} dF(x) = F(M) - F(-M) \geq 1 - \epsilon.$$

Since

$$F_n(M) - F_n(-M) \to F(M) - F(-M) \quad \text{as} \quad n \to \infty,$$

it follows that

$$\int_{-M}^{M} d F_n(x) = F_n(M) - F_n(-M) \geq 1 - 2\epsilon$$

for sufficiently large n. Hence,

$$\left| \int_{-\infty}^{\infty} f(x)[d F_n(x) - d F(x)] \right| \leq \left| \int_{-M}^{M} f(x)[d F_n(x) - d F(x)] \right| + \sup_x |f(x)| 3\epsilon.$$

The same argument used in case 1 can be used to shown that

$$\left| \int_{-M}^{M} f(x)[d F_n(x) - d F(x)] \right| \to 0 \quad \text{as } n \to \infty;$$

since ϵ is arbitrary, it follows that

$$\left| \int_{-\infty}^{\infty} f(x)[d F_n(x) - d F(x)] \right| \to 0$$

as $n \to \infty$, proving the first part of the theorem.

Now suppose that $E[f(X_n)]$ converges to $E[f(X)]$ for all real-valued, bounded, continuous f. Define

$$h(x) = \begin{cases} 1 & \text{if } x < 0 \\ 1 - x & \text{if } 0 \leq x \leq 1 \\ 0 & \text{if } x > 1 \end{cases}$$

and for any $t > 0$ define $h_t(x) = h(tx)$.

Note that, for fixed t, h_t is a real-valued, bounded, continuous function. Hence, for all $t > 0$,

$$\lim_{n \to \infty} E[h_t(X_n)] = E[h_t(X)].$$

For fixed x,

$$I_{\{u \leq x\}} \leq h_t(u - x) \leq I_{\{u \leq x + 1/t\}}$$

for all u, t. Hence,

$$F_n(x) \leq \int_{-\infty}^{\infty} h_t(u - x) d F_n(u) = E[h_t(X_n - x)]$$

and, for any value of $t > 0$,

$$\limsup_{n \to \infty} F_n(x) \leq \lim_{n \to \infty} E[h_t(X_n - x)] = E[h_t(X - x)] \leq F(x + 1/t).$$

It follows that, if F is continuous at x,

$$\limsup_{n \to \infty} F_n(x) \leq F(x). \tag{11.1}$$

Similarly, for fixed x,

$$I_{\{u \leq x - 1/t\}} \leq h_t(u - x + 1/t) \leq I_{\{u \leq x\}}$$

for all $u, t > 0$. Hence,

$$F_n(x) \geq \int_{-\infty}^{\infty} h_t(u - x + 1/t) d F_n(u) = E[h_t(X_n - x + 1/t)]$$

and, for any value of $t > 0$,

$$\liminf_{n \to \infty} F_n(x) \geq \lim_{n \to \infty} E[h_t(X_n - x + 1/t)] = E[h_t(X_n - x + 1/t)] \geq F(x + 1/t).$$

It follows that

$$\liminf_{n \to \infty} F_n(x) \geq F(x - 1/t), \quad t > 0$$

and, hence, that

$$\liminf_{n \to \infty} F_n(x) \geq F(x) \tag{11.2}$$

provided that F is continuous at x.

Combining (11.1) and (11.2), it follows that

$$\lim_{n \to \infty} F_n(x) = F(x)$$

at all continuity points x of F, proving the theorem. ■

It can be shown that the function h_t used in the proof of Theorem 11.1 is not only continuous, it is uniformly continuous. Hence, $X_n \overset{\mathcal{D}}{\to} X$ as $n \to \infty$ provided only that

$$E[f(X_n)] \to E[f(X)] \quad \text{as} \quad n \to \infty$$

for all bounded, uniformly continuous, real-valued functions f. Since the class of all bounded, uniformly continuous functions is smaller than the class of all bounded, continuous functions, this gives a slightly weaker condition for convergence in distribution that is sometimes useful. The details of the argument are given in following corollary.

Corollary 11.1. *Let $X_1, X_2, \ldots$ denote a sequence of real-valued random variables and let X denote a real-valued random variable. If*

$$E[f(X_n)] \to E[f(X)] \quad as \quad n \to \infty$$

for all bounded, uniformly continuous, real-valued functions f, then

$$X_n \overset{\mathcal{D}}{\to} X \quad as \quad n \to \infty.$$

Proof. Suppose that $E[f(X_n)]$ converges to $E[f(X)]$ for all real-valued, bounded, uniformly continuous f. As in the proof of Theorem 11.1, define

$$h(x) = \begin{cases} 1 & \text{if } x < 0 \\ 1 - x & \text{if } 0 \leq x \leq 1 \\ 0 & \text{if } x > 1 \end{cases}$$

and for any $t > 0$ define $h_t(x) = h(tx)$.

The function h_t is uniformly continuous. To see this, note that

$$|h(x_1) - h(x_2)| = \begin{cases} |x_1 - x_2| & \text{if } 0 \leq x_1 \leq 1 \text{ and } 0 \leq x_2 \leq 1 \\ 1 & \text{if } \min(x_1, x_2) < 0 \text{ and } \max(x_1, x_2) > 1 \\ \max(x_1, x_2) & \text{if } \min(x_1, x_2) < 0 \text{ and } 0 < \max(x_1, x_2) \leq 1 \\ 0 & \text{otherwise} \end{cases}.$$

Hence, for all x_1, x_2,

$$|h(x_1) - h(x_2)| \leq |x_1 - x_2|$$

so that, for a given value of t and a given $\epsilon > 0$,

$$|h_t(x_1) - h_t(x_2)| \leq \epsilon$$

whenever

$$|x_1 - x_2| \leq \epsilon/t.$$

It follows that h_t is uniformly continuous and, hence, for all $t > 0$,

$$\lim_{n \to \infty} \mathrm{E}[h_t(X_n)] = \mathrm{E}[h_t(X)].$$

The proof of the corollary now follows as in the proof of Theorem 11.1. ∎

The requirements in Theorem 11.1 that f is bounded and continuous are crucial for the conclusion of the theorem. The following examples illustrate that convergence of the expected values need not hold if these conditions are not satisfied.

Example 11.4 (*Convergence of a sequence of degenerate random variables*). As in Example 11.2, let $X_1, X_2, \ldots$ denote a sequence of random variables such that

$$\Pr(X_n = 1/n) = 1, \quad n = 1, 2, \ldots.$$

We have seen that $X_n \overset{D}{\to} 0$ as $n \to \infty$.
  Let

$$f(x) = \begin{cases} 0 & \text{if } x \leq 0 \\ 1 & \text{if } x > 0 \end{cases};$$

note that f is bounded, but it is discontinuous at $x = 0$. It is easy to see that $\mathrm{E}[f(X_n)] = 1$ for all $n = 1, 2, \ldots$ so that $\lim_{n \to \infty} \mathrm{E}[f(X_n)] = 1$; however, $\mathrm{E}[f(0)] = 0$. □

Example 11.5 (*Pareto distribution*). For each $n = 1, 2, \ldots$, suppose that X_n is a real-valued random variable with an absolutely continuous distribution with density function

$$p_n(x) = \frac{1}{n(1+n)^{\frac{1}{n}}} \frac{1}{x^{2+\frac{1}{n}}}, \quad x > \frac{1}{n+1}.$$

Let F_n denote the distribution function of X_n; then

$$F_n(x) = \begin{cases} 0 & \text{if } x < \frac{1}{1+n} \\ 1 - [(n+1)x]^{-(1+\frac{1}{n})} & \text{if } \frac{1}{n+1} \leq x < \infty. \end{cases}$$

Hence,

$$\lim_{n \to \infty} F_n(x) = \begin{cases} 0 & \text{if } x \leq 0 \\ 1 & \text{if } x > 0 \end{cases}$$

so that $X_n \overset{D}{\to} 0$ as $n \to \infty$.

Consider the function $f(x) = x$, which is continuous, but unbounded. It is straight forward to show that

$$E[f(X_n)] = \frac{1}{n(1+n)^{\frac{1}{n}}} \int_{\frac{1}{n+1}}^{\infty} \frac{1}{x^{1+\frac{1}{n}}} \, dx = 1, \quad n = 1, 2, \ldots$$

while $E[f(0)] = 0$. $\square$

A useful consequence of Theorem 11.1 is the result that convergence in distribution is preserved under continuous transformations. This result is stated in the following corollary; the proof is left as an exercise.

Corollary 11.2. *Let* $X, X_1, X_2, \ldots$ *denote real-valued random variables such that*

$$X_n \overset{D}{\to} X \quad as \ n \to \infty.$$

Let $f : \mathcal{X} \to \mathbf{R}$ *denote a continuous function, where* $\mathcal{X} \subset \mathbf{R}$ *satisfies* $\Pr[X_n \in \mathcal{X}] = 1$, $n = 1, 2, \ldots$ *and* $\Pr(X \in \mathcal{X}) = 1$. *Then*

$$f(X_n) \overset{D}{\to} f(X) \quad as \ n \to \infty.$$

***Example* 11.6 (*Minimum of uniform random variables*).** As in Example 11.3, let $Y_1, Y_2, \ldots$ denote a sequence of independent identically distributed, each with a uniform distribution on $(0, 1)$ and let $X_n = n \min(Y_1, \ldots, Y_n)$, $n = 1, 2, \ldots$. In Example 11.3, it was shown that that $X_n \overset{D}{\to} X$ as $n \to \infty$, where X is a random variable with a standard exponential distribution.

Let $W_n = \exp(X_n)$, $n = 1, 2, \ldots$. Since $\exp(\cdot)$ is a continuous function, it follows from Corollary 11.2 that $W_n \overset{D}{\to} W$ as $n \to \infty$, where $W = \exp(X)$. It is straightforward to show that W has an absolutely continuous distribution with density

$$\frac{1}{w^2}, \quad w \geq 1. \qquad \qquad \square$$

An important result is that convergence in distribution may be characterized in terms of convergence of characteristic functions. The usefulness of this result is due to the fact that the characteristic function of a sum of independent random variables is easily determined from the characteristic functions of the random variables making up the sum. This approach is illustrated in detail in Chapter 12.

Theorem 11.2. *Let* $X_1, X_2, \ldots$ *denote a sequence of real-valued random variables and let* X *denote a real-valued random variable. For each* $n = 1, 2, \ldots$, *let* φ_n *denote the characteristic function of* X_n *and let* φ *denote the characteristic function of* X. *Then*

$$X_n \overset{D}{\to} X \quad as \ n \to \infty$$

if and only if

$$\lim_{n \to \infty} \varphi_n(t) = \varphi(t), \quad for \ all \ \ -\infty < t < \infty.$$

Proof.   Applying Theorem 11.1 to the real and imaginary parts of $\exp\{itx\}$ shows that $X_n \xrightarrow{\mathcal{D}} X$ implies that $\varphi_n(t) \to \varphi(t)$ for all t.

Hence, suppose that

$$\lim_{n \to \infty} \varphi_n(t) = \varphi(t), \quad \text{for all} \quad -\infty < t < \infty.$$

Let g denote an arbitrary bounded uniformly continuous function such that

$$|g(x)| \leq M, \quad -\infty < x < \infty.$$

If it can be shown that

$$\lim_{n \to \infty} \mathrm{E}[g(X_n)] = \mathrm{E}[g(X)] \quad \text{as} \quad n \to \infty,$$

then, by Corollary 11.1, $X_n \xrightarrow{\mathcal{D}} X$ as $n \to \infty$ and the theorem follows.

Given $\epsilon > 0$, choose δ so that

$$\sup_{x,y:|x-y|<\delta} |g(x) - g(y)| < \epsilon.$$

Let Z denote a standard normal random variable, independent of $X, X_1, X_2, \ldots$, and consider $|g(X_n + Z/t) - g(X_n)|$ where $t > 0$. Whenever $|Z|/t$ is less than δ,

$$|g(X_n + Z/t) - g(X_n)| < \epsilon;$$

whenever $|Z|/t \geq \epsilon$, we still have

$$|g(X_n + Z/t) - g(X_n)| \leq |g(X_n + Z/t)| + |g(X_n)| \leq 2M.$$

Hence, for any $t > 0$,

$$\mathrm{E}[|g(X_n + Z/t) - g(X_n)|] \leq \epsilon \Pr(|Z| < t\delta) + 2M\Pr(|Z| > t\delta).$$

For sufficiently large t,

$$\Pr(|Z| > t\delta) \leq \frac{\epsilon}{2M}$$

so that

$$\mathrm{E}[|g(X_n + Z/t) - g(X_n)|] \leq 2\epsilon.$$

Similarly, for sufficiently large t,

$$\mathrm{E}[|g(X + Z/t) - g(X)|] \leq 2\epsilon$$

so that

$$\begin{aligned}
\mathrm{E}[|g(X_n) - g(X)|] &\leq \mathrm{E}[|g(X_n) - g(X_n + Z/t)|] + \mathrm{E}[|g(X_n + Z/t) - g(X + Z/t)|] \\
&\quad + \mathrm{E}[|g(X + Z/t) - g(X)|] \leq 4\epsilon + \mathrm{E}[|g(X_n + Z/t) - g(X + Z/t)|].
\end{aligned}$$

Recall that, by Example 3.2,

$$\varphi_Z(t) = (2\pi)^{\frac{1}{2}} \phi(t),$$

where ϕ denotes the standard normal density.

Hence,

$$
\begin{aligned}
\mathrm{E}[g(X_n + Z/t)] &= \int_{-\infty}^{\infty} \int_{-\infty}^{\infty} g(x + z/t)\phi(z)\,dz\,dF_n(x) \\
&= \frac{1}{(2\pi)^{\frac{1}{2}}} \int_{-\infty}^{\infty} \int_{-\infty}^{\infty} g(x + z/t)\varphi(z)\,dz\,dF_n(x) \\
&= \frac{1}{(2\pi)^{\frac{1}{2}}} \int_{-\infty}^{\infty} \int_{-\infty}^{\infty} g(x + z/t)\mathrm{E}[\exp\{izZ\}]\,dz\,dF_n(x) \\
&= \frac{1}{(2\pi)^{\frac{1}{2}}} \int_{-\infty}^{\infty} \int_{-\infty}^{\infty} \int_{-\infty}^{\infty} g(x + z/t)\exp\{izy\}\phi(y)\,dy\,dz\,dF_n(x).
\end{aligned}
$$

Consider the change-of-variable $u = x + z/t$. Then

$$
\begin{aligned}
\mathrm{E}[g(X_n + Z/t)] &= \frac{1}{t(2\pi)^{\frac{1}{2}}} \int_{-\infty}^{\infty} \int_{-\infty}^{\infty} \int_{-\infty}^{\infty} g(u)\exp\{iyt(u - x)\}\phi(y)\,dy\,du\,dF_n(x) \\
&= \frac{1}{t(2\pi)^{\frac{1}{2}}} \int_{-\infty}^{\infty} \int_{-\infty}^{\infty} g(u)\exp\{-iytu\}\phi(y)\int_{-\infty}^{\infty} \exp\{-iytx\}dF_n(x)\,dy\,du \\
&= \frac{1}{t(2\pi)^{\frac{1}{2}}} \int_{-\infty}^{\infty} \int_{-\infty}^{\infty} g(u)\exp\{-ityu\}\varphi(y)\varphi_n(-ty)\,dy\,du.
\end{aligned}
$$

Similarly,

$$
\mathrm{E}[g(X_n + Z/t)] = \frac{1}{t(2\pi)^{\frac{1}{2}}} \int_{-\infty}^{\infty} \int_{-\infty}^{\infty} g(u)\exp\{-ityu\}\varphi(y)\varphi(-ty)\,dy\,du.
$$

By assumption,

$$
\lim_{n\to\infty} \varphi_n(-ty) = \varphi(-ty) \quad \text{for all} \quad -\infty < y < \infty.
$$

Since $\varphi_n(-ty)$ is bounded, it follows from the dominated convergence theorem (considering the real and imaginary parts separately) that

$$
\lim_{n\to\infty} \mathrm{E}[g(X_n + Z/t)] = \mathrm{E}[g(X + Z/t)] \quad \text{for all} \quad t > 0.
$$

Hence, for sufficiently large t and n,

$$
|\mathrm{E}[g(X_n) - g(X)]| \le 5\epsilon.
$$

Since ϵ is arbitrary, the result follows. ∎

***Example* 11.7.** Let $X_1, X_2, \ldots$ denote a sequence of real-valued random variables such that $X_n \xrightarrow{D} X$ as $n \to \infty$ for some random variable X. Let Y denote a real-valued random variable such that, for each $n = 1, 2, \ldots$, X_n and Y are independent.

Let φ_n denote the characteristic function of X_n, $n = 1, 2, \ldots$, let φ_X denote the characteristic function of X, and let φ_Y denote the characteristic function of Y. Then $X_n + Y$ has characteristic function $\varphi_n(t)\varphi_Y(t)$. Since $X_n \xrightarrow{D} X$ as $n \to \infty$, it follows from Theorem 11.2 that, for each $t \in \mathbf{R}$, $\varphi_n(t) \to \varphi_X(t)$ as $n \to \infty$. Hence,

$$
\lim_{n\to\infty} \varphi_n(t)\varphi_Y(t) = \varphi_X(t)\varphi_Y(t), \quad t \in \mathbf{R}.
$$

It follows that from Theorem 11.2

$$X_n + Y \xrightarrow{D} X + Y_0 \quad \text{as} \quad n \to \infty,$$

where Y_0 denotes a random variable that is independent of X and has the same marginal distribution as Y.   □

***Example* 11.8 (*Normal approximation to the Poisson distribution*).** Let $Y_1, Y_2, \ldots$ denote a sequence of real-valued random variables such that, for each $n = 1, 2, \ldots, Y_n$ has a Poisson distribution with mean n and let

$$X_n = \frac{Y_n - n}{\sqrt{n}}, \quad n = 1, 2, \ldots.$$

Since the characteristic function of a Poisson distribution with mean λ is given by

$$\exp\{\lambda[\exp(it) - 1]\}, \quad -\infty < t < \infty,$$

it follows that the characteristic function of X_n is

$$\varphi_n(t) = \exp\{n \exp(it/\sqrt{n}) - n - \sqrt{n}it\}.$$

By Lemma A2.1 in Appendix 2,

$$\exp\{it\} = \sum_{j=0}^{n} \frac{(it)^j}{j!} + R_n(t)$$

where

$$|R_n(t)| \leq \min\{|t|^{n+1}/(n+1)!, \ 2|t|^n/n!\}.$$

Hence,

$$\exp(it/\sqrt{n}) = 1 + it/\sqrt{n} - \frac{1}{2}t^2/n + R_2(t)$$

where

$$|R_2(t)| \leq \frac{1}{6}t^3/n^{\frac{3}{2}}.$$

It follows that

$$\varphi_n(t) = \exp\{-t^2/2 + nR_2(t)\}$$

and that

$$\lim_{n \to \infty} nR_2(t) = 0, \quad -\infty < t < \infty.$$

Hence,

$$\lim_{n \to \infty} \varphi_n(t) = \exp(-t^2/2), \quad -\infty < t < \infty,$$

the characteristic function of the standard normal distribution.

Let Z denote a random variable with a standard normal distribution. Then, by Theorem 11.2, $X_n \xrightarrow{D} Z$ as $n \to \infty$.

Thus, probabilities of the form $\Pr(X_n \leq z)$ can be approximated by $\Pr(Z \leq z)$; these approximations have the property that the approximation error approaches 0 as $n \to \infty$.

Table 11.1. *Exact probabilities in Example 11.8.*

			α		
n	0.10	0.25	0.50	0.75	0.90
10	0.126	0.317	0.583	0.803	0.919
20	0.117	0.296	0.559	0.788	0.914
50	0.111	0.278	0.538	0.775	0.909
100	0.107	0.270	0.527	0.768	0.907
200	0.105	0.264	0.519	0.763	0.905
500	0.103	0.259	0.512	0.758	0.903

Let Q_Z denote the quantile function of the standard normal distribution. Table 11.1 contains probabilities of the form $\Pr(X_n \le Q_Z(\alpha))$ for various values of n and α; note that a probability of this form can be approximated by α. For each value of α given, it appears that the exact probabilities are converging to the approximation, although the convergence is, in some cases, quite slow; for instance, for $\alpha = 0.50$, the relative error of the approximation is about 2.3% even when $n = 500$. □

Uniformity in convergence in distribution

Convergence in distribution requires only pointwise convergence of the sequence of distributions. However, because of the special properties of distribution functions, in particular, the facts that they are nondecreasing and all have limit 1 at ∞ and limit 0 at $-\infty$, pointwise convergence is equivalent to uniform convergence whenever the limiting distribution function is continuous.

Theorem 11.3. *Let $X, X_1, X_2, \ldots$ denote real-valued random variables such that*

$$X_n \xrightarrow{\mathcal{D}} X \quad as \quad n \to \infty.$$

For $n = 1, 2, \ldots$, let F_n denote the distribution function of X_n and let F denote the distribution function of X. If F is continuous, then

$$\sup_x |F_n(x) - F(x)| \to 0 \quad as \quad n \to \infty.$$

Proof. Fix $\epsilon > 0$. Let $x_1, x_2, \ldots, x_m$ denote a partition of the real line, with $x_0 = -\infty$, $x_{m+1} = \infty$, and

$$F(x_j) - F(x_{j-1}) < \epsilon/2, \quad j = 1, \ldots, m+1.$$

Let $x \in \mathbf{R}$, $x_{j-1} \le x \le x_j$ for some $j = 1, \ldots, m+1$. Since $F_n(x_j) \to F(x_j)$ as $n \to \infty$,

$$F_n(x) - F(x) \le F_n(x_j) - F(x_{j-1}) < F(x_j) + \epsilon/2 - F(x_{j-1})$$

for sufficiently large n, say $n \ge N_j$. Similarly,

$$F_n(x) - F(x) \ge F_n(x_{j-1}) - F(x_j) > F(x_{j-1}) - \epsilon/2 - F(x_j)$$

for $n \ge N_{j-1}$; note that if $j = 1$, the $F_n(x_{j-1}) = F(x_{j-1})$ so that $N_0 = 1$.

Since $F(x_j) - F(x_{j-1}) < \epsilon/2$ and $F(x_{j-1}) - F(x_j) > -\epsilon/2$, it follows that, for $n \geq$ max(N_j, N_{j-1}),

$$F_n(x) - F(x) \leq \epsilon$$

and

$$F_n(x) - F(x) \geq -\epsilon.$$

It follows that, for $n \geq$ max(N_j, N_{j-1}),

$$|F_n(x) - F(x)| < \epsilon.$$

Let $N = $ max$(N_0, N_1, N_2, \ldots, N_{m+1})$. Then, for any value of x,

$$|F_n(x) - F(x)| < \epsilon$$

for $n \geq N$; since the right-hand side of this inequality does not depend on x, it follows that, given ϵ, there exists an N such that

$$\sup_x |F_n(x) - F(x)| \leq \epsilon$$

for $n \geq N$, proving the result. ∎

Example* 11.9 *(Convergence of $F_n(x_n)$). Suppose that a sequence of random variables $X_1, X_2, \ldots$ converges in distribution to a random variable X. Let F_n denote the distribution function of X_n, $n = 1, 2, \ldots$, and let F denote the distribution function of X, where F is continuous on **R**. Then, for each x, $\lim_{n\to\infty} F_n(x) = F(x)$.

Let $x_1, x_2, \ldots$ denote a sequence of real numbers such that $x = \lim_{n\to\infty} x_n$ exists. Then

$$|F_n(x_n) - F(x)| \leq |F_n(x_n) - F(x_n)| + |F(x_n) - F(x)|.$$

Since F is continuous,

$$\lim_{n\to\infty} |F(x_n) - F(x)| = 0;$$

by Theorem 11.3,

$$\lim_{n\to\infty} |F_n(x_n) - F(x_n)| \leq \lim_{n\to\infty} \sup_x |F_n(x) - F(x)| = 0.$$

Hence,

$$\lim_{n\to\infty} |F_n(x_n) - F(x)| = 0;$$

that is, the sequence $F_n(x_n)$ converges to $F(x)$ as $n \to \infty$. □

Convergence in distribution of random vectors

We now consider convergence in distribution of random vectors. The basic definition is a straightforward extension of the definition used for real-valued random variables. Let $X_1, X_2, \ldots$ denote a sequence of random vectors, each of dimension d, and let X denote a random vector of dimension d. For each $n = 1, 2, \ldots$, let F_n denote the distribution function of X_n and let F denote the distribution function of X. We say that X_n converges in distribution to X as $n \to \infty$, written

$$X_n \xrightarrow{\mathcal{D}} X \quad \text{as} \quad n \to \infty$$

provided that

$$\lim_{n \to \infty} F_n(x) = F(x)$$

for all $x \in \mathbf{R}^d$ at which F is continuous.

Many of the properties of convergence in distribution, proven in this section for sequences of real-valued random variables, extend to the case of random vectors. Several of these extensions are presented below without proof; for further discussion and detailed proofs, see, for example, Port (1994, Chapters 50 and 51).

The following result considers convergence in distribution of random vectors in terms of convergence of expected values of bounded functions and generalizes Theorem 11.1 and Corollaries 11.1 and 11.2.

Theorem 11.4. *Let $X_1, X_2, \ldots$ denote a sequence of d-dimensional random vectors and let X denote a d-dimensional random vector.*

   (i) If

$$X_n \xrightarrow{\mathcal{D}} X \quad as \quad n \to \infty$$

   then

$$\mathrm{E}[f(X_n)] \to \mathrm{E}[f(X)] \quad as \quad n \to \infty$$

   for all bounded, continuous, real-valued functions f.

   (ii) If

$$\mathrm{E}[f(X_n)] \to \mathrm{E}[f(X)] \quad as \quad n \to \infty$$

   for all bounded, uniformly continuous, real-valued functions f, then

$$X_n \xrightarrow{\mathcal{D}} X \quad as \quad n \to \infty.$$

   (iii) If $X_n \xrightarrow{\mathcal{D}} X$ as $n \to \infty$, and g is a continuous function, then $g(X_n) \xrightarrow{\mathcal{D}} g(X)$.

Theorem 11.5 below generalizes Theorem 11.2 on the convergence of characteristic functions.

Theorem 11.5. *Let $X_1, X_2, \ldots$ denote a sequence of d-dimensional random vectors and let X denote a d-dimensional random vector.*

Let φ_n denote the characteristic function of X_n and let φ denote the characteristic function of X.

$$X_n \xrightarrow{\mathcal{D}} X \quad as \quad n \to \infty$$

if and only if

$$\lim_{n \to \infty} \varphi_n(t) = \varphi(t), \quad for\ all\ \ t \in \mathbf{R}^d.$$

Recall that, when discussing the properties of characteristic functions of random vectors, it was noted that two random vectors X and Y have the same distribution if and only if $t^T X$

and $t^T Y$ have the same distribution for any vector t. Similarly, a very useful consequence of Theorem 11.5 is the result that convergence in distribution of a sequence of d-dimensional random vectors $X_1, X_2, \ldots$ to a random vector X may be established by showing that, for any vector $t \in \mathbf{R}^d$, $t^T X_1, t^T X_2, \ldots$ converges in distribution to $t^T X$. Thus, a multidimensional problem may be converted to a class of one-dimensional problems. This often called the *Cramér–Wold device.*

Theorem 11.6. *Let* $X, X_1, X_2, \ldots$ *denote d-dimensional random vectors. Then*

$$X_n \xrightarrow{\mathcal{D}} X \quad as \quad n \to \infty$$

if and only if

$$t^T X_n \xrightarrow{\mathcal{D}} t^T X \quad as \quad n \to \infty$$

for all $t \in \mathbf{R}^d$.

Proof. Let φ_n denote the characteristic function of X_n and let φ denote the characteristic function of X. Then $t^T X_n$ has characteristic function

$$\tilde{\varphi}_n(s) = \varphi_n(st), \quad s \in \mathbf{R}$$

and $t^T X$ has characteristic function

$$\tilde{\varphi}(s) = \varphi(st), \quad s \in \mathbf{R}.$$

Suppose $X_n \xrightarrow{\mathcal{D}} X$. Then

$$\varphi_n(t) \to \varphi(t) \quad \text{for all } t \in \mathbf{R}^d$$

so that, fixing t,

$$\varphi_n(st) \to \varphi(st) \quad \text{for all } s \in \mathbf{R},$$

proving that $t^T X_n \xrightarrow{\mathcal{D}} t^T X$.

Now suppose that $t^T X_n \xrightarrow{\mathcal{D}} t^T X$ for all $t \in \mathbf{R}^d$. Then

$$\varphi_n(st) \to \varphi(st) \quad \text{for all } s \in \mathbf{R}, \ t \in \mathbf{R}^d.$$

Taking $s = 1$ shows that

$$\varphi_n(t) \to \varphi(t) \quad \text{for all } t \in \mathbf{R}^d,$$

proving the result. ∎

Thus, according to Theorem 11.6, convergence in distribution of the component random variables of a random vector is a necessary, but not sufficient, condition for convergence in distribution of the random vector. This is illustrated in the following example.

Example 11.10. Let Z_1 and Z_2 denote independent standard normal random variables and, for $n = 1, 2, \ldots$, let $X_n = Z_1 + \alpha_n Z_2$ and $Y_n = Z_2$, where $\alpha_1, \alpha_2, \ldots$ is a sequence of real numbers. Clearly, $Y_n \xrightarrow{\mathcal{D}} Z_2$ as $n \to \infty$ and, if $\alpha_n \to \alpha$ as $n \to \infty$, for some real number α, then $X_n \xrightarrow{\mathcal{D}} N(0, 1 + \alpha^2)$ as $n \to \infty$. Furthermore, since for any $(t_1, t_2) \in \mathbf{R}^2$, $t_1 X_n + t_2 Y_n = t_1 Z_1 + (t_2 + \alpha_n t_1) Z_2$ has characteristic function

$$\exp\{-[t_1^2 + (t_2 + \alpha_n t_1)^2]s^2/2\}, \quad s \in \mathbf{R},$$

which converges to

$$\exp\{-[t_1^2 + (t_2 + \alpha t_1)^2]s^2/2\}, \quad s \in \mathbf{R},$$

it follows that $t_1 X_n + t_2 Y_n$ converges in distribution to a random variable with a normal distribution with mean 0 and variance $t_1^2 + (t_2 + \alpha t_1)^2$. Hence, by Theorem 11.6,

$$\begin{pmatrix} X_n \\ Y_n \end{pmatrix} \xrightarrow{\mathcal{D}} W,$$

where W has a bivariate normal distribution with mean vector 0 and covariance matrix

$$\begin{pmatrix} 1 + \alpha^2 & \alpha \\ \alpha & 1 \end{pmatrix}.$$

Now suppose that $\alpha_n = (-1)^n$, $n = 1, 2, \ldots$. Clearly, $Y_n \xrightarrow{\mathcal{D}} Z_2$ as $n \to \infty$ still holds. Furthermore, since $X_n = Z_1 + (-1)^n Z_2$, it follows that, for each $n = 1, 2, \ldots$, X_n has a normal distribution with mean 0 and variance 2; hence, $X_n \xrightarrow{\mathcal{D}} N(0, 2)$ as $n \to \infty$. However, consider the distribution of $X_n + Y_n = Z_1 + (1 + (-1)^n)Z_2$. This distribution has characteristic function

$$\exp\{-[1 + (1 + (-1)^n)^2]s^2/2\} = \begin{cases} \exp(-s^2/2) & \text{for } n = 1, 3, 5, \ldots \\ \exp(-5s^2/2) & \text{for } n = 2, 4, 6, \ldots \end{cases}, \quad s \in \mathbf{R}.$$

Hence, by Theorem 11.2, $X_n + Y_n$ does not converge in distribution so that, by Theorem 11.6, the random vector (X_n, Y_n) does not converge in distribution. $\quad\square$

11.3   Convergence in Probability

A sequence of real-valued random variables $X_1, X_2, \ldots$ converges in distribution to a random variable X if the distribution functions of $X_1, X_2, \ldots$ converge to that of X. It is important to note that this type of convergence says nothing about the relationship between the random variables X_n and X.

Suppose that the sequence $X_1, X_2, \ldots$ is such that $|X_n - X|$ becomes small with high probability as $n \to \infty$; in this case, we say that X_n *converges in probability* to X. More precisely, a sequence of real-valued random variables $X_1, X_2, \ldots$ converges in probability to a real-valued random variable X if, for any $\epsilon > 0$,

$$\lim_{n \to \infty} \Pr(|X_n - X| \geq \epsilon) = 0.$$

We will denote this convergence by $X_n \xrightarrow{P} X$ as $n \to \infty$. Note that for this definition to make sense, for each n, X and X_n must be defined on the same underlying sample space, a requirement that did not arise in the definition of convergence in distribution.

Example 11.11 (Sequence of Bernoulli random variables). Let $X_1, X_2, \ldots$ denote a sequence of real-valued random variables such that

$$\Pr(X_n = 1) = 1 - \Pr(X_n = 0) = \theta_n, \quad n = 1, 2, \ldots$$

where $\theta_1, \theta_2, \ldots$ denotes a sequence of constants each taking values in the interval $(0, 1)$. For any $\epsilon > 0$,

$$\Pr(|X_n| \geq \epsilon) = \Pr(X_n = 1) = \theta_n;$$

hence, $X_n \xrightarrow{p} 0$ provided that $\lim_{n \to \infty} \theta_n = 0$.   □

Example 11.12 *(Normal random variables).*

Let $Z, Z_1, Z_2, \ldots$ denote independent random variables, each with a standard normal distribution, and let $\alpha_1, \alpha_2, \ldots$ denote a sequence of real numbers satisfying $0 < \alpha_n < 1$ for $n = 1, 2, \ldots$ and $\alpha_n \to 1$ as $n \to \infty$. Let

$$X_n = (1 - \alpha_n)Z_n + \alpha_n Z, \quad n = 1, 2, \ldots$$

and let $X = Z$. Then, for any $m = 1, 2, \ldots, (X_1, \ldots, X_m)$ has a multivariate normal distribution with mean vector and covariance matrix with (i, j)th element given by $\alpha_i \alpha_j$, if $i \neq j$ and by $(1 - \alpha_j)^2 + \alpha_j^2$ if $i = j$.

Note that $X_n - X = (1 - \alpha_n)(Z_n - Z)$ so that, for any $\epsilon > 0$,

$$\Pr\{|X_n - X| \geq \epsilon\} = \Pr\{|Z_n - Z| \geq \epsilon/(1 - \alpha_n)\}$$

so that, by Markov's inequality, together with the fact that $E[|Z_n - Z|^2] = 2$,

$$\Pr\{|X_n - X| \geq \epsilon\} \leq \frac{2(1 - \alpha_n)^2}{\epsilon^2}, \quad n = 1, 2, \ldots.$$

It follows that $X_n \xrightarrow{p} X$ as $n \to \infty$.   □

As noted above, an important distinction between convergence of a sequence $X_n, n = 1, 2, \ldots$, to X in distribution and in probability is that convegence in distribution depends only on the marginal distribution functions of X_n and of X, while convergence in probability is concerned with the distribution of $|X_n - X|$. Hence, for convergence in probability, the joint distribution of X_n and X is relevant. This is illustrated in the following example.

Example 11.13 *(Sequence of Bernoulli random variables).* Let $X_1, X_2, \ldots$ denote a sequence of real-valued random variables such that, for each $n = 1, 2, \ldots$,

$$\Pr(X_n = 1) = 1 - \Pr(X_n = 0) = \frac{1}{2} \frac{n + 1}{n}$$

and let X denote a random variable satisfying

$$\Pr(X = 1) = \Pr(X = 0) = 1/2.$$

Then, by Example 11.1, $X_n \xrightarrow{D} X$ as $n \to \infty$.

However, whether or not X_n converges in X in probability will depend on the joint distributions of $(X, X_1), (X, X_2), \ldots$. For instance, if, for each n, X_n and X are independent,

then

$$\Pr(|X_n - X| \geq 1/2) = \Pr(X_n = 1 \cap X = 0) + \Pr(X_n = 0 \cap X = 1)$$
$$= \frac{1}{4}\frac{n+1}{n} + \frac{1}{4}\frac{n-1}{n}$$
$$= \frac{1}{2};$$

it follows that X_n does not converge to X in probability.

On the other hand, suppose that

$$\Pr(X_n = 1 | X = 1) = 1 \quad \text{and} \quad \Pr(X_n = 1 | X = 0) = \frac{1}{n}.$$

Note that

$$\Pr(X_n = 1) = \Pr(X_n = 1 | X = 1)\Pr(X = 1) + \Pr(X_n = 1 | X = 0)\Pr(X = 0)$$
$$= \frac{1}{2} + \frac{1}{n}\frac{1}{2}$$
$$= \frac{1}{2}\frac{n+1}{n},$$

as stated above. In this case, for any $\epsilon > 0$,

$$\Pr(|X_n - X| \geq \epsilon) = \Pr(X_n = 1 \cap X = 0) + \Pr(X_n = 0 \cap X = 1) = \frac{1}{2n}$$

so that $X_n \overset{p}{\to} X$ as $n \to \infty$.   □

The preceding example shows that convergence in distribution does not necessarily imply convergence in probability. The following result shows that convergence in probability does imply convergence in distribution. Furthermore, when the limiting random variable is a constant with probability 1, then convergence in probability is equivalent to convergence in distribution.

Corollary 11.3. *Let $X, X_1, X_2, \ldots$ denote real-valued random variables.*

   (i) *If $X_n \overset{p}{\to} X$ as $n \to \infty$ then $X_n \overset{D}{\to} X$ as $n \to \infty$.*

   (ii) *If $X_n \overset{D}{\to} X$ as $n \to \infty$ and $\Pr(X = c) = 1$ for some constant c, then $X_n \overset{p}{\to} X$ as $n \to \infty$.*

Proof. Suppose that $X_n \overset{p}{\to} X$. Consider the event $X_n \leq x$ for some real-valued x. If $X_n \leq x$, then, for every $\epsilon > 0$, either $X \leq x + \epsilon$ or $|X_n - X| > \epsilon$. Hence,

$$\Pr(X_n \leq x) \leq \Pr(X \leq x + \epsilon) + \Pr(|X_n - X| > \epsilon).$$

Let F_n denote the distribution function of X_n and let F denote the distribution function of F. Then, for all $\epsilon > 0$,

$$\limsup_{n \to \infty} F_n(x) \leq F(x + \epsilon).$$

Similarly, if, for some $\epsilon > 0$, $X \leq x - \epsilon$, then either $X_n \leq x$ or $|X_n - X| > \epsilon$. Hence,

$$F(x - \epsilon) \leq F_n(x) + \Pr(|X_n - X| > \epsilon)$$

so that, for all $\epsilon > 0$,

$$\liminf_{n \to \infty} F_n(x) \geq F(x - \epsilon).$$

That is, for all $\epsilon > 0$,

$$F(x - \epsilon) \leq \liminf_{n \to \infty} F_n(x) \leq \limsup_{n \to \infty} F_n(x) \leq F(x + \epsilon).$$

Suppose F is continuous at x. Then $F(x + \epsilon) - F(x - \epsilon) \to 0$ as $\epsilon \to 0$. It follows that

$$\lim_{n \to \infty} F_n(x)$$

exists and is equal to $F(x)$ so that $X_n \overset{D}{\to} X$. This proves part (i) of the theorem.

Now suppose that $X_n \overset{D}{\to} X$ and that $X = c$ with probability 1. Then X has distribution function

$$F(x) = \begin{cases} 0 & \text{if } x < c \\ 1 & \text{if } x \geq c \end{cases}.$$

Let F_n denote the distribution function of X_n. Since F is not continuous at $x = c$, it follows that

$$\lim_{n \to \infty} F_n(x) = \begin{cases} 0 & \text{if } x < c \\ 1 & \text{if } x > c \end{cases}.$$

Fix $\epsilon > 0$. Then

$$\Pr(|X_n - c| \geq \epsilon) = \Pr(X_n \leq c - \epsilon) + \Pr(X_n \geq c + \epsilon)$$
$$\leq F_n(c - \epsilon) + 1 - F_n(c + \epsilon/2) \to 0 \quad \text{as} \quad n \to \infty.$$

Since this holds for all $\epsilon > 0$, we have

$$\lim_{n \to \infty} \Pr\{|X_n - c| \geq \epsilon\} = 0 \quad \text{for all} \quad \epsilon > 0;$$

it follows that $X_n \overset{p}{\to} c$ as $n \to \infty$, proving part (ii) of the theorem. ∎

Convergence in probability to a constant

We now consider convergence in probability of a sequence $X_1, X_2, \ldots$ to a constant. Without loss of generality we may take this constant to be 0; convergence to a constant c may be established by noting that X_n converges in probability to c if and only if $X_n - c$ converges in probability to 0.

Since convergence in probability to a constant is equivalent to convergence in distribution, by Corollary 11.2, if $X_n \overset{p}{\to} 0$ and f is a continuous function, then $f(X_n) \overset{p}{\to} f(0)$. Since, in this case, the distribution of X_n becomes concentrated near 0 as $n \to \infty$, convergence of $f(X_n)$ to $f(0)$ holds provided only that f is continuous at 0. The details are given in the following theorem.

Theorem 11.7. *Let $X_1, X_2, \ldots$ denote a sequence of real-valued random variables such that $X_n \overset{p}{\to} 0$ as $n \to \infty$. Let $f : \mathbf{R} \to \mathbf{R}$ denote a function that is continuous at 0. Then*

$$f(X_n) \overset{p}{\to} f(0) \quad as \quad n \to \infty.$$

Proof. Fix ϵ and consider $\Pr(|f(X_n) - f(0)| < \epsilon)$. By the continuity of f, there exists a δ such that $|f(x) - f(0)| < \epsilon$ whenever $|x| < \delta$. Hence,

$$\Pr(|f(X_n) - f(0)| < \epsilon) = \Pr(|X_n| < \delta).$$

The result now follows from the fact that, for any $\delta > 0$,

$$\lim_{n \to \infty} \Pr(|X_n| < \delta) = 1. \qquad \blacksquare$$

According to Theorem 11.2, a necessary and sufficient condition for convergence in distribution is convergence of the characteristic functions. Since the characteristic function of the random variable 0 is 1 for all $t \in \mathbf{R}$, it follows that $X_n \xrightarrow{p} 0$ as $n \to \infty$ if and only if $\varphi_1, \varphi_2, \ldots$, the characteristic functions of $X_1, X_2, \ldots$, respectively, satisfy

$$\lim_{n \to \infty} \varphi_n(t) = 1, \quad t \in \mathbf{R}.$$

Example 11.14 (*Gamma random variables*). Let $X_1, X_2, \ldots$ denote a sequence of real-valued random variables such that, for each $n = 1, 2, \ldots$, X_n has a gamma distribution with parameters α_n and β_n, where $\alpha_n > 0$ and $\beta_n > 0$; see Example 3.4 for further details regarding the gamma distribution. Assume that

$$\lim_{n \to \infty} \alpha_n = \alpha \quad \text{and} \quad \lim_{n \to \infty} \beta_n = \beta$$

for some α, β.

Let φ_n denote the characteristic function of X_n. Then, according to Example 3.4,

$$\log \varphi_n(t) = \alpha_n \log \beta_n - \alpha_n \log(\beta_n - it), \quad t \in \mathbf{R}.$$

Hence, $X_n \xrightarrow{p} 0$ as $n \to \infty$ provided that $\alpha = 0$, $\beta < \infty$, and

$$\lim_{n \to \infty} \alpha_n \log \beta_n = 0. \qquad \square$$

Example 11.15 (*Weak law of large numbers*). Let $Y_n, n = 1, 2, \ldots$ denote a sequence of independent, identically distributed real-valued random variables such that $E(Y_1) = 0$. Let

$$X_n = \frac{1}{n}(Y_1 + \cdots + Y_n), \quad n = 1, 2, \ldots.$$

The characteristic function of X_n is given by

$$\varphi_n(t) = \varphi(t/n)^n$$

where φ denotes the characteristic function of Y_1 and, hence,

$$\log \varphi_n(t) = n \, \log \varphi(t/n).$$

Since $E(Y_1) = 0$, by Theorem 3.5,

$$\varphi(t) = 1 + o(t) \quad \text{as} \quad t \to 0 \qquad \left(1 + \sum_{j=1}^{m} \frac{(it)^j}{j!} E(X^j) + o(t^m)\right)$$

and, hence,

$$\log \varphi(t) = o(t) \quad \text{as} \quad t \to 0;$$

it follows that

$$\log \varphi_n(t) = n \, o\left(\frac{t}{n}\right) \to 0 \quad \text{as} \quad n \to \infty.$$

Hence, by Theorem 11.2, $X_n \xrightarrow{D} 0$ as $n \to \infty$; it now follows from Corollary 11.3 that $X_n \xrightarrow{P} 0$ as $n \to \infty$. This is one version of the *weak law of large numbers*. $\square$

***Example* 11.16 (*Mean of Cauchy random variables*).** Let Y_n, $n = 1, 2, \ldots$ denote a sequence of independent, identically distributed random variables such that each Y_j has a standard Cauchy distribution; recall that the mean of this distribution does not exist so that the result in Example 11.15 does not apply.

Let

$$X_n = \frac{1}{n}(Y_1 + \cdots + Y_n), \quad n = 1, 2, \ldots.$$

The characteristic function of the standard Cauchy distribution is $\exp(-|t|)$ so that the characteristic function of X_n is given by

$$\varphi_n(t) = \exp(-|t|/n)^n = \exp(-|t|).$$

Hence, X_n does not converge in probability to 0; in fact, X_n also has a standard Cauchy distribution. $\square$

Although convergence in probability of X_n to 0 may be established by considering characteristic functions, it is often more convenient to use the connection between probabilities and expected values provided by Markov's inequality (Theorem 1.14). Such a result is given in the following theorem; the proof is left as an exercise.

***Theorem* 11.8.** *Let* $X_1, X_2, \ldots$ *denote a sequence of real-valued random variables. If, for some* $r > 0$,

$$\lim_{n \to \infty} \mathrm{E}(|X_n|^r) = 0$$

then $X_n \xrightarrow{P} 0$.

***Example* 11.17 (*Weak law of large numbers*).** Let $Y_n, n = 1, 2, \ldots$ denote a sequence of real-valued random variables such that $\mathrm{E}(Y_n) = 0$, $n = 1, 2, \ldots$, $\mathrm{E}(Y_n^2) = \sigma_n^2 < \infty$, $n = 1, 2, \ldots$, and $\mathrm{Cov}(Y_i, Y_j) = 0$ for all $i \neq j$.

Let

$$X_n = \frac{1}{n}(Y_1 + \cdots + Y_n), \quad n = 1, 2, \ldots.$$

Then

$$\mathrm{E}(X_n^2) = \mathrm{Var}(X_n) = \frac{1}{n^2} \sum_{j=1}^{n} \sigma_j^2.$$

Hence, $X_n \overset{p}{\to} 0$ as $n \to 0$ provided that

$$\lim_{n \to 0} \frac{1}{n^2} \sum_{j=1}^{n} \sigma_j^2 = 0.$$

This is another version of the weak law of large numbers. □

The weak laws of large numbers given in Examples 11.15 and 11.17 give conditions under which the sample means of a sequence of real-valued random variables, each with mean 0, converges in probability to 0. In Example 11.15, the random variables under consideration were taken to be independent, while in Example 11.17, they were taken to be uncorrelated. The following example shows that a similar result holds, under some conditions, even when the random variables are correlated.

Example 11.18 (Weak law of large numbers for correlated random variables). Let $X_1, X_2, \ldots$ denote a sequence of real-valued random variables such that $E(X_j) = 0$, $\text{Var}(X_j) = 1$, $j = 1, \ldots, n$, and suppose that

$$\text{Cov}(X_i, X_j) = R(i - j), \quad i, j = 1, \ldots, n$$

for some function R. In the language of Chapter 6, $\{X_t : t \in \mathbf{Z}\}$ is a discrete-time covariance-stationary stochastic process and R is the autocovariance function of the process; however, the results of Chapter 6 are not needed for this example. We will show that if $R(j) \to 0$ as $j \to \infty$, then

$$\frac{1}{n} \sum_{j=1}^{n} X_j \overset{p}{\to} 0 \quad \text{as} \quad n \to \infty.$$

Note that the condition that $R(j) \to 0$ as $j \to \infty$ is the condition that the correlation between two random variables in the sequence $X_1, X_2, \ldots$ decreases to 0 as the distance between their indices increases.

By Chebychev's inequality, the result follows provided that

$$\text{Var}\left(\frac{1}{n} \sum_{j=1}^{n} X_j\right) \to 0 \quad \text{as} \quad n \to \infty.$$

Note that

$$\text{Var}\left(\frac{1}{n} \sum_{j=1}^{n} X_j\right) = \frac{1}{n^2} \sum_{i=1}^{n} \sum_{j=1}^{n} R(|i - j|) = \frac{1}{n^2}\left[nR(0) + 2\sum_{j=1}^{n-1}(n - j)R(j)\right]$$

$$\le \frac{2}{n^2} \sum_{j=0}^{n-1}(n - j)R(j).$$

Fix $\epsilon > 0$. Under the assumption that $R(j) \to 0$ as $j \to \infty$, there exists an integer N such that $|R(j)| < \epsilon$ for all $j > N$. Hence, for $n \ge N$,

$$\left|\sum_{j=0}^{n-1}(n - j)R(j)\right| \le \sum_{j=0}^{N}(n - j)|R(j)| + (n - N)^2\epsilon$$

so that

$$\mathrm{Var}\left(\frac{1}{n}\sum_{j=1}^{n}X_j\right) \le \frac{2}{n}\sum_{j=0}^{N}\left(1-\frac{j}{n}\right)|R(j)| + 2\left(1-\frac{N}{n}\right)^2\epsilon.$$

It follows that

$$\limsup_{n\to\infty}\mathrm{Var}\left(\frac{1}{n}\sum_{j=1}^{n}X_j\right) \le 2\epsilon.$$

Since $\epsilon > 0$ is arbitrary, it follows that

$$\lim_{n\to\infty}\mathrm{Var}\left(\frac{1}{n}\sum_{j=1}^{n}X_j\right) = 0,$$

proving the result. □

Convergence in probability of random vectors and random matrices

Convergence in probability can also be applied to random vectors. Let $X_1, X_2, \ldots$ be a sequence of d-dimensional random vectors and let X denote a d-dimensional random vector. Then X_n converges in probability to X as $n \to \infty$, written $X_n \xrightarrow{p} X$ as $n \to \infty$, provided that, for any $\epsilon > 0$,

$$\lim_{n\to\infty}\Pr\{||X_n - X|| \ge \epsilon\} = 0;$$

here $||\cdot||$ denotes Euclidean distance on $\mathbf{R}^d$.

The following result shows that convergence in probability of a sequence of random variables is equivalent to convergence in probability of the sequences of component random variables. Thus, convergence in probability of random vectors does not include any ideas beyond those contained in convergence in probability of real-valued random variables.

Theorem 11.9. *Let $X_1, X_2, \ldots$ denote a sequence of d-dimensional random vectors and, for each $n = 1, 2, \ldots$, let $X_n(j)$, $j = 1, \ldots, d$ denote the components of X_n so that*

$$X_n = \begin{pmatrix} X_n(1) \\ \vdots \\ X_n(d) \end{pmatrix}, \quad n = 1, 2, \ldots.$$

Let X denote a d-dimensional random vector with components $X(1), \ldots, X(d)$. Then $X_n \xrightarrow{p} X$ as $n \to \infty$ if and only if for each $j = 1, 2, \ldots, d$

$$X_n(j) \xrightarrow{p} X(j) \quad \text{as } n \to \infty.$$

Proof. For simplicity, we consider the case in which $d = 2$; the general case follows along similar lines. Since

$$||X_n - X||^2 = |X_n(1) - X(1)|^2 + |X_n(2) - X(2)|^2,$$

for any $\epsilon > 0$,

$$\Pr\{||X_n - X|| \geq \epsilon\} = \Pr\{||X_n - X||^2 \geq \epsilon^2\} \leq \Pr\{|X_n(1) - X(1)|^2 \geq \epsilon^2/2\}$$
$$+ \Pr\{|X_n(2) - X(2)|^2 \geq \epsilon^2/2\}$$
$$= \Pr\{|X_n(1) - X(1)| \geq \epsilon/\sqrt{2}\} + \Pr\{|X_n(2) - X(2)| \geq \epsilon/\sqrt{2}\}.$$

Hence, if $X_n(j) \xrightarrow{P} X(j)$ as $n \to \infty$ for $j = 1, 2$, it follows that $X_n \xrightarrow{P} X$ as $n \to \infty$. Now suppose that $X_n \xrightarrow{P} X$ as $n \to \infty$. Note that

$$|X_n(1) - X(1)|^2 = ||X_n - X||^2 - |X_n(2) - X(2)|^2 \leq ||X_n - X||^2.$$

It follows that, for any $\epsilon > 0$,

$$\Pr\{|X_n(1) - X(1)| \geq \epsilon\} \leq \Pr\{||X_n - X|| \geq \epsilon\}$$

and, hence, that $X_n(1) \xrightarrow{P} X$ as $n \to \infty$; the same argument applies to $X_n(2)$. $\blacksquare$

In statistics it is often useful to also consider random matrices; for instance, we may be interested in an estimator of a covariance matrix. For the purpose of defining convergence in probability, a $d_1 \times d_2$ random matrix can be viewed as a random vector of length $d_1 d_2$. Hence, by Theorem 11.9, a sequence of random matrices $Y_1, Y_2, \ldots$ converges in probability to a matrix Y, if and only if for each (i, j) the sequence formed by taking the (i, j)th element of $Y_1, Y_2, \ldots$ converges in probability to the (i, j)th element of Y.

11.4   Convergence in Distribution of Functions of Random Vectors

For many statistics arising in statistical applications, it is difficult to establish convergence in distribution directly using the results given in Sections 11.1–11.3. However, in some of these cases, the statistic in question may be written as a function of statistics whose convergence properties can be determined. For instance, suppose that we are interested in a statistic $T_n = f(X_n, Y_n)$; it is often possible to establish the convergence in distribution of $T_1, T_2, \ldots$ by first establishing the convergence in distribution of $X_1, X_2, \ldots$ and $Y_1, Y_2, \ldots$ and using properties of the function f. Some basic results of this type are given in this section; further results are discussed in Section 13.2.

The main technical result of this section is the following.

Theorem 11.10. *Let* $X_1, X_2, \ldots$ *denote a sequence of d-dimensional random vectors such that*

$$X_n \xrightarrow{D} X \quad as \ n \to \infty.$$

Let $Y_1, Y_2, \ldots$ *denote a sequence of m-dimensional random vectors such that*

$$Y_n \xrightarrow{P} c \quad as \ n \to \infty$$

for some constant $c \in \mathbf{R}^m$. *Then*

$$\begin{pmatrix} X_n \\ Y_n \end{pmatrix} \xrightarrow{D} \begin{pmatrix} X \\ c \end{pmatrix} \quad as \ n \to \infty.$$

Proof.   Let g denote a bounded, uniformly continuous, real-valued function on $\mathbf{R}^d \times \mathbf{R}^m$. Given $\epsilon > 0$ there exists a δ such that

$$|g(x_1, y_1) - g(x_2, y_2)| \le \epsilon$$

whenever

$$||(x_1, y_1) - (x_2, y_2)|| \le \delta.$$

Note that

$$|E[g(X_n, Y_n) - g(X, c)]| \le |E[g(X_n, Y_n) - g(X_n, c)]| + |E[g(X_n, c) - g(X, c)]|.$$

First consider $E[g(X_n, Y_n) - g(X_n, c)]$. Whenever $||y - c|| \le \delta$,

$$|g(x, y) - g(x, c)| \le \epsilon.$$

Hence,

$$
\begin{aligned}
&|E[g(X_n, Y_n) - g(X_n, c)]| \\
&\le E[|g(X_n, Y_n) - g(X_n, c)| \, I_{\{||Y_n - c|| \le \delta\}}] + E[|g(X_n, Y_n) - g(X_n, c)| I_{\{||Y_n - c|| > \delta\}}] \\
&\le \epsilon + 2G \Pr(||Y_n - c|| > \delta)
\end{aligned}
$$

where

$$G = \sup_{x, y} |g(x, y)|.$$

It follows that

$$\limsup_{n \to \infty} |E[g(X_n, Y_n) - g(X_n, c)]| \le \epsilon.$$

Now consider $E[g(X_n, c) - g(X, c)]$. Define $\bar{g}(x) = g(x, c)$. Then $\bar{g}$ is a bounded continuous function on $\mathbf{R}^m$. Hence,

$$\lim_{n \to \infty} |E[\bar{g}(X_n) - \bar{g}(X)]| \equiv \lim_{n \to \infty} |E[g(X_n, c) - g(X, c)]| = 0.$$

It follows that, for any $\epsilon > 0$,

$$\limsup_{n \to \infty} |E[g(X_n, Y_n) - g(X, c)]| \le \epsilon;$$

the result follows.   ∎

Theorem 11.10 may be used to establish the following result, which is known as *Slutsky's theorem*, and which is often used in statistics. The proof is straightforward and is left as an exercise.

Corollary **11.4.** *Let* $X_1, X_2, \ldots$ *and* $Y_1, Y_2, \ldots$ *denote sequences of d-dimensional random vectors such that*

$$X_n \xrightarrow{\mathcal{D}} X \quad as \quad n \to \infty$$

and

$$Y_n \xrightarrow{P} c \quad as \quad n \to \infty$$

for some constant $c \in \mathbf{R}^d$. Then

$$X_n + Y_n \overset{D}{\to} X + c \quad as \quad n \to \infty$$

and

$$Y_n^T X_n \overset{D}{\to} c^T X \quad as \quad n \to \infty.$$

Example 11.19 (Standardization by a random scale factor). Let $X_1, X_2, \ldots$ denote a sequence of real-valued random variables such that $X_n \overset{D}{\to} X$ as $n \to \infty$ where X has a normal distribution with mean 0 and variance σ^2. Let $Y_1, Y_2, \ldots$ denote a sequence of real-valued, nonzero, random variables such that $Y_n \overset{P}{\to} \sigma$ as $n \to \infty$. Then, by Corollary 11.4, X_n / Y_n converges in distribution to a standard normal random variable.   □

Example 11.20 (Sequences of random variables with correlation approaching 1). Let $X_1, X_2, \ldots$ and $Y_1, Y_2, \ldots$ denote sequences of real-valued random variables, each with mean 0 and standard deviation 1, such that $X_n \overset{D}{\to} X$ as $n \to \infty$ for some random variable X. For each $n = 1, 2, \ldots$, let ρ_n denote the correlation of X_n and Y_n and suppose that $\lim_{n \to \infty} \rho_n = 1$.

Note that we may write

$$Y_n = X_n + (Y_n - X_n)$$

and, since

$$\mathrm{Var}(Y_n - X_n) = 2 - 2\rho_n,$$

it follows that $Y_n - X_n \overset{P}{\to} 0$. Hence, by Corollary 11.4, $Y_n \overset{D}{\to} X$ as $n \to \infty$.   □

As noted previously, random matrices often arise in statistics. The following result gives some properties of convergence in probability of random matrices.

Lemma 11.1. *Let $Y_1, Y_2, \ldots$, denote random $m \times m$ matrices and let C denote an $m \times m$ matrix of constants.*

(i) *$Y_n \overset{P}{\to} C$ as $n \to \infty$ if and only if $a^T Y_n b \overset{P}{\to} a^T C b$ as $n \to \infty$ for all $a, b \in \mathbf{R}^m$*

(ii) *Let $|M|$ denote the determinant of a square matrix M. If $Y_n \overset{P}{\to} C$ as $n \to \infty$, then*

$$|Y_n| \overset{P}{\to} |C| \quad as \quad n \to \infty.$$

(iii) *Suppose each Y_n is invertible and that C is invertible. $Y_n^{-1} \overset{P}{\to} C^{-1}$ as $n \to \infty$ if and only if $Y_n \overset{P}{\to} C$ as $n \to \infty$.*

Proof.   For each $n = 1, 2$, and each $i, j = 1, \ldots, m$, let Y_{nij} denote the (i, j)th element of Y_n and let C_{ij} denote the (i, j)th element of C. Assume that $Y_n \overset{P}{\to} C$ as $n \to \infty$. Then, for each i, j,

$$Y_{nij} \overset{P}{\to} C_{ij} \quad as \quad n \to \infty.$$

Since

$$a^T Y_n b = \sum_{i=1}^{m} \sum_{j=1}^{m} a_i b_j Y_{nij},$$

it follows that

$$a^T Y_n b \xrightarrow{P} a^T C b \quad \text{as} \quad n \to \infty.$$

Now suppose that

$$a^T Y_n b \xrightarrow{P} a^T C b \quad \text{as} \quad n \to \infty$$

for any $a, b \in \mathbf{R}^m$. It follows that

$$Y_{nij} \xrightarrow{P} C_{ij} \quad \text{as} \quad n \to \infty$$

for each i, j, proving part (i) of the theorem.

Part (ii) follows from the fact that the determinant of a matrix is a continuous function of the elements of the matrix.

Part (iii) follows from part (ii), using the fact that the elements of the inverse of a matrix may be written as ratios of determinants. ∎

Corollary 11.4 may now be extended to random matrices.

Corollary 11.5. *Let $Y_1, Y_2, \ldots$ denote random $m \times m$ matrices and let C denote an $m \times m$ matrix of constants. Assume that*

$$Y_n \xrightarrow{P} C \quad \text{as} \quad n \to \infty.$$

Let $X_1, X_2, \ldots$ denote a sequence of m-dimensional random vectors such that

$$X_n \xrightarrow{D} X \quad \text{as} \quad n \to \infty.$$

If each Y_n, $n = 1, 2, \ldots$, is invertible and C is invertible, then

$$Y_n^{-1} X_n \xrightarrow{D} C^{-1} X \quad \text{as} \quad n \to \infty.$$

Proof. Let $a \in \mathbf{R}^m$ and let $W_n^T = a^T Y_n^{-1}$ and $w^T = a^T C$. From Lemma 11.1,

$$W_n \xrightarrow{P} w \quad \text{as} \quad n \to \infty.$$

Since

$$a^T Y_n^{-1} X_n = W_n^T X_n$$

the result then follows from Corollary 11.4. ∎

11.5 Convergence of Expected Values

According to Theorem 11.1, if a sequence of real-valued random variables $X_1, X_2, \ldots,$ converges in distribution to X and there exists a constant M such that $|X_n| \le M$,

$n = 1, 2, \ldots$, then

$$\lim_{n \to \infty} E(X_n) = E(X).$$

However, if the X_n are not bounded in this manner, convergence of the expected values does not necessarily hold. An example of this has already been given in Example 11.5.

Hence, in order for $X_n \overset{D}{\to} X$ as $n \to \infty$ to imply that $E(X_n) \to E(X)$ as $n \to \infty$, additional conditions are needed on the sequence $X_1, X_2, \ldots$. Suppose that

$$\sup_n E[|X_n| I_{\{|X_n| \geq c\}}] = 0 \quad \text{for some} \quad c > 0. \tag{11.3}$$

Then $X_1, X_2, \ldots$ are uniformly bounded by c, for if there exists an N such that $\Pr\{|X_N| \geq c\} > 0$, then

$$E[|X_N| I_{\{|X_N| \geq c\}}] \geq c \Pr\{|X_N| \geq c\} > 0;$$

hence, condition (11.3) implies that $\lim_{n \to \infty} E(X_n) = E(X)$.

The condition of *uniform integrability* is a weaker version of (11.3) that is still strong enough to imply that $\lim_{n \to \infty} E(X_n) = E(X)$. A sequence of real-valued random variables $X_1, X_2, \ldots$ is said to be *uniformly integrable* if

$$\lim_{c \to \infty} \sup_n E[|X_n| I_{\{|X_n| \geq c\}}] = 0.$$

Theorem 11.11. *Let $X, X_1, X_2, \ldots$ denote real-valued random variables such that $X_n \overset{D}{\to} X$ as $n \to \infty$. If $X_1, X_2, \ldots$ is uniformly integrable and $E(|X|) < \infty$, then*

$$\lim_{n \to \infty} E(X_n) = E(X).$$

Proof. For $c > 0$, define

$$g_c(x) = \begin{cases} x & \text{if } |x| < c \\ \text{sgn}(x)c & \text{if } |x| \geq c \end{cases}.$$

Hence,

$$x - g_c(x) = \begin{cases} 0 & \text{if } |x| < c \\ x - \text{sgn}(x)c & \text{if } |x| \geq c \end{cases}.$$

It follows that, for any random variable Z,

$$E(Z) = E[g_c(Z)] + E[Z I_{\{|Z| > c\}}] - c E[\text{sgn}(Z) I_{\{|Z| > c\}}].$$

Hence,

$$\begin{aligned} \left| E(Z) - E[g_c(Z)] \right| &= \left| E[Z\, I_{\{|Z| \geq c\}}] - c E[\text{sgn}(Z)\, I_{\{|Z| \geq c\}}] \right| \\ &\leq E[|Z|\, I_{\{|Z| \geq c\}}] + c E[I_{\{|Z| \geq c\}}] \\ &\leq 2 E[|Z|\, I_{\{|Z| \geq c\}}]. \end{aligned}$$

It follows that

$$|E(X_n) - E(X)| \leq \left| E[g_c(X_n)] - E[g_c(X)] \right| + 2E[|X_n| I_{\{|X_n| \geq c\}}] + 2E[|X| I_{\{|X| \geq c\}}].$$

$$\left| E(X_n) - E[g_c(X_n)] + E[g_c(X_n)] - E[g(X)] + E[g(X)] - E(X) \right|$$

Fix $\epsilon > 0$. Note that, by the dominated convergence theorem, using the fact that $E[|X|] < \infty$,

$$\lim_{c \to \infty} E[|X|I_{\{|X| \geq c\}}] = 0.$$

Hence,

$$\lim_{c \to \infty} \sup_n \{E[|X_n| \, I_{\{|X_n| \geq c\}}] + E[|X|I_{\{|X| \geq c\}}]\} = 0.$$

Therefore, given $\epsilon > 0$, there exists a $c > 0$ such that

$$|E(X_n) - E(X)| \leq |E[g_c(X_n)] - E[g_c(X)]| + \epsilon$$

for all $n = 1, 2, \ldots$.

Since g_c is a bounded continuous function,

$$\limsup_{n \to \infty} |E(X_n) - E(X)| \leq \lim_{n \to \infty} |E[g_c(X_n)] - E[g_c(X)]| + \epsilon = \epsilon.$$

Since $\epsilon > 0$ is arbitrary, the result follows. ■

In fact, the assumption that $E(|X|) < \infty$ is not needed in Theorem 11.11; it may be shown that if $X_1, X_2, \ldots$ is uniformly integrable, then $E(|X|) < \infty$. See, for example, Billingsley (1995, Section 25). However, in typical applications, it is a relatively simple matter to verify directly that $E(|X|) < \infty$.

***Example* 11.21 (*Minimum of uniform random variables*).** As in Example 11.3, let $Y_1, Y_2, \ldots$ denote a sequence of independent identically distributed, each with a uniform distribution on $(0, 1)$ and let $X_n = n \min(Y_1, \ldots, Y_n)$. Recall that $X_n \overset{D}{\to} X$ as $n \to \infty$, where X denotes a random variable with a standard exponential distribution.

Consider

$$E[|X_n|I_{\{|X_n| \geq c\}}] = n \int_{\frac{c}{n}}^1 tn(1-t)^{n-1} \, dt = \frac{n}{n+1}\left(1 - \frac{c}{n}\right)^n (1+c).$$

Note that $n/(n+1)$ and $(1 - c/n)^n$ are both increasing functions of n; it follows that

$$\sup_n E[|X_n|I_{\{|X_n| \geq c\}}] = \lim_{n \to \infty} \frac{n}{n+1}\left(1 - \frac{c}{n}\right)^n (1+c) = (1+c)\exp(-c)$$

and, hence, that

$$\lim_{c \to \infty} \sup_n E[|X_n|I_{\{|X_n| \geq c\}}] = 0$$

so that $X_1, X_2, \ldots$ is uniformly integrable.

Therefore,

$$\lim_{n \to \infty} E(X_n) = E(X) = 1. \qquad \square$$

***Example* 11.22 (*Mixture of normal distributions*).** For each $n = 1, 2, \ldots$, define a random variable X_n as follows. With probability $(n-1)/n$, X_n has a standard normal distribution; with probability $1/n$, X_n has a normal distribution with mean n and standard deviation 1. Hence, X_n has distribution function

$$F_n(x) = \frac{n-1}{n}\Phi(x) + \frac{1}{n}\Phi(x - n), \quad -\infty < x < \infty$$

where Φ denotes the distribution function of the standard normal distribution.

Note that, for any $-\infty < x < \infty$,

$$0 \leq \frac{1}{n}\Phi(x-n) \leq \frac{1}{n}$$

so that

$$\lim_{n\to\infty} F_n(x) = \Phi(x)$$

and, hence, $X_n \overset{D}{\to} Z$ as $n \to \infty$, where Z denotes a random variable with a standard normal distribution.

Consider

$$E[|X_n|\, I_{\{|X_n|\geq c\}}] = \frac{n-1}{n}E[|Z|\, I_{\{|Z|\geq c\}}] + \frac{1}{n}E[|Z+n|\, I_{\{|Z+n|\geq c\}}].$$

Let ϕ denote the standard normal density function; note that

$$E[|Z+n|\, I_{\{|Z+n|\geq c\}}] = \int_{c-n}^{\infty}(z+n)\phi(z)\,dz - \int_{-\infty}^{-(c+n)}(z+n)\phi(z)\,dz$$

$$= n[\Phi(c+n) - \Phi(c-n)] + \int_{c-n}^{\infty}z\phi(z)\,dz + \int_{c+n}^{\infty}z\phi(z)\,dz$$

$$\geq n[\Phi(c+n) - \Phi(c-n)].$$

Hence,

$$E[|X_n|I_{\{|X_n|\geq c\}}] \geq \frac{1}{n}E[|Z+n|I_{\{|Z+n|\geq c\}}] \geq \Phi(c+n) - \Phi(c-n)$$

and

$$\sup_n E[|X_n|\, I_{\{|X_n|\geq c\}}] \geq 1$$

for all $c > 0$. It follows that $X_1, X_2, \ldots$ is not uniformly integrable.

In fact, here $E(X_n) = 1$ for all $n = 1, 2, \ldots$, while $E(Z) = 0$. $\square$

The following result is sometimes useful in showing that a given sequence of random variables is uniformly integrable.

Theorem 11.12. *Let $X_1, X_2, \ldots$ denote real-valued random variables. The sequence $X_1, X_2, \ldots$ is uniformly integrable provided that either of the two following conditions holds:*

(i) $\sup_n E(|X_n|^{1+\epsilon}) < \infty$ for some $\epsilon > 0$

(ii) There exists a uniformly integrable sequence $Y_1, Y_2, \ldots$ such that

$$\Pr(|X_n| \leq |Y_n|) = 1, \quad n = 1, 2, \ldots.$$

Proof. The sufficiency of condition (i) follows from the fact that, for any $\epsilon > 0$ and $c > 0$,

$$E(|X_n|\, I_{\{|X_n|\geq c\}}) \leq \frac{1}{c^{\epsilon}}E(|X_n|^{1+\epsilon}).$$

The sufficiency of condition (ii) follows from the fact that, for any $c > 0$,

$$E[|X_n| \, I_{\{|X_n| \geq c\}}] \leq E[|X_n| \, I_{\{|X_n| \geq c, \, |X_n| \leq |Y_n|\}}] + E[|X_n| \, I_{\{|X_n| \geq c, \, |X_n| > |Y_n|\}}]$$
$$\leq E[|Y_n| \, I_{\{|Y_n| \geq c\}}]. \qquad \blacksquare$$

Example 11.23 (*Minimum of uniform random variables*). Let $Y_1, Y_2, \ldots$ denote a sequence of independent, identically distributed random variables, each with a uniform distribution on $(0, 1)$ and let $X_n = n \min(Y_1, \ldots, Y_n)$. Recall that $X_n \overset{D}{\to} X$ as $n \to \infty$, where X denotes a random variable with a standard exponential distribution; see Example 11.3. Since

$$E(X_n^2) = n^3 \int_0^1 t^2 (1 - t)^{n-1} \, dt = \frac{2n^2}{(n+2)(n+1)},$$

it follows that condition (i) of Theorem 11.12 holds with $\epsilon = 1$. Hence, $X_1, X_2, \ldots$ is uniformly integrable. $\square$

A condition for the convergence of higher-order moments follows easily from Theorems 11.11 and 11.12 and is given in the following corollary; the proof is left as an exercise.

Corollary 11.6. *Let $X_1, X_2, \ldots$ denote real-valued random variables such that $X_n \overset{D}{\to} X$ as $n \to \infty$ for some random variable X. If, for some $r = 1, 2, \ldots$, there exists an $\epsilon > 0$ such that*

$$\sup_n E(|X_n|^{r+\epsilon}) < \infty$$

then

$$\lim_{n \to \infty} E(X_n^r) = E(X^r).$$

Example 11.24 (*Minimum of uniform random variables*). Consider the framework considered in Example 11.23. Note that

$$E(X_n^r) = n^{r+1} \int_0^1 t^r (1-t)^{n-1} \, dt$$
$$= n^{r+1} \frac{\Gamma(r+1)\Gamma(n)}{\Gamma(n+r+1)}$$
$$= n^{r+1} \frac{r!}{(n+r)(n+r-1)\cdots n}, \qquad r = 1, 2, \ldots.$$

Hence,

$$\sup_n E(X_n^r) = r!, \qquad r = 1, 2, \ldots.$$

It follows that

$$\lim_{n \to \infty} E(X_n^r) = \int_0^\infty x^r \exp(-x) \, dx = r!, \qquad r = 1, 2, \ldots. \qquad \square$$

11.6 O_p and o_p Notation

Consider sequences of real numbers, $a_1, a_2, \ldots$ and $b_1, b_2, \ldots$. In describing the relationship between these sequences, it is often convenient to use the O and o notation; see Appendix 3. A similar notation is used for random variables. Let $X_1, X_2, \ldots$ denote a sequence of real-valued random variables. If, for any $\epsilon > 0$, there exists a constant M such that

$$\Pr(|X_n| \geq M) \leq \epsilon, \quad n = 1, 2, \ldots$$

we write $X_n = O_p(1)$; such a sequence is said to be *bounded in probability*. If $X_n \xrightarrow{p} 0$ as $n \to \infty$, we write $X_n = o_p(1)$.

This notation can be extended to consider the relationship between two sequences of random variables. Let $Y_1, Y_2, \ldots$ denote a sequence of real-valued random variables, $Y_n > 0$, $n = 1, 2, \ldots$. If

$$\frac{X_n}{Y_n} = O_p(1) \quad \text{as} \quad n \to \infty$$

we write $X_n = O_p(Y_n)$. This notation is often used when the sequence $Y_1, Y_2, \ldots$ is deterministic; for example, $X_n = O_p(n)$ means

$$\frac{X_n}{n} = O_p(1).$$

If

$$\frac{X_n}{Y_n} = o_p(1) \quad \text{as} \quad n \to \infty$$

we write $X_n = o_p(Y_n)$; again, this notation is often used when $Y_1, Y_2, \ldots$ is deterministic.

Finally, the O_p and o_p notation can be applied when $X_1, X_2, \ldots$ is a sequence of random vectors or random matrices. Let $\mathcal{X}$ be a set such that $\Pr(X_n \in \mathcal{X}) = 1$, $n = 1, 2, \ldots$, and let $|| \cdot ||$ denote a norm on $\mathcal{X}$. We write $X_n = O_p(Y_n)$ if $||X_n|| = O_p(Y_n)$ and $X_n = o_p(Y_n)$ if $||X_n|| = o_p(Y_n)$.

There are a number of simple rules for working with these symbols that makes their use particularly convenient. Several of these are given in the following theorem; other results along these lines can be established using the same general approach.

Theorem 11.13. *Let* $W_1, W_2, \ldots,$ $X_1, X_2, \ldots,$ $Y_1, Y_2, \ldots,$ *and* $Z_1, Z_2, \ldots$ *denote sequences real-valued random variables such that* $Y_n > 0$ *and* $Z_n > 0$, $n = 1, 2, \ldots$.

  (i) *If* $X_n = o_p(1)$ *as* $n \to \infty$, *then* $X_n = O_p(1)$ *as* $n \to \infty$.

 (ii) *If* $W_n = O_p(1)$ *and* $X_n = O_p(1)$ *as* $n \to \infty$, *then* $W_n + X_n = O_p(1)$ *and* $X_n W_n = O_p(1)$ *as* $n \to \infty$; *that is,* $O_p(1) + O_p(1) = O_p(1)$ *and* $O_p(1)O_p(1) = O_p(1)$.

(iii) *If* $W_n = O_p(1)$ *and* $X_n = o_p(1)$ *as* $n \to \infty$, *then* $W_n X_n = o_p(1)$ *and* $W_n + X_n = O_p(1)$ *as* $n \to \infty$; *that is,* $O_p(1)o_p(1) = o_p(1)$ *and* $O_p(1) + o_p(1) = O_p(1)$.

 (iv) *If* $X_n = O_p(Y_n)$ *and* $W_n = O_p(Z_n)$ *as* $n \to \infty$, *then*

$$W_n X_n = O_p(Y_n Z_n) \quad and \quad W_n + X_n = O_p(\max(Z_n, Y_n)) \quad as \quad n \to \infty;$$

that is,

$$O_p(Y_n)O_p(Z_n) = O_p(Y_n Z_n) \quad and \quad O_p(Y_n) + O_p(Z_n) = O_p(\max(Y_n, Z_n)).$$

Proof.   Suppose that $X_n = o_p(1)$. Fix $\epsilon > 0$ and $M_0 > 0$; there exists an N such that

$$\Pr(|X_n| \geq M_0) \leq \epsilon \quad \text{for all} \quad n \geq N.$$

For each $n = 1, 2, \ldots, N - 1$ let M_n denote a constant such that

$$\Pr(|X_n| \geq M_n) \leq \epsilon;$$

since

$$\lim_{M \to \infty} \Pr(|X_n| \geq M) = 0$$

such a constant must exist. Let

$$M = \max\{M_0, M_1, \ldots, M_{N-1}\}.$$

Then

$$\Pr(|X_n| \geq M) \leq \epsilon, \quad n = 1, 2, \ldots.$$

This proves part (i).

Suppose $W_n = O_p(1)$ and $X_n = O_p(1)$. Fix $\epsilon > 0$. Choose M_1 and M_2 so that

$$\Pr(|W_n| \geq M_1) \leq \epsilon/2, \quad n = 1, 2, \ldots$$

and

$$\Pr(|X_n| \geq M_2) \leq \epsilon/2, \quad n = 1, 2, \ldots.$$

Take $M = 2 \max(M_1, M_2)$. Then

$$
\begin{aligned}
\Pr(|W_n + X_n| \geq M) &\leq \Pr(|W_n| + |X_n| \geq M) \\
&\leq \Pr(|W_n| \geq M/2 \ \cup \ |X_n| \geq M/2) \\
&\leq \Pr(|W_n| \geq M_1) + \Pr(|X_n| \geq M_2) \leq \epsilon;
\end{aligned}
$$

hence, $X_n + W_n = O_p(1)$. Similarly,

$$
\begin{aligned}
\Pr\{|X_n W_n| \geq M_1 M_2\} &\leq \Pr\{|X_n| \geq M_1 \ \cup \ |W_n| \geq M_2\} \\
&\leq \Pr\{|X_n| \geq M_1\} + \Pr\{|W_n| \geq M_2\} \leq \epsilon.
\end{aligned}
$$

It follows that $X_n W_n = O_p(1)$, proving part (ii).

Suppose that $W_n = O_p(1)$ and $X_n = o_p(1)$. For any $t > 0$ and any $M > 0$,

$$
\begin{aligned}
\Pr(|W_n X_n| \geq t) &= \Pr(|W_n X_n| \geq t \ \cap \ |W_n| \geq M) + \Pr(|W_n X_n| \geq t \ \cap \ |W_n| < M) \\
&\leq \Pr(|W_n| \geq M) + \Pr(|X_n| \geq t/M).
\end{aligned}
$$

Fix $\epsilon > 0$ and $\delta > 0$. Choose M so that

$$\Pr(|W_n| \geq M) \leq \epsilon, \quad n = 1, 2, \ldots.$$

Then, for any $t > 0$,

$$\Pr(|W_n X_n| \geq t) \leq \epsilon + \Pr(|X_n| \geq t/M)$$

and, since $X_n \xrightarrow{p} 0$ as $n \to \infty$,

$$\limsup_{n \to \infty} \Pr(|W_n X_n| \geq t) \leq \epsilon.$$

Since ϵ is arbitrary,

$$\lim_{n\to\infty} \Pr(|W_n X_n| \geq t) = 0,$$

proving the first part of part (iii). The second part of part (iii) follows immediately from parts (i) and (ii).

Suppose $X_n = O_p(Y_n)$ and $W_n = O_p(Z_n)$. Then $X_n/Y_n = O_p(1)$ and $W_n/Z_n = O_p(1)$; it follows from part (ii) of the theorem that

$$\frac{X_n}{Y_n}\frac{W_n}{Z_n} = \frac{X_n Y_n}{Y_n Z_n} = O_p(1),$$

so that $W_n X_n = O_p(Y_n Z_n)$.

By part (ii) of the theorem,

$$\frac{|X_n|}{Y_n} + \frac{|W_n|}{Z_n} = O_p(1).$$

Since

$$\frac{|X_n + W_n|}{\max(Y_n, Z_n)} \leq \frac{|X_n| + |W_n|}{\max(Y_n, Z_n)} \leq \frac{|X_n|}{Y_n} + \frac{|W_n|}{Z_n},$$

it follows that

$$\frac{|X_n + W_n|}{\max(Y_n, Z_n)} = O_p(1)$$

and, hence, that $X_n + W_n = O_p(\max(Y_n, Z_n))$. This proves part (iv). ∎

The important thing to keep in mind when working with the O_p, o_p symbols is that each occurrence of these symbols in a given expression refers to a different sequence of random variables. For instance, we cannot conclude that $O_p(1) - O_p(1) = 0$; instead, all we can say is that $O_p(1) - O_p(1) = O_p(1)$.

The following result gives some conditions that can be used to show that a sequence $X_1, X_2, \ldots$ is $O_p(1)$.

Theorem 11.14. *Let $X_1, X_2, \ldots$ denote a sequence of real-valued random variables. Assume that one of the following conditions is satisfied:*

(i) *$X_n \overset{D}{\to} X$ as $n \to \infty$ for some random variable X.*
(ii) *There exists an increasing function $g : [0, \infty) \to \mathbf{R}$ such that*

$$\sup_n \mathrm{E}[g(|X_n|)] < \infty.$$

Then $X_n = O_p(1)$ as $n \to \infty$.

Proof. Suppose that $X_n \overset{D}{\to} X$ as $n \to \infty$ and let F denote the distribution function of X. Fix $\epsilon > 0$. Choose $M_0 > 0$ such M_0 and $-M_0$ are continuity points of F, such that

$$F(M_0) - F(-M_0) < \epsilon/2.$$

Then

$$\limsup_{n\to\infty} \Pr(|X_n| \geq M_0) \leq \epsilon$$

so that there exists $N > 0$ such that

$$\Pr(|X_n| \geq M_0) \leq \epsilon \quad \text{for all } n \geq N.$$

For each $n = 1, 2, \ldots, N - 1$, let M_n denote a constant such that

$$\Pr(|X_n| \geq M_n) \leq \epsilon.$$

Let

$$M = \max\{M_0, M_1, \ldots; M_{N-1}\}.$$

Then

$$\Pr(|X_n| \geq M) \leq \epsilon, \quad n = 1, 2, \ldots$$

so that $X_n = O_p(1)$.

Now suppose that condition (ii) of the theorem is satisfied. Given $\epsilon > 0$, choose M_0 so that

$$M_0 > \frac{\sup_n \mathrm{E}[g(|X_n|)]}{\epsilon}.$$

Then, by Markov's inequality,

$$\Pr\{g(|X_n|) \geq M_0\} \leq \frac{\mathrm{E}[g(|X_n|)]}{M_0} < \epsilon, \quad n = 1, 2, \ldots.$$

Note that, since g is an increasing function, it is invertible. Let $M = g^{-1}(M_0)$. Then

$$\Pr\{|X_n| \geq M\} = \Pr\{g(|X_n|) \geq g(M)\} = \Pr\{g(|X_n|) \geq M_0\} < \epsilon, \quad n = 1, 2, \ldots.$$

It follows that $X_n = O_p(1)$. ■

The advantage of the O_p, o_p notation is that these symbols can be used in algebraic expressions; this is illustrated in the following examples.

Example 11.25 (Sample mean). Let $X_1, X_2, \ldots$ denote independent, identically distribution random variables such that $\mathrm{E}(|X_1|) < \infty$ and let $\mu = \mathrm{E}(X_1)$. Using the weak law of large numbers given in Example 11.15,

$$\frac{1}{n} \sum_{j=1}^{n} X_j \xrightarrow{p} \mu \quad \text{as } n \to \infty.$$

Hence,

$$\frac{1}{n} \sum_{j=1}^{n} X_j = \mu + o_p(1) \quad \text{as } n \to \infty.$$

Now suppose that $\sigma^2 \equiv \mathrm{E}[(X_1 - \mu)^2] < \infty$. By Markov's inequality, for any $M > 0$,

$$\Pr\left\{\left|\frac{1}{\sqrt{n}} \sum_{j=1}^{n} X_j - \mu\right| \geq M\right\} \leq \frac{1}{M^2} \mathrm{E}\left\{\frac{1}{n}\left|\sum_{j=1}^{n}(X_j - \mu)\right|^2\right\}.$$

Since $\sum_{j=1}^{n}(X_j - \mu)$ has mean 0 and variance $n\sigma^2$,

$$\mathrm{E}\left\{\left|\sum_{j=1}^{n}(X_j - \mu)\right|^2\right\} = n\sigma^2$$

and

$$\Pr\left\{\left|\frac{1}{\sqrt{n}}\sum_{j=1}^{n}X_j - \mu\right| \geq M\right\} \leq \frac{\sigma^2}{M}.$$

It follows that, given $\epsilon > 0$, we can always find $M > 0$ such that

$$\Pr\left\{\left|\frac{1}{\sqrt{n}}\sum_{j=1}^{n}X_j - \mu\right| \geq M\right\} \leq \epsilon;$$

that is,

$$\frac{1}{\sqrt{n}}\left[\sum_{j=1}^{n}X_j - \mu\right] = O_p(1).$$

Written another way, we have that

$$\frac{1}{n}\sum_{j=1}^{n}X_j = \mu + O_p\left(\frac{1}{\sqrt{n}}\right) \quad \text{as} \quad n \to \infty. \qquad \square$$

***Example* 11.26 (*Approximation of a function of a random variable*).** Let $X_1, X_2, \ldots$ denote a sequence of real-valued random variables such that

$$\sqrt{n}X_n \xrightarrow{D} Z \quad \text{as} \quad n \to \infty$$

for some real-valued random variable Z. Note that $X_n = o_p(1)$ as $n \to \infty$.

Let $\mathcal{X}_n$ denote the range of X_n, $n = 1, 2, \ldots$, and let $\mathcal{X} = \cup_{n=1}^{\infty}\mathcal{X}_n$. Let $f : \mathcal{X} \to \mathbf{R}$ denote a twice-continuously-differentiable function such that $|f'(0)| > 0$. Let $Y_n = f(X_n)$, $n = 1, 2, \ldots$. Then, using a Taylor's series expansion,

$$Y_n = f(X_n) = f(0) + f'(0)X_n + \int_0^{X_n}(X_n - t)f''(t)\,dt.$$

Note that

$$\left|\int_0^{X_n}(X_n - t)f''(t)\,dt\right| \leq |X_n|\left|\int_0^{X_n}f''(t)\,dt\right|$$

and, since

$$\int_0^{x}f''(t)\,dt$$

is a continuous function of x,

$$\int_0^{X_n}f''(t)\,dt \xrightarrow{P} 0 \quad \text{as} \quad n \to \infty.$$

It follows that

$$Y_n = f(0) + f'(0)X_n + o_p(1).$$

However, a stronger statement is possible. Since

$$\sqrt{n}|X_n| = O_p(1),$$

it follows that

$$\sqrt{n}|X_n| \left| \int_0^{X_n} f''(t)\,dt \right| \xrightarrow{p} 0 \quad \text{as} \quad n \to \infty$$

and, hence, that

$$Y_n = f(0) + f'(0)X_n + O_p(n^{-\frac{1}{2}}).$$

In fact, by the mean-value theorem for integrals, there exists X_n^*, $|X_n^*| \le |X_n|$, such that

$$\int_0^{X_n} (X_n - t)f''(t)\,dt = \frac{1}{2}X_n^2 f''(X_n^*).$$

Since $|X_n^*| \le |X_n|$ and $X_n = o_p(1)$, it follows that $X_n^* = o_p(1)$ and, hence, that $f''(X_n^*) = f''(0) + o_p(1)$. We may therefore write

$$Y_n = f(0) + f'(0)X_n + \frac{1}{2}X_n^2[f''(0) + o_p(1)].$$

Since $X_n^2 = O_p(n^{-1})$, it follows that

$$Y_n = f(0) + f'(0)X_n + \frac{1}{2}f''(0)X_n^2 + o_p(n^{-1}) \quad \text{as} \quad n \to \infty. \qquad \square$$

11.7 Exercises

11.1 For each n, let X_n denote a discrete random variable taking values in the set $\{1, 2, \dots\}$ and let X denote a discrete random variable also taking values in $\{1, 2, \dots\}$. Let $p_n(x)$ denote the frequency function of X_n and let $p(x)$ denote the frequency function of X. Show that

$$X_n \xrightarrow{D} X \quad \text{as} \quad n \to \infty$$

if and only if

$$\lim_{n \to \infty} p_n(x) = p(x) \quad \text{for each } x = 1, 2, \dots.$$

11.2 Let $X_1, X_2, \dots$ denote independent, identically distributed random variables such that X_1 is continuously distributed with density $p(x)$ and distribution function $F(x)$ where $F(0) = 0$, $F(x) > 0$ for all $x > 0$, and $p(0) > 0$. Let

$$Y_n = n \min\{X_1, \dots, X_n\}.$$

Then there exists a random variable Y such that

$$Y_n \xrightarrow{D} Y \quad \text{as} \quad n \to \infty.$$

Find the distribution function of Y.

11.3 Let $Y_1, Y_2, \dots$ denote a sequence of Poisson random variables such that Y_n has mean $\lambda_n > 0$ for all $n = 1, 2, \dots$.
Give conditions on the sequence λ_n so that there exist sequences a_n and b_n such that

$$a_n Y_n + b_n \xrightarrow{D} Z \quad \text{as} \quad n \to \infty$$

where Z denotes a standard normal random variable. Give expressions for a_n and b_n.

11.4 Let X_n denote a random variable with a Γ distribution with parameters α_n, β_n so that X_n has density function

$$\frac{\Gamma(\alpha_n)}{\beta_n^{\alpha_n}} x^{\alpha_n-1} \exp\{-\beta_n x\}, \quad x > 0$$

where $\alpha_n > 0$ and $\beta_n > 0$. Find conditions on α_n and β_n such that there exist sequences of constants c_n, d_n satisfying

$$\frac{X_n - d_n}{c_n} \xrightarrow{D} N(0, 1).$$

Express c_n and d_n in terms of α_n and β_n.

11.5 For each $n = 1, 2, \ldots$, suppose that X_n is a discrete random variable with range $\{1/n, 2/n, \ldots, 1\}$ and

$$\Pr(X_n = j/n) = \frac{2j}{n(n+1)}, \quad j = 1, \ldots, n.$$

Does $X_1, X_2, \ldots$ converge in distribution to some random variable X? If so, find the distribution of X.

11.6 Let $X_1, X_2, \ldots$ denote independent random variables such that X_j has a Poisson distribution with mean λt_j where $\lambda > 0$ and $t_1, t_2, \ldots$ are known positive constants.

(a) Find conditions on $t_1, t_2, \ldots$ so that

$$Y_n = \frac{\sum_{j=1}^n X_j / \sum_{j=1}^n t_j - \lambda}{\mathrm{Var}(\sum_{j=1}^n X_j / \sum_{j=1}^n t_j)}$$

converges in distribution to a standard normal random variable.

(b) Suppose that, for each $j = 1, \ldots, t_j$ lies in the interval (a, b) where $0 < a < b < \infty$. Does it follow that Y_n converges in distribution to a standard normal random variable?

(c) Suppose that $t_j = j, j = 1, \ldots$. Does it follow that Y_n converges in distribution to a standard normal random variable?

11.7 Consider a discrete distribution with frequency function $p(x; n, \theta), x = 0, 1, \ldots$, where n is a nonnegative integer and $\theta \in \Theta$ for some set $\Theta \subset \mathbf{R}$. Let $\theta_1, \theta_2, \ldots$ denote a sequence in Θ and let X_n denote a random variable with frequency function $p(x; n, \theta_n), n = 1, 2, \ldots$. For the two choices of p given below, find conditions on the sequence $\theta_1, \theta_2, \ldots$ so that

$$X_n \xrightarrow{D} X \quad \text{as } n \to \infty,$$

where X has a Poisson distribution.

(a)

$$p(x; n, \theta) = \binom{n}{x} \theta^x (1 - \theta)^{n-x}, \quad x = 0, 1, \ldots, n; \quad \Theta = (0, 1)$$

(b)

$$p(x; n, \theta) = \binom{n + x - 1}{n} \theta^n (1 - \theta)^x, \quad x = 0, 1, \ldots, ; \quad \Theta = (0, 1).$$

In each case give examples of sequences satisfying your condition.

11.8 Let $X_1, X_2, \ldots$ denote a sequence of real-valued random variables such that

$$X_n \xrightarrow{D} X \quad \text{as } n \to \infty$$

for some real-valued random variable X and suppose that, for some set $B \subset \mathbf{R}$,

$$\Pr(X_n \in B) = 1 \quad \text{for all } n = 1, 2, \ldots.$$

Does it follow that $\Pr(X \in B) = 1$?

11.9 Let $X_1, X_2, \ldots$ denote a sequence of real-valued random variables such that, for each $n = 1, 2, \ldots$, X_n has a discrete distribution with frequency function p_n. Let X denote a discrete random variable with frequency function p such that

$$X_n \xrightarrow{D} X \quad \text{as } n \to \infty.$$

Does it follow that

$$\lim_{n \to \infty} p_n(x) = p(x), \quad x \in \mathbf{R}?$$

11.10 For each $n = 1, 2, \ldots$, define real-valued random variables X_n and Y_n as follows. Let $T_1, T_2, \ldots, T_n$ denote independent, identically distributed random variables, each uniformly distributed on $(0, 1)$ and let $T_{(1)}, \ldots, T_{(n)}$ denote the order statistics. Define

$$X_n = T_{(n)} - T_{(n-1)} \quad \text{and} \quad Y_n = T_{(n)} - T_{(n-2)}.$$

Find sequences of constants $a_1, a_2, \ldots$ and $b_1, b_2, \ldots$ such that, as $n \to \infty$,

$$a_n X_n \xrightarrow{D} X \quad \text{and} \quad b_n Y_n \xrightarrow{D} Y$$

for some non-degenerate random variables X and Y. Find the distributions of X and Y.

11.11 Let $X_1, X_2, \ldots$ and $Y_1, Y_2, \ldots$ denote sequences of real-valued random variables such that

$$X_n \xrightarrow{D} X \quad \text{as } n \to \infty$$

for some real-valued random variable X and that

$$\lim_{n \to \infty} \mathrm{E}[(X_n - Y_n)^2] = 0.$$

Show that

$$Y_n \xrightarrow{D} X \quad \text{as } n \to \infty.$$

Consider the following converse to this result. Suppose that

$$\begin{pmatrix} X_n \\ Y_n \end{pmatrix} \xrightarrow{D} \begin{pmatrix} X \\ X \end{pmatrix} \quad \text{as } n \to \infty$$

and that $\mathrm{E}(X_n^2) < \infty$ and $\mathrm{E}(Y_n^2) < \infty$ for all n. Does it follow that

$$\lim_{n \to \infty} \mathrm{E}[(X_n - Y_n)^2] = 0?$$

11.12 Let X_n and Y_n, $n = 1, 2, \ldots$, denote sequences of real-valued random variables such that $E(Y_n) = E(X_n) = 0$ and $\mathrm{Var}(Y_n) = \mathrm{Var}(X_n) = 1$. Let $\rho_n = \mathrm{Cov}(X_n, Y_n)$. Suppose that, as $n \to \infty$,

(a) $X_n \xrightarrow{D} N(0, 1)$

(b) $\rho_n \to 1$

(c) $Y_n \xrightarrow{D} Y$ for some random variable Y.

Find the distribution of Y or show that the distribution cannot be determined from the information given.

11.13 Let $X_1, X_2, \ldots$ denote a sequence of real-valued random variables such that, for each $n = 1, 2, \ldots$, X_n has an absolutely continuous distribution. Let X denote a real-valued random variable with a non-degenerate distribution such that

$$X_n \xrightarrow{D} X \quad \text{as } n \to \infty.$$

Does it follow that X has an absolutely continuous distribution?

11.14 For each $n = 1, 2, \ldots$, let X_n denote a random variable with a noncentral chi-squared distribution with m degrees of freedom and noncentrality parameter Δ_n, where $\Delta_n \to 0$ as $n \to \infty$. Let X denote a random variable with a (central) chi-squared distribution with m degrees of freedom. Does it follow that $X_n \xrightarrow{D} X$ as $n \to \infty$? Are any conditions needed on the sequence $\Delta_1, \Delta_2, \ldots$?

11.15 Let $X, X_1, X_2, \ldots$ denote real-valued random variables such that, for each $n = 1, 2, \ldots$, X_n has an absolutely continuous distribution with density p_n and X has an absolutely continuous distribution with density p. Suppose that

$$\lim_{n \to \infty} p_n(x) = p(x), \quad x \in \mathbf{R}.$$

Show that

$$X_n \xrightarrow{D} X \quad \text{as } n \to \infty.$$

11.16 Let X denote a random variable such that

$$\Pr(X = x_j) = \theta_j, \quad j = 1, 2, \ldots,$$

where $x_1, x_2, \ldots$ is an increasing sequence of real numbers and $\theta_1, \theta_2, \ldots$ is a sequence of nonnegative real numbers such that

$$\sum_{j=1}^{\infty} \theta_j = 1.$$

Let $X_1, X_2, \ldots$ denote a sequence of real-valued random variables, each with an absolutely continuous distribution, such that $X_n \xrightarrow{D} X$ as $n \to \infty$. For each $n = 1, 2, \ldots$, let F_n denote the distribution function of X_n and let F denote the distribution function of X.

(a) Find $\alpha > 0$ such that

$$\sup_x |F_n(x) - F(x)| \geq \alpha, \quad n = 1, 2, \ldots.$$

(b) Let a, b be real numbers such that

$$x_{j-1} < a < b < x_j$$

for some $j = 1, 2, \ldots$, where $x_0 = -\infty$. Does it follow that

$$\lim_{n \to \infty} \sup_{a \leq x \leq b} |F_n(x) - F(x)| = 0?$$

11.17 Let $X, X_1, X_2, \ldots$ denote real-valued random variables such that

$$X_n \xrightarrow{P} c \quad \text{as } n \to \infty$$

for some constant c. For each $n = 1, 2, \ldots$, let F_n denote the distribution function of X_n and let $a_1, a_2, \ldots$ denote a sequence such that $a_n \to a$ as $n \to \infty$ for some $a \in \mathbf{R}$. Does it follow that

$$\lim_{n \to \infty} F_n(a_n) = \begin{cases} 0 & \text{if } a < c \\ 1 & \text{if } a > c \end{cases} ?$$

11.18 Let $X_n, n = 1, \ldots$ denote a sequence of real-valued random variables; X_n is said to *converge in mean* to a random variable X if

$$\lim_{n \to \infty} E[|X_n - X|] = 0.$$

(a) Show that if X_n converges to X in mean then X_n converges to X in probability.

(b) Show that if X_n converges to X in mean and $E[|X|] < \infty$ then

$$\lim_{n \to \infty} E[X_n] = E[X].$$

(c) Give a counter example to show that if X_n converges to X in probability then X_n does not necessarily converge to X in mean.

11.19 Prove Corollary 11.2.

11.20 Let $X_1, X_2, \ldots$ denote a sequence of real-valued random variables such that

$$X_n \overset{D}{\to} X \quad \text{as } n \to \infty$$

for some real-valued random variable X and let Y denote a real-valued random variable. Does it follow that

$$\begin{pmatrix} X_n \\ Y \end{pmatrix} \overset{D}{\to} \begin{pmatrix} X \\ Y \end{pmatrix} \quad \text{as } n \to \infty?$$

11.21 Let X_n, $n = 1, 2, \ldots$, and Y_n, $n = 1, 2, \ldots$, denote sequences of real-valued random variables such that, for each $n = 1, 2, \ldots$, X_n and Y_n are independent. Suppose that there exist independent random variables X and Y such that, as $n \to \infty$,

$$X_n \overset{D}{\to} X \quad \text{and} \quad Y_n \overset{D}{\to} Y.$$

Does it follow that

$$\begin{pmatrix} X_n \\ Y_n \end{pmatrix} \overset{D}{\to} \begin{pmatrix} X \\ Y \end{pmatrix} \quad \text{as } n \to \infty?$$

11.22 Prove Corollary 11.4.

11.23 Let $X_1, X_2, \ldots$ and $Y_1, Y_2, \ldots$ denote sequences of real-valued random variables, each of which is uniformly integrable. Does it follow that $X_1 + Y_1, X_2 + Y_2, \ldots$ is uniformly integrable?

11.24 Let $X_1, X_2, \ldots$ denote a uniformly integrable sequence of real-valued random variables. Show that

$$\sup_n E(|X_n|) < \infty.$$

11.25 Find a sequence of random variables $X_1, X_2, \ldots$ that is uniformly integrable, but

$$\sup_n E(|X_n|^{1+\epsilon}) = \infty \quad \text{for all } \epsilon > 0.$$

Thus, the condition given in part (i) of Theorem 11.12 is sufficient, but not necessary.

11.26 Let $X_1, X_2, \ldots$ and $Y_1, Y_2, \ldots$ denote sequences of real-valued random variables such that $Y_1, Y_2, \ldots$ is uniformly integrable and

$$\Pr(|X_n| \le |Y_n|) \ge \alpha_n, \quad n = 1, 2, \ldots$$

where $0 < \alpha_n < 1, n = 1, 2, \ldots$ and $\alpha_n \to 1$ as $n \to \infty$.
Give conditions on the sequence $\alpha_1, \alpha_2, \ldots$ so that $X_1, X_2, \ldots$ is uniformly integrable or show that such conditions do not exist. Does your answer change if it is also assumed that $\sup_n E[|X_n|] < \infty$?

11.27 Prove Theorem 11.8.

11.28 Let $X_1, X_2, \ldots$ denote a sequence of real-valued random variables such that

$$\lim_{n \to \infty} \text{Var}(X_n) = 0.$$

Does it follow that

$$X_n = \mathrm{E}(X_n) + o_p(1) \quad \text{as} \quad n \to \infty?$$

11.29 Let $X_1, X_2, \ldots$ denote a sequence of real-valued random variables such that $X_n = O_p(1)$ as $n \to \infty$. Let $f : \mathbf{R} \to \mathbf{R}$ denote a function and suppose that there exists a set B such that f is continuous on B and

$$\Pr(X_n \in B) = 1, \quad n = 1, 2, \ldots.$$

Does it follow that $f(X_n) = O_p(1)$ as $n \to \infty$?

11.30 Let $X_1, X_2, \ldots$ denote a sequence of real-valued random variables. Suppose that $X_n = O_p(1)$ as $n \to \infty$. Does it follow that $\mathrm{E}(X_n) = O(1)$ as $n \to \infty$? Suppose that $\mathrm{E}(X_n) = O(1)$ as $n \to \infty$. Does it follow that $X_n = O_p(1)$ as $n \to \infty$?

11.31 Suppose that $X_1, X_2, \ldots$ is a sequence of real-valued random variable such that $X_1, X_2, \ldots$ is uniformly integrable. Does it follow that $X_n = O_p(1)$ as $n \to \infty$?

11.32 Let $X_1, X_2, \ldots$ denote a sequence of real-valued random variables. Suppose that, given $\epsilon > 0$, there exists an M and a positive integer N such that

$$\Pr(|X_n| \geq M) \leq \epsilon, \quad n = N + 1, N + 2, \ldots.$$

Does it follow that $X_n = O_p(1)$ as $n \to \infty$?

11.8   Suggestions for Further Reading

Convergence in distribution is covered in many books on probability theory; see, for example, Billingsley (1995, Chapter 5) and Port (1994, Chapter 50). Billingsley (1968) gives a general treatment of convergence in distribution in metric spaces. Useful books on the theory and application of large-sample methods in statistics include Ferguson (1996), Lehmann (1999), Sen and Singer (1993), and van der Vaart (1998).

12

Central Limit Theorems

12.1 Introduction

In this chapter, we consider the asymptotic distribution of the sample mean. For instance, let $\bar{X}_n$ denote the sample mean based on random variables $X_1, \ldots, X_n$. In this chapter, it is shown that, under a wide variety of conditions, a suitably scaled and shifted version of $\bar{X}_n$ converges in distribution to a standard normal random variable. Such a result is called a central limit theorem. The simplest case is when $X_1, \ldots, X_n$ are independent and identically distributed; this is discussed in Section 12.2. The remaining sections of this chapter consider more general settings, including cases in which the random variables under consideration are not identically distributed or are not independent.

12.2 Independent, Identically Distributed Random Variables

We begin by considering the asymptotic behavior of sample means of independent, identically distributed random variables. These results rely heavily on the fact that convergence in distribution may be established by showing convergence of characteristic functions. This fact is particularly useful in the present context since the characteristic function of a sum of independent random variables is simply the product of the characteristic functions of the component random variables.

When discussing limiting distributions of this type, it is convenient to use the symbol $N(0, 1)$ to denote a standard normal random variable.

Theorem 12.1. *Let* $X_1, X_2, \ldots$ *denote independent, identically distributed real-valued random variables with each mean* μ *and variance* $\sigma^2 < \infty$ *and let*

$$\bar{X}_n = \frac{X_1 + X_2 + \cdots + X_n}{n}, \quad n = 1, 2, \ldots.$$

Then

$$\frac{\sqrt{n}(\bar{X}_n - \mu)}{\sigma} \xrightarrow{D} N(0, 1) \quad as \quad n \to \infty.$$

Proof. Let φ denote the common characteristic function of the X_j and let φ_n denote the characteristic function of

$$\frac{\sqrt{n}(\bar{X}_n - \mu)}{\sigma} = \frac{X_1 + X_2 + \cdots + X_n - n\mu}{\sigma\sqrt{n}}.$$

365

Then, for each $t \in \mathbf{R}$,

$$\varphi_n(t) = \exp\{-it\mu/(\sigma/\sqrt{n})\}\varphi(t/(\sigma\sqrt{n}))^n.$$

Recall that, by Theorem 4.13,

$$\log \varphi(t) = it\mu - \frac{t^2}{2}\sigma^2 + o\left(t^2\right) \quad \text{as} \quad t \to 0.$$

Hence, for each t,

$$\log \varphi_n(t) = n\left(it\frac{\mu}{\sigma\sqrt{n}} - \frac{t^2}{2n} + o\left(n^{-1}\right)\right) - \frac{i\mu\sqrt{n}}{\sigma\sqrt{n}} = -\frac{t^2}{2} + o(1)$$

as $n \to \infty$. The result now follows from Theorem 11.2. ∎

The conclusion of Theorem 12.1 is often stated "$\bar{X}_n$ is asymptotically distributed according to a normal distribution with mean μ and variance σ^2/n." This statement should not be taken to mean that $\bar{X}_n$ converges in distribution to a normally distributed random variable with mean μ and variance σ^2/n, which would make no sense since the limiting distribution would then depend on n. Instead this statement should be viewed as an informal expression of the result that

$$\frac{\sqrt{n}(\bar{X}_n - \mu)}{\sigma} \xrightarrow{\mathcal{D}} N(0, 1) \quad \text{as} \quad n \to \infty.$$

In spite of the lack of precision, the informal statement is still useful and will be used here in the examples; however, in the theorems, results will be stated in terms of convergence in distribution.

Example 12.1 (Sample mean of Bernoulli random variables). Let $X_1, X_2, \ldots$ denote independent, identically distributed random variables such that, for each $j = 1, 2, \ldots$,

$$\Pr(X_j = 1) = 1 - \Pr(X_j = 0) = \theta$$

where $0 < \theta < 1$. It is straightforward to show that

$$\mu = \mathrm{E}(X_1) = \theta \quad \text{and} \quad \mathrm{Var}(X_1) = \theta(1 - \theta).$$

It follows that $\bar{X}_n = (X_1 + \cdots + X_n)/n$ is asymptotically distributed according to a normal distribution with mean θ and variance $\theta(1 - \theta)/n$.

Since $X_1 + \cdots + X_n$ has a binomial distribution with parameters n and θ, this result also shows that the binomial distribution may be approximated by the normal distribution when the index n is large. □

Example 12.2 (Normal approximation to the chi-squared distribution). Let $Z_1, Z_2, \ldots$ denote independent, identically distributed standard normal random variables and let $X_j = Z_j^2$, $j = 1, 2, \ldots$ Consider the distribution of

$$S_n = \sum_{j=1}^{n} X_j = \sum_{j=1}^{n} Z_j^2.$$

Recall that it was shown in Chapter 8 that the exact distribution of S_n is a chi-squared distribution with n degrees of freedom. Here we consider an approximation to the distribution of S_n, and hence, to the chi-squared distribution, based on the central limit theorem.

Table 12.1. *Normal approximation to the*
chi-squared distribution.

		n	
α	5	10	20
0.01	0.0799	0.0481	0.0317
0.05	0.111	0.0877	0.0740
0.10	0.142	0.125	0.116
0.20	0.200	0.196	0.196
0.80	0.765	0.779	0.787
0.90	0.910	0.910	0.908
0.95	0.973	0.968	0.964
0.99	0.999	0.998	0.997

Since the X_j are independent and identically distributed, each with mean 1 and variance 2, it follows that

$$\frac{\sqrt{n}(S_n/n - 1)}{\sqrt{2}} \xrightarrow{\mathcal{D}} N(0, 1) \quad \text{as} \quad n \to \infty.$$

Thus, S_n is approximately normally distributed with mean n and variance $2n$ when n, the degrees of freedom of the chi-squared distribution, is large; that is, for large n, $\Pr(S_n \leq s)$ can be approximated by $\Phi((s - n)/\sqrt{(2n)})$, where Φ denotes the distribution function of the standard normal distribution.

Table 12.1 contains approximations of the form $\Phi((s_{n\alpha} - n)/\sqrt{(2n)})$, where $s_{n\alpha}$ satisfies

$$\Pr(S_n \leq s_{n\alpha}) = \alpha,$$

for several choices of n and α. These results show that the normal approximation is generally quite accurate, except when n and α are both small.

Using an argument based on the central limit theorem, the normal approximation to the chi-squared distribution holds only when n, the degrees of freedom, is an integer. A more direct proof based on an expansion of the characteristic function of the chi-squared distribution shows that the normal approximation holds in general for large degrees-of-freedom. $\square$

12.3 Triangular Arrays

Theorem 12.1, which applies only to independent, identically distributed random variables, is too limited for many applications. A more general version of the central limit theorem applies to *triangular arrays*. Consider a collection of random variables of the form $\{X_{nk}, k = 1, \ldots, n, n = 1, \ldots\}$. Hence, when written as a two-dimensional array, this collection has the representation

$$\begin{matrix} X_{11} & & & \\ X_{21} & X_{22} & & \\ X_{31} & X_{32} & X_{33} & \\ \vdots & \vdots & \vdots & \vdots \end{matrix}$$

which leads to the term "triangular array." In such a triangular array the random variables in any row are taken to be independent; however, they do not need to be identically distributed. Furthermore, the distributions may change from row to row.

More generally, instead of having n elements in the nth row of the triangular array, we can consider a triangular array with r_n elements in the nth row, where $r_1, r_2, \ldots$ is a given deterministic sequence. The following result considers the asymptotic distribution of the sample mean in such a setting.

Theorem 12.2. *Let $r_1, r_2, \ldots$ be a given sequence such that $r_n \to \infty$ as $n \to \infty$. For each $n = 1, 2, \ldots$, let $X_{n1}, \ldots, X_{nr_n}$ denote independent random variables and let*

$$\mu_{nk} = \mathrm{E}(X_{nk}), \qquad \sigma_{nk}^2 = \mathrm{Var}(X_{nk}), \quad k = 1, \ldots, r_n$$

For $\alpha > 0$, define

$$\gamma_{nk}(\alpha) = \mathrm{E}\left[|X_{nk} - \mu_{nk}|^{2+\alpha}\right], \quad k = 1, \ldots, r_n; \ n = 1, 2, \ldots.$$

Assume that, for some $\alpha > 0$,

$$\lim_{n \to \infty} \frac{\sum_{k=1}^{r_n} \gamma_{nk}(\alpha)}{\left[\sum_{k=1}^{r_n} \sigma_{nk}^2\right]^{1+\frac{\alpha}{2}}} = 0. \tag{12.1}$$

Let

$$\bar{X}_n = \frac{1}{r_n} \sum_{k=1}^{r_n} X_{nk} \quad and \quad \bar{\mu}_n = \frac{1}{r_n} \sum_{k=1}^{r_n} \mu_{nk}.$$

Then

$$\frac{\sqrt{r_n}(\bar{X}_n - \bar{\mu}_n)}{\left[\sum_{k=1}^{r_n} \sigma_{nk}^2/r_n\right]^{\frac{1}{2}}} \xrightarrow{\mathcal{D}} N(0, 1) \quad as \ n \to \infty.$$

The proof uses the following lemma.

Lemma 12.1. *Let Y denote a real-valued random variable with distribution function F. Suppose that, for some $\alpha > 0$,*

$$\mathrm{E}(|Y|^{2+\alpha}) < \infty.$$

Then, for all $\epsilon > 0$,

$$\mathrm{E}[\min\{|tY|^2, |tY|^3\}] \leq \epsilon |t|^3 \mathrm{E}(Y^2) + \frac{t^2}{\epsilon^\alpha} \mathrm{E}[|Y|^{2+\alpha}].$$

Proof. Note that, for all $\epsilon > 0$,

$$\mathrm{E}[\min\{|tY|^2, |tY|^3\}] \leq |t|^3 \int_{\{|y|<\epsilon\}} |y|^3 \, dF(y) + t^2 \int_{\{|y|\geq\epsilon\}} |y|^2 \, dF(y).$$

Using the facts that

$$\int_{\{|y|<\epsilon\}} |y|^3 \, dF(y) \leq \epsilon \int_{\{|y|<\epsilon\}} |y|^2 \, dF(y) \leq \epsilon \mathrm{E}(|Y|^2)$$

and

$$\int_{\{|y|\geq\epsilon\}} |y|^2 \, dF(y) \leq \frac{1}{\epsilon^\alpha} \int_{\{|y|\geq\epsilon\}} |y|^{2+\alpha} \, dF(y) \leq \frac{1}{\epsilon^\alpha} \mathrm{E}(|Y|^{2+\alpha}),$$

the result follows.   ∎

Proof of Theorem 12.2.   Without loss of generality we may take $\mu_{nk} = 0$ for all n and k. For $n = 1, 2, \ldots$ and $k = 1, 2, \ldots, r_n$, let φ_{nk} denote the characteristic function of X_{nk} and let φ_n denote the characteristic function of

$$\frac{\sqrt{r_n}\bar{X}_n}{\left[\sum_{k=1}^{r_n} \sigma_{nk}^2/r_n\right]^{\frac{1}{2}}} = \frac{X_{n1} + \cdots + X_{nr_n}}{\left[\sum_{k=1}^{r_n} \sigma_{nk}^2\right]^{\frac{1}{2}}}.$$

Then, for each $t \in \mathbf{R}$,

$$\varphi_n(t) = \prod_{k=1}^{r_n} \varphi_{nk}(t/S_n),$$

where

$$S_n^2 = \sum_{k=1}^{r_n} \sigma_{nk}^2.$$

To prove the result, we show that, for each $t \in \mathbf{R}$,

$$\lim_{n\to\infty} \varphi_n(t) = \exp\{-t^2/2\}.$$

Recall that, by Lemma A2.1,

$$\exp\{it\} = \sum_{j=0}^{2} (it)^j/j! + R(t)$$

where

$$|R(t)| \leq \min\left\{\frac{1}{2}|t|^2, \frac{1}{6}|t|^3\right\}.$$

Hence,

$$\left|\varphi_{nk}(t) - \left(1 - \frac{t^2}{2}\sigma_{nk}^2\right)\right| \leq \mathrm{E}\left[\min\left\{|t|^2 X_{nk}^2, |t|^3|X_{nk}|^3\right\}\right].$$

It follows from Lemma 12.1 that, for all $\epsilon > 0$,

$$\left|\varphi_{nk}(t/S_n) - \left(1 - \frac{t^2}{2}\frac{\sigma_{nk}^2}{S_n^2}\right)\right| \leq \epsilon|t|^3 \frac{\sigma_{nk}^2}{S_n^2} + \frac{t^2}{\epsilon^\alpha}\frac{\gamma_{nk}(\alpha)}{S_n^{1+\alpha/2}}.$$

Note that, for any complex numbers $a_1, \ldots, a_n; b_1, \ldots, b_n$ with modulus at most 1,

$$|a_1 \cdots a_n - b_1 \cdots b_n| \leq \sum_{j=1}^{n} |a_j - b_j|.$$

Hence, for all $\epsilon > 0$,

$$\left|\varphi_n(t) - \prod_{k=1}^{r_n}\left(1 - \frac{t^2}{2}\frac{\sigma_{nk}^2}{S_n^2}\right)\right| = \left|\prod_{k=1}^{r_n} \varphi_{nk}\frac{t}{S_n} - \prod_{k=1}^{r_n}\left(1 - \frac{t^2}{2}\frac{\sigma_{nk}^2}{S_n^2}\right)\right| \leq \epsilon|t|^3 + \frac{t^2}{\epsilon^\alpha}\sum_{k=1}^{r_n}\frac{\gamma_{nk}(\alpha)}{S_n^{2+\alpha}}.$$

It follows from (12.1) that, for each $t \in \mathbf{R}$,

$$\limsup_{n\to\infty} \left| \varphi_n(t) - \prod_{k=1}^{r_n} \left(1 - \frac{t^2}{2} \frac{\sigma_{nk}^2}{S_n^2} \right) \right| \le \epsilon |t|^3.$$

Since $\epsilon > 0$ is arbitrary,

$$\lim_{n\to\infty} \left| \varphi_n(t) - \prod_{k=1}^{r_n} \left(1 - \frac{t^2}{2} \frac{\sigma_{nk}^2}{S_n^2} \right) \right| = 0$$

for each $t \in \mathbf{R}$.

The result now holds provided that

$$\lim_{n\to\infty} \prod_{k=1}^{r_n} \left(1 - \frac{t^2}{2} \frac{\sigma_{nk}^2}{S_n^2} \right) = \exp\left\{ \frac{-t^2}{2} \right\}.$$

By Taylor's theorem, for $x > 0$,

$$\exp(-x) = 1 - x + \frac{1}{2} \exp(-x^*) x^2$$

for some $x^*, 0 \le x^* \le x$. Hence,

$$\left| \exp\left\{ -\frac{t^2}{2} \frac{\sigma_{nk}^2}{S_n^2} \right\} - \left(1 - \frac{t^2}{2} \frac{\sigma_{nk}^2}{S_n^2} \right) \right| \le \frac{1}{8} t^4 \frac{\sigma_{nk}^4}{S_n^4}.$$

Fix t. Note that, since

$$\mathrm{E}\left[|X_{nk}|^2 \right]^{1+\frac{\alpha}{2}} \le \mathrm{E}\left[|X_{nk}|^{2+\alpha} \right],$$

$\sigma_{nk}^{2+\alpha} \le \gamma_{nk}(\alpha)$. Hence,

$$\left(\frac{\sup_{1\le k\le r_n} \sigma_{nk}}{S_n} \right)^{2+\alpha} = \frac{\sup_{1\le k\le r_n} \sigma_{nk}^{2+\alpha}}{S_n^{2+\alpha}} \le \frac{\sum_{k=1}^{r_n} \sigma_{nk}^{2+\alpha}}{S_n^{2+\alpha}} \le \frac{\sum_{k=1}^{r_n} \gamma_{nk}(\alpha)}{S_n^{2+\alpha}}.$$

It follows that

$$\lim_{n\to\infty} \frac{\sup_{1\le k\le r_n} \sigma_{nk}^2}{S_n^2} = 0.$$

Hence, for sufficiently large n,

$$\sup_{1\le k\le r_n} \left| 1 - \frac{t^2}{2} \frac{\sigma_{nk}^2}{S_n^2} \right| \le 1.$$

It follows that

$$\left| \prod_{k=1}^{r_n} \exp\left\{ -\frac{t^2}{2} \frac{\sigma_{nk}^2}{S_n^2} \right\} - \prod_{k=1}^{r_n} \left(1 - \frac{t^2}{2} \frac{\sigma_{nk}^2}{S_n^2} \right) \right| \le \sum_{k=1}^{r_n} \left| \exp\left\{ -\frac{t^2}{2} \frac{\sigma_{nk}^2}{S_n^2} \right\} - \left(1 - \frac{t^2}{2} \frac{\sigma_{nk}^2}{S_n^2} \right) \right|$$

$$\le \sum_{k=1}^{r_n} \frac{t^4}{8} \frac{\sigma_{nk}^4}{S_n^4} \le \sup_{1\le k\le r_n} \frac{\sigma_{nk}^2}{S_n^2} \frac{t^4}{8}$$

so that, for each $t \in \mathbf{R}$,

$$\lim_{n\to\infty} \left| \prod_{k=1}^{r_n} \exp\left\{ -\frac{t^2}{2} \frac{\sigma_{nk}^2}{S_n^2} \right\} - \prod_{k=1}^{r_n} \left(1 - \frac{t^2}{2} \frac{\sigma_{nk}^2}{S_n^2} \right) \right| = 0.$$

The result now follows from the fact that

$$\prod_{k=1}^{r_n} \exp\left\{-\frac{t^2}{2}\frac{\sigma_{nk}^2}{S_n^2}\right\} = \exp\left\{\frac{-t^2}{2}\right\}$$

so that

$$\lim_{n\to\infty}\prod_{k=1}^{r_n}\left(1 - \frac{t^2}{2}\frac{\sigma_{nk}^2}{S_n^2}\right) = \exp\left\{\frac{-t^2}{2}\right\},$$

as required.   ∎

Condition (12.1) of the theorem, known as *Lyapounov's condition*, may be weakened to the *Lindeberg condition*: for each $\epsilon > 0$,

$$\lim_{n\to\infty}\sum_{k=1}^{n}\frac{1}{S_n^2}\mathrm{E}[X_{nk}^2 \mathrm{I}_{\{|X_{nk}\geq\epsilon S_n\}}] = 0.$$

See, for example, Billingsley (1995, Section 27).

***Example* 12.3 (*Sum of Bernoulli random variables*).** Let $X_1, X_2, \ldots$ denote independent random variables such that, for each $n = 1, 2, \ldots,$

$$\Pr(X_n = 1) = 1 - \Pr(X_n = 0) = \theta_n,$$

where $\theta_1, \theta_2, \ldots$ is a sequence in $(0, 1)$, and consider the asymptotic distribution of $\bar{X}_n = \sum_{j=1}^{n} X_j/n$. Thus, $X_1, X_2, \ldots$ are independent, but not identically distributed.

Note that $X_1, X_2, \ldots$ can be viewed as a triangular array by taking $r_n = n$ and $X_{nk} = X_k$, $k = 1, \ldots, n, n = 1, 2, \ldots$. Using this representation,

$$\mu_{nk} = \theta_k \quad\text{and}\quad \sigma_{nk}^2 = \theta_k(1 - \theta_k), \qquad k = 1, 2, \ldots, n; \ n = 1, 2, \ldots.$$

Since, for $k = 1, 2, \ldots, n$ and $n = 1, 2, \ldots,$

$$\gamma_{nk}(\alpha) = \theta_k(1 - \theta_k)\left[\theta_k^{1+\alpha} + (1 - \theta_k)^{1+\alpha}\right],$$

it follows that (12.1) is satisfied if

$$\sum_{k=1}^{n}\theta_k(1 - \theta_k) \to \infty \quad\text{as}\quad n \to \infty.$$

Under this condition,

$$\frac{\sqrt{n}(\bar{X}_n - \sum_{k=1}^{n}\theta_k/n)}{\left[\sum_{k=1}^{n}\theta_k(1 - \theta_k)/n\right]^{\frac{1}{2}}} \xrightarrow{\mathcal{D}} N(0, 1) \quad\text{as}\quad n \to \infty.$$

That is, $\bar{X}_n$ follows the central limit theorem provided that

$$\sum_{k=1}^{n}\theta_k(1 - \theta_k) \to \infty \quad\text{as}\quad n \to \infty.$$

Although this type of analysis is useful for establishing formal limit theorems, it is not very useful for determining when a normal distribution will be a useful approximation to the distribution of $\bar{X}_n$ for a given value of n. For instance, suppose that we wish to approximate the distribution of $\bar{X}_n$ for $n = 50$, for given values of $\theta_1, \ldots, \theta_{50}$. Note that the terms $\theta_1, \ldots, \theta_{50}$ can be embedded both in a sequence satisfying

$$\sum_{k=1}^{n} \theta_k(1 - \theta_k) \to \infty \quad \text{as} \quad n \to \infty$$

and in a sequence not satisfying this condition. Furthermore, the magnitude of the sum of the first 50 terms in a sequence may be a poor guide as to the behavior of the infinite sum. Thus, the formal limit theorem cannot be used directly to assess the usefulness of the normal approximation. However, Theorem 12.2, as applied in this setting, suggests that the larger

$$\sum_{k=1}^{50} \theta_k(1 - \theta_k)$$

is, the more accuracy we can expect from the normal approximation to the distribution of $\bar{X}_n$. $\square$

Example 12.4 (*Linear regression model*). For each $n = 1, 2, \ldots$, let $Y_1, \ldots, Y_n$ denote independent random variables such that

$$Y_j = \theta z_j + \epsilon_j, \quad j = 1, \ldots, n$$

where $\epsilon_1, \epsilon_2, \ldots$ are independent identically distributed random variables with mean 0, variance σ^2 and

$$\gamma \equiv \mathrm{E}\left[|\epsilon_1|^3\right] < \infty.$$

Here $z_1, z_2, \ldots$ are fixed constants.

For $n = 1, 2, \ldots$ define

$$T_n = \frac{\sum_{j=1}^{n} z_j Y_j}{\sum_{j=1}^{n} z_j^2};$$

T_n is the least-squares estimator of θ in the model for $Y_1, \ldots, Y_n$. Our goal is to determine the asymptotic distribution of T_n. Note that

$$T_n = \frac{1}{n} \sum_{j=1}^{n} \frac{z_j}{\sum_{j=1}^{n} z_j^2/n} Y_j$$

so that T_n is the sample mean of the random variables

$$\frac{z_j}{\sum_{j=1}^{n} z_j^2/n} Y_j, \quad j = 1, \ldots, n.$$

Since the distribution of

$$\frac{z_j}{\sum_{j=1}^{n} z_j^2/n} Y_j$$

depends on n, we need to use the central limit theorem for a triangular array.

For $n = 1, 2, \ldots$ and $j = 1, 2, \ldots, n$, define

$$X_{nj} = \frac{z_j}{\sum_{j=1}^{n} z_j^2/n} Y_j;$$

then $T_n = \sum_{j=1}^{n} X_{nj}/n$. Note that X_{nj} has mean

$$\mu_{nj} = \theta \frac{z_j^2}{\sum_{j=1}^{n} z_j^2/n},$$

variance

$$\sigma_{nj}^2 = \frac{z_j^2}{\left[\sum_{j=1}^{n} z_j^2/n\right]^2} \sigma^2,$$

and

$$\gamma_{nj} \equiv \gamma_{nj}(1) = \mathrm{E}\left[|X_{nj} - \mu_{nj}|^3\right] = \frac{|z_j|^3}{\left[\sum_{j=1}^{n} z_j^2/n\right]^3} \gamma,$$

where $\gamma = \mathrm{E}\left[|\epsilon_1|^3\right]$.

Then

$$\frac{\sum_{j=1}^{n} \gamma_{nj}}{\left[\sum_{k=1}^{n} \sigma_{nk}^2\right]^{\frac{3}{2}}} = \frac{\sum_{j=1}^{n} |z_j|^3}{\left[\sum_{j=1}^{n} z_j^2\right]^{\frac{3}{2}}} \frac{\gamma}{\sigma^3}.$$

Hence, condition (12.1) of Theorem 12.2 is satisfied provided that the sequence $z_1, z_2, \ldots$ satisfies

$$\lim_{n \to \infty} \frac{\sum_{j=1}^{n} |z_j|^3}{\left[\sum_{j=1}^{n} z_j^2\right]^{\frac{3}{2}}} = 0. \tag{12.2}$$

This holds, for instance, if for all j, $|z_j| \leq M$ for some M and $\sum_{j=1}^{n} z_j^2$ diverges to ∞ as $n \to \infty$. Then

$$\frac{\sum_{j=1}^{n} |z_j|^3}{\left[\sum_{j=1}^{n} z_j^2\right]^{\frac{3}{2}}} \leq M \frac{\sum_{j=1}^{n} |z_j|^2}{\left[\sum_{j=1}^{n} z_j^2\right]^{\frac{3}{2}}} = \frac{M}{\left[\sum_{j=1}^{n} z_j^2\right]^{\frac{1}{2}}}$$

which approaches 0 as $n \to \infty$.

When (12.2) holds, T_n is asymptotically normally distributed with mean

$$\frac{1}{n} \sum_{j=1}^{n} \mu_{nj} = \theta$$

and variance

$$\frac{1}{n^2} \sum_{j=1}^{n} \sigma_{nj}^2 = \frac{\sigma^2}{\sum_{j=1}^{n} z_j^2}. \qquad \square$$

Example 12.5 (*Normal approximation to the binomial distribution*). Let $r_1, r_2, \ldots$ be a given sequence of nonnegative integers such that $r_n \to \infty$ as $n \to \infty$ and, for each $n = 1, 2, \ldots$, let Y_n have a binomial distribution with parameters r_n and θ_n, where $\theta_1, \theta_2, \ldots$ is a sequence in $(0, 1)$. We will consider the conditions under which

$$\frac{Y_n - r_n \theta_n}{\sqrt{[r_n \theta_n (1 - \theta_n)]}}$$

converges in distribution to a standard normal random variable. That is, we will consider the conditions under which a binomial distribution with parameters r_n and θ_n can be approximated by a normal distribution when n is large.

For each $n = 1, 2, \ldots$, let $X_{n1}, \ldots, X_{nr_n}$ denote independent, identically distributed Bernoulli random variables such that

$$\Pr(X_{nj} = 1) = \theta_n, \quad j = 1, \ldots, r_n.$$

Then Y_n has the same distribution as

$$\sum_{j=1}^{r_n} X_{nj}.$$

Thus, Theorem 12.2 can be used to study the asymptotic distribution of the standardized version of Y_n.

Consider condition (12.1). Clearly, X_{nj} has mean $\mu_{nj} = \theta_n$ and variance $\sigma_{nj}^2 = \theta_n(1 - \theta_n)$. Since

$$\mathrm{E}\left[|X_{n1} - \theta_n|^{2+\alpha}\right] = \theta_n(1 - \theta_n)\left[\theta_n^{1+\alpha} + (1 - \theta_n)^{1+\alpha}\right],$$

$$\frac{r_n \mathrm{E}\left[|X_{n1} - \theta_n|^{2+\alpha}\right]}{[r_n\theta_n(1 - \theta_n)]^{1+\alpha/2}} = \frac{1}{r_n^{\frac{\alpha}{2}}} \frac{\theta_n^{1+\alpha} + (1 - \theta_n)^{1+\alpha}}{[\theta_n(1 - \theta_n)]^{\frac{\alpha}{2}}}$$

$$= \frac{\theta_n^{1+\alpha/2}}{[r_n(1 - \theta_n)]^{\frac{\alpha}{2}}} + \frac{(1 - \theta_n)^{1+\alpha/2}}{[r_n\theta_n]^{\frac{\alpha}{2}}}.$$

It follows that a sufficient condition for condition (12.1) is that

$$r_n(1 - \theta_n) \to \infty \quad \text{and} \quad r_n\theta_n \to \infty$$

as $n \to \infty$.

Thus, we expect that a normal approximation to the distribution of Y_n will be accurate whenever $r_n(1 - \theta_n)$ and $r_n\theta_n$ are both large, a criterion that is often given in elementary statistics textbooks. $\quad \square$

Example 12.6 (First-order autoregressive process). Let $Z_1, Z_2, \ldots$ denote independent, identically distributed random variables each with mean 0, variance σ^2, and $\gamma = \mathrm{E}[|Z_1|^3] < \infty$. For $j = 1, 2, \ldots$ let

$$Y_j = \begin{cases} \frac{1}{\sqrt{(1-\rho^2)}} Z_1 & \text{if } j = 1 \\[2mm] \rho Y_{j-1} + Z_j & \text{if } j = 2, 3, \ldots \end{cases}$$

where $|\rho| < 1$. This is a first-order autoregressive process; such a process was considered in Example 6.2 under the additional assumption that $Z_1, Z_2, \ldots$ each have a normal distribution.

Consider the asymptotic distribution of $\sum_{j=1}^{n} Y_j/n$. Note that we may write

$$Y_j = \sum_{i=0}^{j-2} \rho^i Z_{j-i} + \frac{\rho^{j-1}}{\sqrt{(1-\rho^2)}} Z_1.$$

Hence,

$$\sum_{j=1}^{n} Y_j = \sum_{i=0}^{n-1} \frac{\rho^i}{\sqrt{(1-\rho^2)}} Z_1 + \sum_{i=0}^{n-2} \rho^i Z_2 + \sum_{i=0}^{n-3} \rho^i Z_3 + \cdots + Z_n.$$

Let

$$a_{nj} = \begin{cases} \frac{1-\rho^n}{(1-\rho)\sqrt{(1-\rho^2)}} & \text{if } j = 1 \\[2mm] \frac{1-\rho^{n+1-j}}{1-\rho} & \text{if } j = 2, 3, \ldots, n \end{cases}.$$

Then

$$\sum_{j=1}^{n} Y_j = \sum_{j=1}^{n} a_{nj} Z_j.$$

Define $X_{nj} = a_{nj} Z_j$. Then

$$\frac{1}{n} \sum_{j=1}^{n} Y_j = \frac{1}{n} \sum_{j=1}^{n} X_{nj}$$

where X_{nj}, $j = 1, \ldots, n$, $n = 1, 2, \ldots$ forms a triangular array.

Note that each X_{nj} has mean 0, variance

$$\sigma_{nj}^2 = a_{nj}^2 \sigma^2,$$

and

$$\gamma_{nj} \equiv \gamma_{nj}(1) = \mathrm{E}\left[|X_{nj}|^3\right] = |a_{nj}|^3 \gamma.$$

Hence, $\sum_{j=1}^{n} Y_j/n$ is asymptotically normally distributed provided that

$$\lim_{n \to \infty} \frac{\sum_{j=1}^{n} |a_{nj}|^3}{\left[\sum_{j=1}^{n} a_{nj}^2\right]^{\frac{3}{2}}} = 0. \tag{12.3}$$

Using the expression for the sum of a geometric series, it is straightforward to show that

$$\sum_{j=1}^{n} a_{nj}^2 = \frac{(n-1) - 2\frac{\rho-\rho^n}{1-\rho} + \frac{\rho^2-\rho^{2n}}{1-\rho^2}}{(1-\rho)^2} + \frac{(1-\rho^n)^2}{(1-\rho)^2(1-\rho^2)}.$$

Hence,

$$\sum_{j=1}^{n} a_{nj}^2 = O(n) \quad \text{as} \quad n \to \infty.$$

Clearly, there exists an M, depending on ρ, such that

$$\sup_{n} \sup_{j=1,\ldots,n} |a_{nj}| \le M.$$

It follows that (12.3) holds and, hence, $\sum_{j=1}^{n} Y_j/n$ is asymptotically normally distributed with mean 0 and variance

$$\frac{1}{n^2} \sum_{j=1}^{n} a_{nj}^2 \sigma^2.$$

Since

$$\lim_{n \to \infty} \frac{1}{n} \sum_{j=1}^{n} a_{nj}^2 = \frac{1}{(1-\rho)^2},$$

by Corollary 11.4, $\sum_{j=1}^{n} Y_j/n$ is asymptotically normally distributed with mean 0 and variance

$$\frac{1}{n} \frac{1}{(1-\rho)^2}. \qquad \square$$

12.4 Random Vectors

Both versions of the central limit theorem considered thus far in this chapter may be extended to the case of random vectors. Here we consider only the independent, identically distributed case; see, for example, van der Vaart (1998, Chapter 2) for the case of a triangular array.

Theorem 12.3. *Let $X_1, X_2, \ldots$ denote independent, identically distributed d-dimensional random vectors each with mean vector μ and covariance matrix Σ, where Σ is nonnegative definite and $|\Sigma| < \infty$. Let*

$$\bar{X}_n = \frac{1}{n} \sum_{j=1}^{n} X_j, \quad n = 1, 2, \ldots.$$

Then

$$\sqrt{n}(\bar{X}_n - \mu) \overset{\mathcal{D}}{\to} Z \quad as \quad n \to \infty,$$

where Z has a d-dimensional multivariate normal distribution with mean vector 0 and covariance matrix Σ.

Proof. Let a denote an arbitrary element of $\mathbf{R}^d$ such that $a^T \Sigma a > 0$. Then $a^T X_1, a^T X_2, \ldots$ are independent, identically distributed real-valued random variables each with mean $a^T \mu$ and variance $0 < a^T \Sigma a < \infty$. It follows from Theorem 12.1 that

$$\frac{a^T X_1 + a^T X_2 + \cdots + a^T X_n - n a^T \mu}{[a^T \Sigma a]^{\frac{1}{2}} \sqrt{n}} \overset{\mathcal{D}}{\to} Z_0 \quad as \quad n \to \infty,$$

where Z_0 has a standard normal distribution. That is,

$$\frac{a^T X_1 + a^T X_2 + \cdots + a^T X_n - n a^T \mu}{\sqrt{n}} \overset{\mathcal{D}}{\to} Z_1 \quad as \quad n \to \infty,$$

where Z_1 has a normal distribution with mean 0 and variance $a^T \Sigma a$. The result now follows from Theorem 11.6.

Now suppose that $a \in \mathbf{R}^d$ satisfies $a^T \Sigma a = 0$. Then $a^T X_1, a^T X_2, \ldots$ are independent, identically distributed real-valued random variables each with mean $a^T \mu$ and variance 0. Hence,

$$\frac{a^T X_1 + a^T X_2 + \cdots + a^T X_n - n a^T \mu}{[a^T \Sigma a]^{\frac{1}{2}} \sqrt{n}} \xrightarrow{\mathcal{D}} a^T \mu \quad \text{as} \quad n \to \infty.$$

The result now follows from the fact that $a^T \mu$ may be viewed as a random variable with a normal distribution with mean $a^T \mu$ and variance 0.   ∎

***Example* 12.7 *(Multinomial distribution)*.** Let $Y = (Y_1, \ldots, Y_m)$ denote a random vector with a multinomial distribution with parameters n and $(\theta_1, \ldots, \theta_{m+1})$. Recall that this distribution has frequency function

$$p(x_1, \ldots, x_m) = \binom{n}{x_1, x_2, \ldots, x_m} \theta_1^{x_1} \theta_2^{x_2} \cdots \theta_m^{x_m} \theta_{m+1}^{(n - x_1 - \cdots - x_m)},$$

for $x_j = 0, 1, \ldots, n$, $j = 1, \ldots, m$, such that $\sum_{j=1}^m x_j \le n$; here $\theta_{m+1} = 1 - \sum_{j=1}^m \theta_j$. We will consider an approximation to the distribution of Y that is valid for large n.

For $j = 1, \ldots, m$ let $e_j \in \mathbf{R}^m$ denote the vector with a 1 in the jth position and zeros in all other positions, so that $\{e_1, \ldots, e_m\}$ is the usual set of basis vectors for $\mathbf{R}^m$, and let e_{m+1} be a vector of all zeros. Define a random variable X as follows. Let T denote a random variable such that

$$\Pr(T = j) = \theta_j, \quad j = 1, \ldots, m + 1$$

and let $X \equiv X(T) = e_T$. For instance, if $T = 1$, then

$$X = \begin{pmatrix} 1 \\ 0 \\ \vdots \\ 0 \end{pmatrix}.$$

Then, for any $x_1, \ldots, x_m$, each taking the values 0 or 1 such that $\sum_{j=1}^m x_j \le 1$,

$$\Pr(X = (x_1, \ldots, x_m)) = \theta_1^{x_1} \cdots \theta_m^{x_m} \theta_{m+1}^{1 - (x_1 + \cdots + x_m)};$$

it follows that X has a multinomial distribution with parameters $(\theta_1, \ldots, \theta_{m+1})$ and $n = 1$. It is straightforward to show that the mean vector of X is

$$\mu = \begin{pmatrix} \theta_1 \\ \vdots \\ \theta_m \end{pmatrix}$$

and, since at most one component of X is nonzero, it follows that X has covariance matrix Σ with (i, j)th element

$$\Sigma_{ij} = \begin{cases} \theta_j(1 - \theta_j) & \text{if } i = j \\ -\theta_i \theta_j & \text{if } i \ne j \end{cases},$$

$i, j = 1, \ldots, m$.

Let $X_1, \ldots, X_n$ denote independent, identically distributed random vectors such that each X_j has the same distribution as X. Then $\sum_{j=1}^n X_j$ has a multinomial distribution with parameters $(\theta_1, \ldots, \theta_m)$ and n. It follows from Theorem 12.3 that $\sum_{j=1}^n X_j / n$ is asymptotically distributed according to a multivariate normal distribution with mean μ and covariance matrix Σ/n. That is, the multinomial distribution can be approximated by a multivariate normal distribution in the same way that a binomial distribution is approximated by a univariate normal distribution. $\quad\square$

12.5   Random Variables with a Parametric Distribution

In statistical applications, the random variables of interest are often distributed according to a distribution depending on a parameter whose value is unknown. Let $X_1, X_2, \ldots$ denote independent, identically distributed random variables each distributed according to a distribution depending on a parameter $\theta \in \Theta$ and consider an approximation to the distribution of

$$\bar{X}_n = \frac{X_1 + \cdots + X_n}{n}.$$

Let

$$\mu(\theta) = \mathrm{E}(X_1; \theta) \quad \text{and} \quad \sigma^2(\theta) = \mathrm{Var}(X_1; \theta), \qquad \theta \in \Theta.$$

Under the assumption that $\sigma^2(\theta) < \infty$ for $\theta \in \Theta$, Theorem 12.1 may be used to show that

$$\lim_{n\to\infty} \Pr\left\{ \frac{\sqrt{n}(\bar{X}_n - \mu(\theta))}{\sigma(\theta)} \le t; \theta \right\} = \Phi(t), \quad -\infty < t < \infty$$

and, hence, the distribution of $\bar{X}_n$ may be approximated by the normal distribution with mean $\mu(\theta)$ and standard deviation $\sigma(\theta)/\sqrt{n}$.

When the random variables are distributed according to the distribution with parameter θ, the approximation error decrease to 0 as $n \to \infty$ for any value of θ. However, the accuracy of the approximation may depend heavily on the value of θ under consideration. In some cases, for any value of n, it may be possible to find a value of θ, θ_n, such that the normal approximation to the distribution of $\bar{X}_n$, when the data are distributed according to the distribution with parameter value θ_n, is inaccurate. The following example illustrates this possibility.

***Example* 12.8** *(Normal approximation to the binomial distribution).* Let $Y_1, Y_2, \ldots$ denote independent, identically distributed random variables such that

$$\Pr(Y_1 = 1; \theta) = 1 - \Pr(Y_1 = 0) = \theta$$

where $0 < \theta < 1$. Since Y_1 has mean θ and variance $\theta(1 - \theta)$, we know that for any $0 < \theta < 1$,

$$\lim_{n\to\infty} \Pr\left\{ \frac{\sqrt{n}(\bar{Y}_n - \theta)}{[\theta(1 - \theta)]^{\frac{1}{2}}} \le t; \theta \right\} = \Phi(t), \quad -\infty < t < \infty.$$

Suppose $t < -1$ and, for $n = 1, 2, \ldots$, let $\theta_n = 1/(n + 1)$. Then,

$$\Pr\left\{ \frac{\sqrt{n}(\bar{Y}_n - \theta_n)}{[\theta_n(1 - \theta_n)]^{\frac{1}{2}}} \le t; \theta_n \right\} = \Pr\left\{ n\bar{Y}_n \le \frac{n}{n + 1}(t + 1); \theta_n \right\}.$$

Since $n\bar{Y}_n$ has a binomial distribution and $t + 1 < 0$, it follows that

$$\Pr\left\{n\bar{Y}_n \le \frac{n}{n+1}(t+1)\right\} = 0.$$

Hence, although the error in the normal approximation to the distribution of $\bar{Y}_n$ decreases to 0 as $n \to \infty$, for any value of θ, for any given value of n, there is a value of θ for which the error in the normal approximation is at least $\Phi(t)$ for $t < -1$. $\square$

Thus, it may be the case that even though

$$\lim_{n \to \infty} \Pr\left\{\frac{\sqrt{n}(\bar{X}_n - \mu(\theta))}{\sigma(\theta)} \le t; \theta\right\} = \Phi(t), \quad -\infty < t < \infty,$$

for each $\theta \in \Theta$, the maximum error in the normal approximation over all θ does not converge to 0 as $n \to \infty$. In order to be aware of this type of situation, we can consider the maximum error in the normal approximation for $\theta \in \Theta$. For instance, we might require that

$$\lim_{n \to \infty} \sup_{\theta \in \Theta} \left|\Pr\left\{\frac{\sqrt{n}(\bar{X}_n - \mu(\theta))}{\sigma(\theta)} \le t; \theta\right\} - \Phi(t)\right| = 0, \quad -\infty < t < \infty.$$

In this case, we say that

$$\frac{\sqrt{n}(\bar{X}_n - \mu(\theta))}{\sigma(\theta)}$$

converges in distribution to a standard normal distribution *uniformly in θ* for $\theta \in \Theta$. If this type of uniform convergence holds, then, given $\epsilon > 0$ there is an n_0 such that

$$\sup_{\theta \in \Theta} \left|\Pr\left\{\frac{\sqrt{n}(\bar{X}_n - \mu(\theta))}{\sigma(\theta)} \le t; \theta\right\} - \Phi(t)\right| < \epsilon, \quad \text{for} \quad n \ge n_0;$$

that is, for sufficiently large n, the approximation error is less than ϵ for all values of θ.

Theorem 12.4 gives a version of the central limit theorem that can be used to establish this type of uniform convergence. Note that, in this result, we allow the random variables $X_1, X_2, \ldots$ to also depend on the parameter θ.

Theorem 12.4. *For each $\theta \in \Theta$, let $X_1(\theta), X_2(\theta), \ldots$ denote independent, identically distributed real-valued random variables such that*

$$\mathrm{E}[X_1(\theta); \theta] = 0$$

and let

$$\sigma^2(\theta) \equiv \mathrm{Var}[X_1(\theta); \theta], \quad \theta \in \Theta.$$

Suppose there exists $\delta > 0$ such that

$$\sup_{\theta \in \Theta} \mathrm{E}\left[\left|\frac{X_1(\theta)}{\sigma(\theta)}\right|^{2+\delta}; \theta\right] < \infty.$$

Then

$$\lim_{n \to \infty} \sup_{\theta \in \Theta} \left|\Pr\left\{\frac{\sum_{j=1}^{n} X_j(\theta)/\sqrt{n}}{\sigma(\theta)} \le t; \theta\right\} - \Phi(t)\right| = 0$$

for each $t \in \mathbf{R}$.

The proof of the theorem relies on the following lemma.

Lemma 12.2. *Define* $X_1(\theta)$, $X_2(\theta)$, ... *as in Theorem 12.4. If*

$$\lim_{n \to \infty} \sup_{\theta \in \Theta} \left| E\left[f\left(\frac{\sum_{j=1}^{n} X_j(\theta)}{\sigma(\theta)\sqrt{n}} \right); \theta \right] - E[f(Z)] \right| = 0$$

for all bounded, continuous functions f having bounded derivatives of each order, then

$$\lim_{n \to \infty} \sup_{\theta \in \Theta} \left| \Pr\left\{ \frac{\sum_{j=1}^{n} X_j(\theta)/\sqrt{n}}{\sigma(\theta)} \leq t; \theta \right\} - \Phi(t) \right| = 0$$

for each $t \in \mathbf{R}$.

Proof. Let $S_n(\theta) = \sum_{j=1}^{n} X_j(\theta)$, let Z be a random variable with a standard normal distribution that is independent of $X_1(\theta)$, $X_2(\theta)$, ..., and let f denote a function satisfying the conditions of the lemma. Then

$$\left| \sup_{\theta \in \Theta} E\left[f\left(\frac{S_n(\theta)}{\sigma(\theta)\sqrt{n}} \right); \theta \right] - E[f(Z)] \right| \leq \sup_{\theta \in \Theta} \left| E\left[f\left(\frac{S_n(\theta)}{\sigma(\theta)\sqrt{n}} \right); \theta \right] - E[f(Z)] \right|$$

and

$$\left| \inf_{\theta \in \Theta} E\left[f\left(\frac{S_n(\theta)}{\sigma(\theta)\sqrt{n}} \right); \theta \right] - E[f(Z)] \right| \leq \sup_{\theta \in \Theta} \left| E\left[f\left(\frac{S_n(\theta)}{\sigma(\theta)\sqrt{n}} \right); \theta \right] - E[f(Z)] \right|.$$

Hence,

$$\lim_{n \to \infty} \sup_{\theta \in \Theta} E\left[f\left(\frac{S_n(\theta)}{\sigma(\theta)\sqrt{n}} \right); \theta \right] = \lim_{n \to \infty} \inf_{\theta \in \Theta} E\left[f\left(\frac{S_n(\theta)}{\sigma(\theta)\sqrt{n}} \right); \theta \right] = E[f(Z)]. \quad (12.4)$$

Define

$$q(z) = \begin{cases} 1 & \text{if } z \leq 0 \\ \int_z^1 \exp\{-1/(t(1-t))\}\, dt / \int_0^1 \exp\{-1/(t(1-t))\}\, dt & \text{if } 0 < z < 1 \\ 0 & \text{if } z \geq 1 \end{cases}$$

and, for each $u > 0$, let

$$q_u(z) = q(uz), \quad z \in \mathbf{R}.$$

Note that, for all $u > 0$, q_u is a bounded, continuous function having bounded derivatives of each order and

$$I_{\{z \leq 0\}} \leq q_u(z) \leq I_{\{z \leq 1/u\}}, \quad z \in \mathbf{R};$$

hence, for any random variable Y and any $u > 0$,

$$\Pr(Y \leq y) \leq E[q_u(Y - y)] \leq \Pr(Y \leq y + 1/u), \quad -\infty < y < \infty. \quad (12.5)$$

Let

$$F_n(t; \theta) = \Pr\left\{ \frac{S_n(\theta)}{\sigma(\theta)\sqrt{n}} \leq t; \theta \right\}.$$

Then, by (12.5) for each $u > 0$,

$$F_n(t; \theta) \leq E\left[q_u\left(\frac{S_n(\theta)}{\sigma(\theta)\sqrt{n}} - t \right); \theta \right].$$

Since q_u satisfies the conditions of the lemma, by (12.4) and (12.5),

$$\lim_{n\to\infty} \inf_{\theta\in\Theta} \mathrm{E}\left[q_u\left(\frac{S_n(\theta)}{\sigma(\theta)\sqrt{n}} - t\right); \theta\right] = \lim_{n\to\infty} \sup_{\theta\in\Theta} \mathrm{E}\left[q_u\left(\frac{S_n(\theta)}{\sigma(\theta)\sqrt{n}} - t\right); \theta\right]$$
$$= \mathrm{E}[q_u(Z - t)] \le \Phi(t + 1/u).$$

It follows that, for all $u > 0$,

$$\limsup_{n\to\infty} \sup_{\theta\in\Theta} F_n(t; \theta) \le \Phi(t + 1/u)$$

and

$$\limsup_{n\to\infty} \inf_{\theta\in\Theta} F_n(t; \theta) \le \Phi(t + 1/u);$$

hence,

$$\limsup_{n\to\infty} \sup_{\theta\in\Theta} F_n(t; \theta) \le \Phi(t)$$

and

$$\limsup_{n\to\infty} \inf_{\theta\in\Theta} F_n(t; \theta) \le \Phi(t).$$

Similarly,

$$F_n(t; \theta) \ge \mathrm{E}\left[q_u\left(\frac{S_n(\theta)}{\sigma(\theta)\sqrt{n}} - t - 1/u\right); \theta\right]$$

so that

$$\liminf_{n\to\infty} \sup_{\theta\in\Theta} F_n(t; \theta) \ge \lim_{n\to\infty} \sup_{\theta\in\Theta} \mathrm{E}\left[q_u\left(\frac{S_n(\theta)}{\sigma(\theta)\sqrt{n}} - t - 1/u\right); \theta\right]$$
$$= \mathrm{E}[q_u(Z - t - 1/u)] \ge \Phi(t - 1/u)$$

and

$$\liminf_{n\to\infty} \inf_{\theta\in\Theta} F_n(t; \theta) \ge \lim_{n\to\infty} \inf_{\theta\in\Theta} \mathrm{E}\left[q_u\left(\frac{S_n(\theta)}{\sigma(\theta)\sqrt{n}} - t - 1/u\right); \theta\right]$$
$$= \mathrm{E}[q_u(Z - t - 1/u)] \ge \Phi(t - 1/u).$$

Since this holds for all $u > 0$,

$$\liminf_{n\to\infty} \sup_{\theta\in\Theta} F_n(t; \theta) \ge \Phi(t)$$

and

$$\liminf_{n\to\infty} \inf_{\theta\in\Theta} F_n(t; \theta) \ge \Phi(t).$$

It follows that

$$\lim_{n\to\infty} \sup_{\theta\in\Theta} F_n(t; \theta) = \Phi(t)$$

and

$$\lim_{n\to\infty} \inf_{\theta\in\Theta} F_n(t;\theta) = \Phi(t).$$

The result now follows from the fact that

$$\sup_{\theta\in\Theta} |F_n(t;\theta) - \Phi(t)| \le |\sup_{\theta\in\Theta} F_n(t;\theta) - \Phi(t)| + |\inf_{\theta\in\Theta} F_n(t;\theta) - \Phi(t)|. \quad \blacksquare$$

Proof of Theorem 12.4. Let $S_n(\theta) = \sum_{j=1}^{n} X_j(\theta)$. Using Lemma 12.2, the result follows provided that, for every bounded continuous function f having bounded derivatives of each order,

$$\sup_{\theta\in\Theta} \left| E\left[f\left(\frac{S_n(\theta)}{\sigma(\theta)\sqrt{n}} \right); \theta \right] - E[f(Z); \theta] \right| \to 0$$

as $n \to \infty$, where Z denotes a random variable with a standard normal distribution that is independent of $X_1(\theta), X_2(\theta), \ldots$.

Fix $f(\cdot)$ and $\theta \in \Theta$. For $h \in \mathbf{R}$, define

$$g(h) = \sup_x \left| f(x + h) - f(x) - f'(x)h - \frac{1}{2}f''(x)h^2 \right|.$$

Then

$$|g(h)| \le \frac{1}{6}\sup_x |f'''(x)| \, |h|^3 \le K_1 \, |h|^3,$$

for some constant K_1, and

$$|g(h)| \le \sup_x |f''(x)| \, |h|^2 \le K_2 \, |h|^2,$$

for some constant K_2. Also note that for $h_1, h_2 \in \mathbf{R}$,

$$\left| f(x + h_1) - f(x + h_2) - f'(x)(h_1 - h_2) - \frac{1}{2}f''(x)\left(h_1^2 - h_2^2 \right) \right| \le g(h_1) + g(h_2).$$

Let $Z_j \equiv Z_j(\theta)$, $j = 1, 2, \ldots$, denote independent random variables, each normally distributed according to a normal distribution with mean 0 and standard deviation $\sigma(\theta)$, that are independent of $X_j \equiv X_j(\theta)$, $j = 1, 2, \ldots$. For $k = 1, \ldots, n$, let

$$W_k \equiv W_k(\theta) = \sum_{j=1}^{k-1} X_j(\theta) + \sum_{j=k+1}^{n} Z_j(\theta)$$

so that

$$W_k + X_k = \sum_{j=1}^{k} X_j + \sum_{j=k+1}^{n} Z_j,$$

$W_k + Z_k = W_{k-1} + X_{k+1}$, $W_n + X_n = S_n$, and $W_1 + Z_1$ has a normal distribution with mean 0 and variance $n\sigma(\theta)^2$. Then

$$\left| \mathrm{E}\left[f\left(\frac{S_n(\theta)}{\sigma(\theta)\sqrt{n}} \right); \theta \right] - \mathrm{E}[f(Z); \theta] \right| = \left| \mathrm{E}\left[f\left(\frac{W_n + X_n}{\sigma(\theta)\sqrt{n}} \right); \theta \right] - \mathrm{E}\left[f\left(\frac{W_1 + Z_1}{\sigma(\theta)\sqrt{n}} \right); \theta \right] \right|$$

$$= \left| \mathrm{E}\left\{ \sum_{j=2}^{n} \left[f\left(\frac{W_j + X_j}{\sigma(\theta)\sqrt{n}} \right) - f\left(\frac{W_{j-1} + X_{j-1}}{\sigma(\theta)\sqrt{n}} \right) \right] \right. \right.$$

$$\left. \left. + f\left(\frac{W_1 + X_1}{\sigma(\theta)\sqrt{n}} \right) - f\left(\frac{W_1 + Z_1}{\sigma(\theta)\sqrt{n}} \right); \theta \right\} \right|$$

$$= \left| \mathrm{E}\left\{ \sum_{j=1}^{n} \left[f\left(\frac{W_j + X_j}{\sigma(\theta)\sqrt{n}} \right) - f\left(\frac{W_j + Z_j}{\sigma(\theta)\sqrt{n}} \right) \right]; \theta \right\} \right|$$

$$\leq \sum_{j=1}^{n} \left| \mathrm{E}\left\{ \left[f\left(\frac{W_j + X_j}{\sigma(\theta)\sqrt{n}} \right) - f\left(\frac{W_j + Z_j}{\sigma(\theta)\sqrt{n}} \right) \right]; \theta \right\} \right|$$

$$\leq \sum_{j=1}^{n} \mathrm{E}\left[g\left(\frac{X_j}{\sigma(\theta)\sqrt{n}} \right) + g\left(\frac{Z_j}{\sigma(\theta)\sqrt{n}} \right); \theta \right]$$

since

$$\mathrm{E}\left[f'\left(\frac{W_j}{\sigma(\theta)\sqrt{n}} \right) \frac{X_j - Z_j}{\sigma(\theta)\sqrt{n}}; \theta \right] = \mathrm{E}\left[f''\left(\frac{W_j}{\sigma(\theta)\sqrt{n}} \right) \frac{X_j^2 - Z_j^2}{\sigma(\theta)\sqrt{n}}; \theta \right] = 0$$

by the independence of W_j, X_j, Z_j.

Hence, for each θ,

$$\left| \mathrm{E}\left[f\left(\frac{S_n(\theta)}{\sigma(\theta)\sqrt{n}} \right); \theta \right] - \mathrm{E}[f(Z); \theta] \right| \leq n \mathrm{E}\left[g\left(\frac{X_1}{\sigma(\theta)\sqrt{n}} \right) + g\left(\frac{Z_1}{\sigma(\theta)\sqrt{n}} \right); \theta \right]$$

where g depends only on the function f.

Consider

$$\mathrm{E}\left[g\left(\frac{X_1}{\sigma(\theta)\sqrt{n}} \right); \theta \right].$$

Let F_θ denote the distribution function of $X_1(\theta)$ for a particular value of $\theta \in \Theta$. For $\epsilon > 0$ and $n = 1, 2, \ldots$, let $A_{n,\epsilon}(\theta)$ denote the event that

$$|X_1(\theta)| \leq \epsilon \sigma(\theta)\sqrt{n}.$$

Then

$$\mathrm{E}\left[g\left(\frac{X_1}{\sigma(\theta)\sqrt{n}} \right); \theta \right] = \int_{A_{n,\epsilon}(\theta)} g\left(\frac{x}{\sigma(\theta)\sqrt{n}} \right) dF_\theta(x) + \int_{A_{n,\epsilon}(\theta)^c} g\left(\frac{x}{\sigma(\theta)\sqrt{n}} \right) dF_\theta(x)$$

$$\leq K_1 \int_{A_{n,\epsilon}(\theta)} \frac{|x|^3}{\sigma(\theta)^3 n^{\frac{3}{2}}} dF_\theta(x) + K_2 \int_{A_{n,\epsilon}(\theta)^c} \frac{|x|^2}{\sigma(\theta)^2 n} dF_\theta(x)$$

$$\leq \frac{K_1 \epsilon}{n} + \frac{K_2}{n} \int_{A_{n,\epsilon}(\theta)^c} \left| \frac{x}{\sigma(\theta)} \right|^{2+\delta} \left| \frac{\sigma(\theta)}{x} \right|^\delta dF_\theta(x)$$

$$\leq \frac{K_1 \epsilon}{n} + \frac{K_2}{n} \frac{1}{(\epsilon\sqrt{n})^\delta} \mathrm{E}\left[\left| \frac{X_1(\theta)}{\sigma(\theta)} \right|^{2+\delta}; \theta \right].$$

Note that this holds for all $\theta \in \Theta$ and K_1, K_2 depend only on f. Also,

$$\mathrm{E}\left[g\left(\frac{Z_1}{\sigma(\theta)\sqrt{n}}\right);\theta\right] \le K_1 \mathrm{E}\left[\left|\frac{Z_1}{\sigma(\theta)\sqrt{n}}\right|^3;\theta\right] = \frac{3K_1}{n^{\frac{3}{2}}}.$$

Hence, for every $\epsilon > 0$ and every $\theta \in \Theta$,

$$\left|\mathrm{E}\left[f\left(\frac{S_n(\theta)}{\sigma(\theta)\sqrt{n}}\right);\theta\right] - \mathrm{E}[f(Z)]\right| \le K_1\epsilon + \frac{K_2}{\epsilon^\delta n^{\frac{\delta}{2}}}\mathrm{E}\left[\left|\frac{X_1(\theta)}{\sigma(\theta)}\right|^{2+\delta};\theta\right] + \frac{3K_1}{\sqrt{n}}.$$

It follows that

$$\sup_{\theta\in\Theta}\left|\mathrm{E}\left[f\left(\frac{S_n(\theta)}{\sigma(\theta)\sqrt{n}}\right);\theta\right] - \mathrm{E}[f(Z)]\right| \le K_1\epsilon + \frac{K_2}{\epsilon^\delta n^{\frac{\delta}{2}}}\sup_{\theta\in\Theta}\mathrm{E}\left[\left|\frac{X_1(\theta)}{\sigma(\theta)}\right|^{2+\delta};\theta\right] + \frac{3K_1}{\sqrt{n}}.$$

for every $\epsilon > 0$. Hence, for all $\epsilon > 0$,

$$\limsup_{n\to\infty}\sup_{\theta\in\Theta}\left|\mathrm{E}\left[f\left(\frac{S_n(\theta)}{\sigma(\theta)\sqrt{n}}\right);\theta\right] - \mathrm{E}[f(Z)]\right| \le K_1\epsilon$$

so that

$$\lim_{n\to\infty}\sup_{\theta\in\Theta}\left|\mathrm{E}\left[f\left(\frac{S_n(\theta)}{\sigma(\theta)\sqrt{n}}\right);\theta\right] - \mathrm{E}[f(Z)]\right| = 0,$$

proving the result. ∎

***Example* 12.9 (*t-distribution*).** Let $X_1, X_2, \ldots$ denote independent, identically distributed random variables, each distributed according to the absolutely continuous distribution with density function

$$c(\theta)\left(1 + \frac{x^2}{\theta}\right)^{\frac{-(\theta+1)}{2}}, \quad -\infty < x < \infty$$

where $c(\theta)$ is a constant and $\theta \ge 3$. This is a standard t-distribution with θ degrees of freedom; recall that the variance of this distribution is finite only when $\theta > 2$.

Here

$$\mathrm{E}(Y_1;\theta) = 0 \quad \text{and} \quad \sigma^2(\theta) \equiv \mathrm{Var}(X_1;\theta) = \frac{\theta}{\theta-2};$$

it is straightforward to show that, for $0 < \delta < 1$,

$$\mathrm{E}[|X_1|^{2+\delta}] = \frac{\Gamma\left(\frac{3+\delta}{2}\right)}{\Gamma\left(\frac{1}{2}\right)}\theta^{\frac{2+\delta}{2}}\frac{\Gamma(\theta/2-1-\delta/2)}{\Gamma(\theta/2)}, \quad \theta \ge 3.$$

Hence,

$$\sup_{\theta\ge3}\mathrm{E}\left[\left|\frac{X_1}{\sigma(\theta)}\right|^{2+\delta}\right] = \frac{\Gamma\left(\frac{3+\delta}{2}\right)}{\Gamma\left(\frac{1}{2}\right)}\sup_{\theta\ge3}(\theta-2)^{\frac{\delta}{2}}\frac{\Gamma(\theta/2-1-\delta/2)}{\Gamma(\theta/2-1)}.$$

Let

$$H(\theta) = (\theta-2)^{\frac{\delta}{2}}\frac{\Gamma(\theta/2-1-\delta/2)}{\Gamma(\theta/2-1)}.$$

Note that $H(\cdot)$ is continuous on $[3, \infty)$ and

$$H(3) = \frac{\Gamma(1/2 - \delta/2)}{\Gamma(1/2)}$$

which is finite for $\delta < 1$. Consider the properties of $H(\theta)$ as $\theta \to \infty$. According to the asymptotic expansion for the ratio of gamma functions given in Example 9.5,

$$H(\theta) = (\theta - 2)^{\frac{\delta}{2}} \left\{ \frac{1}{(\theta/2 - 1)^{\frac{\delta}{2}}} + O\left(\frac{1}{(\theta/2 - 1)^{1+\delta/2}} \right) \right\} \quad \text{as} \quad \theta \to \infty.$$

It follows that

$$\lim_{\theta \to \infty} H(\theta)$$

exists and, hence,

$$\sup_{\theta \geq 3} H(\theta) < \infty.$$

According to Theorem 12.4,

$$\lim_{n \to \infty} \sup_{\theta \geq 3} \left| \Pr\left\{ \frac{\sum_{j=1}^{n} X_j / \sqrt{n}}{\sigma(\theta)} \leq t; \theta \right\} - \Phi(t) \right| = 0$$

for each $t \in \mathbf{R}$. $\quad\square$

In many cases in which

$$\lim_{n \to \infty} \sup_{\theta \in \Theta} \left| \Pr\left\{ \frac{\sum_{j=1}^{n} X_j(\theta) / \sqrt{n}}{\sigma(\theta)} \leq t; \theta \right\} - \Phi(t) \right| = 0$$

does not hold for the entire parameter space Θ, the convergence is uniform in a subset of Θ_0; it follows from Theorem 12.4 that if there exists a subset $\Theta_0 \subset \Theta$ such that

$$\sup_{\theta \in \Theta_0} \mathrm{E}\left[\left| \frac{X_1(\theta)}{\sigma(\theta)} \right|^{2+\delta} \right] < \infty$$

for some $\delta > 0$, then

$$\lim_{n \to \infty} \sup_{\theta \in \Theta_0} \left| \Pr\left\{ \frac{\sum_{j=1}^{n} X_j(\theta) / \sqrt{n}}{\sigma(\theta)} \leq t; \theta \right\} - \Phi(t) \right| = 0.$$

***Example* 12.10 (*Normal approximation to the binomial distribution*).** As in Example 12.8, let $Y_1, Y_2, \ldots$ denote independent, identically distributed random variables such that

$$\Pr(Y_1 = 1; \theta) = 1 - \Pr(Y_1 = 0) = \theta,$$

where $0 < \theta < 1$, and take $X_j(\theta) = Y_j - \theta$, $j = 1, 2, \ldots, n$. For $\delta > 0$,

$$\mathrm{E}\left[|X_1(\theta)|^{2+\delta}; \theta \right] = \theta^{2+\delta}(1 - \theta) + (1 - \theta)^{2+\delta}\theta$$

and

$$\mathrm{E}\left[\frac{|X_1(\theta)|^{2+\delta}}{\sigma(\theta)^{2+\delta}}; \theta \right] = \frac{\theta^{1+\delta/2}}{(1-\theta)^{\frac{\delta}{2}}} + \frac{(1-\theta)^{1+\delta/2}}{\theta^{\frac{\delta}{2}}}.$$

Hence,

$$\sup_{0<\theta<1} \mathrm{E}\left[\frac{|X_1(\theta)|^{2+\delta}}{\sigma(\theta)^{2+\delta}};\theta\right] = \infty.$$

However, let Θ_0 denote any compact subset of $(0, 1)$; such a subset must have a minimum value $a > 0$ and a maximum value $b < 1$. It follows that

$$\sup_{\theta\in\Theta_0} \mathrm{E}\left[\frac{|X_1(\theta)|^{2+\delta}}{\sigma(\theta)^{2+\delta}};\theta\right] < \infty$$

and, hence, that

$$\lim_{n\to\infty} \sup_{\theta\in\Theta_0} \left| \Pr\left\{ \frac{\sum_{j=1}^n X_j(\theta)/\sqrt{n}}{\sigma(\theta)} \le t;\theta \right\} - \Phi(t) \right| = 0$$

for all $-\infty < t < \infty$.   □

12.6   Dependent Random Variables

In this section, we consider the asymptotic distribution of a sample mean $\sum_{j=1}^n X_j/n$ for cases in which $X_1, X_2, \ldots$ are dependent random variables. In many cases, sample means based on dependent random variables follow the central limit theorem. However, the conditions required depend on the exact nature of the dependence and, hence, there are many versions of the central limit theorem for dependent random variables. Here we present two examples of these results. In order to keep the proofs as simple as possible, the regularity conditions required are relatively strong; in both cases, similar results are available under weaker conditions.

The first result applies to a sequence of real-valued random variables $X_1, X_2, \ldots$ such that X_i and X_j are independent if $|i - j|$ is sufficiently large. Specifically, the stochastic process $\{X_t : t \in \mathbf{Z}\}$ is said to be *m-dependent* if there exists a positive integer m such that, for positive integers r and s, $s > r$, and any $n = 1, 2, \ldots,$

$$(X_1, X_2, \ldots, X_r) \quad \text{and} \quad (X_s, X_{s+1}, \ldots, X_{s+n})$$

are independent whenever $s - r > m$.

Recall that the autocovariance function of a stationary process is given by

$$R(j) = \mathrm{Cov}(X_1, X_{1+j}), \quad j = 0, 1, \ldots$$

and the autocorrelation function is given by

$$\rho(j) = R(j)/R(0), \quad j = 0, 1, \ldots.$$

Hence, if the process is m-dependent, $R(j) = \rho(j) = 0$ for $j = m + 1, m + 2, \ldots.$

***Example* 12.11.** Let $Y_1, Y_2, \ldots$ be independent, identically distributed random variables each with range $\mathcal{Y}$. Let m denote an integer, let $f : \mathcal{Y}^{m+1} \to \mathbf{R}$ denote a function and let

$$X_j = f(Y_j, Y_{j+1}, \ldots, Y_{j+m}), \quad j = 1, 2, \ldots.$$

The process $\{X_t : t \in \mathbf{Z}\}$ is clearly m-dependent.   □

The following result gives a central limit theorem for an *m*-dependent process.

Theorem 12.5. *Let* $\{X_t : t \in \mathbf{Z}\}$ *denote an m-dependent stationary stochastic process such that* $\mathrm{E}(X_1) = 0$ *and* $\mathrm{E}(|X_1|^3) < \infty$*. Then*

$$\frac{1}{\tau} \frac{1}{\sqrt{n}} \sum_{j=1}^{n} X_j \xrightarrow{\mathcal{D}} N(0, 1),$$

$$\tau^2 = \left\{ 1 + 2 \sum_{j=1}^{m} \rho(j) \right\} \sigma^2$$

where $\sigma^2 = \mathrm{Var}(X_1)$ *and* $\rho(\cdot)$ *is the autocorrelation function of the process.*

Proof. The basic idea of the proof is that we can divide the sequence $X_1, X_2, \ldots, X_n$ into large and small blocks such that large blocks are separated by $m + 1$ units of time and, hence, are independent. The sample mean $(X_1 + \cdots + X_n)/n$ can be written as the mean of the large-block means together with a remainder term based on the small blocks. The large-block means are independent and, hence, their average follows the central limit theorem. Provided that the small blocks are small enough, their contribution to the sum is negligible and does not affect the limiting distribution. We now consider the details of the argument.

Let $k_n, n = 1, 2, \ldots$, denote an increasing sequence of integers such that $k_n > m$ for all n and $k_n = o(n^{\frac{1}{3}})$ as $n \to \infty$. For each $n = 1, 2, \ldots$, let d_n denote the largest integer such that $d_n \le n/k_n$; hence

$$n = k_n d_n + r_n, \quad n = 1, 2, \ldots,$$

where $0 \le r_n \le k_n$.

Let $S_n = X_1 + \cdots + X_n$. For each $j = 1, \ldots, d_n$, define

$$S_{nj} = (X_{jk_n - k_n + 1} + \cdots + X_{jk_n - m})/\sqrt{k_n}.$$

Hence,

$$S_{n1} = (X_1 + \cdots + X_{k_n - m})/\sqrt{k_n},$$

$$S_{n2} = (X_{k_n + 1} + \cdots + X_{2k_n - m})/\sqrt{k_n},$$

and so on.

Define

$$T_{nj} = X_{jk_n - m + 1} + \cdots + X_{jk_n}, \quad j = 1, \ldots, d_n - 1$$

and

$$T_{nd_n} = X_{k_n d_n - m + 1} + \cdots + X_n.$$

Hence,

$$S_n = \sqrt{k_n}(S_{n1} + \cdots + S_{nd_n}) + (T_{n1} + \cdots + T_{nd_n})$$

where $S_{n1}, \ldots, S_{nd_n}$ are independent and $T_{n1}, \ldots, T_{nd_n}$ are also independent.

It is straightforward to show that, for $j = 1, \ldots, d_n$,

$$E(S_{nj}) = 0 \quad \text{and} \quad \text{Var}(S_{nj}) = \tau_n^2$$

where

$$k_n \tau_n^2 = (k_n - m)\text{Var}(X_1) + 2 \sum_{1 \le i < j \le k_n - m} \text{Cov}(X_i, X_j)$$

$$= \left\{ (k_n - m) + 2 \sum_{j=1}^{k_n - m - 1} (k_n - m - j)\rho(j) \right\} \sigma^2.$$

Since the process is m-dependent,

$$\rho(j) = 0 \quad \text{for} \quad j \ge m + 1$$

so that, for $k_n \ge 2m + 1$,

$$\tau_n^2 = \left\{ \frac{k_n - m}{k_n} + 2 \sum_{j=1}^{m} \frac{k_n - m - j}{k_n}\rho(j) \right\} \sigma^2.$$

Note that

$$\lim_{n \to \infty} \tau_n^2 = \tau^2,$$

where τ^2 is given in the statement of the theorem.

We may write

$$\frac{1}{\sqrt{n}} S_n = \left(\frac{d_n k_n}{n} \right)^{\frac{1}{2}} \frac{1}{\sqrt{d_n}} \sum_{j=1}^{d_n} S_{nj} + \frac{1}{\sqrt{n}} \sum_{j=1}^{d_n} T_{nj}.$$

Recall that $d_n k_n = n - r_n$, where $r_n \le k_n = o(n^{\frac{1}{3}})$; hence,

$$\lim_{n \to \infty} \frac{d_n k_n}{n} = 1.$$

It follows from Corollary 11.4 that the theorem holds if it can be shown that

$$\frac{1}{\sqrt{d_n}} \sum_{j=1}^{d_n} \frac{S_{nj}}{\tau_n} \overset{D}{\to} N(0, 1) \quad \text{as} \quad n \to \infty \tag{12.6}$$

and

$$\frac{1}{\sqrt{n}} \sum_{j=1}^{d_n} T_{nj} \overset{P}{\to} 0 \quad \text{as} \quad n \to \infty. \tag{12.7}$$

Equation (12.7) can be established by showing that

$$\lim_{n \to \infty} \frac{\sum_{j=1}^{d_n} \text{Var}(T_{nj})}{n} = 0. \tag{12.8}$$

Note that, for $j = 1, \ldots, d_n - 1$,

$$
\begin{aligned}
\mathrm{Var}(T_{nj}) &= \left\{ (m-1) + 2 \sum_{j=1}^{m-2} (m-1-j)\rho(j) \right\} \sigma^2 \\
&\leq \left\{ (m-1) + 2 \sum_{j=1}^{m-2} (m-1-j) \right\} \sigma^2 \\
&\leq (m-1)^2 \sigma^2
\end{aligned}
$$

and

$$
\mathrm{Var}(T_{nd_n}) \leq (n + m - k_n d_n)^2 \sigma^2.
$$

Hence,

$$
\frac{\sum_{j=1}^{d_n} \mathrm{Var}(T_{nj})}{n} \leq \left\{ \frac{(m-1)^2 + (n - k_n d_n + m)^2}{n} \right\} \sigma^2 = \frac{(m-1)^2 + (r_n + m)^2}{n};
$$

(12.8) now follows from the fact that $r_n \leq k_n = o(n^{\frac{1}{3}})$. Hence, (12.7) holds.

To show (12.6), we use Theorem 12.2. The random variables S_{nj}, $j = 1, \ldots, d_n$, $n = 1, 2, \ldots$ form a triangular array such that, for each $n = 1, 2, \ldots, S_{n1}, \ldots, S_{nd_n}$ are independent with mean 0, standard deviation τ_n. Using the Hölder inequality for sums (see Appendix 3)

$$
\mathrm{E}\left[\left| \sum_{j=1}^{n} X_j \right|^3 \right\} \leq \mathrm{E}\left[\left(\sum_{j=1}^{n} |X_j| \right)^3 \right] \leq n^{\frac{2}{3}} \mathrm{E}\left[\sum_{j=1}^{n} |X_j|^3 \right] = n^{\frac{5}{3}} \mathrm{E}[|X_1|^3]
$$

so that

$$
\mathrm{E}\{|S_{nj}|^3\} \leq \frac{(k_n - m)^{\frac{5}{3}}}{k_n^{\frac{3}{2}}} \mathrm{E}\{|X_1|^3\}. \tag{12.9}
$$

Using Theorem 12.2 with $\alpha = 1$, the result follows provided that condition (12.1) is satisfied, that is, provided that

$$
\lim_{n \to \infty} \frac{\sum_{j=1}^{d_n} \mathrm{E}\{|S_{nj}|^3\}}{[d_n \tau_n^2]^{\frac{1}{2}}} = 0.
$$

Using (12.9),

$$
\frac{\sum_{j=1}^{d_n} \mathrm{E}\{|S_{nj}|^3\}}{[d_n \tau_n^2]^{\frac{1}{2}}} \leq \frac{\frac{(k_n - m)^{\frac{5}{3}}}{k_n^{\frac{3}{2}}} \mathrm{E}[|X_1|^3]}{d_n^{\frac{3}{2}} \tau_n^3}.
$$

Since $\tau_n \to \tau$ as $n \to \infty$, the result holds provided that

$$
\lim_{n \to \infty} \frac{k_n^{\frac{1}{6}}}{d_n^{\frac{3}{2}}} = 0
$$

which follows from the facts that $k_n = o(n^{\frac{1}{3}})$ and $d_n k_n = O(n)$. ∎

***Example* 12.12 (*Sample autocovariances under independence*).** Let $Y_1, Y_2, \ldots$ denote independent, identically distributed random variables each with mean 0, standard deviation 1 and assume that $E\{|Y_1|^3\} < \infty$. Consider the statistic

$$\frac{1}{n} \sum_{j=1}^{n} Y_j Y_{j+m}$$

for some fixed integer $m > 0$. This may be viewed a sample version of the autocovariance of order m of the process $Y_1, Y_2, \ldots$ based on the observation of $Y_1, \ldots, Y_{n+m}$.

   Let

$$X_j = Y_j Y_{j+m}, \quad j = 1, \ldots.$$

Clearly, the process $X_1, X_2, \ldots$ is m-dependent and, since

$$E\{|X_j|^3\} = E\{|Y_1|^3\}^2 < \infty,$$

the conditions of Theorem 12.5 are satisfied. Note that $E(X_1) = 0$, $\sigma^2 = \text{Var}(X_1) = 1$, and the autocorrelation function of the process $X_1, X_2, \ldots$ is given by

$$\begin{aligned}
\rho(j) &= \text{Cov}(X_1, X_{1+j}) \\
&= \text{Cov}(Y_1 Y_m, Y_{1+j} Y_{m+j}) = E(Y_1 Y_m Y_{1+j} Y_{m+j}) \\
&= 0, \quad j = 1, 2, \ldots.
\end{aligned}$$

Hence,

$$\frac{1}{\sqrt{n}} \sum_{j=1}^{n} Y_j Y_{j+m}$$

converges in distribution to a standard normal random variable. $\square$

***Example* 12.13 (*Finite moving-average process*).** Let $\ldots, Z_{-1}, Z_0, Z_1, \ldots$ denote a sequence of independent, identically distributed random variables with

$$E(Z_0) = 0 \quad \text{and} \quad \text{Var}(Z_0) = \sigma^2 < \infty$$

and $E(|Z_0|^3) < \infty$. Let $\alpha_0, \alpha_1, \ldots, \alpha_m$ denote constants and let

$$X_j = \sum_{i=0}^{m} \alpha_i Z_{j-i}, \quad j = 1, 2, \ldots.$$

Clearly, the process $X_1, X_2, \ldots$ is m-dependent with autocovariance function $R(\cdot)$ given by

$$\begin{aligned}
R(j) &= \text{Cov}(X_1, X_{1+j}) = \text{Cov}\left(\sum_{i=0}^{m} \alpha_i Z_{-i}, \sum_{i=0}^{m} \alpha_i Z_{j-i}\right) \\
&= \text{Cov}\left(\sum_{i=0}^{m} \alpha_i Z_{j-i}, \sum_{i=-j}^{m-j} \alpha_{j+i} Z_{-i}\right) \\
&= \sum_{i=0}^{m-j} \alpha_i \alpha_{j+i} \sigma^2, \quad j = 0, \ldots, m;
\end{aligned}$$

for $j > m$, $R(j) = 0$.

It is straightforward to show that $E[|Z_0|^3] < \infty$ implies that $E[|X_1|^3] < \infty$ so that the conditions of Theorem 12.5 are satisfied and, hence,

$$\frac{1}{\tau} \frac{1}{\sqrt{n}} \sum_{j=1}^{n} X_j \overset{\mathcal{D}}{\to} N(0, 1) \quad \text{as} \quad n \to \infty$$

where

$$\tau^2 = \left\{ 1 + 2 \sum_{j=1}^{m} \rho(j) \right\} R(0) = R(0) + 2 \sum_{j=1}^{m} R(j)$$

$$= \left[\sum_{i=0}^{m} \alpha_i^2 + 2 \sum_{j=1}^{m} \sum_{i=0}^{m-j} \alpha_i \alpha_{j+i} \right] \sigma^2.$$

Note that

$$\sum_{j=1}^{m} \sum_{i=0}^{m-j} \alpha_i \alpha_{j+i} = \sum_{i<j} \alpha_i \alpha_j$$

so that

$$\tau^2 = \left[\sum_{i=0}^{m} \alpha_i^2 + 2 \sum_{i<j} \alpha_i \alpha_j \right] \sigma^2 = \left(\sum_{i=0}^{m} \alpha_i \right)^2 \sigma^2. \qquad \square$$

Theorem 12.5 applies to stationary stochastic processes in which X_j and X_{j+r} are independent for sufficiently large r. The following result considers a different scenario in which nothing is assumed regarding the dependence between X_j and X_{j+r}. Instead we impose conditions upon the extent to which the conditional mean and variance of X_n given $X_0, X_1, \ldots, X_{n-1}$ depend on $X_0, X_1, \ldots, X_{n-1}$.

Theorem 12.6. *Let $X_0, X_1, \ldots$ denote a sequence of real-valued random variables such that, for each $n = 1, 2, \ldots$,*

$$E(X_n \mid X_0, X_1, \ldots, X_{n-1}) = 0$$

and

$$E(X_n^4) < \infty.$$

Let

$$\sigma_n^2 = \text{Var}(X_n), \quad n = 1, 2, \ldots$$

and

$$s_n^2 = \text{Var}(X_n \mid X_0, X_1, \ldots, X_{n-1}), \quad n = 1, 2, \ldots.$$

Assume that

$$\lim_{n \to \infty} \frac{\sum_{j=1}^{n} E(|X_j|^4)}{\left(\sum_{1}^{n} \sigma_j^2 \right)^2} = 0 \qquad (12.10)$$

and

$$\lim_{n \to \infty} \frac{\sum_{j=1}^{n} \mathrm{Var}(s_j^2)^{\frac{1}{2}}}{\sum_1^n \sigma_j^2} = 0. \tag{12.11}$$

Let

$$\bar{X}_n = \frac{1}{n} \sum_{j=1}^{n} X_j, \quad n = 1, 2, \ldots.$$

Then

$$\frac{\sqrt{n}\,\bar{X}_n}{\left[\sum_1^n \sigma_j^2/n\right]^{\frac{1}{2}}} \xrightarrow{\mathcal{D}} N(0, 1) \quad as \ \ n \to \infty.$$

Proof. For $n = 1, 2, \ldots$, let

$$V_n^2 = \sum_{j=1}^{n} \sigma_j^2, \quad V_0 = 0,$$

$$Z_{nj} = \frac{1}{V_n} \sum_{i=1}^{j} X_i, \quad j = 1, \ldots, n, \quad \text{and} \quad Z_{n0} = 0.$$

Define

$$\varphi_n(t) = \mathrm{E}\{\exp(it Z_{nn})\}, \quad t \in \mathbf{R}, \quad n = 1, 2, \ldots.$$

The result follows provided that, for each $t \in \mathbf{R}$,

$$\lim_{n \to \infty} \varphi_n(t) = \exp(-t^2/2).$$

For each $n = 1, 2, \ldots$, let

$$h_{nk}(t) = \exp\left\{-\frac{V_n^2 - V_k^2}{V_n^2} t^2/2\right\}, \quad k = 1, \ldots, n, \ \ h_{n0}(t) = 1, \ \ t \in \mathbf{R}$$

and

$$g_{nk}(t) = h_{nk}(t)\mathrm{E}\{\exp(it Z_{nk})\}, \quad k = 0, \ldots, n, \ \ t \in \mathbf{R}.$$

Note that

$$\varphi_n(t) - \exp\left(\frac{-t^2}{2}\right) = g_{nn}(t) - g_{n0}(t) = \sum_{k=1}^{n} [g_{nk}(t) - g_{n,k-1}(t)].$$

It is straightforward to show that

$$h_{n,k-1}(t) = \exp\left(-\frac{\sigma_k^2}{2V_n^2} t^2\right) h_{nk}(t)$$

and, using a Taylor's series expansion for the exponential function,

$$\left| \exp\left(-\frac{\sigma_k^2}{2V_n^2} t^2\right) - \left(1 - \frac{\sigma_k^2}{2V_n^2} t^2\right) \right| \leq \frac{\sigma_k^4}{8V_n^4} t^4.$$

Using the expansion (A2.1) for $\exp(ix)$, together with the fact that

$$E(X_k \mid X_0, \ldots, X_{k-1}) = 0$$

for all $k = 1, 2, \ldots$, it follows that

$$
\begin{aligned}
E[\exp(it\,Z_{nk})] &= E[\exp(it\,Z_{n,k-1})\exp(it\,X_k/V_n)] \\
&= E\Big[\exp(it\,Z_{n,k-1})\big\{1 + it\,X_k/V_n - t^2 X_k^2/(2V_n^2) + u(t\,X_k/V_n)\big\}\Big] \\
&= E\Big[\exp(it\,Z_{n,k-1})\Big\{1 - \frac{1}{2}\frac{t^2}{V_n^2}s_k^2 + u\Big(\frac{t\,X_k}{V_n}\Big)\Big\}\Big]
\end{aligned}
$$

where $|u(t)| \le t^3/6$.

Hence,

$$
\begin{aligned}
g_{nk}(t) - g_{n,k-1}(t) = h_{nk}(t)\Big\{ &E\Big[\exp(it\,Z_{n,k-1})\Big(1 - \frac{1}{2}\frac{s_k^2}{V_n^2}t^2 + u\Big(\frac{t\,X_k}{V_n}\Big)\Big)\Big] \\
&- \Big(1 - \frac{1}{2}\frac{\sigma_k^2}{V_n^2}t^2 + r_n\frac{\sigma_k^4 t^4}{8V_n^4}\Big)E[\exp(it\,Z_{n,k-1})]\Big\}
\end{aligned}
$$

where $|r_n| \le 1$. It follows that

$$
\begin{aligned}
|g_{nk}(t) - g_{n,k-1}(t)| \le |h_{nk}(t)|\,&\Big|E\big[\exp(it\,Z_{n,k-1})\big(s_k^2 - \sigma_k^2\big)\big]\Big|\frac{t^2}{2V_n^2} \\
&+ \Big|E\Big[\exp(it\,Z_{n,k-1})u\Big(\frac{t\,X_k}{V_n}\Big)\Big]\Big| + \frac{\sigma_k^4 t^4}{8V_n^4}
\end{aligned}
$$

so that

$$
\begin{aligned}
|\varphi_n(t) - \exp(-t^2/2)| \le \frac{t^2}{2V_n^2}\sum_{k=1}^{n}\Big|E\big[&\exp(it\,Z_{n,k-1})\big(s_k^2 - V_k^2\big)\big]\Big| \\
&+ \sum_{k=1}^{n}E\Big[\Big|u\Big(\frac{t\,X_k}{V_n}\Big)\Big|\Big] + \frac{\sum_{k=1}^{n}\sigma_k^4}{8V_n^4}t^4 \equiv M_{n1} + M_{n2} + M_{n3}.
\end{aligned}
$$

The result follows provided that M_{n1}, M_{n2}, and M_{n3} all approach 0 as $n \to \infty$.

By the Cauchy-Schwarz inequality,

$$
\Big|E\big[\exp(it\,Z_{n,k-1})\big(s_k^2 - V_k^2\big)\big]\Big| \le [E\{\exp(2it\,Z_{n,k-1})\}]^{\frac{1}{2}}\big[E\{\big(s_k^2 - V_k^2\big)^2\}\big]^{\frac{1}{2}} \le \mathrm{Var}\big(s_k^2\big)^{\frac{1}{2}}.
$$

Hence, it follows from (12.11) that

$$\lim_{n\to\infty} M_{n1} = 0.$$

Again using the Cauchy-Schwarz inequality,

$$
E\Big[\Big|\frac{X_k}{V_n}\Big|^3\Big] \le \frac{E(X_k^4)^{\frac{1}{2}}}{V_n^2}\frac{\sigma_k}{V_n}
$$

and

$$
\sum_{k=1}^{n}\big[E\big(X_k^4\big)\big]^{\frac{1}{2}}\sigma_k \le V_n\Big[\sum_{k=1}^{n}E(X_k^4)\Big]^{\frac{1}{2}}.
$$

Hence,

$$M_{n2} = \sum_{k=1}^{n} E\left[\left|u\left(\frac{tX_k}{V_n}\right)\right|\right] \le \frac{|t|^3}{6V_n^3} \sum_{k=1}^{n} E[|X_{nk}|^3] \le \frac{|t|^3}{6} \frac{\left[\sum_{k=1}^{n} E(X_k^4)\right]^{\frac{1}{2}}}{V_n^2}$$

and, by condition (12.10) of the theorem, $M_{n2} \to 0$ as $n \to \infty$.

Finally, since $\sum_{k=1}^{n} \sigma_k^4 = \sum_{k=1}^{n} E(X_k^2)^2 \le \sum_{k=1}^{n} E(X_k^4)$, it follows from (12.10) that

$$\lim_{n\to\infty} M_{n3} = \lim_{n\to\infty} \frac{t^4}{8} \frac{\sum_{k=1}^{n} \sigma_k^4}{V_n^4} = 0.$$

The result follows. ∎

Example 12.14 *(A generalization of the central limit theorem for independent random variables).* Let $X_0, X_1, X_2, \ldots$ denote a sequence of real-valued random variables such that

$$E(X_n \mid X_0, X_1, \ldots, X_{n-1}) = 0, \quad n = 1, 2, \ldots$$

$$\sup_n E(X_n^4) < \infty$$

and

$$\mathrm{Var}(X_n \mid X_0, X_1, \ldots, X_{n-1}) = \sigma^2$$

for some constant $\sigma^2 > 0$. Hence, using the notation of Theorem 12.6, $\sigma_n^2 = \sigma^2$, $n = 1, 2, \ldots$ and $\mathrm{Var}(s_n^2) = 0$ so that

$$\lim_{n\to\infty} \frac{\sum_{j=1}^{n} \mathrm{Var}(s_j^2)^{\frac{1}{2}}}{\sum_1^n \sigma_j^2} = 0.$$

It follows that

$$\lim_{n\to\infty} \frac{\sum_{j=1}^{n} E(|X_j|^4)}{\left(\sum_1^n \sigma_j^2\right)^2} \le \frac{M}{n\sigma^2},$$

for some constant M, so that (12.10) holds. Hence,

$$\frac{\sum_{j=1}^{n} X_j}{\sigma\sqrt{n}} \xrightarrow{\mathcal{D}} N(0, 1) \quad \text{as} \quad n \to \infty.$$

That is, if $E(X_n \mid X_0, \ldots, X_{n-1}) = 0$, $\mathrm{Var}(X_n \mid X_0, \ldots, X_{n-1})$ is constant and the X_n have bounded fourth moments, then $\bar{X}_n$ is asymptotically normally distributed without any other conditions on the dependence structure of $X_1, X_2, \ldots$. □

Example 12.15 *(Martingale differences).* Let $Y_0, Y_1, \ldots$ be a martingale; then

$$E[Y_{n+1} \mid Y_0, \ldots, Y_n] = Y_n$$

with probability 1. Let $X_0 = Y_0$ and

$$X_n = Y_n - Y_{n-1}, \quad n = 1, 2, \ldots;$$

$X_0, X_1, \ldots$ are sometimes called *martingale differences*. Note that, for each $n = 1, 2, \ldots,$ $(X_0, \ldots, X_n)$ is a one-to-one function of $(Y_0, \ldots, Y_n)$:

$$Y_n = X_0 + \cdots + X_n, \quad n = 1, 2, \ldots.$$

Hence,

$$\mathrm{E}[X_n \mid X_0, \ldots, X_{n-1}] = \mathrm{E}[Y_n - Y_{n-1} \mid Y_0, \ldots, Y_{n-1}] = 0$$

by the martingale property.

Let $\sigma_n^2 = \mathrm{Var}(X_n)$, $n = 1, 2, \ldots$. Note that, for $m < n$,

$$\mathrm{E}(X_m X_n) = \mathrm{E}[\mathrm{E}(X_m X_n \mid X_0, \ldots, X_{n-1})] = \mathrm{E}[X_m \mathrm{E}(X_n \mid X_0, \ldots, X_{n-1})] = 0,$$

$$\sum_{j=1}^{n} \sigma_j^2 = \mathrm{Var}(Y_n - Y_0).$$

It follows that, under the conditions of Theorem 12.6,

$$\frac{Y_n - Y_0}{\sqrt{\mathrm{Var}(Y_n - Y_0)}} \xrightarrow{\mathcal{D}} N(0, 1) \quad \text{as} \quad n \to \infty.$$

In particular, this result holds if

$$\sup_n \mathrm{E}(X_n^4) < \infty, \quad \lim_{n \to \infty} \mathrm{Var}(Y_n - Y_0) = 0$$

and

$$\mathrm{Var}(Y_n \mid Y_0, \ldots, Y_{n-1}) = \mathrm{Var}(Y_n), \quad n = 1, 2, \ldots. \qquad \square$$

12.7 Exercises

12.1 Let X_n denote the sample mean of n independent, identically distributed random variables, each with an exponential distribution with rate parameter λ, i.e., with mean λ^{-1}. Note that the exact distribution of X_n is available using properties of the gamma distribution. Let $F(x; \lambda) = \Pr(X_n \le x; \lambda)$.
   (a) Give an approximation to $F(x; \lambda)$ based on the central limit theorem.
   (b) Consider the case $\lambda = 1/2$ and $n = 9$. For this choice of λ, nX_n has a chi-square distribution with $2n$ degrees of freedom. For $x = 2.5289, 2.8877, 3.2077,$ and 3.8672 approximate $F(x; \lambda)$ using the approximation derived in part (a) and compare the results to the exact values.

12.2 Let $X_1, X_2, \ldots, X_n$ denote independent, identically distributed random variables, each distributed according to the discrete distribution with frequency function

$$(1 - \theta)\theta^x, \quad x = 0, 1, \ldots$$

where $0 < \theta < 1$. Find a normal approximation to the distribution of $\bar{X}$. For the case $\theta = 1/3$ and $n = 9$, approximate $\Pr(\bar{X} \le 5/12)$.

12.3 Let (X_j, Y_j), $j = 1, 2, \ldots, n$, denote independent, identically distributed pairs of random variables such that X_1 is uniformly distributed in the interval $[0, 1]$ and that the conditional distribution of Y_1 given X_1 is an exponential distribution with mean βX_1, where $\beta > 0$. Find the asymptotic distribution of $T = \sum_{j=1}^{n} X_j Y_j / \sum_{j=1}^{n} X_j^2$.

12.4 Let $X_1, X_2, \ldots$ denote independent, identically distributed random variables each with mean μ and standard deviation σ. Let

$$\bar{X}_n = \frac{1}{n} \sum_{j=1}^{n} X_j \quad \text{and} \quad S_n^2 = \frac{1}{n} \sum_{j=1}^{n} (X_j - \bar{X}_n)^2.$$

Find the asymptotic distribution of

$$\frac{\sqrt{n}(\bar{X}_n - \mu)}{S_n}.$$

12.5 Let $X_1, X_2, \ldots$ denote independent, identically distributed random variables each with mean μ and standard deviation σ and let

$$\bar{X}_n = \frac{1}{n} \sum_{j=1}^{n} X_j.$$

Let t_n, $n = 1, 2, \ldots$, denote a sequence of real numbers and consider approximation of the probability $\Pr(\bar{X}_n \le t_n)$ using the approximation to the distribution of $\bar{X}_n$ given by the central limit theorem.

(a) Suppose $t_n \to t$ as $n \to \infty$, where $t \ne \mu$. Find

$$\lim_{n \to \infty} \Pr(\bar{X}_n \le t_n).$$

(b) Suppose that $t = \mu + c/\sqrt{n} + o(1/\sqrt{n})$, as $n \to \infty$. Find

$$\lim_{n \to \infty} \Pr(\bar{X}_n \le t_n).$$

12.6 Let $X_1, X_2, \ldots$ denote independent, identically distributed random variables each with mean μ and standard deviation σ and suppose that each X_j takes values in the set $\{0, 1, \ldots\}$. Let

$$S_n = \sum_{j=1}^{n} X_j$$

and consider approximation of $\Pr(S_n \le s)$, where s is a nonnegative integer, using an approximation based on the central limit theorem.

(a) Let $F(s)$ denote the central-limit-theorem-based approximation to $\Pr(S_n \le s)$ and let $G(s)$ denote the central-limit-theorem-based approximation to $\Pr(S_n \ge s + 1)$. Show that

$$F(s) + G(s) < 1.$$

Note that

$$\Pr(S_n \le s) + \Pr(S_n \ge s + 1) = 1.$$

(b) Note that $\Pr(S_n \le s) = \Pr(S_n \le s + \delta)$, for all $0 \le \delta < 1$. Hence, $\Pr(S_n \le s)$ can be approximated by $F(s + \delta)$; similarly, $\Pr(S_n \ge s + 1)$ can be approximated by $G(s + 1 - \delta)$ for $0 \le \delta < 1$. Find conditions on δ^* so that

$$F(s + \delta^*) + G(s + 1 - \delta^*) = 1.$$

The approximations $F(s + \delta^*)$ and $G(s + 1 - \delta^*)$ are known as the *continuity-corrected* approximations.

12.7 Let $X_1, X_2, \ldots$ denote independent random variables and let $\mu_1, \mu_2, \ldots$ denote a sequence of real-valued, positive constants. Let p denote a density function on the real line such that $p(x) = 0$, for $x < 0$,

$$\int_0^\infty x \, p(x) \, dx = 1, \quad \int_0^\infty x^2 \, p(x) \, dx = 2,$$

and

$$\int_0^\infty x^3 \, p(x) \, dx < \infty.$$

Assume that, for each $j = 1, 2, \ldots$, X_j has an absolutely continuous distribution with density function

$$\frac{1}{\mu_j} p(x/\mu_j).$$

(a) Find conditions on $\mu_1, \mu_2, \ldots$ such that

$$\frac{\sum_{j=1}^n (X_j - \mu_j)}{\left(\sum_{j=1}^n \mu_j^2\right)^{\frac{1}{2}}} \xrightarrow{D} N(0, 1) \quad \text{as} \quad n \to \infty.$$

(b) Suppose that $\mu_j = j^\beta$, $j = 1, 2, \ldots$, for some constant β, $-\infty < \beta < \infty$. Find conditions on β so that your conditions in part (a) are satisfied.

12.8 For each $n = 1, 2, \ldots$, let $X_{n1}, \ldots, X_{nn}$ denote independent, identically distributed random variables such that

$$\Pr(X_{nj} = -c_n) = \Pr(X_{nj} = c_n) = \frac{1}{2}.$$

Find conditions on $c_1, c_2, \ldots$ so that

$$\frac{\sum_{j=1}^n X_{nj}}{\left[\sum_{j=1}^n \text{Var}(X_{nj})\right]^{\frac{1}{2}}} \xrightarrow{D} N(0, 1) \quad \text{as} \quad n \to \infty.$$

12.9 Let $Y_1, Y_2, \ldots$ denote independent random variables and for $j = 1, 2, \ldots$, let μ_j, σ_j denote the mean and standard deviation of Y_j. Thus, the random variables are independent, but not identically distributed. Find conditions on the distribution of $Y_1, Y_2, \ldots$ such that

$$\frac{\sum_{j=1}^n (Y_j - \mu_j)}{\left[\sum_{j=1}^n \sigma_j^2\right]^{\frac{1}{2}}} \xrightarrow{D} N(0, 1) \quad \text{as} \quad n \to \infty.$$

12.10 Let $r_1, r_2, \ldots$ denote a given sequence such that $r_n \to \infty$ as $n \to \infty$ and for each $n = 1, 2, \ldots$, let $X_{n1}, \ldots, X_{nr_n}$ denote independent random variables. Suppose that all X_{nj} are bounded by a constant M. Give conditions under which condition (12.1) of Theorem 12.2 is satisfied.

12.11 Let $X_1, X_2, \ldots$ denote independent, identically distributed random variables, each with a standard exponential distribution. Find the asymptotic distribution of

$$\frac{1}{n} \left(\frac{\sum_{j=1}^n X_j}{\sum_{j=1}^n \log X_j} \right).$$

12.12 Let $X_1, X_2, \ldots$ denote independent, identically distributed random variables such that X_1 takes values in the set $\{0, 1, \ldots\}$. Let σ denote the standard deviation of X_1 and let p denote $\Pr(X_1 = 0)$; assume that $\sigma > 0$ and $p > 0$. Find the asymptotic distribution of $(\bar{X}_n, \hat{p}_n)$, where

$$\bar{X}_n = \frac{1}{n} \sum_{j=1}^n X_j \quad \text{and} \quad \hat{p}_n = \frac{1}{n} \sum_{j=1}^n I_{\{X_j = 0\}}.$$

12.13 Let $(X_1, Y_1), (X_2, Y_2), \ldots$ denote independent, identically distributed random vectors, taking values in $\mathbf{R}^2$, and each with mean vector $(0, 0)$ and covariance matrix Σ. Find

$$\lim_{n \to \infty} \Pr(\bar{X}_n \leq \bar{Y}_n)$$

where

$$\bar{X}_n = \frac{1}{n} \sum_{j=1}^{n} X_j \quad \text{and} \quad \bar{Y}_n = \frac{1}{n} \sum_{j=1}^{n} Y_j.$$

12.14 Let $X_1, X_2, \ldots$ denote independent, identically distributed, real-valued random variables such that $E(X_1) = 0$ and $E(X_1^2) < \infty$. Show that Lindeberg's condition holds (with $r_n = n$), but that Lyapounov's condition does not hold without further conditions.

12.15 Let $Y_1, Y_2, \ldots$ denote independent, identically distributed random variables, each with a Poisson distribution with mean θ, $\theta > 0$. Is the convergence of

$$\frac{\sum_{j=1}^{n}(Y_j - \theta)}{\sqrt{(n\theta)}}$$

to a standard normal distribution uniform in $\theta \in (0, \infty)$?

12.16 Let $Y_1, Y_2, \ldots$ denote independent, identically distributed real-valued random variables such that each Y_j has an absolutely continuous distribution with density $p(y; \theta)$, $\theta \in \Theta$. Suppose that $p(y; \theta)$ is of the form

$$p(y; \theta) = \frac{1}{\sigma} f\left(\frac{y - \mu}{\sigma}\right), \quad \theta = (\mu, \sigma) \in \mathbf{R} \times (0, \infty).$$

Hence, μ is a location parameter and σ is a scale parameter.

Show that, if

$$\lim_{n \to \infty} \Pr\left\{\frac{\sum_{j=1}^{n}(Y_j - \mu_0)/\sqrt{n}}{\sigma_0} \leq t; (\mu_0, \sigma_0)\right\} = \Phi(t), \quad t \in \mathbf{R}$$

for some $\mu_0 \in \mathbf{R}$ and $\sigma_0 > 0$, then

$$\lim_{n \to \infty} \sup_{\mu \in \mathbf{R}, \sigma > 0} \left| \Pr\left\{\frac{\sum_{j=1}^{n}(Y_j - \mu)/\sqrt{n}}{\sigma} \leq t; (\mu, \sigma)\right\} - \Phi(t) \right| = 0, \quad t \in \mathbf{R}.$$

12.17 For each $\theta \in \Theta$, let $X_1(\theta), X_2(\theta), \ldots$ denote independent, identically distributed real-valued random variables such that

$$E[X_1(\theta); \theta] = 0$$

and let

$$\sigma^2(\theta) \equiv \text{Var}[X_1(\theta)], \quad \theta \in \Theta.$$

Suppose that

$$\lim_{n \to \infty} \sup_{\theta \in \Theta} \left| \Pr\left\{\frac{\sum_{j=1}^{n} X_j(\theta)/\sqrt{n}}{\sigma(\theta)} \leq t; \theta\right\} - \Phi(t) \right| = 0$$

for each $t \in \mathbf{R}$.

Let $\theta_1, \theta_2, \ldots$ denote a sequence in Θ such that $\theta_n \to \theta$ as $n \to \infty$ for some $\theta \in \Theta$. Show that

$$\lim_{n \to \infty} \Pr\left\{\frac{\sum_{j=1}^{n} X_j(\theta_n)/\sqrt{n}}{\sigma(\theta_n)} \leq t; \theta_n\right\} = \Phi(t).$$

12.18 For each $\theta \in \Theta$, let $X_1(\theta), X_2(\theta), \ldots$ denote independent, identically distributed real-valued random variables such that

$$E[X_1(\theta); \theta] = 0$$

and let

$$\sigma^2(\theta) \equiv \text{Var}[X_1(\theta)], \quad \theta \in \Theta.$$

Suppose that there exists constants M and m such that

$$\sup_{\theta \in \Theta} |X_1(\theta)| \leq M$$

with probability one and suppose that

$$\inf_{\theta \in \Theta} \sigma(\theta) > m.$$

Does it follow that

$$\limsup_{n \to \infty} \sup_{\theta \in \Theta} \left| \Pr\left\{ \frac{\sum_{j=1}^n X_j(\theta)/\sqrt{n}}{\sigma(\theta)} \leq t \right\} - \Phi(t) \right| = 0$$

for each $t \in \mathbf{R}$? Why or why not?

12.19 Let $Y_1, Y_2, \ldots$ denote independent, identically distributed, random variables, each uniformly distributed on the interval $(0, 1)$ and let

$$X_n = \max(Y_n, Y_{n+1}), \quad n = 1, 2, \ldots.$$

Find the asymptotic distribution of

$$\frac{\sum_{j=1}^n X_n}{n}.$$

12.8 Suggestions for Further Reading

The references given in the previous chapter all contain various versions of the central limit theorem. Although Theorem 12.4 is new, the technique used in its proof is based on the proof of Theorem 7.2 of Billingsley (1968). Theorem 12.5 is based on Hoeffding and Robbins (1948); see also Ferguson (1996, Chapter 7). Theorem 12.6 is based on Ibragimov (1963).

There are many different versions of the central limit theorem for dependent random variables; the results given in this chapter give just two examples of these. Central limit theorems for martingales are given by Billingsley (1961), Brown (1971), Doob (1953, p. 382), and McLeish (1974), among others. Chernoff and Teicher (1958) give a central limit for exchangeable random variables. Central limit theorems for random variables with a moving average structure are given in Anderson (1975) and Fuller (1976). Port (1994, Chapter 61) gives a central limit theorem for Markov chains.

13

Approximations to the Distributions of More General Statistics

13.1 Introduction

In Chapter 12, the focus was on the asymptotic distribution of the sample mean and several versions of the central limit theorem were presented. However, in statistical applications, we often encounter a wider range of statistics. The purpose of this chapter is to consider the asymptotic distributions for several other types of statistics.

For instance, the statistic of interest may not be a sample mean, but it may be written as a function of a sample mean, or of several sample means. In many of these cases, the asymptotic distribution of the statistic may be determined from the asymptotic distribution of the sample means involved. Other possibilities are statistics that are functions of the order statistics or ranks.

13.2 Nonlinear Functions of Sample Means

Suppose that $X_1, X_2, \ldots$ are independent, identically distributed random variables each with mean μ and variance σ^2. Then $\bar{X}_n = \sum_{j=1}^{n} X_j / n$ is asymptotically normally distributed with mean μ and variance σ^2/n. Suppose the statistic of interest may be written as $g(\bar{X}_n)$ where g is a smooth function. If $g(\cdot)$ is of the form $g(x) = ax + b$ for some constants a and b, then clearly $g(\bar{X}_n)$ will be asymptotically normally distributed with mean $a\mu + b$ and variance $a^2\sigma^2/n$.

Of course, this function is a very special case. However, we know that, for large n, $\bar{X}_n$ will be very close to μ with high probability. Given that g is a smooth function, if $g'(\mu) \neq 0$, then we expect that g may be well-approximated by a function of the form $ax + b$ for x near μ. Hence, using this approach, we may be able to determine the asymptotic distribution of $\sqrt{n}(g(\bar{X}_n) - g(\mu))$.

Theorem 13.1 gives a formal result of this type; this result is sometimes called the δ-*method*. Here we use the symbol $N_d(0, \Sigma)$ to denote a d-dimensional random vector with a multivariate normal distribution with mean vector 0 and covariance matrix Σ.

Theorem 13.1. *Let $X_1, X_2, \ldots$ denote a sequence of d-dimensional random vectors such that, for some vector μ,*

$$\sqrt{n}(X_n - \mu) \overset{\mathcal{D}}{\to} N_d(0, \Sigma) \quad as \quad n \to \infty,$$

where Σ is a positive definite matrix with $|\Sigma| < \infty$.

Let $g: \mathbf{R}^d \to \mathbf{R}^k$ denote a continuously differentiable function and let $g'(x) = \partial g(x)/\partial x$ denote the $k \times d$ matrix of partial derivatives of g with respect to x. Then

$$\sqrt{n}(g(X_n) - g(\mu)) \xrightarrow{\mathcal{D}} N_k(0, g'(\mu)\Sigma g'(\mu)^T) \quad as \quad n \to \infty.$$

Proof. Let a denote an arbitrary vector in $\mathbf{R}^k$ and let $h: \mathbf{R}^d \to \mathbf{R}$ denote the function given by $h = a^T g$. By Taylor's theorem, for any $x \in \mathbf{R}^d$,

$$h(x) - h(\mu) = h'(tx + (1-t)\mu)(x - \mu)$$

for some $0 \le t \le 1$; here $h' = a^T g'$ is a $1 \times d$ vector of derivatives. It follows that, for $n = 1, 2, \ldots,$

$$\sqrt{n}(h(X_n) - h(\mu)) = h'(t_n X_n + (1-t_n)\mu)\sqrt{n}(X_n - \mu)$$
$$= h'(\mu)\sqrt{n}(X_n - \mu) + [h'(t_n X_n + (1-t_n)\mu) - h'(\mu)]\sqrt{n}(X_n - \mu),$$

where $0 \le t_n \le 1$; note that t_n generally depends on X_n.

Since $X_n \xrightarrow{p} \mu$ as $n \to \infty$,

$$t_n X_n + (1-t_n)\mu \xrightarrow{p} \mu \quad as \quad n \to \infty.$$

Since

$$\sqrt{n}(X_n - \mu) \xrightarrow{\mathcal{D}} N_d(0, \Sigma) \quad as \quad n \to \infty,$$

it follows that $\sqrt{n}(X_n - \mu) = O_p(1)$ and, hence, by the continuity of h',

$$h'(t_n X_n + (1-t_n)\mu) - h'(\mu) \xrightarrow{p} 0 \quad as \quad n \to \infty,$$

and

$$h'(\mu)\sqrt{n}(X_n - \mu) \xrightarrow{\mathcal{D}} N(0, h'(\mu)\Sigma h'(\mu)^T) \quad as \quad n \to \infty.$$

That is, for any $a \in \mathbf{R}^k$, $a^T \sqrt{n}(g(X_n) - g(\mu))$ converges in the distribution to a normal random variable with mean 0 and variance

$$h'(\mu)\Sigma h'(\mu)^T = a^T g'(\mu)\Sigma g'(\mu)^T a.$$

It now follows from Theorem 11.6 that

$$\sqrt{n}(g(X_n) - g(\mu)) \xrightarrow{\mathcal{D}} N_k(0, g'(\mu)\Sigma g'(\mu)^T) \quad as \quad n \to \infty,$$

proving the result. ■

Example* 13.1 *(Variance-stabilizing transformations). Suppose that, for each $n = 1, 2, \ldots,$ X_n is the sample mean of n independent, identically distributed real-valued random variables, each with mean θ and variance σ^2. Suppose further that σ^2 is a function of θ, $\sigma^2(\theta)$. Then

$$\frac{\sqrt{n}(X_n - \theta)}{\sigma(\theta)} \xrightarrow{\mathcal{D}} N(0, 1) \quad as \quad n \to \infty.$$

Let g denote a real-valued, continuously differentiable function defined on the range of $X_1, X_2, \ldots$. Then, according to Theorem 13.1,

$$\frac{\sqrt{n}[g(X_n) - g(\theta)]}{|g'(\theta)|\sigma(\theta)} \xrightarrow{D} N(0, 1) \quad \text{as} \quad n \to \infty.$$

provided that $g'(\theta) \neq 0$.

Suppose that g is chosen so that $|g'(\theta)|\sigma(\theta)$ is a constant not depending on θ; then the asymptotic variance of X_n does not depend on the parameter θ. Such a function g is said to be a *variance-stabilizing transformation*.

For instance, suppose that the underlying random variables have a Poisson distribution with mean θ so that $\sigma^2(\theta) = \theta$. A variance-stabilizing transformation g then satisfies $g'(\theta)\sqrt{\theta} = c$ for some constant c; that is, $g'(\theta) = c\theta^{-\frac{1}{2}}$. Hence, we may take $g(\theta) = \sqrt{\theta}$ so that

$$\frac{\sqrt{n}(\sqrt{X_n} - \sqrt{\theta})}{1/2} \xrightarrow{D} N(0, 1) \quad \text{as} \quad n \to \infty.$$

That is, $\sqrt{X_n}$ is asymptotically distributed according to a normal distribution with mean $\sqrt{\theta}$ and variance $1/(4n)$. $\square$

Example 13.2 (*Ratio of correlated sample means*). Let $Y_1, Y_2, \ldots$ denote independent, identically distributed two-dimensional random vectors each with mean vector $\mu = (\mu_1, \mu_2)$ and covariance matrix

$$\Sigma = \begin{pmatrix} \sigma_1^2 & \rho\sigma_1\sigma_2 \\ \rho\sigma_1\sigma_2 & \sigma_2^2 \end{pmatrix}.$$

Let $X_n = \sum_{j=1}^n Y_j / n$ and consider the function

$$g(X_n) = \frac{a^T X_n}{b^T X_n},$$

where a and b are given elements of $\mathbf{R}^2$ such that

$$\frac{a}{||a||} \neq \frac{b}{||b||}$$

and $b^T \mu \neq 0$.

Then

$$g'(x) = \frac{1}{b^T x}\left(a^T - \frac{a^T x}{b^T x}b^T\right) = \frac{1}{b^T x}[a - g(x)b]^T.$$

Let $c = a - g(\mu)b$. Then

$$\sqrt{n}(g(X_n) - g(\mu)) \xrightarrow{D} Z_1 \quad \text{as} \quad n \to \infty,$$

where Z_1 has a normal distribution with mean 0 and variance

$$\sigma^2 = \frac{c^T \Sigma c}{(b^T \mu)^2}.$$

For instance, suppose that $a = (1, 0)$ and $b = (0, 1)$ and write $X_n = (X_{n1}, X_{n2})$. Then

$$\sqrt{n}\left(\frac{X_{n1}}{X_{n2}} - \frac{\mu_1}{\mu_2}\right) \xrightarrow{D} Z \quad \text{as} \quad n \to \infty,$$

where Z has a normal distribution with mean 0 and variance

$$\sigma^2 = \frac{\sigma_1^2 - 2\rho\sigma_1\sigma_2\mu_1/\mu_2 + (\mu_1/\mu_2)^2\sigma_2^2}{\mu_2^2}.$$ □

***Example* 13.3 (*Joint asymptotic distribution of the sample mean and variance*).** Let $Y_1, Y_2, \ldots$ denote independent, identically distributed real-valued random variables such that $E(Y_1^4) < \infty$. Suppose we wish to determine the asymptotic distribution of the vector $(\bar{Y}, S^2)$, suitably normalized, where $\bar{Y} = \sum_{j=1}^n Y_j/n$ and $S^2 = \sum_{j=1}^n (Y_j - \bar{Y})^2/n$.

For each $n = 1, 2, \ldots$ let

$$X_n = \begin{pmatrix} \frac{1}{n}\sum_{j=1}^n Y_j \\ \frac{1}{n}\sum_{j=1}^n Y_j^2 \end{pmatrix}.$$

For $x \in \mathbf{R}^2$, define

$$g(x) = \begin{pmatrix} x_1 \\ x_2 - x_1^2 \end{pmatrix};$$

then

$$\begin{pmatrix} \bar{Y} \\ S^2 \end{pmatrix} = g(X_n).$$

Let $\mu = E(Y_1)$, $\sigma^2 = \mathrm{Var}(Y_1)$ and

$$\mu_r = E[(Y_1 - \mu)^r], \quad r = 3, 4.$$

Then, according to Theorem 12.3,

$$\sqrt{n}\left(X_n - \begin{pmatrix} \mu \\ \mu^2 + \sigma^2 \end{pmatrix}\right) \xrightarrow{\mathcal{D}} N_2(0, \Sigma)$$

where

$$\Sigma = \begin{pmatrix} \sigma^2 & \mathrm{Cov}(Y_1, Y_1^2) \\ \mathrm{Cov}(Y_1, Y_1^2) & \mathrm{Var}(Y_1^2) \end{pmatrix}.$$

Since

$$g'(\mu, \mu^2 + \sigma^2) = \begin{pmatrix} 1 & 0 \\ -2\mu & 1 \end{pmatrix},$$

it follows from Theorem 13.1 that

$$\sqrt{n}\begin{pmatrix} \bar{Y} - \mu \\ S^2 - \sigma^2 \end{pmatrix}$$

is asymptotically normally distributed with mean vector zero and covariance matrix given by

$$\begin{pmatrix} 1 & 0 \\ -2\mu & 1 \end{pmatrix} \Sigma \begin{pmatrix} 1 & -2\mu \\ 0 & 1 \end{pmatrix} = \begin{pmatrix} \sigma^2 & \mu_3 \\ \mu_3 & \mu_4 - \sigma^4 \end{pmatrix}.$$

Thus, $\bar{Y}$ and S^2 are asymptotically independent if and only if $\mu_3 = 0$. □

Note that Theorem 13.1 holds even if $g'(0) = 0$; however, in that case, the limiting distribution of $\sqrt{n}(g(X_n) - g(\mu))$ is degenerate. A formal statement of this result is given in the following corollary; the proof is left as an exercise.

Corollary 13.1. *Let* $X_1, X_2, \ldots$ *denote a sequence of d-dimensional random vectors satisfying the conditions of Theorem 13.1.*

Let $g : \mathbf{R}^d \to \mathbf{R}^k, k \leq d,$ *denote a continuously differentiable function. If* $g'(\mu) = 0,$ *then*

$$\sqrt{n}(g(X_n) - g(\mu)) \xrightarrow{p} 0 \quad as \quad n \to \infty.$$

13.3 Order Statistics

In this section, we consider approximations to the distributions of order statistics. When studying the asymptotic theory of order statistics, there are essentially two cases to consider. The first is the case of *central* order statistics. These are order statistics, such as the sample median, that converge to a quantile of the underlying distribution as the sample size increases. The asymptotic properties of the central order statistics are similar in many respects to those of a sample mean; for instance, a suitably normalized central order statistic is asymptotically normally distributed.

The second case is that of the *extreme* order statistics, such as the minimum or maximum. The asymptotic properties of the extreme order statistics differ considerably from those of the sample mean; for instance, the extreme order statistics have non-normal limiting distributions.

Central order statistics

First consider the properties of central order statistics based on independent, identically distributed random variables $X_1, X_2, \ldots, X_n$ each with distribution function F. Let $X_{n1}, \ldots, X_{nn}$ denote the order statistics based on $X_1, \ldots, X_n$. The central order statistics are of the form $X_{n(k_n)}$ where k_n/n converges to a constant in $(0, 1)$. That is, the relative position of $X_{n(k_n)}$ in the set of order statistics $X_{n1}, \ldots, X_{nn}$ stays, roughly, constant.

When studying the properties of order statistics, the following approach is often useful. First, the properties may be established for the case in which the random variables have a uniform distribution on $(0, 1)$. The corresponding result for the general case can then be established by noting that the random variables $F(X_1), \ldots, F(X_n)$ are uniformly distributed on $(0, 1)$ and then using the δ-method, as described in the previous section.

Theorem 13.2. *Let* $X_1, X_2, \ldots, X_n$ *denote independent, identically distributed random variables, each distributed according to a uniform distribution on* $(0, 1)$. *Let* $k_n, n = 1, 2, \ldots$ *denote a sequence of integers such that* $k_n \leq n$ *for all n and*

$$\sqrt{n}\left(\frac{k_n}{n} - q\right) \to 0 \quad as \quad n \to \infty$$

where $0 < q < 1$. *Let* $X_{n(k_n)}$ *denote the* k_n*th order statistic of* $X_1, \ldots, X_n$. *Then*

$$\sqrt{n}(X_{n(k_n)} - q) \xrightarrow{D} N(0, q(1 - q)) \quad as \quad n \to \infty.$$

Proof. Fix $0 < t < 1$. Define

$$Z_j(t) = \begin{cases} 0 & \text{if } X_j > t \\ 1 & \text{if } X_j \leq t, \end{cases} \quad j = 1, 2, \ldots.$$

Note that, for each $n = 1, 2, \ldots,$ $Z_1(t), \ldots, Z_n(t)$ are independent identically distributed Bernoulli random variables with

$$\Pr(Z_1(t) = 1) = t.$$

Then

$$\Pr(X_{n(k_n)} \leq t) = \Pr\left(\sum_{j=1}^n Z_j(t) \geq k_n\right)$$

and

$$\Pr(\sqrt{n}(X_{n(k_n)} - q) \leq t) = \Pr\left(\frac{1}{\sqrt{n}}\left[\sum_{j=1}^n Z_j(q + t/\sqrt{n}) - n(q + t/\sqrt{n})\right]\right.$$
$$\left.\geq \frac{k_n - nq - t\sqrt{n}}{\sqrt{n}}\right).$$

Define

$$Z_{nj} = Z_j(q + t/\sqrt{n}) - (q + t/\sqrt{n}), \quad j = 1, \ldots, n, \quad n = 1, 2, \ldots.$$

This is a triangular array.

Let

$$\gamma_{nj} = \mathrm{E}[|Z_{nj}|^3] \quad \text{and} \quad \sigma_{nj}^2 = \mathrm{Var}(Z_{nj}), \qquad j = 1, \ldots, n, \ n = 1, 2, \ldots.$$

It is straightforward to show that γ_{nj} are uniformly bounded in n and j and that

$$\sigma_{nj}^2 = \left(q + \frac{t}{\sqrt{n}}\right)\left(1 - q - \frac{t}{\sqrt{n}}\right), \quad j = 1, \ldots, n$$

so that

$$\sum_{j=1}^n \sigma_{nj}^2 = nq(1 - q) + \sqrt{n}t(1 - 2q) - t^2.$$

Hence,

$$\frac{\sum_{j=1}^n \gamma_{nj}}{\left[\sum_{j=1}^n \sigma_{nj}^2\right]^{\frac{3}{2}}} \to 0 \quad \text{as} \quad n \to \infty.$$

It follows from Theorem 12.2 that

$$\frac{\sum_{j=1}^n Z_j(q + t/\sqrt{n}) - n(q + t/\sqrt{n})}{\left[nq(1 - q) + \sqrt{n}t(1 - 2q) - t^2\right]^{\frac{1}{2}}} \xrightarrow{\mathcal{D}} N(0, 1) \quad \text{as} \quad n \to \infty$$

so that, by Corollary 11.4,

$$\frac{1}{\sqrt{n}} \frac{\sum_{j=1}^n Z_j(q + t/\sqrt{n}) - n(q + t/\sqrt{n})}{[q(1 - q)]^{\frac{1}{2}}} \xrightarrow{\mathcal{D}} N(0, 1) \quad \text{as} \quad n \to \infty.$$

Hence,

$$\lim_{n \to \infty} \Pr[\sqrt{n}(X_{n(k_n)} - q) \leq t] = 1 - \Phi\left(\lim_{n \to \infty} \frac{k_n - nq - t\sqrt{n}}{[nq(1 - q)]^{\frac{1}{2}}}\right).$$

It is straightforward to show that

$$\lim_{n \to \infty} \frac{k_n - nq - t\sqrt{n}}{[nq(1-q)]^{\frac{1}{2}}} = -\frac{t}{[q(1-q)]^{\frac{1}{2}}},$$

proving the result. ■

The result in Theorem 13.2 can now be extended to random variables with a wider range of absolutely continuous distributions.

Corollary 13.2. *Let $X_1, X_2, \ldots, X_n$ denote independent, identically distributed random variables, each with an absolutely continuous distribution with distribution function F and density function p. Let $k_n, n = 1, 2, \ldots$ denote a sequence of integers such that $k_n \le n$ for all n and*

$$\sqrt{n}\left(\frac{k_n}{n} - q\right) \to 0 \quad as \quad n \to \infty$$

where $0 < q < 1$. Assume that there exists a unique $x_q \in \mathbf{R}$ such that $F(x_q) = q$ and that p is strictly positive in a neighborhood of x_q.
   Then

$$\sqrt{n}(X_{n(k_n)} - x_q) \overset{\mathcal{D}}{\to} N(0, \sigma^2),$$

where

$$\sigma^2 = \frac{q(1-q)}{p(x_q)^2}.$$

Proof. Let $Z_j = F(X_j)$. Then $Z_1, Z_2, \ldots, Z_n$ are independent, uniformly distributed random variables on $(0, 1)$ and

$$Z_{n(j)} = F(X_{n(j)}), \quad j = 1, \ldots, n, \quad n = 1, 2, \ldots$$

By Theorem 13.2,

$$\sqrt{n}(Z_{n(k_n)} - q) \overset{\mathcal{D}}{\to} N(0, q(1-q)) \quad as \quad n \to \infty.$$

Consider

$$\Pr\{\sqrt{n}(X_{n(k_n)} - x_q) \le t\} = \Pr\{X_{n(k_n)} \le x_q + t/\sqrt{n}\}.$$

Since p is strictly positive in a neighborhood of x_q, F is strictly increasing in a neighborhood of x_q. Hence, there exists an integer N such that

$$\Pr\{X_{n(k_n)} \le x_q + t/\sqrt{n}\} = \Pr\{F(X_{n(k_n)}) \le F(x_q + t/\sqrt{n})\}$$
$$= \Pr\{\sqrt{n}(Z_{n(k_n)} - q) \le \sqrt{n}(F(x_q + t/\sqrt{n}) - q)\}.$$

As $n \to \infty$,

$$\sqrt{n}(F(x_q + t/\sqrt{n}) - q) \to p(x_q)t.$$

Hence, by Theorem 11.3 and Example 11.9, together with the asymptotic normality of $Z_{n(k_n)}$,

$$\lim_{n \to \infty} \Pr\{\sqrt{n}(X_{n(k_n)} - x_q) \le t\} = \Pr\{Z \le p(x_q)t/\sqrt{[q(1-q)]}\},$$

where Z has a standard normal distribution. The result follows. ■

***Example* 13.4 (*Order statistics from an exponential distribution*).** Let $X_1, X_2, \ldots, X_n$ denote independent, identically distributed random variables each distributed according to the distribution with density

$$p(x; \lambda) = \lambda^{-1} \exp(-x/\lambda), \quad x > 0,$$

where $\lambda > 0$ is an unknown constant. Let k_n, $n = 1, 2, \ldots$ denote a sequence of integers such that

$$\sqrt{n}\left(\frac{k_n}{n} - q\right) \to 0 \quad \text{as} \quad n \to \infty,$$

where $0 < q < 1$.

Note that $F(x) = 1 - \exp(-x/\lambda)$, $x > 0$ so that $x_q = -\log(1 - q)\lambda$. Hence,

$$\sqrt{n}(X_{n(k_n)} + \log(1 - q)\lambda) \xrightarrow{\mathcal{D}} N(0, \sigma^2),$$

where

$$\sigma^2 = \lambda^2 \frac{q}{1 - q}. \qquad \qquad \Box$$

***Example* 13.5 (*Sample median*).** Let $X_1, X_2, \ldots, X_n$ denote independent, identically distributed random variables each with an absolutely continuous distribution with common density p. Assume that $\Pr(X_1 \leq 0) = 1/2$ and that, for some $\epsilon > 0$,

$$p(x) > 0 \quad \text{for all} \quad |x| < \epsilon.$$

For $n = 1, 2, \ldots$, let $\hat{m}_n$ denote the sample median, given by

$$\hat{m}_n = \begin{cases} X_{n\left(\frac{n+1}{2}\right)} & \text{if } n \text{ is odd} \\ \frac{X_{n\left(\frac{n}{2}\right)} + X_{n\left(\frac{n}{2}+1\right)}}{2} & \text{if } n \text{ is even} \end{cases}.$$

Fix t and consider $\Pr(\sqrt{n}\hat{m}_n \leq t)$. If n is odd,

$$\Pr(\sqrt{n}\hat{m}_n \leq t) = \Pr\left(\sqrt{n}X_{n\left(\frac{n+1}{2}\right)} \leq t\right);$$

if n is even,

$$\Pr\left(\sqrt{n}X_{n\left(\frac{n}{2}+1\right)} \leq t\right) \leq \Pr(\sqrt{n}\hat{m}_n \leq t) \leq \Pr\left(\sqrt{n}X_{n\left(\frac{n}{2}\right)} \leq t\right).$$

Let k_n denote the smallest integer greater than or equal to $n/2$. Then, for all n,

$$\Pr(\sqrt{n}X_{n(k_n+1)} \leq t) \leq \Pr(\sqrt{n}\hat{m}_n \leq t) \leq \Pr(\sqrt{n}X_{n(k_n)} \leq t).$$

Note that

$$\frac{1}{2} \leq \frac{k_n}{n} \leq \frac{1}{2} + \frac{1}{2n}, \quad n = 1, 2, \ldots$$

so that, as $n \to \infty$,

$$\sqrt{n}\left(\frac{k_n}{n} - \frac{1}{2}\right) \to 0 \quad \text{and} \quad \sqrt{n}\left(\frac{k_n + 1}{n} - \frac{1}{2}\right) \to 0.$$

It now follows from Corollary 13.2 that

$$\lim_{n \to \infty} \Pr(\sqrt{n}X_{n(k_n+1)} \leq t) = \lim_{n \to \infty} \Pr(\sqrt{n}X_{n(k_n)} \leq t) = \Pr(Z \leq t),$$

where Z has a normal distribution with mean 0 and variance $1/4p(0)^2$; hence,

$$\sqrt{n}\hat{m}_n \overset{D}{\to} N\left(0, \frac{1}{4p(0)^2}\right) \quad \text{as} \quad n \to \infty. \qquad \square$$

Let $X_1, X_2, \ldots, X_n$ denote independent, identically distributed random variables and let F denote the distribution function of X_1. Define

$$\hat{F}(t) = \frac{1}{n}\sum_{j=1}^{n}\mathrm{I}_{\{X_j \le t\}}, \quad -\infty < t < \infty.$$

Hence, this is a random function on $\mathbf{R}$; if Ω denotes the sample space of the underlying experiment, then, for each $t \in \mathbf{R}$,

$$\hat{F}(t)(\omega) = \frac{1}{n}\sum_{j=1}^{n}\mathrm{I}_{\{X_j(\omega) \le t\}}, \quad \omega \in \Omega.$$

Note that, for each $\omega \in \Omega$, $\hat{F}(\cdot)(\omega)$ is a distribution function on $\mathbf{R}$, called the *empirical distribution function* based on $X_1, \ldots, X_n$.

Define the *qth sample quantile* by

$$\hat{X}_{nq} = \inf\{x : \hat{F}_n(x) \ge q\}.$$

Clearly, the sample quantiles are closely related to the order statistics and it is straightforward to use Corollary 13.2 to determine the asymptotic distribution of a sample quantile. The result is given in the following corollary; the proof is left as an exercise.

Corollary 13.3. *Let $X_1, X_2, \ldots, X_n$ denote independent, identically distributed random variables, each with an absolutely continuous distribution with density p. Fix $0 < q < 1$. Assume that there is a unique $x_q \in \mathbf{R}$ such that $F(x_q) = q$ and that p is strictly positive in a neighborhood of x_q.*

For each $n = 1, 2, \ldots$, define $\hat{X}_{nq}$ as above. Then

$$\sqrt{n}(\hat{X}_{nq} - x_q) \overset{D}{\to} N(0, \sigma^2),$$

where

$$\sigma^2 = \frac{q(1-q)}{p(x_q)^2}.$$

Pairs of central order statistics

The same approach used in Theorem 13.2 for a single order statistic can be applied to the joint asymptotic distribution of several order statistics; in Theorem 13.3, we consider the case of two order statistics.

Theorem 13.3. *Let $X_1, X_2, \ldots, X_n$ denote independent, identically distributed random variables, each distributed according to a uniform distribution on $(0, 1)$. Let k_n, and m_n, $n = 1, 2, \ldots$ denote sequences of integers such that $k_n \le n$ and $m_n \le n$ for all n and*

$$\sqrt{n}\left(\frac{k_n}{n} - q_1\right) \to 0 \quad \text{as} \quad n \to \infty,$$

and

$$\sqrt{n}\left(\frac{m_n}{n} - q_2\right) \to 0 \quad as \quad n \to \infty,$$

where $0 < q_1 < 1$ and $0 < q_2 < 1$. Assume that $q_1 < q_2$.

Let $X_{n(k_n)}$ and $X_{n(m_n)}$ denote the k_nth and m_nth order statistic, respectively, of $X_1, \ldots, X_n$. Then

$$\sqrt{n}\begin{pmatrix} X_{n(k_n)} - q_1 \\ X_{n(m_n)} - q_2 \end{pmatrix} \xrightarrow{\mathcal{D}} N_2(0, \Sigma), \quad as \quad n \to \infty,$$

where

$$\Sigma = \begin{pmatrix} q_1(1 - q_1) & q_1(1 - q_2) \\ q_1(1 - q_2) & q_2(1 - q_2) \end{pmatrix}.$$

Proof.   The proof is very similar to that of Theorem 13.2; however, since the multivariate version of the central limit theorem for triangular arrays is not given in Chapter 12, some additional details are needed.

Fix $0 < t < 1$ and $0 < s < 1$. Define

$$Z_j(t) = \begin{cases} 0 & \text{if } X_j > t \\ 1 & \text{if } X_j \le t \end{cases}$$

and

$$W_j(s) = \begin{cases} 0 & \text{if } X_j > s \\ 1 & \text{if } X_j \le s \end{cases}.$$

Then

$$\Pr(X_{n(k_n)} \le t, \ X_{n(m_n)} \le s) = \Pr\left[\sum_{j=1}^n Z_j(t) \ge k_n, \ \sum_{j=1}^n W_j(s) \ge m_n\right]$$

and

$$\Pr(\sqrt{n}(X_{n(k_n)} - q_1) \le t, \ \sqrt{n}(X_{n(m_n)} - q_2) \le s)$$

$$= \Pr\left(\frac{1}{\sqrt{n}}\left[\sum_{j=1}^n Z_j(q_1 + t/\sqrt{n}) - n(q_1 + t/\sqrt{n})\right] \ge \frac{k_n - nq_1 - t\sqrt{n}}{\sqrt{n}}, \right.$$

$$\left. \frac{1}{\sqrt{n}}\left[\sum_{j=1}^n W_j(q_2 + s/\sqrt{n}) - n(q_2 + s/\sqrt{n})\right] \ge \frac{m_n - nq_2 - s\sqrt{n}}{\sqrt{n}}\right).$$

Define

$$Z_{nj} = Z_j(q_1 + t/\sqrt{n}) - (q_1 + t/\sqrt{n}), \quad j = 1, \ldots, n$$

and

$$W_{nj} = W_j(q_2 + s/\sqrt{n}) - (q_2 + s/\sqrt{n}), \quad j = 1, \ldots, n.$$

Let a and b denote arbitrary real-valued constants and let

$$V_{nj} = aZ_{nj} + bW_{nj}, \quad j = 1, 2, \ldots, n, \ n = 1, 2, \ldots;$$

this is a triangular array with $E(V_{nj}) = 0$,

$$\sigma_n^2 \equiv \text{Var}(V_{nj})$$
$$= a^2(q_1 + t/\sqrt{n})(1 - q_1 - t/\sqrt{n}) + b^2(q_2 + s/\sqrt{n})(1 - q_2 - s/\sqrt{n})$$
$$+ 2ab(q_1 + t/\sqrt{n})(q_2 + s/\sqrt{n}),$$

and $E(|V_{nj}|^3) \le |a + b|^3$, for all j, n.

Since

$$\sum_{j=1}^{n} \text{Var}(V_{nj}) = n\sigma_n^2 = O(n) \quad \text{as} \quad n \to \infty$$

and $E(|V_{nj}|^3)$ is bounded, it follows that the conditions of Theorem 12.2 are satisfied and

$$\frac{\sum_{j=1}^{n} V_{nj}}{\sigma_n \sqrt{n}} \xrightarrow{D} N(0, 1) \quad \text{as} \quad n \to \infty.$$

Note that

$$\lim_{n \to \infty} \sigma_n^2 = a^2 q_1(1 - q_1) + b^2 q_2(1 - q_2) + 2ab q_1(1 - q_2);$$

since this holds for any a, b, by Theorem 11.6,

$$\frac{1}{\sqrt{n}} \begin{pmatrix} \sum_{j=1}^{n} Z_{nj} \\ \sum_{j=1}^{n} W_{nj} \end{pmatrix} \xrightarrow{D} N_2(0, \Sigma)$$

where Σ is given in the statement of the theorem.

It follows that

$$\Pr\{\sqrt{n}(X_{n(k_n)} - q_1) \le t, \quad \sqrt{n}(X_{n(m_n)} - q_2) \le s\}$$
$$= \Pr\left\{\frac{1}{\sqrt{n}} \sum_{j=1}^{n} Z_{nj} - (k_n - nq_1)/\sqrt{n} \ge -t, \quad \frac{1}{\sqrt{n}} \sum_{j=1}^{n} W_{nj} - (m_n - nq_2)/\sqrt{n} \ge -s\right\}$$

and, hence, by Corollary 11.4, that

$$\lim_{n \to \infty} \Pr[\sqrt{n}(X_{n(k_n)} - q_1) \le t, \quad \sqrt{n}(X_{n(m_n)} - q_2) \le s]$$
$$= \lim_{n \to \infty} \Pr\left\{\frac{1}{\sqrt{n}} \sum_{j=1}^{n} Z_{nj} \ge -t, \quad \frac{1}{\sqrt{n}} \sum_{j=1}^{n} W_{nj} \ge -s\right\} = \Pr(Z \ge -t, \quad W \ge -s)$$

where (Z, W) has a bivariate normal distribution with mean 0 and covariance matrix Σ.

Note that

$$\Pr(Z \ge -t, \quad W \ge -s) = \Pr(-Z \le t, \quad -W \le s)$$

and that $(-Z, -W)$ has the same distribution as (Z, W); the result follows. ∎

As was done in the case of a single central order statistic, the result in Theorem 13.3 for the uniform distribution can now be extended to a wider class of distributions. The result is given in the following corollary; the proof is left as an exercise.

Corollary 13.4. *Let* $X_1, X_2, \ldots, X_n$ *denote independent, identically distributed random variables, each distributed according to an absolutely continuous distribution with* *distribution function F and density p. Let k_n, and m_n, $n = 1, 2, \ldots$ denote sequences of integers such that $k_n \leq n$ and $m_n \leq n$ for all n and*

$$\sqrt{n}\left(\frac{k_n}{n} - q_1\right) \to 0 \quad as \quad n \to \infty$$

and

$$\sqrt{n}\left(\frac{m_n}{n} - q_2\right) \to 0 \quad as \quad n \to \infty,$$

where $0 < q_1 < q_2 < 1$.

Assume that there exist unique $x_{q_1}, x_{q_2} \in \mathbf{R}$ such that $F(x_{q_1}) = q_1$ and $F(x_{q_2}) = q_2$ and that p is strictly positive in neighborhoods of q_1 and q_2.

Let $X_{n(k_n)}$ and $X_{n(m_n)}$ denote the k_nth and m_nth order statistic, respectively, of $X_1, \ldots, X_n$. Then

$$\sqrt{n}\begin{pmatrix} X_{n(k_n)} - x_{q_1} \\ X_{n(m_n)} - x_{q_2} \end{pmatrix} \xrightarrow{D} N_2(0, \Sigma), \quad as \quad n \to \infty,$$

where

$$\Sigma = \begin{pmatrix} \dfrac{q_1(1 - q_1)}{p(x_{q_1})^2} & \dfrac{q_1(1 - q_2)}{p(x_{q_1})p(x_{q_2})} \\ \dfrac{q_1(1 - q_2)}{p(x_{q_1})p(x_{q_2})} & \dfrac{q_2(1 - q_2)}{p(x_{q_2})^2} \end{pmatrix}.$$

***Example* 13.6 (*Interquartile range*).** Let $X_1, X_2, \ldots, X_n$ denote independent, identically distributed random variables, each distributed according to a distribution with density p that is positive on $\mathbf{R}$ and symmetric about 0. Let k_n denote the smallest integer greater than or equal to $n/4$ and let m_n denote the smallest integer greater than or equal to $3n/4$. Then

$$\sqrt{n}\left(\frac{k_n}{n} - \frac{1}{4}\right) \to 0 \quad and \quad \sqrt{n}\left(\frac{m_n}{n} - \frac{3}{4}\right) \to 0$$

as $n \to \infty$. The order statistic $X_{n(k_n)}$ is sometimes called the *lower quartile* of the sample and $X_{n(m_n)}$ is sometimes called the *upper quartile* of the sample; the *interquartile range* is given by

$$\hat{Q}_n = X_{n(m_n)} - X_{n(k_n)}$$

and is a measure of the dispersion in the data.

Let c denote the constant satisfying

$$\Pr(X_1 \leq -c) = \frac{1}{4};$$

note that, since the distribution of X_1 is symmetric about 0,

$$\Pr(X_1 \leq c) = \frac{3}{4}.$$

Hence,

$$\sqrt{n}\begin{pmatrix} X_{n(k_n)} + c \\ X_{n(m_n)} - c \end{pmatrix} \xrightarrow{D} N_2(0, \Sigma),$$

where

$$\Sigma = \frac{1}{16p(c)^2} \begin{pmatrix} 3 & 1 \\ 1 & 3 \end{pmatrix}.$$

It follows that

$$\sqrt{n}(\hat{Q}_n - 2c) \xrightarrow{D} N(0, \sigma^2),$$

where

$$\sigma^2 = \frac{1}{4p(c)^2}. \qquad \qquad \square$$

Sample extremes

We now consider the asymptotic properties of the sample mean and sample minimum. As noted earlier, the asymptotic theory of these extreme order statistics is quite different from the asymptotic theory of the central order statistics described above.

Let $X_1, X_2, \ldots, X_n$ denote independent, identically distributed scalar random variables each with distribution function F, which is assumed to be absolutely continuous. Let

$$Y_n = \max_{1 \le j \le n} X_j$$

and

$$Z_n = \min_{1 \le j \le n} X_j.$$

The asymptotic properties of Y_n and Z_n follow from the following general result.

Theorem 13.4. *For each $n = 1, 2, \ldots$, let Y_n and Z_n be defined as above and let W denote a random variable with a standard exponential distribution. Then*

$$n(1 - F(Y_n)) \xrightarrow{D} W \quad as \quad n \to \infty,$$

$$nF(Z_n)) \xrightarrow{D} W \quad as \quad n \to \infty,$$

and $n(1 - F(Y_n))$ and $nF(Z_n)$ are asymptotically independent.

Proof. Consider the probability

$$\Pr\{n(1 - F(Y_n)) \le y, \ nF(Z_n) \le x\} = \Pr\{F(Y_n) \ge 1 - y/n, \ F(Z_n) \le x/n\}$$
$$= 1 - \Pr\{F(Y_n) \le 1 - y/n \ \cup \ F(Z_n) \ge x/n\}$$
$$= 1 - \Pr\{F(Y_n) \le 1 - y/n\} - \Pr\{F(Z_n) \ge x/n\}$$
$$+ \Pr\{F(Y_n) \le 1 - y/n, \ F(Z_n) \ge x/n\}.$$

It is straightforward to show that if x or y is negative, then this probability converges to 0 as $n \to \infty$; hence, assume that $x \ge 0$ and $y \ge 0$.

Note that $F(Z_n)$ and $F(Y_n)$ may be viewed as the first and nth order statistic, respectively, based on n independent, identically distributed random variables each uniformly distributed on the interval $(0, 1)$. Hence,

$$\Pr\{F(Y_n) \le 1 - y/n\} = (1 - y/n)^n, \quad 0 \le y \le n,$$

$$\Pr\{F(Z_n) \ge x/n\} = (1 - x/n)^n, \quad 0 \le x \le n,$$

and

$$\Pr\{F(Y_n) \le 1 - y/n, \ F(Z_n) \ge x/n\} = (1 - y/n - x/n)^n, \quad 0 \le x + y \le n.$$

It follows that

$$\lim_{n\to\infty} \Pr\{n[1 - F(Y_n)] \le y, \ nF(Z_n) \le x\} = 1 - \exp(-y) - \exp(-x) + \exp(-(x + y))$$
$$= [1 - \exp(-y)] [1 - \exp(-x)],$$

proving the result.   ■

The asymptotic behavior of the minimum and maximum of n independent identically distributed random variables will therefore depend on the properties of the distribution function F. Some of the possibilities are illustrated by the following examples; in each case,

$$Y_n = \max_{1 \le j \le n} X_j$$

and

$$Z_n = \min_{1 \le j \le n} X_j,$$

as in Theorem 13.4.

***Example* 13.7 (*Beta distribution*).** Let $X_1, X_2, \ldots, X_n$ denote independent, identically distributed random variables each distributed according to the distribution with density function

$$p(x; \alpha) = \alpha x^{\alpha - 1}, \quad 0 < x < 1$$

where $\alpha > 0$ is a constant. The corresponding distribution function is given by

$$F(x; \alpha) = x^\alpha, \quad 0 < x < 1.$$

Then, according to Theorem 13.4,

$$\lim_{n\to\infty} \Pr\left(n\left(1 - Y_n^\alpha\right) \le t\right) = 1 - \exp(-t), \quad t > 0$$

and

$$\lim_{n\to\infty} \Pr\left(nZ_n^\alpha \le t\right) = 1 - \exp(-t), \quad t > 0.$$

First consider the asymptotic distribution of Y_n. Since

$$\Pr\left[n\left(1 - Y_n^\alpha\right) \le t\right] = \Pr\left[Y_n \ge (1 - t/n)^{\frac{1}{\alpha}}\right],$$

$$\lim_{n\to\infty} \Pr\{n(Y_n - 1)/\alpha \le n[(1 - t/n)^{\frac{1}{\alpha}} - 1]/\alpha\} = \exp(-t), \quad t > 0.$$

Since

$$\lim_{n\to\infty} n[(1 - t/n)^{\frac{1}{\alpha}} - 1]/\alpha = -t,$$

it follows that

$$\lim_{n\to\infty} \Pr\{n(Y_n - 1)/\alpha \le -t\} = \exp(-t), \quad t > 0$$

or, equivalently,

$$\lim_{n \to \infty} \Pr\{n(Y_n - 1)/\alpha \le t\} = \exp(t), \quad t < 0.$$

Now consider Z_n. Since

$$\Pr(nZ_n^\alpha \le t) = \Pr(Z_n \le (t/n)^{\frac{1}{\alpha}}),$$

it follows that

$$\lim_{n \to \infty} \Pr[Z_n \le (t/n)^{\frac{1}{\alpha}}] = 1 - \exp(-t), \quad t > 0.$$

Hence,

$$\lim_{n \to \infty} \Pr[n^{\frac{1}{\alpha}} Z_n \le t^{\frac{1}{\alpha}}] = 1 - \exp(-t), \quad t > 0,$$

or, equivalently,

$$\lim_{n \to \infty} \Pr(n^{\frac{1}{\alpha}} Z_n \le t) = 1 - \exp(-t^\alpha), \quad t > 0. \qquad \square$$

***Example* 13.8 (*Exponential distribution*).** Let $X_1, X_2, \ldots, X_n$ denote independent, identically distributed random variables each distributed according to a standard exponential distribution. Then, according to Theorem 13.4,

$$\lim_{n \to \infty} \Pr[n \exp(-Y_n) \le t] = 1 - \exp(-t), \quad t > 0$$

and

$$\lim_{n \to \infty} \Pr\{n[1 - \exp(-Z_n)] \le t\} = 1 - \exp(-t), \quad t > 0.$$

First consider the distribution of Y_n. Note that

$$\lim_{n \to \infty} \Pr\{Y_n \le -\log(t/n)\} = \exp(-t), \quad t > 0.$$

It follows that

$$\lim_{n \to \infty} \Pr[Y_n - \log(n) \le -\log(t)] = \exp(-t), \quad t > 0,$$

or, equivalently,

$$\lim_{n \to \infty} \Pr[Y_n - \log(n) \le t)] = \exp\{-\exp(-t)\}, \quad -\infty < t < \infty.$$

Now consider Z_n. Note that

$$\lim_{n \to \infty} \Pr[Z_n \le -\log(1 - t/n)] = 1 - \exp(-t), \quad t > 0$$

so that

$$\lim_{n \to \infty} \Pr[nZ_n \le -n \log(1 - t/n)] = 1 - \exp(-t), \quad t > 0.$$

Since

$$\lim_{n \to \infty} -n \log(1 - t/n) = t,$$

it follows that

$$\lim_{n\to\infty} \Pr(nZ_n \le t) = 1 - \exp(-t), \quad t > 0. \qquad \square$$

Example* 13.9 *(Pareto distribution). Let $X_1, X_2, \ldots, X_n$ denote independent, identically distributed random variables each distributed according to the distribution with density function

$$p(x) = \frac{1}{x^2}, \quad x > 1;$$

the corresponding distribution function is given by

$$F(x) = 1 - \frac{1}{x}, \quad x > 1.$$

It follows from Theorem 13.4 that

$$\lim_{n\to\infty} \Pr(n/Y_n \le t) = 1 - \exp(-t), \quad t > 0$$

and

$$\lim_{n\to\infty} \Pr[n(1 - 1/Z_n) \le t] = 1 - \exp(-t), \quad t > 0.$$

Hence,

$$\lim_{n\to\infty} \Pr(Y_n/n \le t) = \exp(-1/t), \quad t > 0$$

and

$$\lim_{n\to\infty} \Pr\{n(Z_n - 1) \le t/(1 - t/n)\} = 1 - \exp(-t), \quad t > 0$$

so that

$$\lim_{n\to\infty} \Pr\{n(Z_n - 1) \le t\} = 1 - \exp(-t), \quad t > 0. \qquad \square$$

13.4   U-Statistics

Let $X_1, X_2, \ldots, X_n$ denote independent, identically distributed real-valued random variables, each with range $\mathcal{X}$. Let h denote a real-valued function defined on $\mathcal{X}$ such that $E[h(X_1)^2] < \infty$. Then the sample mean $\sum_{j=1}^{n} h(X_j)/n$ follows the central limit theorem; that is, a suitably normalized form of the statistic is asymptotically distributed according to a normal distribution.

As a generalization of this type of statistic, suppose the function h is defined on $\mathcal{X}^2$ and suppose that h is symmetric in its arguments. Consider the statistic

$$U = \frac{1}{\binom{n}{2}} \sum_{\beta} h(X_{\beta_1}, X_{\beta_2}),$$

where the sum is over all unordered pairs of integers $\beta = (\beta_1, \beta_2)$ chosen from $\{1, 2, \ldots, n\}$. A statistic of this form is called a *U-statistic* of order 2; the function h is called the *kernel* of the statistic.

***Example* 13.10 (*Sample variance*).** Consider the kernel

$$h(x_1, x_2) = \frac{1}{2}(x_1 - x_2)^2.$$

The corresponding U-statistic is given by

$$S^2 = \frac{1}{2\binom{n}{2}} \sum_\beta (X_{\beta_1} - X_{\beta_2})^2.$$

Note that

$$\sum_\beta (X_{\beta_1} - X_{\beta_2})^2 = (n-1)\sum_{j=1}^n X_j^2 - 2\sum_{j<k} X_j X_k.$$

Since

$$\left(\sum_{j=1}^n X_j\right)^2 = \sum_{j=1}^n X_j^2 + 2\sum_{j<k} X_j X_k,$$

$$\sum_\beta (X_{\beta_1} - X_{\beta_2})^2 = n\sum_{j=1}^n X_j^2 - \left(\sum_{j=1}^n X_j\right)^2.$$

Hence,

$$S^2 = \frac{1}{n-1}\sum_{j=1}^n X_j^2 - \frac{n}{(n-1)}\left(\sum_{j=1}^n X_j/n\right)^2 = \frac{1}{n-1}\sum_{j=1}^n (X_j - \bar{X})^2,$$

where $\bar{X} = \sum_{j=1}^n X_j/n$; that is, S^2 is the sample variance with divisor $n-1$. □

This type of statistic may be generalized to allow h to be a real-valued function on $\mathcal{X}^r$ for some $r = 1, 2, \ldots$. Again, we assume that h is symmetric in its arguments. Then a *U-statistic with kernel h* is a statistic of the form

$$U = \frac{1}{\binom{n}{r}} \sum h(X_{\beta_1}, \ldots, X_{\beta_r}),$$

where the sum is over all unordered sets of r integers chosen from $\{1, 2, \ldots, n\}$. Note that, since the random variables appearing in $h(X_{\beta_1}, \ldots, X_{\beta_r})$ are always independent and identically distributed,

$$\mathrm{E}(U) = \mathrm{E}[h(X_1, \ldots, X_r)];$$

hence, in statistical terminology, U is an *unbiased estimator* of $\mathrm{E}[h(X_1, \ldots, X_r)]$. That is, if we are interested in estimating $\mathrm{E}[h(X_1, \ldots, X_r)]$ based on $X_1, \ldots, X_n$, the statistic U has the property that its expected value is exactly the quantity we are attempting to estimate.

***Example* 13.11 (*Integer power of a mean*).** For a given positive integer r, consider the kernel

$$h(x_1, \ldots, x_r) = x_1 x_2 \cdots x_r.$$

This leads to the U-statistic

$$U = \frac{1}{\binom{n}{r}} \sum_{\beta} X_{\beta_1} X_{\beta_2} \cdots X_{\beta_r}.$$

Hence, U has expected value μ^r where $\mu = \mathrm{E}(X_1)$.

For instance, for $r = 2$,

$$U = \frac{2}{n(n-1)} \sum_{j<k} X_j X_k = \frac{1}{n(n-1)} \left[\left(\sum_{j=1}^{n} X_j \right)^2 - \sum_{j=1}^{n} X_j^2 \right]$$

$$= \frac{1}{n-1} \left[n\bar{X}^2 - \frac{1}{n} \sum_{j=1}^{n} X_j^2 \right]. \qquad \square$$

Clearly, a U-statistic of order r is an average of the values $h(X_{\beta_1}, \ldots, X_{\beta_r})$. Furthermore, since $X_1, \ldots, X_n$ have an exchangeable distribution and h is a symmetric function, the variables $h(X_{\beta_1}, \ldots, X_{\beta_r})$ as β varies are identically distributed. However, they are not independent so that it does not follow immediately that U follows the central limit theorem.

The asymptotic normality of a U-statistic may be established by a general technique known as the *projection method*. Consider a real-valued random variable T and consider a collection of real-valued random variables $\mathcal{V}$. The projection of T onto $\mathcal{V}$ is that element of $\hat{V}$ of $\mathcal{V}$ that minimizes $\mathrm{E}[(T - V)^2]$ over $V \in \mathcal{V}$. Hence, $T - \hat{V}$ is as small as possible, in a certain sense. Writing $T = \hat{V} + T - \hat{V}$, the asymptotic properties of T may be obtained from those of $\hat{V}$, provided that $T - \hat{V}$ is negligible. Clearly, for this approach to be effective, the class of random variables $\mathcal{V}$ must be chosen so that the asymptotic properties of $\hat{V}$ are available. For instance, a commonly used class of random variables are those that are sums of independent identically distributed random variables; the resulting projection is known as the *Hájek projection*. See, for example, van der Vaart (1998, Chapter 11) for further details on projections.

Let U denote a U-statistic of order r based on a kernel h and let

$$\theta = \mathrm{E}[h(X_1, \ldots, X_r)].$$

Let

$$\hat{U} = \sum_{j=1}^{n} \mathrm{E}(U \mid X_j);$$

note that

$$\hat{U} - n\theta = \sum_{j=1}^{n} [\mathrm{E}(U \mid X_j) - \theta]$$

is of the form $\sum_{j=1}^{n} g(X_j)$ with g taken to be $g(x) = \mathrm{E}(U \mid X_1 = x) - \theta$.

The following argument shows that $\hat{U} - n\theta$ is the projection of $U - \theta$ onto the space of random variables of the form $\sum_{j=1}^{n} g(X_j)$ for some function g such that $\mathrm{E}[g(X_1)^2] < \infty$. First note that $(U - \theta) - (\hat{U} - n\theta)$ is uncorrelated with any statistic of the form

$\sum_{j=1}^{n} g_0(X_j)$, where g_0 is a function on $\mathcal{X}$ such that $\mathrm{E}[g_0(X_1)^2] < \infty$:

$$\mathrm{Cov}\left[(U - \theta) - (\hat{U} - n\theta), \sum_{j=1}^{n} g_0(X_j) \right]$$

$$= \sum_{j=1}^{n} \mathrm{E}[(U - \theta)g_0(X_j)] - \sum_{j=1}^{n} \mathrm{E}\{[\mathrm{E}(U \mid X_j) - \theta]g_0(X_j)\}$$

$$= \sum_{j=1}^{n} \{\mathrm{E}[U g_0(X_j)] - \mathrm{E}[\mathrm{E}(U \mid X_j)g_0(X_j)]\}$$

$$= \sum_{j=1}^{n} \{\mathrm{E}[U g_0(X_j)] - \mathrm{E}[U g_0(X_j)]\} = 0.$$

It follows that, for any random variable W of the form $\sum_{j=1}^{n} g_0(X_j)$, where g_0 satisfies $\mathrm{E}[g_0(X_j)^2] < \infty$,

$$\mathrm{E}\{[(U - \theta) - W]^2\} = \mathrm{E}\{[(U - \theta) - (\hat{U} - n\theta) + (\hat{U} - n\theta) - W]^2\}$$

$$= \mathrm{E}\{[(U - \theta) - (\hat{U} - n\theta)]^2\} + \mathrm{E}\{[(\hat{U} - n\theta) - W]^2\}.$$

Hence,

$$\mathrm{E}\{[(U - \theta) - (\hat{U} - n\theta)]^2\} \le \mathrm{E}\{[(U - \theta) - W]^2$$

so that, in terms of expected squared distance, $\hat{U} - n\theta$ is closer to $U - \theta$ than is W.

***Example* 13.12 (*Sample mean*).** Consider a U-statistic based on the kernel

$$h(x_1, x_2) = \frac{1}{2}(x_1 + x_2).$$

Then $U = \bar{X}$ and $\theta = \mathrm{E}(X_1)$. Clearly, the U-statistic approach is not needed here and this example is included for illustration only.

Writing

$$U = \frac{1}{n(n-1)} \sum_{j<k} (X_j + X_k),$$

we see that

$$\mathrm{E}(X_j + X_k \mid X_i) = \begin{cases} X_i + \theta & \text{if } i = j \text{ or } i = k \\ 2\theta & \text{otherwise} \end{cases}.$$

Hence,

$$\mathrm{E}(U \mid X_i) = \frac{1}{n(n-1)}[(n-1)(X_i + \theta) + (n-1)^2 2\theta] = \frac{1}{n}X_i + (n-1)\theta$$

and

$$\hat{U} = \sum_{j=1}^{n} \mathrm{E}(U \mid X_j) = \bar{X} + (n-1)\theta.$$

It follows that

$$\hat{U} - n\theta = \bar{X} - \theta = U - \theta;$$

this is not surprising since $U - \theta$ itself is of the form $\sum_{j=1}^{n} g(X_j)$. $\quad\square$

Since $E(U \mid X_j)$ is a function of X_j, $\hat{U}$ is a sum of independent identically distributed random variables; hence, under standard conditions, $\hat{U}$ follows the central limit theorem. The asymptotic normality of U follows provided that $U - \hat{U}$ is sufficiently small in an appropriate sense. The following theorem gives the details of this approach.

Theorem 13.5. *Let $X_1, X_2, \ldots, X_n$ denote independent, identically distributed real-valued random variables and let*

$$U = \frac{1}{\binom{n}{r}} \sum h(X_{\beta_1}, \ldots, X_{\beta_r})$$

denote a U-statistic with kernel h, where

$$E[h(X_1, X_2, \ldots, X_r)^2] < \infty.$$

Then

$$\sqrt{n}(U - \theta) \xrightarrow{\mathcal{D}} N(0, \sigma^2) \quad \text{as} \quad n \to \infty;$$

here

$$\theta = E[h(X_1, X_2, \ldots, X_r)]$$

and

$$\sigma^2 = r^2 \operatorname{Cov}[h(X_1, X_2, \ldots, X_r), h(X_1, \tilde{X}_2, \ldots, \tilde{X}_r)],$$

where $X_1, X_2, \ldots, X_r, \tilde{X}_2, \ldots, \tilde{X}_r$ are independent and identically distributed.

Proof.   Let

$$\hat{h}(x) = E[h(x, X_2, \ldots, X_r)].$$

Then, since h is symmetric in its arguments,

$$E[h(X_{\beta_1}, \ldots, X_{\beta_r}) \mid X_i = x] = \begin{cases} \hat{h}(x) & \text{if } i \in \{\beta_1, \ldots, \beta_r\} \\ \theta & \text{otherwise} \end{cases}.$$

There are $\binom{n}{r}$ terms in the sum defining U and, for any $i = 1, \ldots, n$, $\binom{n-1}{r-1}$ of them include X_i. Hence,

$$E[U \mid X_i = x] = \frac{\binom{n-1}{r-1}}{\binom{n}{r}} \hat{h}(x) + \frac{\binom{n}{r} - \binom{n-1}{r-1}}{\binom{n}{r}} \theta$$

$$= \frac{r}{n} \hat{h}(x) + \frac{n-r}{n} \theta.$$

It follows that

$$\hat{U} = \frac{r}{n} \sum_{i=1}^{n} \hat{h}(X_i) + (n - r)\theta,$$

or,

$$\hat{U} - n\theta = \frac{r}{n} \sum_{i=1}^{n} [\hat{h}(X_i) - \theta].$$

Note that

$$\hat{h}(X_1) - \theta, \ldots, \hat{h}(X_n) - \theta$$

are independent, indentically distributed random variables each with mean 0 and variance

$$\sigma_1^2 = E\{[\hat{h}(X_1) - \theta]^2\}.$$

Note that

$$
\begin{aligned}
E\{[\hat{h}(X_1) - \theta]^2\} &= E[(E[h(X_1, X_2, \ldots, X_r)|X_1] - \theta)^2] \\
&= E\{E[h(X_1, X_2, \ldots, X_r) - \theta|X_1]E[h(X_1, \tilde{X}_2, \ldots, \tilde{X}_r) - \theta|X_1]\} \\
&= \mathrm{Cov}[h(X_1, X_2, \ldots, X_r), h(X_1, \tilde{X}_2, \ldots, \tilde{X}_r)] \equiv \sigma^2/r^2.
\end{aligned}
$$

Hence,

$$\sqrt{n}(\hat{U} - n\theta) \xrightarrow{D} N(0, \sigma^2) \quad \text{as} \quad n \to \infty.$$

The result now follows provided that

$$\sqrt{n}[(\hat{U} - n\theta) - (U - \theta)] \xrightarrow{P} 0 \quad \text{as} \quad n \to \infty. \tag{13.1}$$

Note that

$$E[(\hat{U} - n\theta) - (U - \theta)] = 0;$$

hence, (13.1) holds, provided that

$$n\,\mathrm{Var}(U - \hat{U}) \to 0 \quad \text{as} \quad n \to \infty.$$

Now,

$$\mathrm{Cov}(U, \hat{U}) = \mathrm{Cov}\left(U, \sum_{i=1}^{n} E(U|X_i)\right) = \sum_{i=1}^{n} \mathrm{Cov}(U, E(U|X_i))$$

and

$$
\begin{aligned}
\mathrm{Cov}(U, E(U|X_i)) &= E[U E(U|X_i)] - E(U)^2 = E\{E[U E(U|X_i)|X_i]\} - E[E(U|X_i)]^2 \\
&= E[E(U|X_i)^2] - E[E(U|X_i)]^2 \\
&= \mathrm{Var}[E(U|X_i)]
\end{aligned}
$$

so that

$$\mathrm{Cov}(U, \hat{U}) = \mathrm{Var}(\hat{U}).$$

It follows that

$$\mathrm{Var}(U - \hat{U}) = \mathrm{Var}(U) - \mathrm{Var}(\hat{U}).$$

Since

$$U = \frac{1}{\binom{n}{r}} \sum_{\beta} h(X_{\beta_1}, \ldots, X_{\beta_r}),$$

$$\mathrm{Var}(U) = \frac{1}{\binom{n}{r}^2} \sum_{\beta} \sum_{\tilde{\beta}} \mathrm{Cov}[h(X_{\tilde{\beta}_1}, \ldots, X_{\tilde{\beta}_r}), h(X_{\beta_1}, \ldots, X_{\beta_r})].$$

Note that $\text{Cov}[h(X_{\tilde{\beta}_1}, \ldots, X_{\tilde{\beta}_r}), h(X_{\beta_1}, \ldots, X_{\beta_r})]$ depends on how many indices appear in both

$$\{\tilde{\beta}_1, \ldots, \tilde{\beta}_r\} \quad \text{and} \quad \{\beta_1, \ldots, \beta_r\}.$$

Of the $\binom{n}{r}^2$ total subsets of $\{\tilde{\beta}_1, \ldots, \tilde{\beta}_r\}$ and $\{\beta_1, \ldots, \beta_r\}$ under consideration, the number with m indices in common is

$$\binom{n}{r}\binom{r}{m}\binom{n-r}{r-m}.$$

This result may be obtained by noting that there are $\binom{n}{r}$ ways to choose the first subset and, given the first subset, the second must be chosen in such a way that m indices are selected from the r indices in the first subset.

Let $\sigma_m = \text{Cov}[h(X_{\tilde{\beta}_1}, \ldots, X_{\tilde{\beta}_r}), h(X_{\beta_1}, \ldots, X_{\beta_r})]$ when there are m indices in common. Then

$$\text{Var}(U) = \sum_{m=1}^{r} \frac{\binom{r}{m}\binom{n-r}{r-m}}{\binom{n}{r}} \sigma_m \equiv \sum_{m=1}^{r} Q_m \sigma_m.$$

For $m = 1$,

$$Q_1 = \frac{\binom{r}{m}\binom{n-r}{r-m}}{\binom{n}{r}} = \frac{r^2[(n-r)!]^2}{n!(n-2r+1)!} = r^2 \frac{(n-r)\cdots(n-r-(r-2))}{n\cdots(n-(r-1))}.$$

Since there are $r-1$ terms in the numerator of this ratio and r terms in the denominator, $Q_1 = O(1/n)$. For $m = 2$,

$$Q_2 = \frac{\binom{r}{2}\binom{n-r}{r-2}}{\binom{n}{r}} = \frac{\binom{r}{2}\binom{n-r}{r-2}}{\binom{r}{1}\binom{n-r}{r-1}} \left[\frac{\binom{r}{1}\binom{n-r}{r-1}}{\binom{n}{r}}\right] = \frac{(r-1)^2}{2} \frac{1}{n-2r+2} Q_1 = O(n^{-2}).$$

In general, it is straightforward to show that

$$\frac{Q_{m+1}}{Q_m} = \frac{(r-m)^2}{(m+1)} \frac{1}{n-2r+m+1}$$

so that $Q_m = O(n^{-m})$, $m = 1, \ldots, r$. It follows that

$$\text{Var}(U) = \frac{r\binom{n-r}{r-1}}{\binom{n}{r}} \sigma_1 + O(n^{-2}) = r^2 \frac{[(n-r)!]^2}{n!(n-2r+1)!} \sigma_1 + O(n^{-2}).$$

Note that

$$\frac{[(n-r)!]^2}{n!(n-2r+1)!} = \frac{(n-r)(n-r-1)\cdots(n-2r+2)}{n(n-1)\cdots(n-r+1)}$$

$$= \frac{1}{n} \frac{n-r}{n-1} \frac{n-r-1}{n-2} \frac{n-r-(r-2)}{n-(r-1)}.$$

Hence,

$$\lim_{n\to\infty} n\text{Var}(U) = r^2\sigma_1 \equiv \text{Var}(\hat{U}).$$

It follows that

$$n\text{Var}(U - \hat{U}) = n[\text{Var}(U) - \text{Var}(\hat{U})] \to 0 \quad \text{as} \quad n \to \infty,$$

proving (13.1). The result follows. ■

***Example* 13.13** *(Sample variance).* Consider the second-order kernel

$$h(x_1, x_2) = \frac{1}{2}(x_1 - x_2)^2.$$

Recall that the corresponding U-statistic is the sample variance,

$$S^2 = \frac{1}{n-1} \sum_{j=1}^{n} (X_j - \bar{X})^2$$

and $\theta = \text{Var}(X_1)$; see Example 13.10. Assume that $\text{E}(X_1^4) < \infty$; it follows that

$$\text{E}[h(X_1, X_2)^2] < \infty.$$

Then $\sqrt{n}(S^2 - \theta)$ converges in distribution to a random variable with a normal distribution with mean 0 and variance

$$\begin{aligned}
\sigma^2 &= \text{Cov}[(X_1 - X_2)^2, (X_1 - X_3)^2] \\
&= \text{Cov}[(X_1 - \mu - X_2 + \mu)^2 + (X_1 - \mu - X_3 + \mu)^2] \\
&= \text{Cov}[(X_1 - \mu)^2, (X_1 - \mu)^2] \\
&= \text{E}[(X_1 - \mu)^4] - \tau^4,
\end{aligned}$$

where $\mu = \text{E}(X_1)$ and $\tau^2 = \text{Var}(X_1)$. □

***Example* 13.14** *(Integer power of a mean).* For some $r = 1, 2, \ldots$, consider the kernel of order r given by

$$h(x_1, \ldots, x_r) = x_1 x_2 \cdots x_r.$$

Then $\theta = \mu^r$ where $\mu = \text{E}(X_1)$. Assume that $\text{E}(X_1^2) < \infty$; it follows that

$$\text{E}[h(X_1, \ldots, X_r)^2] < \infty.$$

It follows from Theorem 13.5 that the corresponding U-statistic is such that $\sqrt{n}(U - \theta)$ converges in distribution to a random variable with a normal distribution with mean 0 and variance

$$\sigma^2 = r^2 \text{Cov}(X_1 X_2 \cdots X_r, X_1 \tilde{X}_2 \cdots \tilde{X}_r) = r^2 \left[\text{E}\left(X_1^2\right) \mu^{2(r-1)} - \mu^{2r} \right] = r^2 \tau^2 \mu^{2(r-1)},$$

where $\tau^2 = \text{Var}(X_1)$. □

13.5   Rank Statistics

Let $X_1, X_2, \ldots, X_n$ denote independent, real-valued random variables and denote the ranks of the sample by $R_1, R_2, \ldots, R_n$. Recall that, under the assumption that the X_j are identically distributed with an absolutely continuous distribution, the vector of ranks is uniformly distributed over the set of all permutations of $\{1, 2, \ldots, n\}$. Since this is true for any absolutely continuous distribution of X_1, statistical procedures based on $R_1, R_2, \ldots, R_n$ are *distribution-free* in the sense that the properties of the procedures do not require additional assumptions regarding the distribution of X_1.

***Example* 13.15** (*Rank correlation statistic*). Let $X_1, X_2, \ldots, X_n$ denote independent, random variables and let $t_1, \ldots, t_n$ denote a sequence of unique constants; without loss of generality we may assume that $t_1 < t_2 < \cdots < t_n$. In order to investigate the possibility of a relationship between the sequence $X_1, X_2, \ldots, X_n$ and the sequence $t_1, t_2, \ldots, t_n$ we might consider the statistic $\sum_{j=1}^{n} jR_j$ where $R_1, R_2, \ldots, R_n$ denote the ranks of $X_1, \ldots, X_n$. Note that the vector $(1, 2, \ldots, n)$ is the vector of ranks corresponding to $t_1, t_2, \ldots, t_n$.

Note that since $\sum_{j=1}^{n} j, \sum_{j=1}^{n} j^2, \sum_{j=1}^{n} R_j$, and $\sum_{j=1}^{n} R_j^2$ are all deterministic functions of n, $\sum_{j=1}^{n} jR_j$ is a deterministic function of the correlation between the vector of ranks $(R_1, \ldots, R_n)$ and the vector $(1, \ldots, n)$. □

In this section, we consider the asymptotic distribution of linear functions of the ranks based on independent identically distributed random variables. For each fixed n, these ranks constitute a random permutation of the integers from 1 to n; hence, in studying the asymptotic properties of such statistics, we need to use a triangular array in order to express the ranks. Here we will use $Y_{n1}, \ldots, Y_{nn}$ to denote the ranks based on a sample of size n.

We are often interested in statistics of the form $\sum_{j=1}^{n} a_{nj} Y_{nj}$ where, for each $n = 1, 2, \ldots,$ $a_{n1}, \ldots, a_{nn}$ are known constants. More generally, we consider a statistic of the form

$$S_n = \sum_{j=1}^{n} a_{nj} g\left(\frac{Y_{nj}}{n}\right),$$

where g is a known function on $[0, 1]$. The following theorem gives a central limit theorem for this type of statistic.

***Theorem* 13.6.** *Let* $(Y_{n1}, Y_{n2}, \ldots, Y_{nn})$ *denote a random permutation of* $(1, 2, \ldots, n)$. *For each* $n = 1, 2, \ldots,$ *let* $a_{nj}, j = 1, 2, \ldots, n$ *denote a sequence of constants satisfying*

$$\lim_{n \to \infty} \frac{1}{n} \sum_{j=1}^{n} |a_{nj} - \bar{a}_n|^2 = c^2,$$

for some constant c, and

$$\lim_{n \to \infty} \frac{\sum |a_{nj} - \bar{a}_n|^3}{\left[\sum (a_{nj} - \bar{a}_n)^2\right]^{\frac{3}{2}}} = 0, \tag{13.2}$$

where

$$\bar{a}_n = \frac{1}{n} \sum_{j=1}^{n} a_{nj}.$$

Let g denote a bounded function on $[0, 1]$ *such that g is continuous almost everywhere. Define*

$$S_n = \sum_{j=1}^{n} a_{nj} g\left(\frac{Y_{nj}}{n}\right).$$

Then

$$\frac{S_n - \mathrm{E}(S_n)}{[\mathrm{Var}(S_n)]^{\frac{1}{2}}} \to N(0, 1) \quad as \quad n \to \infty.$$

We will prove this theorem through a series of lemmas. The first lemma considers a statistic of the form S_n, except that Y_{nj}/n is replaced by a random variable uniformly distributed on $(0, 1)$. Note that Y_{nj}/n is uniformly distributed on the set $\{1/n, 2/n, \ldots, 1\}$.

Lemma 13.1. *Let $U_1, U_2, \ldots, U_n$ denote independent, random variables each uniformly distributed on* $[0, 1]$ *and let*

$$T_n = \sum_{j=1}^{n}(a_{nj} - \bar{a}_n)g(U_j) + n\gamma_n\bar{a}_n,$$

where

$$\gamma_n = \frac{1}{n}\sum_{j=1}^{n}g(j/n).$$

Then, under the conditions of Theorem 13.6,

$$\frac{T_n - \mathrm{E}(T_n)}{[\mathrm{Var}(T_n)]^{\frac{1}{2}}} \xrightarrow{D} N(0, 1) \quad as \quad n \to \infty.$$

Proof. Let σ^2 denote the variance of $g(U_1)$. Then

$$\mathrm{Var}(T_n) = \sigma^2 \sum(a_{nj} - \bar{a}_n)^2.$$

Define

$$Z_{nj} = (a_{nj} - \bar{a}_n)(g(U_j) - \mathrm{E}[g(U_j)]), \quad j = 1, \ldots, n.$$

Then

$$T_n - \mathrm{E}(T_n) = \sum_{j=1}^{n} Z_{nj}.$$

Since g is bounded,

$$|Z_{nj}| \le 2M|a_{nj} - \bar{a}_n|,$$

where $M = \sup_{0 \le t \le 1}|g(t)|$, and

$$\mathrm{E}[|Z_{nj}|^3] \le 8M^3|a_{nj} - \bar{a}|^3.$$

Hence,

$$\frac{\sum_{j=1}^{n}\mathrm{E}[|Z_{nj}|^3]}{\left[\sum_{j=1}^{n}\mathrm{Var}(Z_{nj})\right]^{\frac{3}{2}}} \le \frac{8M^3}{\sigma^3}\frac{\sum_{j=1}^{n}|a_{nj} - \bar{a}_n|^3}{\left[\sum_{j=1}^{n}|a_{nj} - \bar{a}|^2\right]^{\frac{3}{2}}} \to 0$$

by condition (13.2). The result now follows from the central limit theorem for triangular arrays (Theorem 12.2). ∎

The second lemma shows that, as $n \to \infty$, the random variable $g(R_1/n)$, where R_1 is uniformly distributed on $\{1, \ldots, n\}$, is well-approximated by $g(U_1)$, where U_1 is uniformly distributed on $(0, 1)$.

Lemma 13.2. *Let* $U_1, U_2, \ldots, U_n$ *denote independent, random variables, each uniformly distributed on* $(0, 1)$*, and let* R_1 *denote the rank of* U_1*. Then, under the conditions of Theorem 13.6,*

$$\mathrm{E}\left[\left|g(U_1) - g\left(\frac{R_1}{n}\right)\right|^2\right] \to 0 \quad as \quad n \to \infty.$$

Proof. Note that

$$\mathrm{E}\left[\left|g(U_1) - g\left(\frac{R_1}{n}\right)\right|^2\right] = \mathrm{E}[g(U_1)^2] + \mathrm{E}\left[g\left(\frac{R_1}{n}\right)^2\right] - 2\mathrm{E}\left[g(U_1)g\left(\frac{R_1}{n}\right)\right],$$

$$\mathrm{E}[g(U_1)^2] = \int_0^1 g(x)^2 \, dx$$

and

$$\mathrm{E}\left[g\left(\frac{R_1}{n}\right)^2\right] = \frac{1}{n}\sum_{j=1}^n g(j/n)^2.$$

Under the conditions of Theorem 13.6, g is Riemann-integrable so that

$$\lim_{n \to \infty} \mathrm{E}\left[g\left(\frac{R_1}{n}\right)^2\right] = \lim_{n \to \infty} \frac{1}{n}\sum_{j=1}^n g(j/n)^2 = \int_0^1 g(x)^2 \, dx.$$

Hence, the result follows provided that

$$\lim_{n \to \infty} \mathrm{E}[g(U_1)g(R_1/n)] = \int_0^1 g(x)^2 \, dx.$$

Note that

$$\mathrm{E}\left[g(U_1)g\left(\frac{R_1}{n}\right)\right] = \mathrm{E}\left\{\mathrm{E}\left[g(U_1)g\left(\frac{R_1}{n}\right)|R_1\right]\right\} = \mathrm{E}\left\{g\left(\frac{R_1}{n}\right)\mathrm{E}[g(U_1)|R_1]\right\}.$$

Given that $R_1 = r$, U_1 is the rth-order statistic from $U_1, \ldots, U_n$. Hence,

$$\mathrm{E}[g(U_1)|R_1 = r] = \mathrm{E}[g(U_{(r)})|R_1 = r].$$

By Theorem 7.11, the order statistics and ranks are independent. Hence,

$$\mathrm{E}[g(U_{(r)})|R_1 = r] = \mathrm{E}[g(U_{(r)})] = \int_0^1 \frac{n!}{(r-1)!(n-r)!} x^{r-1}(1-x)^{n-r} g(x) \, dx$$

so that $\mathrm{E}[g(U_{(r)})] = \mathrm{E}[g(B_{r,n-r+1})]$, where $B_{r,s}$ denotes a random variable with a beta distribution with parameters r and s; see, Example 7.24. It follows that

$$\mathrm{E}\left[g(U_1)g\left(\frac{R_1}{n}\right)\right] = \mathrm{E}\left[g(B_{R_1,n-R_1+1})g\left(\frac{R_1}{n}\right)\right].$$

Since R_1 is uniformly distributed on $\{1, \ldots, n\}$,

$$
\begin{aligned}
E\left[g(U_1)g\left(\frac{R_1}{n}\right)\right] &= \frac{1}{n}\sum_{j=1}^{n} g\left(\frac{j}{n}\right) E[g(B_{j,n-j+1})] \\
&= \frac{1}{n}\sum_{j=1}^{n} g\left(\frac{j}{n}\right) \int_0^1 \frac{n!}{(j-1)!(n-j)!} x^{j-1}(1-x)^{n-j} g(x)\, dx \\
&= \int_0^1 \left\{ \sum_{j=1}^{n} \frac{1}{n}\frac{n!}{(j-1)!(n-j)!} x^{j-1}(1-x)^{n-j} g\left(\frac{j}{n}\right) \right\} g(x)\, dx \\
&\equiv \int_0^1 g_n(x)g(x)\, dx,
\end{aligned}
$$

where

$$
g_n(x) = \sum_{j=1}^{n} \frac{1}{n}\frac{n!}{(j-1)!(n-j)!} x^{j-1}(1-x)^{n-j} g\left(\frac{j}{n}\right).
$$

Note that

$$
g_n(x) = \sum_{j=0}^{n-1} \binom{n-1}{j} x^j (1-x)^{n-1-j} g\left(\frac{j+1}{n}\right) = E\left[g\left(\frac{Y_n(x)+1}{n}\right)\right],
$$

where $Y_n(x)$ has a binomial distribution with parameters $n-1$ and x. It is straightforward to show that

$$
\frac{Y_n(x)+1}{n} \xrightarrow{p} x \quad \text{as} \quad n \to \infty
$$

for all $0 \le x \le 1$. Let x_0 denote a point in $[0, 1]$ at which g is continuous. Then, by Theorem 11.7,

$$
g\left(\frac{Y_n(x_0)+1}{n}\right) \xrightarrow{p} g(x_0) \quad \text{as} \quad n \to \infty,
$$

which, since $g(x_0)$ is a constant, is equivalent to

$$
g\left(\frac{Y_n(x_0)+1}{n}\right) \xrightarrow{D} g(x_0) \quad \text{as} \quad n \to \infty.
$$

Since g is a bounded function, it follows that

$$
\lim_{n\to\infty} E\left[g\left(\frac{Y_n(x_0)+1}{n}\right)\right] = g(x_0).
$$

That is,

$$
\lim_{n\to\infty} g_n(x) = g(x)
$$

for all x at which g is continuous.

Recall that g is continuous almost everywhere so that $g_n(x) \to g(x)$ as $n \to \infty$, for almost all x. Furthermore,

$$
\sup_{x\in[0,1]} |g_n(x)| \le \sup_{x\in[0,1]} |g(x)| < \infty.
$$

It now follows from the dominated convergence theorem (see Appendix 1) that

$$\lim_{n\to\infty} \int_0^1 g_n(x)g(x)\,dx = \int_0^1 g(x)^2\,dx.$$

The result follows. ∎

Lemma 13.2 can now be used to show that a function of $(R_1/n, \ldots, R_n/n)$, where $(R_1, \ldots, R_n)$ is a random permutation of $(1, \ldots, n)$, can be approximated by an analogous statistic with $(R_1/n, \ldots, R_n/n)$ replaced by $(U_1, \ldots, U_n)$, where $U_1, \ldots, U_n$ are independent, identically distributed random variables, each uniformly distributed on $(0, 1)$.

Lemma 13.3. *Define $U_1, U_2, \ldots, U_n$ as in Lemma 13.2 and let the ranks of $U_1, U_2, \ldots, U_n$ be denoted by $R_1, \ldots, R_n$. Let γ_n be defined as in Lemma 13.1, let*

$$\tilde{S}_n = \sum_{j=1}^n (a_{nj} - \bar{a}_n)g\left(\frac{R_j}{n}\right) + n\gamma_n\bar{a}_n$$

and define T_n as in the statement of Lemma 13.1. Then, under the conditions of Theorem 13.6,

$$\lim_{n\to\infty} \frac{E[(T_n - \tilde{S}_n)^2]}{\mathrm{Var}(T_n)} = 0.$$

Proof. Since

$$T_n - \tilde{S}_n = \sum_{j=1}^n (a_{nj} - \bar{a}_n)\left[g(U_j) - g\left(\frac{R_j}{n}\right)\right],$$

$$|T_n - \tilde{S}_n|^2 \leq \left[\sum_{j=1}^n |a_{nj} - \bar{a}_n|^2\right]^{\frac{1}{2}} \left[\sum_{j=1}^n \left|g(U_j) - g\left(\frac{R_j}{n}\right)\right|^2\right]^{\frac{1}{2}}$$

and, using the fact that $E(|X|^{\frac{1}{2}}) \leq E(|X|)^{\frac{1}{2}}$,

$$E[|T_n - \tilde{S}_n|^2] \leq \left[\sum_{j=1}^n |a_{nj} - \bar{a}_j|^2\right]^{\frac{1}{2}} E\left\{\left[\sum_{j=1}^n \left|g(U_j) - g\left(\frac{R_j}{n}\right)\right|^2\right]^{\frac{1}{2}}\right\}$$

$$\leq \left[\sum_{j=1}^n |a_{nj} - \bar{a}_j|^2\right]^{\frac{1}{2}} E\left\{\left[\sum_{j=1}^n \left|g(U_j) - g\left(\frac{R_j}{n}\right)\right|^2\right]^{\frac{1}{2}}\right\}.$$

Note that

$$E\left[\sum_{j=1}^n \left|g(U_j) - g\left(\frac{R_j}{n}\right)\right|^2\right] = nE\left[|g(U_1) - g\left(\frac{R_1}{n}\right)|^2\right] \equiv n\epsilon_n$$

where, by Lemma 13.2, $\epsilon_n \to 0$ as $n \to \infty$. Hence,

$$\frac{E[(T_n - \tilde{S}_n)^2]}{\mathrm{Var}(T_n)} \leq \frac{\sqrt{n}\sqrt{\epsilon_n}}{\sigma^2\left[\sum(a_{nj} - \bar{a}_n)^2\right]^{\frac{1}{2}}}.$$

The result now follows from the assumption that

$$\lim_{n \to \infty} \frac{1}{n} \sum_{j=1}^{n} |a_{nj} - \bar{a}_n|^2 = c^2. \quad \blacksquare$$

Lemma 13.4 is a technical result on convergence in distribution. It shows that if a sequence $Z_1, \ldots, Z_n$, appropriately standardized, converges in distribution to a standard normal random variable, and Y_n approximates Z_n in the sense of Lemma 13.3, then Y_n is also asymptotically distributed according to a standard normal distribution.

***Lemma* 13.4.** *Let $Z_1, Z_2, \ldots$ and $Y_1, Y_2, \ldots$ denote sequences of real-valued random variables such that, for all $n = 1, 2, \ldots$, $E(Z_n^2) < \infty$ and $E(Y_n^2) < \infty$.*
   If

$$\frac{Z_n - E(Z_n)}{[\text{Var}(Z_n)]^{\frac{1}{2}}} \xrightarrow{\mathcal{D}} N(0, 1) \quad as \quad n \to \infty$$

and

$$\lim_{n \to \infty} \frac{E[(Z_n - Y_n)^2]}{\text{Var}(Z_n)} = 0,$$

then

$$\frac{Y_n - E(Y_n)}{[\text{Var}(Y_n)]^{\frac{1}{2}}} \xrightarrow{\mathcal{D}} N(0, 1) \quad as \quad n \to \infty.$$

Proof. For each $n = 1, 2, \ldots$, let $D_n = Y_n - Z_n$ so that $Y_n = Z_n + D_n$. Under the conditions of the lemma,

$$\lim_{n \to \infty} \frac{\text{Var}(D_n)}{\text{Var}(Z_n)} = 0$$

and, hence, by the Cauchy–Schwarz inequality,

$$\limsup_{n \to \infty} \frac{\text{Cov}(Z_n, D_n)^2}{\text{Var}(Z_n)^2} \leq \lim_{n \to \infty} \frac{\text{Var}(D_n)}{\text{Var}(Z_n)} = 0.$$

Since

$$\frac{\text{Var}(Y_n)}{\text{Var}(Z_n)} = \frac{\text{Var}(Z_n + D_n)}{\text{Var}(Z_n)} = 1 + \frac{\text{Var}(D_n)}{\text{Var}(Z_n)} + \frac{2\text{Cov}(Z_n, D_n)}{\text{Var}(Z_n)},$$

it follows that

$$\lim_{n \to \infty} \frac{\text{Var}(Y_n)}{\text{Var}(Z_n)} = 1.$$

Hence, by Corollary 11.4, it suffices to show that

$$\frac{Y_n - E(Y_n)}{[\text{Var}(Z_n)]^{\frac{1}{2}}} \xrightarrow{\mathcal{D}} N(0, 1) \quad \text{as} \quad n \to \infty$$

and, since

$$\frac{Y_n - E(Y_n)}{[\text{Var}(Z_n)]^{\frac{1}{2}}} = \frac{Z_n - E(Z_n)}{[\text{Var}(Z_n)]^{\frac{1}{2}}} + \frac{D_n - E(D_n)}{[\text{Var}(Z_n)]^{\frac{1}{2}}},$$

it suffices to show that

$$\frac{D_n - \mathrm{E}(D_n)}{[\mathrm{Var}(Z_n)]^{\frac{1}{2}}} \xrightarrow{p} 0 \quad \text{as} \quad n \to \infty.$$

The result now follows from Chebychev's inequality, together with the fact that

$$\lim_{n \to \infty} \frac{\mathrm{Var}(D_n)}{\mathrm{Var}(Z_n)} = 0. \quad \blacksquare$$

Lemmas 13.1–13.4 can now be used to prove Theorem 13.6. The argument proceeds as follows. Let S_n be defined as in Theorem 13.6. Then $S_n - \mathrm{E}(S_n)$ is simply $\tilde{S}_n$, as defined in Lemma 13.3. By Lemma 13.3, $\tilde{S}_n$ is well-approximated by T_n, as defined in Lemma 13.1. By Lemma 13.1, T_n is asymptotically normally distributed and, by Lemma 13.4, this implies asymptotic normality for $\tilde{S}_n$ and, hence, for S_n. The details are given in the following proof.

Proof of Theorem 13.6.   Note that

$$\mathrm{E}(S_n) = \sum_{j=1}^{n} a_{nj} \mathrm{E}\left[g\left(\frac{Y_{nj}}{n} \right) \right] = \sum_{j=1}^{n} a_{nj} \mathrm{E}\left[g\left(\frac{Y_{n1}}{n} \right) \right] = n\bar{a}_n \frac{1}{n} \sum_{j=1}^{n} g\left(\frac{j}{n} \right) = n\bar{a}_n \gamma_n.$$

Hence,

$$S_n - \mathrm{E}(S_n) = \tilde{S}_n$$

and

$$\mathrm{Var}(S_n) = \mathrm{Var}(\tilde{S}_n),$$

where $\tilde{S}_n$ is given in the statement of Lemma 13.3. Hence, it suffices to show that

$$\frac{\tilde{S}_n}{[\mathrm{Var}(\tilde{S}_n)]^{\frac{1}{2}}} \to N(0, 1) \quad \text{as} \quad n \to \infty.$$

Define T_n as the statement of Lemma 13.1; by Lemmas 13.1 and 13.3,

$$\frac{T_n - \mathrm{E}(T_n)}{[\mathrm{Var}(T_n)]^{\frac{1}{2}}} \xrightarrow{\mathcal{D}} N(0, 1), \ n \to \infty, \quad \text{and} \quad \lim_{n \to \infty} \frac{\mathrm{E}[(T_n - \tilde{S}_n)^2]}{\mathrm{Var}(T_n)} = 0.$$

The result now follows from Lemma 13.4. $\blacksquare$

In the course of proving Theorem 13.6 we have derived an expression for $\mathrm{E}(S_n)$, as well as approximations to $\mathrm{Var}(S_n)$. These are given in Corollary 13.5; the proof is left as an exercise.

Corollary 13.5. *Under the condtions of Theorem 13.6,*

$$\frac{\sqrt{12}(S_n - n\bar{a}_n \gamma_n)}{c\sqrt{n}} \xrightarrow{\mathcal{D}} N(0, 1) \quad \text{as} \quad n \to \infty$$

and

$$\frac{\sqrt{12}(S_n - n\bar{a}_n\gamma_n)}{\left[\sum_{j=1}^{n}(a_{nj} - \bar{a}_n)^2\right]^{\frac{1}{2}}} \xrightarrow{\mathcal{D}} N(0, 1) \quad as \quad n \to \infty,$$

where

$$\gamma_n = \frac{1}{n}\sum_{j=1}^{n}\gamma\left(\frac{j}{n}\right).$$

Example 13.16 (*Rank correlation statistic*). Let $R_1, \ldots, R_n$ denote the ranks of a set of independent, identically distributed random variables, each with an absolutely continuous distribution. Consider the statistic

$$\sum_{j=1}^{n}jR_j,$$

which is a determinstic function of the rank correlation statistic; see Example 13.15. In order to use Theorem 13.6 above, we will consider the equivalent statistic

$$S_n = \frac{1}{n^2}\sum_{j=1}^{n}jR_j,$$

which is of the form

$$\sum_{j=1}^{n}a_{nj}g\left(\frac{Y_{nj}}{n}\right)$$

with $g(u) = u$, $a_{nj} = j/n$, and where, for each n, $Y_{n1}, \ldots, Y_{nn}$ is a random permutation of $\{1, \ldots, n\}$.

It is straightforward to show that

$$\bar{a}_n = \frac{1}{2}\frac{n+1}{n}$$

and that

$$\sum_{j=1}^{n}|a_{nj} - \bar{a}_n|^2 = \frac{n^2 - 1}{12n};$$

hence, $c^2 = 1/12$. Since $|a_{nj} - \bar{a}_n| \le 1/2$, it follows that condition (13.2) is satisfied.

Note that

$$\gamma_n = \frac{1}{n}\sum_{j=1}^{n}\frac{j}{n} = \frac{n+1}{2n}.$$

It follows from Corollary 13.5 that

$$\frac{S_n - (n+1)/2}{\sqrt{n/12}} \xrightarrow{\mathcal{D}} N(0, 1) \quad as \quad n \to \infty. \qquad \square$$

Example 13.17 (*Two-sample rank statistic*). Let $R_1, R_2, \ldots, R_n$ denote the ranks of n independent, identically distributed random variables, where $n = n_1 + n_2$, and consider the statistic

$$S_n = \frac{\sum_{j=1}^{n_1} R_j}{n_1} - \frac{\sum_{j=1}^{n_2} R_{n_1+j}}{n_2}.$$

This statistic may be written in the form

$$\sum_{j=1}^{n} a_{nj} R_j / n,$$

where

$$a_{nj} = \begin{cases} n/n_1 & \text{if } j = 1, \dots, n_1 \\ -n/n_2 & \text{if } j = n_1 + 1, \dots, n_1 + n_2 \end{cases}.$$

Then

$$\sum_{j=1}^{n} a_{nj} = 0$$

and

$$\sum_{j=1}^{n} a_{nj}^2 = \frac{n^2}{n_1} + \frac{n^2}{n_2}.$$

Suppose that

$$\lim_{n \to \infty} \frac{n_1}{n} = q,$$

where $0 < q < 1$. Then the constant c^2, defined in Theorem 13.6, is given by

$$c^2 = \frac{1}{q(1-q)}.$$

Note that

$$\frac{\sum_{j=1}^{n} |a_{nj}|^3}{\left[\sum_{j=1}^{n} a_{nj}^2 \right]^{\frac{3}{2}}} = \frac{n^2/n_1^2 + n^2/n_2^2}{[n/n_1 + n/n_2]^{\frac{3}{2}}} \frac{1}{\sqrt{n}}$$

so that condition (13.2) of Theorem 13.6 holds. It follows that

$$\frac{S_n}{\text{Var}(S_n)^{\frac{1}{2}}} \xrightarrow{\mathcal{D}} N(0, 1) \quad \text{as } n \to \infty.$$

Using Corollary 13.5, we also have that

$$[12q(1-q)]^{\frac{1}{2}} S_n / \sqrt{n} \xrightarrow{\mathcal{D}} N(0, 1) \quad \text{as } n \to \infty$$

and that

$$\left[\frac{12 n_1 n_2}{n^3} \right]^{\frac{1}{2}} S_n \xrightarrow{\mathcal{D}} N(0, 1) \quad \text{as } n \to \infty. \qquad \square$$

13.6   Exercises

13.1 Let $Y_1, Y_2, \ldots$ denote independent, identically distributed random variables such that

$$\Pr(Y_1 = 1) = 1 - \Pr(Y_1 = 0) = \theta$$

for some $0 < \theta < 1$. Find the asymptotic distribution of

$$\log\left(\frac{\bar{Y}_n}{1 - \bar{Y}_n}\right),$$

where $\bar{Y}_n = \sum_{j=1}^{n} Y_j/n$.

13.2 Let $Y_1, Y_2, \ldots$ denote independent, identically distributed random variables each normally distributed with mean μ and standard deviation σ, $\mu \in \mathbf{R}$, $\sigma > 0$. Let $\bar{Y}_n = \sum_{j=1}^{n} Y_j$, $S_n^2 = \sum_{j=1}^{n}(Y_j - \bar{Y})^2/n$ and let z be a fixed real number. Find the asymptotic distribution of

$$\Phi\left(\frac{z - \bar{Y}_n}{S_n}\right).$$

13.3 Prove Corollary 13.1.

13.4 Find the variance-stabilizing transformation for the binomial distribution.

13.5 Find the variance-stabilizing transformation for the exponential distribution.

13.6 Let $Y_1, Y_2, \ldots$ denote independent, identically distributed random variables, each distributed according to a Poisson distribution with mean θ, $\theta > 0$. Let k, j, $k \neq j$ denote nonnegative integers and let $\bar{Y}_n = \sum_{j=1}^{n} Y_j$. Find the asymptotic distribution of

$$\begin{pmatrix} \bar{Y}_n^j \exp(-\bar{Y}_n)/j! \\ \bar{Y}_n^k \exp(-\bar{Y}_n)/k! \end{pmatrix}.$$

13.7 Let $Y_1, Y_2, \ldots$ denote independent, identically distributed random variables each normally distributed with mean μ and standard deviation σ, $\mu > 0$, $\sigma > 0$. Let $\bar{Y}_n = \sum_{j=1}^{n} Y_j$, $S_n^2 = \sum_{j=1}^{n}(Y_j - \bar{Y})^2/n$. Find the asymptotic distribution of $S_n/\bar{Y}_n$.

13.8 Let $X_1, X_2, \ldots$ denote a sequence of d-dimensional random vectors such that, for some vector μ,

$$\sqrt{n}(X_n - \mu) \xrightarrow{\mathcal{D}} N_d(0, \Sigma),$$

where Σ is a positive definite matrix with $|\Sigma| < \infty$.
Let $g : \mathbf{R}^d \to \mathbf{R}^k$, $k \leq d$, denote a continuously differentiable function. Let A denote a nonrandom $k \times k$ matrix. Find conditions on A and $g'(\mu)$ such that

$$n(g(X_n) - g(\mu))^T A(g(X_n) - g(\mu)) \xrightarrow{\mathcal{D}} W,$$

where W has a chi-squared distribution with ν degrees of freedom. Find ν in terms of $g'(\mu)$ and A.

13.9 Let $X_1, \ldots, X_n$ denote independent, identically distributed random variables such that all moments of X_j exist. Let g denote a real-valued, twice-differentiable function defined on the real line. Let $\mu = E(X_j)$. Find a large-sample approximation to the distribution of $h(\bar{X})$ for the case in which $h'(\mu) = 0$ and $h''(\mu) \neq 0$. Make whatever (reasonable) assumptions are required regarding h.

13.10 Let $X_1, X_2, \ldots$ denote independent, standard normal random variables. Find the asymptotic distribution of $\cosh(\bar{X}_n)$, $\bar{X}_n = \sum_{j=1}^{n} X_j/n$.

13.11 Prove Corollary 13.3.

13.12 Let $X_1, X_2, \ldots, X_n$ denote independent, identically distributed random variables such that X_1 has a standard exponential distribution and let $X_{n(1)}, \ldots, X_{n(n)}$ denote the order statistics of

$X_1, \ldots, X_n$. Let k_n and m_n, $n = 1, 2, \ldots$, denote sequences of nonnegative integers such that $k_n \leq n$ and $m_n \leq n$ for all n,

$$\sqrt{n}\left(\frac{k_n}{n} - q_1\right) \to 0 \quad \text{as} \quad n \to \infty$$

and

$$\sqrt{n}\left(\frac{m_n}{n} - q_2\right) \to 0 \quad \text{as} \quad n \to \infty$$

for some q_1, q_2 in $(0, 1)$, $q_1 \neq q_2$. Find the asymptotic distribution of $X_{n(k_n)}/X_{n(m_n)}$.

13.13 Let $X_1, X_2, \ldots, X_n$ denote independent, identically distributed standard normal random variables and let $X_{n(1)}, \ldots, X_{n(n)}$ denote the order statistics of $X_1, \ldots, X_n$. Let k_n, $n = 1, 2, \ldots$, denote a sequence of nonnegative integers such that $k_n \leq n/2$ for all n and

$$\sqrt{n}\left(\frac{k_n}{n} - q\right) \to 0 \quad \text{as} \quad n \to \infty$$

for some $0 < q < 1$. Find the asymptotic distribution of

$$\frac{X_{n(k_n)} + X_{n(m_n)}}{2},$$

where $m_n = n - k_n$, $n = 1, 2, \ldots$.

13.14 Prove Corollary 13.4.

13.15 Let $X_1, X_2, \ldots, X_n$ denote independent, identically distributed random variables each uniformly distributed on $(0, 1)$ and let $X_{n(1)}, \ldots, X_{n(n)}$ denote the order statistics of $X_1, \ldots, X_n$. Let k_n, $n = 1, 2, \ldots$, denote a sequence of nonnegative integers such that $k_n \leq n$ for all n and

$$\sqrt{n}\left(\frac{k_n}{n} - q\right) \to 0 \quad \text{as} \quad n \to \infty$$

for some $0 < q < 1$. Let $g : (0, 1) \to \mathbf{R}$ denote a differentiable, strictly increasing function. Find the asymptotic distribution of $g(X_{n(k_n)})$ first by finding the asymptotic distribution of $X_{n(k_n)}$ and using the δ-method and then by finding the asymptotic distribution of $Y_{n(k_n)}$ where $Y_j = g(X_j)$, $j = 1, \ldots, n$.

13.16 Let $X_1, \ldots, X_n$ denote independent, identically distributed random variables such that X_1 has an absolutely continuous distribution with density function

$$\frac{\exp(-x)}{[1 + \exp(-x)]^2}, \quad -\infty < x < \infty.$$

Let

$$Y_n = \max_{1 \leq j \leq n} X_j.$$

Find a constant α and a sequence of constants $\beta_1, \beta_2, \ldots$ such that $n^\alpha(Y_n - \beta)$ has a nondegenerate limiting distribution.

13.17 Let $X_1, X_2, \ldots$ denote independent, identically distributed real-valued random variables such that X_1 has a standard exponential distribution. Let

$$U = \frac{2}{n(n-1)} \sum_{1 \leq i < j \leq n} |X_i - X_j|;$$

this statistic is known as *Gini's mean difference* and is a measure of dispersion. Find the asymptotic distribution of U.

13.18 Let $X_1, X_2, \ldots$ denote independent, identically distributed real-valued random variables such that the distribution of X_1 is absolutely continuous and symmetric about 0. Let

$$U = \frac{1}{n(n-1)} \sum_{1 \le i < j \le n} I_{\{X_i + X_j \le 0\}}.$$

Find the asymptotic distribution of U.

13.19 Let $X_1, X_2, \ldots$ denote independent, identically distributed real-valued random variables such that $E(X_1^4) < \infty$ and let σ^2 denote the variance of X_1. Construct a U-statistic with expected value equal to σ^4; find the asymptotic distribution of this statistic.

13.20 Let

$$U_1 = \frac{1}{\binom{n}{r}} \sum_{\beta} h_1(X_{\beta_1}, \ldots, X_{\beta_2})$$

and

$$U_2 = \frac{1}{\binom{n}{r}} \sum_{\beta} h_2(X_{\beta_1}, \ldots, X_{\beta_2})$$

denote two U-statistics, of the same order, based on the same set of observations. Find the asymptotic distribution of (U_1, U_2).

13.21 Prove Corollary 13.5.

13.22 For each $n = 1, 2, \ldots$, let $Y_{n1}, \ldots, Y_{nn}$ denote the ranks of n independent, identically distributed real-valued random variables, each with an absolutely continuous distribution. Find the asymptotic distribution of $\sum_{j=1}^{n} j Y_{nj}^2$.

13.7   Suggestions for Further Reading

The δ-method, presented in Section 13.2, is used often in statistics; see, for example, Lehmann (1999, Chapters 2 and 5), Rao (1973, Chapter 6), and Serfling (1980, Chapter 3) for further discussion. The asymptotic properties of order statistics are considered in Ferguson (1996, Chapters 13–15), Sen and Singer (1993, Chapter 4), and Serfling (1980, Chapters 2 and 8); a comprehensive treatment of extreme order statistics is given in Galambos (1978).

   The discussion of U-statistics in Section 13.4 is based on van der Vaart (1998, Chapters 11 and 12); see also Lehmann (1999, Chapter 6) Sen and Singer (1993, Chapter 5), and Serfling (1980, Chapter 5). Section 13.5 on rank statistics is based on Port (1994, Section 61.1); see also Ferguson (1996, Chapter 12) and Serfling (1980, Chapter 9).

14

Higher-Order Asymptotic Approximations

14.1 Introduction

In Chapter 12, approximations to the distributions of sample means were derived; the normal approximation $\hat{F}(x)$ to the distribution function $F_n(x)$ of the normalized sample mean has the property that, for each x,

$$\lim_{n \to \infty} F_n(x) = \hat{F}(x);$$

alternatively, we may write this as

$$F_n(x) = \hat{F}(x) + o(1) \quad \text{as} \quad n \to \infty.$$

That is, the error in the approximation tends to 0 as $n \to \infty$. This approximation is known as a *first-order* asymptotic approximation.

In this chapter, we consider higher-order asymptotic approximations to distribution functions and density functions. For instance, in Section 14.2, an Edgeworth series approximation to the distribution function of a sample mean is derived that has the property that

$$F_n(x) = \hat{F}_n(x) + o\left(\frac{1}{\sqrt{n}}\right) \quad \text{as} \quad n \to \infty.$$

In this case, not only does the error in the approximation tend to 0 as $n \to \infty$, $\sqrt{n}$ times the error approaches 0 as $n \to \infty$. The approximation $\hat{F}_n$ is known as a *second-order* asymptotic approximation. Asymptotic approximations of this type are the subject of this chapter.

14.2 Edgeworth Series Approximations

Let $X_1, \ldots, X_n$ denote independent, identically distributed, real-valued random variables with mean μ, standard deviation σ, and characteristic function $\varphi(t)$. Let $Y_n = \sqrt{n}(\bar{X}_n - \mu)/\sigma$. Then $\varphi_n(t)$, the characteristic function of Y_n, may be expanded

$$\log \varphi_n(t) = -\frac{1}{2}t^2 + \frac{\kappa_3}{6}\frac{(it)^3}{\sqrt{n}} + \cdots \tag{14.1}$$

where κ_3 denotes the third cumulant of X_1. A first-order approximation to the $\log \varphi_n(t)$ is given by $-t^2/2$, which corresponds to the log of the characteristic function of the normal distribution. Hence, to first order, the distribution of Y_n may be approximated by the standard normal distribution.

A higher-order approximation to the distribution of Y_n may be obtained by retaining more terms in the approximation to $\log \varphi_n(t)$. For instance, including the next term in the expansion (14.1) yields the approximation

$$-\frac{1}{2}t^2 + \frac{\kappa_3}{6} \frac{(it)^3}{\sqrt{n}}.$$

We may obtain an approximation to the distribution function of Y_n by finding the distribution function that corresponds to this expansion; the same approach may be used to approximate the density function of Y_n. This is the idea behind the Edgeworth series approximations.

The following theorem gives a formal proof of this result.

Theorem 14.1. *For each $n = 1, 2, \ldots$, let X_n denote a real-valued random variable with characteristic function φ_n satisfying*

$$\log \varphi_n(t) = -\frac{1}{2}t^2 + \frac{\kappa_{n3}}{6\sqrt{n}}(it)^3 + R_n(t)$$

where, for each $t \in \mathbf{R}$,

$$\lim_{n \to \infty} \frac{R_n(t)}{\sqrt{n}} = 0$$

and

$$\lim_{n \to \infty} \kappa_{n3} = \kappa_3$$

for some finite constant κ_3.
Assume that the following conditions hold:

(i) Given $\epsilon > 0$ there exists a $\delta > 0$ and a positive integer N such that

$$|R_n(t)| \le \epsilon \frac{|t|^3}{\sqrt{n}} \quad \text{for } |t| \le \delta\sqrt{n} \quad \text{and } n \ge N.$$

(ii) For any $\delta > 0$,

$$\int_{|t|>\delta\sqrt{n}} |\varphi_n(t)|\, dt = o\left(\frac{1}{\sqrt{n}}\right) \quad \text{as } n \to \infty.$$

Let p_n denote the density function of X_n and let

$$\hat{p}_n(x) = \phi(x) + \frac{\kappa_3}{6\sqrt{n}} H_3(x)\phi(x).$$

Then

$$\sup_x |p_n(x) - \hat{p}_n(x)| = o\left(\frac{1}{\sqrt{n}}\right).$$

Let F_n denote the distribution function of X_n and let

$$\hat{F}_n(x) = \Phi(x) - \frac{\kappa_3}{6} H_2(x)\phi(x)\frac{1}{\sqrt{n}}.$$

Then

$$\sup_x |F_n(x) - \hat{F}_n(x)| = o\left(\frac{1}{\sqrt{n}}\right).$$

Proof.   Consider the first part of the theorem. Note that the Fourier transform of $\hat{p}_n$ is $\hat{\varphi}_n(t)/\sqrt{(2\pi)}$, where

$$\hat{\varphi}_n(t) = \exp\left\{-\frac{1}{2}t^2\right\}\left[1 + \frac{\kappa_3}{6\sqrt{n}}(it)^3\right].$$

Let $d_n = p_n - \hat{p}_n$. Then d_n has Fourier transform $(\varphi_n - \hat{\varphi}_n)/\sqrt{(2\pi)}$. Under the conditions of the theorem

$$\int_{-\infty}^{\infty} |\hat{\varphi}_n(t)|\, dt < \infty \quad \text{and} \quad \int_{-\infty}^{\infty} |\varphi_n(t)|\, dt < \infty$$

for sufficiently large n. It follows that, using the argument used in Theorem 3.8,

$$d_n(x) = \frac{1}{2\pi}\int_{-\infty}^{\infty} [\varphi_n(t) - \hat{\varphi}_n(t)]\, \exp(-itx)\, dt$$

so that

$$|d_n(x)| \le \frac{1}{2\pi}\int_{-\infty}^{\infty} |\varphi_n(t) - \hat{\varphi}_n(t)|\, dt.$$

By assumption, for any $\delta > 0$,

$$\int_{|t|>\delta\sqrt{n}} |\varphi_n(t)|\, dt = o\left(\frac{1}{\sqrt{n}}\right)$$

and since clearly

$$\int_{|t|>\delta\sqrt{n}} |\hat{\varphi}_n(t)|\, dt = o\left(\frac{1}{\sqrt{n}}\right),$$

it follows that, for any $\delta > 0$,

$$\int_{|t|>\delta\sqrt{n}} |\varphi_n(t) - \hat{\varphi}_n(t)|\, dt = o\left(\frac{1}{\sqrt{n}}\right) \quad \text{as} \quad n \to \infty.$$

Hence, it suffices to show that, for some $\delta > 0$,

$$\int_{|t|\le\delta\sqrt{n}} |\varphi_n(t) - \hat{\varphi}_n(t)|\, dt = o\left(\frac{1}{\sqrt{n}}\right) \quad \text{as} \quad n \to \infty.$$

Note that

$$\varphi_n(t) - \hat{\varphi}_n(t) = \exp\left\{-\frac{1}{2}t^2\right\}\left[\exp\left\{\frac{\kappa_{n3}}{6\sqrt{n}}(it)^3 + R_n(t)\right\} - 1 - \frac{\kappa_3}{6\sqrt{n}}(it)^3\right]$$

and, using Lemma A2.2,

$$|\varphi_n(t) - \hat{\varphi}_n(t)| \le \exp\left\{-\frac{1}{2}t^2\right\}\left[\frac{|\kappa_{n3} - \kappa_3|}{6\sqrt{n}}|t|^3 + \frac{\kappa_{n3}^2}{72n}|t|^6\right]\exp\left\{\frac{|\kappa_{n3}|}{6\sqrt{n}}|t|^3 + |R_n(t)|\right\}.$$

Since $\kappa_{n3} \to \kappa_3$ as $n \to \infty$, $|\kappa_{n3}| \le 2|\kappa_3|$ for sufficiently large n so that, for $|t| \le \delta\sqrt{n}$,

$$\exp\left\{|\kappa_{n3}|\frac{|t|}{6\sqrt{n}}t^2\right\} \le \exp\left\{\frac{|\kappa_3|}{3}\delta t^2\right\}.$$

Also, by condition (i), for sufficiently small δ and sufficiently large n,

$$|R_n(t)| \leq \frac{1}{8} \frac{|t|^3}{\sqrt{n}}, \quad |t| \leq \delta\sqrt{n}$$

$$\leq \frac{1}{8}t^2.$$

Hence, there exists a $\delta > 0$ and a positive integer N such that

$$\exp\left\{-\frac{1}{2}t^2 + \frac{|\kappa_{n3}|}{6\sqrt{n}}|t|^3 + |R_n(t)|\right\} \leq \exp\left\{-\frac{1}{4}t^2\right\}, \quad |t| \leq \delta\sqrt{n}.$$

It follows that

$$\int_{|t|\leq\delta\sqrt{n}} |\varphi_n(t) - \hat{\varphi}_n(t)|\, dt \leq \int_{|t|\leq\delta\sqrt{n}} \frac{|\kappa_{n3} - \kappa_3|}{6\sqrt{n}}|t|^3 \exp\left\{-\frac{1}{4}t^2\right\}\, dt$$

$$+ \frac{1}{n}\int_{|t|\leq\delta\sqrt{n}} \frac{\kappa_{n3}^2}{72}|t|^6 \exp\left\{-\frac{1}{4}t^2\right\}\, dt.$$

Since $\kappa_{n3} - \kappa_3 \to 0$ as $n \to \infty$, clearly

$$\int_{|t|\leq\delta\sqrt{n}} \frac{|\kappa_{n3} - \kappa_3|}{6\sqrt{n}} 21|t|^3 \exp\left\{-\frac{1}{4}t^2\right\}\, dt = o\left(\frac{1}{\sqrt{n}}\right);$$

also, it follows immediately that

$$\frac{1}{n}\int_{|t|\leq\delta\sqrt{n}} \frac{\kappa_{n3}^2}{72}|t|^6 \exp\left\{-\frac{1}{4}t^2\right\}\, dt = O\left(\frac{1}{n}\right) = o\left(\frac{1}{\sqrt{n}}\right).$$

Hence, for all sufficiently small $\delta > 0$ and sufficiently large n,

$$\sup_x |p_n(x) - \hat{p}_n(x)| = o\left(\frac{1}{\sqrt{n}}\right), \tag{14.2}$$

proving the first part of the theorem.

Now consider the distribution functions. Let x and x_0 denote continuity points of F_n; note that $\hat{F}_n$ is continuous everywhere. By Theorem 3.3,

$$F_n(x) - F_n(x_0) = \frac{1}{2\pi} \lim_{T\to\infty} \int_{-T}^{T} \frac{\exp(-itx_0) - \exp(-itx)}{it}\varphi_n(t)\, dt.$$

Note that, since $\hat{F}_n$ is not necessarily nondecreasing and nonnegative, it is not necessarily a distribution function. However, the proof of Theorem 3.3 still applies to $\hat{F}_n$ provided that it is bounded, a fact which is easily verified. Hence,

$$\hat{F}_n(x) - \hat{F}_n(x_0) = \frac{1}{2\pi} \lim_{T\to\infty} \int_{-T}^{T} \frac{\exp(-itx_0) - \exp(-itx)}{it}\hat{\varphi}_n(t)\, dt.$$

It follows that

$$\left|[F_n(x) - \hat{F}_n(x)] - [F_n(x_0) - \hat{F}_n(x_0)]\right|$$

$$\leq \frac{1}{2\pi}\left|\lim_{T\to\infty} \int_{-T}^{T} \frac{\exp(-itx_0) - \exp(-itx)}{it}[\varphi_n(t) - \hat{\varphi}_n(t)]\, dt\right|$$

$$\leq \frac{1}{\pi} \lim_{T\to\infty} \int_{-T}^{T} \frac{|\varphi_n(t) - \hat{\varphi}_n(t)|}{|t|}\, dt.$$

Letting x_0 approach $-\infty$, it follows that

$$|F_n(x) - \hat{F}_n(x)| \leq \frac{1}{\pi} \lim_{T \to \infty} \int_{-T}^{T} \frac{|\varphi_n(t) - \hat{\varphi}_n(t)|}{|t|} \, dt.$$

Note that, for any $\delta > 0$,

$$\int_{-T}^{T} \frac{|\varphi_n(t) - \hat{\varphi}_n(t)|}{|t|} \, dt$$

$$\leq \int_{|t| \leq \delta \sqrt{n}} \frac{|\varphi_n(t) - \hat{\varphi}_n(t)|}{|t|} \, dt + \int_{|t| > \delta \sqrt{n}} \frac{|\varphi_n(t)|}{|t|} \, dt + \int_{|t| > \delta \sqrt{n}} \frac{|\hat{\varphi}_n(t)|}{|t|} \, dt$$

$$\leq \int_{|t| \leq \delta \sqrt{n}} \frac{|\varphi_n(t) - \hat{\varphi}_n(t)|}{|t|} \, dt + \frac{1}{\delta \sqrt{n}} \int_{|t| > \delta \sqrt{n}} |\varphi_n(t)| \, dt + \frac{1}{\delta \sqrt{n}} \int_{|t| > \delta \sqrt{n}} |\hat{\varphi}_n(t)| \, dt$$

$$= \int_{|t| \leq \delta \sqrt{n}} \frac{|\varphi_n(t) - \hat{\varphi}_n(t)|}{|t|} \, dt + o\left(\frac{1}{n}\right).$$

Hence,

$$|F_n(x) - \hat{F}_n(x)| \leq \int_{|t| \leq \delta \sqrt{n}} \frac{|\varphi_n(t) - \hat{\varphi}_n(t)|}{|t|} \, dt + o\left(\frac{1}{n}\right).$$

The proof now closely parallels the proof for the densities. Note that, for sufficiently small δ and sufficiently large n,

$$\int_{|t| \leq \delta \sqrt{n}} \frac{|\varphi_n(t) - \hat{\varphi}_n(t)|}{|t|} \, dt \leq \int_{|t| \leq \delta \sqrt{n}} \frac{|\kappa_{n3} - \kappa_3|}{6\sqrt{n}} |t|^2 \exp\left\{-\frac{1}{4} t^2\right\} \, dt$$

$$+ \frac{1}{n} \int_{|t| \leq \delta \sqrt{n}} \frac{\kappa_{n3}^2}{72} |t|^5 \exp\left\{-\frac{1}{4} t^2\right\} \, dt = o\left(\frac{1}{\sqrt{n}}\right).$$

It follows that

$$\lim_{n \to \infty} \sqrt{n} \sup_x |F_n(x) - \hat{F}_n(x)| = 0,$$

proving the second part of the theorem. ∎

Note that the conditions of Theorem 14.1 are not satisfied when X_n, $n = 1, 2, \ldots$, has a lattice distribution. For instance, suppose that the range of X_n is $\{b_n j, \ j = 0, \pm 1, \pm 2, \ldots\}$. Then, according to Theorem 3.11,

$$\varphi_n(t) = \varphi_n(t + 2\pi k/b_n), \quad k = 0, \pm 1, \pm 2, \ldots.$$

Hence, for any $k = 0, \pm 1, \pm 2, \ldots$,

$$\int_{t > \delta \sqrt{n}} |\varphi_n(t)| \, dt = \int_{t > \delta \sqrt{n}} |\varphi_n(t - 2\pi k/b_n)| \, dt$$

$$= \int_{t > \delta \sqrt{n} - 2\pi k/b_n} |\varphi_n(t)| \, dt.$$

By choosing k sufficiently large, it follows that

$$\int_{t > \delta \sqrt{n}} |\varphi_n(t)| \, dt \geq \int_{t > 0} |\varphi_n(t)| \, dt.$$

Since a similar result holds for the integral over the region $t < \delta\sqrt{n}$, condition (ii) of Theorem 14.1 implies that

$$\int_{-\infty}^{\infty} |\varphi_n(t)|\, dt \to 0 \quad \text{as} \quad n \to \infty.$$

However, under Theorem 3.8, this implies that the distribution of X_n is absolutely continuous, a contradiction of the assumption that X_n has a lattice distribution. Hence, application of Theorem 14.1 is restricted to random variables with a nonlattice distribution.

Theorem 14.2 below applies Theorem 14.1 to the case of a normalized sample mean based on independent, identically distributed random variables, each with an absolutely continuous distribution.

Theorem 14.2. *Let $X_1, X_2, \ldots$ denote independent, identically distributed, real-valued random variables such that X_1 has an absolutely continuous distribution with characteristic function φ. Assume that*
   (i) $E(X_1) = 0$, $\text{Var}(X_1) = 1$, and $E(|X_1|^3) < \infty$
   (ii) for some $\alpha \geq 1$,

$$\int_{-\infty}^{\infty} |\varphi(t)|^{\alpha}\, dt < \infty.$$

Let f_n and F_n denote the density and distribution function, respectively, of

$$\sum_{j=1}^{n} X_j / \sqrt{n}.$$

Then

$$\sup_x \left| f_n(x) - \left[\phi(x) + \frac{\kappa_3}{6} H_3(x)\phi(x)\frac{1}{\sqrt{n}} \right] \right| = o\left(\frac{1}{\sqrt{n}}\right)$$

and

$$\sup_x \left| F_n(x) - \left[\Phi(x) - \frac{\kappa_3}{6} H_2(x)\phi(x)\frac{1}{\sqrt{n}} \right] \right| = o\left(\frac{1}{\sqrt{n}}\right)$$

as $n \to \infty$.

Proof. The characteristic function of $\sum_{j=1}^{n} X_j / \sqrt{n}$ is given by $\varphi(t/\sqrt{n})^n$. Hence, the result follows from Theorem 14.1 provided that

$$\int_{|t| > \delta\sqrt{n}} |\varphi(t/\sqrt{n})|^n\, dt = o\left(\frac{1}{\sqrt{n}}\right) \quad \text{as} \quad n \to \infty$$

and

$$n \log \varphi(t/\sqrt{n}) = -\frac{1}{2}t^2 + \frac{\kappa_3}{6\sqrt{n}}(it)^3 + R_n(t) \tag{14.3}$$

where

$$\frac{R_n(t)}{\sqrt{n}} \to 0 \quad \text{as} \quad n \to \infty$$

for each t and given $\epsilon > 0$ there exist $\delta > 0$ and N such that

$$|R_n(t)| \le \epsilon \frac{|t|^3}{\sqrt{n}} \quad \text{for } |t| \le \delta\sqrt{n}, \ n \ge N.$$

Note that, since the distribution of X_1 is absolutely continuous, by Theorem 3.11,

$$|\varphi(t)| < 1 \quad \text{for } |t| \ne 0$$

and, by Theorem 3.9,

$$\varphi(t) \to 0 \quad \text{as } |t| \to \infty.$$

Hence, for each fixed δ there exists a constant $C < 1$ such that

$$|\varphi(t)| < C \quad \text{for } |t| \ge \delta.$$

It follows that

$$\int_{|t|>\delta\sqrt{n}} |\varphi(t/\sqrt{n})|^n \, dt = \sqrt{n} \int_{|t|>\delta} |\varphi(t)|^n \, dt \le \sqrt{n} C^{n-\alpha} |\varphi(t)|^\alpha \, dt = o\left(\frac{1}{n}\right).$$

Let $\gamma(t) = \log\varphi(t), \equiv \gamma_1(t) + i\gamma_2(t)$, where γ_1 and γ_2 are real-valued. Using Taylor's series expansions,

$$\gamma_1(t) = -\frac{1}{2}t^2 + \gamma_1'''(t_1^*)t^3/6$$

and

$$\gamma_2(t) = \gamma_2'''(t_2^*)t^3/6$$

where $|t_j^*| \le |t|$ and γ_j''' is continuous, $j = 1, 2$. Hence,

$$n\log\varphi(t/\sqrt{n}) = n\gamma(t/\sqrt{n}) = -\frac{1}{2}t^2 + \frac{\kappa_3}{6}\frac{(it)^3}{\sqrt{n}} + \{\gamma_1'''(t_1^*) - i[\gamma_2'''(t_2^*) - \gamma_2'''(0)]\}\frac{t^3}{6\sqrt{n}}$$

where $|t_j^*| \le t/\sqrt{n}$.

Note that $\gamma_1'''(0) = 0$. It follows that in the expansion (14.3) we may take

$$R_n(t) = \{[\gamma_1'''(t_1^*) - \gamma_1'''(0)] - i[\gamma_2'''(t_2^*) - \gamma_2'''(0)]\}\frac{t^3}{6\sqrt{n}}.$$

Clearly,

$$\lim_{n\to\infty} \frac{R_n(t)}{\sqrt{n}} = 0.$$

By the continuity of γ_1''' and γ_2''', for any given $\epsilon > 0$, we may choose δ sufficiently small so that

$$|\gamma_j'''(t_j^*) - \gamma_j'''(0)| \le \epsilon, \quad j = 1, 2$$

for $|t_j^*| \le \delta$, $j = 1, 2$. Then, there exists an N and a $\delta > 0$ such that for each $|t| \le \delta\sqrt{n}$,

$$|R_n(t)| \le 2\epsilon\frac{|t|^3}{6\sqrt{n}} \quad \text{for } n \ge N.$$

Hence, (14.3) holds. The result now follows from Theorem 14.1.   ∎

A sufficient condition for condition (ii) of Theorem 14.2 may be given in terms of the density function of X_1.

Lemma 14.1. *Let X denote a real-valued random variable with an absolutely continuous distribution; let p denote the density of this distribution and let φ denote the characteristic function. If there exists a constant $M < \infty$ such that $p(x) \leq M$ for all x, then*

$$\int_{-\infty}^{\infty} |\varphi(t)|^2 \, dt < \infty.$$

Proof. Let X_1 and X_2 denote independent random variables, each with the same distribution as X, and let $Y = X_1 + X_2$. Then Y has density function

$$p_Y(y) = \int_{-\infty}^{\infty} p(y-x)p(x) \, dx$$

and characteristic function

$$\varphi_Y(t) = \varphi(t)^2.$$

Note that

$$p_Y(y) = \int_{-\infty}^{\infty} p(y-x)p(x) \, dx \leq M \int_{-\infty}^{\infty} p(x) \, dx = M.$$

Let $\phi(\cdot)$ denote the standard normal density function. For $\lambda > 0$

$$\int_{-\infty}^{\infty} \varphi_Y(t)\phi(t/\lambda) \, dt = \lambda \int_{-\infty}^{\infty} \int_{-\infty}^{\infty} \exp(ity)p_Y(y) \, dy \frac{1}{\lambda} \phi(t/\lambda) \, dt$$

$$= \lambda \int_{-\infty}^{\infty} \int_{-\infty}^{\infty} \exp(ity) \frac{1}{\lambda} \phi(t/\lambda) \, dt \; p_Y(y) \, dy.$$

Note that the inner integral is simply the characteristic function of the normal distribution with mean 0 and standard deviation λ, evaluated at y. Hence,

$$\int_{-\infty}^{\infty} \varphi_Y(t)\phi(t/\lambda) \, dt = \lambda \int_{-\infty}^{\infty} \exp\left(-\frac{\lambda^2}{2}y^2\right) p_Y(y) \, dy$$

$$\leq M\lambda \int_{-\infty}^{\infty} \exp\left(-\frac{\lambda^2}{2}y^2\right) dy = M\sqrt{(2\pi)}.$$

Therefore, we have shown that for all $\lambda > 0$,

$$\int_{-\infty}^{\infty} \varphi_Y(t)\phi(t/\lambda) \, dt \leq M\sqrt{(2\pi)}.$$

Hence,

$$\limsup_{\lambda \to \infty} \int_{-\infty}^{\infty} \varphi_Y(t)\phi(t/\lambda) \, dt \leq \int_{-\infty}^{\infty} \varphi_Y(t) \limsup_{\lambda \to \infty} \phi(t/\lambda) \, dt = \int_{-\infty}^{\infty} \varphi_Y(t) \, dt \leq M\sqrt{(2\pi)}.$$

That is,

$$\int_{-\infty}^{\infty} |\varphi(t)|^2 \, dt \leq M\sqrt{(2\pi)},$$

proving the lemma. ■

***Example* 14.1 (*Sample mean of random variables with a symmetric distribution*).** Let $Y_1, \ldots, Y_n$ denote independent random variables each distributed according to a distribution with mean 0, standard deviation 1, and $E(|Y_1|^3) < \infty$. Assume that the density of Y_1 is bounded and symmetric about 0. It follows that $\kappa_3 = 0$ and, hence, using a Edgeworth series approximation

$$\Pr(\sqrt{n}\bar{Y} \le t) = \Phi(t) + o\left(\frac{1}{\sqrt{n}}\right);$$

that is, the usual first-order normal approximation has error of order $o(1/\sqrt{n})$ rather than the usual $o(1)$. □

***Example* 14.2 (*Chi-squared distribution*).** Let $Z_1, \ldots, Z_n$ denote independent standard normal random variables and let $S_n = \sum_{j=1}^{n} Z_j^2$. Then S_n has a chi-squared distribution with n degrees of freedom. We will consider approximations to the chi-squared distribution function based on an Edgeworth expansion.

It is straightforward to show that the first three cumulants of Z_j^2 are 1, 2, and 8, respectively. Let

$$Y_n = \sqrt{n}\frac{S_n/n - 1}{\sqrt{2}}.$$

Then $\Pr(Y_n \le y)$ may be approximated by

$$\Phi(y) - \frac{2}{3\sqrt{2}}H_2(y)\phi(y)\frac{1}{\sqrt{n}}.$$

Table 14.1 contains approximations to $\Pr(S_n \le s_{n\alpha})$, based on the Edgeworth series approximation described above, where $s_{n\alpha}$ satisfies

$$\Pr(S_n \le s_{n\alpha}) = \alpha,$$

for several choices of n and α. Recall that corresponding approximations based on the central limit theorem are given in Table 12.1. These results show that the Edgeworth series approximation is generally, but not always, an improvement over the approximation based on the central limit theorem. □

***Example* 14.3 (*Normal approximation evaluated at the mean*).** Let $X_1, X_2, \ldots, X_n$ denote independent, identically distributed random variables satisfying the conditions of Theorem 14.2. Then, according to Theorem 14.2, f_n, the density of $\sum_{j=1}^{n} X_j/\sqrt{n}$, satisfies

$$\sup_x \left| f_n(x) - \left[\phi(x) + \frac{\kappa_3}{6}H_3(x)\phi(x)\frac{1}{\sqrt{n}}\right] \right| = o\left(\frac{1}{\sqrt{n}}\right).$$

Hence,

$$f_n(x) = \phi(x) + O\left(\frac{1}{\sqrt{n}}\right).$$

Table 14.1. *Edgeworth series approximation to the chi-squared distribution.*

	n		
α	5	10	20
0.01	0.0493	0.0217	0.0133
0.05	0.0920	0.0679	0.0579
0.10	0.135	0.116	0.107
0.20	0.218	0.208	0.203
0.80	0.796	0.797	0.798
0.90	0.883	0.890	0.895
0.95	0.937	0.942	0.946
0.99	0.995	0.993	0.991

Consider $f_n(0)$, the density of $\sum_{j=1}^{n} X_j / \sqrt{n}$ evaluated at the mean of the distribution. Since $H_3(0) = 0$,

$$f_n(0) = \phi(0) + o\left(\frac{1}{\sqrt{n}}\right) = \frac{1}{\sqrt{(2\pi)}} + o\left(\frac{1}{\sqrt{n}}\right);$$

that is, the normal approximation to the density, evaluated at 0, has error $o(1/\sqrt{n})$. In general, if the normal approximation to the density is evaluated at x_n, where $x_n = o(1/\sqrt{n})$, the error of the approximation is $o(1/\sqrt{n})$. □

Third- and higher-order approximations

More refined approximations may be obtained by retaining more terms in the expansion of $\log \varphi_n(t)$. Here we briefly describe the more general results; for further discussion and references see Section 14.7.

In general, an approximation to the density function of Y_n is given by

$$\phi(y) \left[1 + \frac{\kappa_3}{6} H_3(y) \frac{1}{\sqrt{n}} + \left(\frac{\kappa_4}{24} H_4(y) + \frac{\kappa_3^2}{72} H_6(y) \right) \frac{1}{n} + \cdots \right].$$

Here the functions $H_j(y)$ are the Hermite polynomials, defined by

$$H_r(y)\phi(y) = (-1)^r \frac{d^r \phi(y)}{dy^r};$$

see, Section 10.3 for further details of the Hermite polynomials. An approximation to the distribution function of Y_n may then be obtained by integrating the approximation to the density. This procedure is simplified by recalling that

$$\int_{-\infty}^{y} H_r(t)\phi(t)\,dt = (-1)^r \int_{-\infty}^{y} \frac{d^r \phi(t)}{dt^r}\,dt = (-1)^r \frac{d^{r-1}\phi(t)}{dt^{r-1}}\Big|_{-\infty}^{y}$$

$$= -H_{r-1}(y)\phi(y);$$

see Theorem 10.8. Hence, an approximation to the distribution function F_n of Y_n is given by

$$\Phi(y) - \phi(y) \left[\frac{\kappa_3}{6} H_2(y) \frac{1}{\sqrt{n}} + \left(\frac{\kappa_4}{24} H_3(y) + \frac{\kappa_3^2}{72} H_5(y) \right) \frac{1}{n} + \cdots \right].$$

Typically, either the $n^{-\frac{1}{2}}$ term or the $n^{-\frac{1}{2}}$ term together with the n^{-1} term are used when approximating the distribution function of Y_n. The error of the approximation is one power of $n^{-\frac{1}{2}}$ greater than the last included term. For instance, if the approximation includes only the $n^{-\frac{1}{2}}$ term, then the error is of order n^{-1}; this is the case considered in Theorem 14.2.

Example 14.4 (*Chi-squared distribution*). Consider approximations to the chi-squared distribution function based on an Edgeworth expansion, as in Example 14.2. The fourth cumulant of Z_j^2 is 48. Hence, the distribution function of

$$Y_n = \sqrt{n}\frac{X_n/n - 1}{\sqrt{2}}$$

may be approximated by

$$\Phi(y) - \frac{\sqrt{2}}{3}H_2(y)\phi(y)\frac{1}{\sqrt{n}} - \left(2H_3(y) + \frac{8}{9}H_5(y)\right)\phi(y)\frac{1}{n}.$$

The error of this approximation is $o(1/n)$.   □

Expansions for quantiles

An Edgeworth expansion may be used to approximate the quantiles of the distribution of a sample mean. That is, suppose $X_1, X_2, \ldots, X_n$ are independent, identically distributed random variables each with mean 0 and standard deviation 1 and suppose we wish to approximate the value x_α satisfying

$$\Pr(\sqrt{n}\bar{X}_n \le x_\alpha) = \alpha,$$

where α is a given number, $0 < \alpha < 1$. Using the central limit theorem to approximate the distribution of $\bar{X}_n$, a first-order approximation to x_α is given by the corresponding quantile of the standard normal distribution, $z_\alpha \equiv \Phi^{-1}(\alpha)$.

If an Edgeworth series approximation is used to approximate the distribution of $\sqrt{n}\bar{X}_n$, we obtain a series approximation for x_α; such an approximation is known as a *Cornish–Fisher inversion* of the Edgeworth series approximation. The following result gives the Cornish–Fisher inversion corresponding to an Edgeworth series approximation of the form given in Theorem 14.1.

Theorem 14.3. *Let X_n, $n = 1, 2, \ldots$, denote a sequence real-valued random variables such that each X_n has an absolutely continuous distribution function F_n and let $x_\alpha^{(n)}$ satisfy $F_n(x_\alpha^{(n)}) = \alpha$, where $0 < \alpha < 1$ is given.*
Suppose that there exists a constant β such that

$$F_n(x) = \Phi(x) - \frac{\beta}{\sqrt{n}}H_2(x)\phi(x) + o\left(\frac{1}{\sqrt{n}}\right)$$

uniformly in x. Then

$$x_\alpha^{(n)} = z_\alpha + \frac{\beta}{\sqrt{n}}H_2(z_\alpha) + o\left(\frac{1}{\sqrt{n}}\right)$$

where z_α satisfies $\Phi(z_\alpha) = \alpha$.

Proof. Consider an expansion for $x_\alpha^{(n)}$ of the form

$$x_\alpha^{(n)} = a_n + b_n/\sqrt{n} + o\left(\frac{1}{\sqrt{n}}\right)$$

where a_n and b_n are $O(1)$. Using the expression for F_n given in the statement of the theorem together with the fact that Φ, H_2, and ϕ are all differentiable,

$$
\begin{aligned}
F_n(x_\alpha^{(n)}) &= F_n\left(a_n + b_n/\sqrt{n} + o\left(\frac{1}{\sqrt{n}}\right)\right) \\
&= \Phi(a_n + b_n/\sqrt{n}) - \frac{\beta}{\sqrt{n}} H_2(a_n)\phi(a_n) + o\left(\frac{1}{\sqrt{n}}\right) \\
&= \Phi(a_n) + \frac{b_n}{\sqrt{n}}\phi(a_n) - \frac{\beta}{\sqrt{n}} H_2(a_n)\phi(a_n) + o\left(\frac{1}{\sqrt{n}}\right).
\end{aligned}
$$

Hence, to achieve

$$F_n\left(x_\alpha^{(n)}\right) = \alpha,$$

we need $\Phi(a_n) = \alpha + o(1/\sqrt{n})$ and

$$\left[\frac{b_n}{\sqrt{n}} - \frac{\beta}{\sqrt{n}} H_2(a_n)\right]\phi(a_n) = o\left(\frac{1}{\sqrt{n}}\right).$$

This implies that $a_n = z_\alpha + o(1/\sqrt{n})$ and

$$b_n = \beta H_2(z_\alpha) + o\left(\frac{1}{\sqrt{n}}\right)$$

so that

$$x_\alpha^{(n)} = z_\alpha + \frac{\beta}{\sqrt{n}} H_2(z_\alpha) + o\left(\frac{1}{\sqrt{n}}\right)$$

as stated in the theorem. ∎

***Example* 14.5 (*Chi-squared distribution*).** Let $\chi_n^2(\alpha)$ denote the α-quantile of the chi-squared distribution with n degrees of freedom, and let X_n denote a chi-squared random variable with n degrees of freedom. Then, using the results of Example 14.2, the distribution of

$$\sqrt{n}\frac{X_n/n - 1}{\sqrt{2}}$$

has α-quantile of the form

$$z_\alpha + \frac{\sqrt{2}}{3\sqrt{n}}\left(z_\alpha^2 - 1\right) + o\left(\frac{1}{\sqrt{n}}\right).$$

That is,

$$\chi_n^2(\alpha) = n + \sqrt{(2n)}z_\alpha + \frac{2}{3}\left(z_\alpha^2 - 1\right) + o(1).$$

Table 14.2. *Exact and approximate quantiles of the chi-squared distribution.*

	Quantile			
	$n = 10$		$n = 20$	
α	Exact	Approx.	Exact	Approx.
0.01	2.56	2.54	8.26	8.23
0.05	3.94	3.78	10.85	10.73
0.10	4.87	4.70	12.44	12.44
0.20	6.18	6.04	14.58	14.58
0.80	13.44	13.57	25.04	25.13
0.90	15.99	16.16	28.41	28.53
0.95	18.31	18.49	31.41	31.54
0.99	23.21	23.35	37.57	37.65

Table 14.2 contains the exact quantiles of the chi-squared distribution with n degrees of freedom, together with the approximation

$$n + \sqrt{(2n)}z_\alpha + \frac{2}{3}\left(z_\alpha^2 - 1\right),$$

for $n = 10$ and $n = 20$. These results indicate that approximation is generally quite accurate. $\square$

14.3 Saddlepoint Approximations

Let $X_1, \ldots, X_n$ denote independent, identically distributed random variables, each an absolutely continuous distribution with density p. Let $K(t)$ denote the cumulant-generating function of X_1, which is assumed to be finite for $t_0 < t < t_1$ for some $t_0, t_1, t_0 < 0 < t_1$ For $t_0 < \lambda < t_1$ define

$$p(x; \lambda) = \exp\{x\lambda - K(\lambda)\}p(x);$$

assume that the distribution of X under $p(x; \lambda)$ is non-degenerate for $\lambda \in (t_0, t_1)$. Note that

$$\int_{-\infty}^{\infty} p(x; \lambda)\,dx = \int_{-\infty}^{\infty} \exp\{x\lambda\}p(x)\,dx\, \exp\{-K(\lambda)\} = 1$$

so that, for each λ, $p(x; \lambda)$ defines a density function. The cumulant generating function of the density $p(x; \lambda)$ is $K(\lambda + s) - K(\lambda)$ and, hence, the cumulants are given by $K'(\lambda)$, $K''(\lambda)$, and so on.

Let $S_n = X_1 + \cdots + X_n$, and let $p_n(s; \lambda)$ denote the density function of S_n under the density $p(x; \lambda)$ for the X_j; then the actual density of S_n is given by $p_n(s) \equiv p_n(s; 0)$.

Note that

$$p_n(s;\lambda) = \int_{-\infty}^{\infty} \cdots \int_{-\infty}^{\infty} p(s - x_2 - \cdots - x_n;\lambda)p(x_2;\lambda)\cdots p(x_n;\lambda)\,dx_2\cdots dx_n$$

$$= \exp(s\lambda - nK(\lambda))\int_{-\infty}^{\infty} \cdots \int_{-\infty}^{\infty} p(s - x_2 - \cdots - x_n)p(x_2)\cdots p(x_n)\,dx_2\cdots dx_n$$

$$= \exp(s\lambda - nK(\lambda))p_n(s);$$

this holds for any λ, $t_0 < \lambda < t_1$. Let $\hat{p}_n(s;\lambda)$ denote an approximation to $p_n(s;\lambda)$. Then an approximation to $p_n(s)$ is given by

$$\exp\{nK(\lambda) - s\lambda\}\hat{p}_n(s;\lambda).$$

The idea behind the saddlepoint approximation is to choose λ so that $\hat{p}_n(s;\lambda)$ is an accurate approximation to $p_n(s;\lambda)$; the value of λ chosen will depend on s. Since we are not directly approximating the density of interest, $p_n(s;0)$, the saddlepoint approximation is often referred to as an *indirect* approximation, in contrast to the Edgeworth series approximation which is referred to as a *direct* approximation.

In Example 14.3 we have seen that the normal approximation to a density function is very accurate when it is evaluated at the mean. Hence, the value of λ is chosen so that the point s at which the density of S is to be evaluated corresponds to the mean of $p(\cdot;\lambda)$. That is, given s, choose $\lambda = \hat{\lambda}_s$ so that

$$s = \mathrm{E}(S_n;\lambda) = n\mathrm{E}(Y_1;\lambda) = nK'(\lambda).$$

It follows that $\hat{\lambda}_s$ satisfies $nK'(\hat{\lambda}_s) = s$. Note that, since $K''(\lambda)$ is the variance of X_j under $p(x;\lambda)$, $K''(\lambda) > 0$ for all λ so that $K(\lambda)$ is a convex function, which implies that the equation $nK'(\hat{\lambda}_s) = s$ has at most one solution. We assume that s is such that a solution exists.

For the approximation $\hat{p}_n(s;\lambda)$ we use the normal approximation given by the central limit theorem; since the evaluation of the density is at the mean of S_n and the variance of the X_j under $p(x;\lambda)$ is $K''(\lambda)$, the approximation is given by

$$[2\pi nK''(\hat{\lambda}_s)]^{-\frac{1}{2}}.$$

It follows that an approximation to $p_n(s)$ is given by

$$\hat{p}_n(s) = \exp\{nK(\hat{\lambda}_s) - s\hat{\lambda}_s\}[2\pi nK''(\hat{\lambda}_s)]^{-\frac{1}{2}}.$$

Let $\bar{X} = S_n/n$. Then an approximation to the density of $\bar{X}$ is given by

$$\exp\{n[K(\hat{\lambda}_x) - x\hat{\lambda}_x]\}[2\pi K''(\hat{\lambda}_x)/n]^{-\frac{1}{2}}$$

where $\hat{\lambda}_x$ satisfies $K'(\hat{\lambda}_x) = x$. Since the normal approximation is evaluated at the mean of the distribution, the error of this approximation is of order $o(n^{-\frac{1}{2}})$; see Example 14.3.

The saddlepoint approximation is known to be extremely accurate, even more accurate than is suggested by the error term of order $o(1/\sqrt{n})$. This is due, at least in part, to the fact that the normal approximation is always evaluated at the mean, a region in which the approximation is generally very accurate.

Theorem 14.4 gives a formal statement of the saddlepoint method.

Theorem 14.4. *Let* $X_1, X_2, \ldots$ *denote independent, identically distributed random variables. Assume that the distribution of* X_1 *is absolutely continuous with bounded density* p *and that the moment-generating function of the distribution,* $M(t)$*, exists for* $t_0 < t < t_1$ *for some* t_0, t_1*,* $t_0 < 0 < t_1$*, and let* $K(t) = \log M(t)$*.*

Let $p_{\bar{X}_n}$ *denote the density of* $\bar{X}_n = \sum_{j=1}^n X_j / n$*. Then, for each* x *such that* $K'(\lambda) = x$ *has a solution in* λ*,* λ_x*,*

$$p_{\bar{X}_n}(x) = \exp(nK(\lambda_x) - n\lambda_x) \frac{\sqrt{n}}{[2\pi K''(\lambda_x)]^{\frac{1}{2}}} \left[1 + o\left(\frac{1}{\sqrt{n}}\right) \right].$$

Proof. Let p denote the density of X_1. For $t_0 < \lambda < t_1$, define

$$p(x; \lambda) = \exp\{\lambda x - K(\lambda)\} p(x)$$

where $K(\lambda) = \log M(\lambda)$. Note that

$$\int_{-\infty}^{\infty} p(x; \lambda) \, dx = 1$$

so that $p(\cdot; \lambda)$ represents a valid probability density function for each $|\lambda| < t_0$.

The moment-generating function corresponding to $p(\cdot; \lambda)$ is given by

$$M(t; \lambda) = \int_{-\infty}^{\infty} \exp\{tx\} p(x; \lambda) \, dx = \exp\{K(t + \lambda) - K(\lambda)\},$$

for all t such that $|t + \lambda| < t_0$. Hence, the cumulants of X_1 under $p(\cdot; \lambda)$ are given by $K'(\lambda)$, $K''(\lambda)$, and so on.

Let $S_n = \sum_1^n X_j$, $\bar{X}_n = S_n / n$, and let p_n denote the density of S_n. When $X_1, X_2, \ldots$ have density $p(\cdot; \lambda)$, then S_n has density

$$p_n(s; \lambda) = \exp\{\lambda s - nK(\lambda)\} p_n(s).$$

For a given value of λ, define

$$Z_\lambda = \sqrt{n} \frac{\bar{X}_n - K'(\lambda)}{[K''(\lambda)]^{\frac{1}{2}}}$$

and let $f_n(\cdot; \lambda)$ denote the density of Z_λ when $X_1, X_2, \ldots$ have density $p(\cdot; \lambda)$. Note that, since the moment-generating function of X_1 exists,

$$\lim_{x \to \pm\infty} \exp(tx) p(x) = 0$$

for all $|t| < t_0$, since otherwise the integral defining $M(t)$ would not be finite. Hence, given ϵ, there exists a $B > 0$ such that

$$\exp(tx) p(x) \le \epsilon \quad \text{for all } |x| > B.$$

Since p is bounded, and $\exp(tx)$ is bounded for $|x| \le B$ for any value of t, it follows that $p(x; \lambda)$ is bounded for any λ, $|\lambda| < t_0$. Hence, by Lemma 14.1, condition (ii) of Theorem 14.2 holds.

It is straightforward to show that, under the conditions of the theorem, the remaining conditions of Theorem 14.2 are satisfied so that

$$f_n(z; \lambda) = \phi(z) + \frac{K'''(\lambda)}{6\sqrt{n}} (z^3 - 3z) \phi(z) + o\left(\frac{1}{\sqrt{n}}\right),$$

uniformly in z.

It follows that $p_n(s; \lambda)$ is of the form

$$\frac{1}{[nK''(\lambda)]^{\frac{1}{2}}}[\phi(z(s)) + \frac{K'''(\lambda)}{6\sqrt{n}}(z(s)^3 - 3z(s))\phi(z(s)) + o\left(\frac{1}{\sqrt{n}}\right),$$

where $z(s) = [s/\sqrt{n} - \sqrt{n}K'(\lambda)]/[K''(\lambda)]^{\frac{1}{2}}$. Since

$$p_n(s; \lambda) = \exp\{s\lambda - nK(\lambda)\}p_n(s),$$

it follows that

$$p_n(s) = \exp\{nK(\lambda) - s\lambda\}\left[\frac{1}{[nK''(\lambda)]^{\frac{1}{2}}}[\phi(z(s))\right.$$
$$\left. + \frac{K'''(\lambda)}{6\sqrt{n}}(z(s)^3 - 3z(s))\phi(z(s)) + o\left(\frac{1}{\sqrt{n}}\right)\right].$$

This result holds for any value of λ. Hence, take $\lambda \equiv \lambda_s$ such that $K'(\lambda_s) = s/n$. Then

$$p_n(s) = \exp\{nK(\lambda_s) - s\lambda_s\}\frac{1}{[nK''(\lambda_s)]^{\frac{1}{2}}}\left[\phi(0) + o\left(\frac{1}{\sqrt{n}}\right)\right]$$
$$= \exp\{nK(\lambda_s) - s\lambda_s\}\frac{1}{[2\pi nK''(\lambda_s)]^{\frac{1}{2}}}\left[1 + o\left(\frac{1}{\sqrt{n}}\right)\right].$$

Now consider the $p_{\bar{X}_n}$, the density of $\bar{X}_n$. Since

$$p_{\bar{X}_n}(x) = np_n(nx),$$

it follows that

$$p_{\bar{X}_n}(x) = \exp\{nK(\lambda_x) - nx\lambda_x\}\frac{\sqrt{n}}{[2\pi K''(\lambda_x)]^{\frac{1}{2}}}\left[1 + o\left(\frac{1}{\sqrt{n}}\right)\right]$$

where λ_x solves $K'(\lambda_x) = x$. ∎

Example 14.6 (*Sample mean of Laplace random variables*). Let $Y_1, \ldots, Y_n$ denote independent random variables, each with a standard Laplace distribution. Since the cumulant-generating function is given by $K(t) = -\log(1 - t^2)$, $|t| < 1$,

$$K'(t) = \frac{2t}{1 - t^2}$$

and the equation $K'(\lambda) = y$ may be reduced to a quadratic equation. This quadratic has two solutions, but only one in the interval $(-1, 1)$,

$$\hat{\lambda}_y = \frac{(1 + y^2)^{\frac{1}{2}} - 1}{y}.$$

The saddlepoint approximation to the density of $\bar{Y}$ is therefore given by

$$\frac{\sqrt{n}\exp\{n\}}{2^n(2\pi)^{\frac{1}{2}}}\frac{|y|^{2n-1}}{\left[(1 + y^2)^{\frac{1}{2}} - 1\right]^{n-\frac{1}{2}}}\frac{\exp\left\{-n(1 + y^2)^{\frac{1}{2}}\right\}}{(1 + y^2)^{\frac{1}{4}}}. \qquad \square$$

Example 14.7 (*Chi-squared distribution*). Let $Z_1, Z_2, \ldots$ denote independent, identically distributed standard normal random variables and consider approximation of the distribution of $Y_n = \sum_{j=1}^{n} Z_j^2$ which, of course, has a chi-squared distribution with n degrees of freedom.

The cumulant-generating function of Z_1^2 is $-\log(1-2t)/2, t < 1/2$ so that the solution to $K'(\lambda) = y$ is given by $\lambda_y = (1 - 1/x)/2$; it follows that $K(\lambda_y) = \log(y)/2$ and $K''(\lambda_y) = 2y^2$. It follows that the saddlepoint approximation to the density of $Y_n = \sum_{j=1}^{n} Z_j^2$ is given by

$$\exp\left\{\frac{n}{2}\log(y) - \frac{1}{2}s\left(1 - \frac{1}{s}\right)\right\}\frac{1}{[4\pi ns^2]^{\frac{1}{2}}} = \frac{\exp(1/2)}{2\sqrt{(\pi n)}}y^{\frac{n}{2}-1}\exp\left(-\frac{1}{2}y\right).$$

Comparing this approximation to the density function of the chi-squared distribution with n degrees of freedom, we see that the saddlepoint approximation is exact, aside from a normalization factor. $\square$

Renormalization of saddlepoint approximations

It is important to note that saddlepoint approximations to densities do not necessarily integrate to 1. Furthermore, unlike Edgeworth series expansions, saddlepoint approximations cannot generally be integrated analytically. Let $\hat{p}_{\bar{X}_n}(x)$ denote a saddlepoint approximation to the density of the sample mean and let

$$\frac{1}{c} = \int_{-\infty}^{\infty} \hat{p}_{\bar{X}_n}(x)\,dx;$$

note that it is often necessary to perform this integration numerically. Then the *renormalized* saddlepoint approximation is given by $c\hat{p}_{\bar{X}_n}(x)$.

***Example* 14.8 (*Sample mean of Laplace random variables*).** Consider the saddlepoint approximation to the density of the sample mean of Laplace random variables, derived in Example 14.6, for the case $n = 5$. The integral of the density function, determined by numerical integration, is approximately $1/1.056$. Hence, the renormalized saddlepoint approximation is given by

$$\frac{1.056\sqrt{5}\exp\{5\}}{2^5(2\pi)^{\frac{1}{2}}}\frac{|y|^9}{\left[(1+y^2)^{\frac{1}{2}} - 1\right]^{\frac{9}{2}}}\frac{\exp\left\{-9(1+y^2)^{\frac{1}{2}}\right\}}{(1+y^2)^{\frac{1}{4}}}. \qquad \square$$

Integration of saddlepoint approximations

Saddlepoint approximations can be used as the basis for approximations to the distribution function of the sample mean; equivalently, we can consider approximations to tail probabilities, which we do here. As noted above, analytical integration of saddlepoint density approximations is not generally possible; hence, to approximate tail probabilities, approximation of the resulting integral is needed. Several such approximations were given in Chapter 9. Here we consider the application of the method described in Section 9.6. Only a brief description is given; for further details on this method see Section 14.7.

Consider the problem of approximating $\Pr(\bar{X} \geq t)$. Using the renormalized saddlepoint approximation for the density of $\bar{X}$, this probability can be approximated by

$$c\int_t^{\infty} \exp\{n[K(\hat{\lambda}_x) - x\hat{\lambda}_x]\}[2\pi K''(\hat{\lambda}_x)/n]^{-\frac{1}{2}}\,dx. \qquad (14.4)$$

To approximate this integral using the approximation derived in Section 9.6, we first need
to write (14.4) in the form

$$\int_z^\infty h_n(t)\sqrt{n}\phi(\sqrt{n}t)\,dt. \tag{14.5}$$

Let

$$r(x) = \text{sgn}(\hat\lambda_x)\{2[x\hat\lambda_x - K(\hat\lambda_x)]\}^{\frac{1}{2}}.$$

Note that, for fixed x, the function $x\lambda - K(\lambda)$ is uniquely maximized at $\lambda = \hat\lambda_x$ and since
this function is 0 at $\lambda = 0$, it follows that

$$x\hat\lambda_x - K(\hat\lambda_x) \geq 0.$$

It follows from the fact that $r(x)^2 = 2[x\hat\lambda_x - K(\hat\lambda_x)]$ that

$$r'(x)r(x) = \hat\lambda_x + [x - K'(\hat\lambda_x)]\frac{d\hat\lambda_x}{dx} = \hat\lambda_x.$$

Hence, $r(x)$ is a strictly increasing function of x. Note that, since $K(0) = 0$, this implies
that $r(x) = 0$ if and only if $\hat\lambda_x = 0$.

The integral in (14.4) may be written

$$c\int_t^\infty \left(\frac{n}{2\pi}\right)^{\frac{1}{2}} \exp\left\{-\frac{n}{2}r(x)^2\right\} e K''(\hat\lambda_x)^{-\frac{1}{2}}\,dx. \tag{14.6}$$

Let $z = r(x)$; note that $dx/dz = z/\hat\lambda_x$. Then (14.6) may be written

$$\bar c\int_{r(t)}^\infty \left(\frac{n}{2\pi}\right)^{\frac{1}{2}} \exp\left\{-\frac{n}{2}z^2\right\} K''(\hat\lambda_x)^{-\frac{1}{2}}\frac{z}{\hat\lambda_x}\,dz$$

where $x = r^{-1}(z)$. This is of the form (14.5) with

$$h(z) = \frac{z}{\hat\lambda_x[K''(\hat\lambda_x)]^{\frac{1}{2}}}, \quad x = r^{-1}(z).$$

Note that

$$h(0) = \frac{1}{K''(0)^{\frac{1}{2}}}\lim_{z\to 0}\frac{z}{\hat\lambda_x} = \frac{1}{K''(0)^{\frac{1}{2}}}\lim_{z\to 0}\frac{1}{d\hat\lambda_x/dz}$$

where

$$\frac{d\hat\lambda_x}{dz} = \frac{d\hat\lambda_x}{dx}\frac{dx}{dz} = \frac{1}{K''(\hat\lambda_x)}\frac{z}{\hat\lambda_x} \tag{14.7}$$

since $K'(\hat\lambda_x) = x$ implies that $d\hat\lambda_x/dx = 1/K''(\hat\lambda_x)$. It follows that from the last part of
(14.7) that

$$h(0) = K''(0)^{\frac{1}{2}}\lim_{z\to 0}\frac{\hat\lambda_x}{z}$$

so that $h(0) = 1/h(0)$. Hence, $h(0) = \pm 1$; however, z and $\hat{\lambda}_x$ have the same sign so that $h(0) > 0$. It follows that $h(0) = 1$. Also,

$$h(r(t)) = \frac{r(t)}{\hat{\lambda}_x K''(\hat{\lambda}_x)^{\frac{1}{2}}}, \quad x = r^{-1}(r(t)) = t$$

$$= \frac{r(t)}{\hat{\lambda}_t K''(\hat{\lambda}_t)^{\frac{1}{2}}}.$$

An approximation to the integral (14.4) is given by Theorem 9.16 with

$$\frac{h(r(t)) - h(0)}{r(t)} = \frac{1}{\hat{\lambda}_t K''(\hat{\lambda}_t)^{\frac{1}{2}}} - \frac{1}{r(t)}.$$

Hence, $\Pr(\bar{X}_n \geq t)$ may be approximated by

$$1 - \Phi(\sqrt{n}r) + \frac{1}{n}\left[\frac{1}{\hat{\lambda}_t K''(\hat{\lambda}_t)^{\frac{1}{2}}} - \frac{1}{r}\right]\sqrt{n}\phi(\sqrt{n}r), \quad r = r(t).$$

This approximation has relative error $O(n^{-\frac{3}{2}})$ for fixed r, corresponding to t of the form $t = \mathrm{E}(\bar{X}) + O(n^{-\frac{1}{2}})$, and relative error $O(n^{-1})$ for $r = O(\sqrt{n})$, corresponding to fixed values of t.

Example 14.9 (Chi-squared distribution). As in Example 14.7, let Y_n denote a chi-squared random variable with n degrees of freedom and consider approximation of $\Pr(Y_n \geq y)$; an Edgeworth series approximation to this probability was given in Example 14.2.

Let $\bar{X}_n = Y_n/n$. Then $\bar{X}_n$ is the sample mean of n independent, identically distributed random variables each with a distribution with cumulant-generating function

$$K(t) = -\frac{1}{2}\log(1 - 2t), \quad t < 1/2.$$

The solution to $K'(t) = x$ is given by $\hat{\lambda}_x = -(x-1)/(2x)$ so that

$$r(x) = \mathrm{sgn}(x-1)\{x - 1 - \log(x)\}^{\frac{1}{2}}$$

and $K''(\hat{\lambda}_x) = 2x^2$. It follows that $\Pr(X_n \geq x)$ may be approximated by

$$1 - \Phi[\sqrt{n}r(x)] + \frac{1}{n}\left[\frac{\sqrt{2}}{x-1} - \frac{1}{r(x)}\right]\sqrt{n}\phi[\sqrt{n}r(x)].$$

Now consider approximation of $\Pr(Y_n \geq y) = \Pr(\bar{X}_n \geq y/n)$. Let

$$r_n = \sqrt{n}r(y/n) = \mathrm{sgn}(y - n)\{y - n - n\,\log(y/n)\}^{\frac{1}{2}}.$$

It follows that $\Pr(Y_n \geq y)$ may be approximated by

$$[1 - \Phi(r_n)] + \left[\frac{\sqrt{2n}}{n - y} - \frac{1}{r_n}\right]\phi(r_n);$$

equivalently, $\Pr(Y_n \leq y)$ may be approximated by

$$\Phi(r_n) - \left[\frac{\sqrt{2n}}{y - n} - \frac{1}{r_n}\right]\phi(r_n).$$

Table 14.3. *Saddlepoint approximation to the chi-squared distribution.*

	n		
α	1	2	5
0.01	0.0114	0.0105	0.0101
0.05	0.0557	0.0516	0.0502
0.10	0.109	0.102	0.100
0.20	0.213	0.203	0.200
0.80	0.802	0.800	0.800
0.90	0.900	0.900	0.900
0.95	0.950	0.950	0.950
0.99	0.990	0.990	0.990

Table 14.3 contains approximations to $\Pr(S_n \leq s_{n\alpha})$ based on the saddlepoint method described above, where $s_{n\alpha}$ satisfies

$$\Pr(S_n \leq s_{n\alpha}) = \alpha,$$

for several choices of n and α. Recall that corresponding approximations based on the central limit theorem and on an Edgeworth series approximation are given in Tables 12.1 and 14.1, respectively. These results show that the saddlepoint approximation is very accurate, even when $n = 1$; for $n = 5$, the error in the approximation is essentially 0. $\quad\square$

14.4 Stochastic Asymptotic Expansions

Let $Y_1, Y_2, \ldots, Y_n$ denote a sequence of random variables. Thus far, when considering approximations to Y_n we have focused on approximations for the distribution function or density function of Y_n. Another approach is to approximate the random variable Y_n directly by other random variables, the properties of which are well-understood. For instance, we might be able to write

$$Y_n = \hat{Y}_0 + \hat{Y}_1 \frac{1}{n} + O_p\left(\frac{1}{n^2}\right)$$

for some random variables $\hat{Y}_0, \hat{Y}_1$. This type of approximation is known as a *stochastic asymptotic expansion*.

A stochastic asymptotic expansion for Y_n can often be used to derive an approximation to the distribution of Y_n by approximating the distribution of the terms in the expansion. For instance, suppose that $Y_n = \hat{Y}_n + o_p(1)$, where $\hat{Y}_n$ is asymptotically distributed according to a standard normal distribution. It follows from Slutsky's theorem that Y_n is also asymptotically distributed according to a standard normal distribution. We have already seen one application of this idea, the δ-method described in Section 13.2.

Now suppose that

$$Y_n = \hat{Y}_n + O_p(n^{-1})$$

where the distribution of $\hat{Y}_n$ has an Edgeworth series expansion of the form

$$\phi(x)\left[1 + \frac{\rho_3}{6\sqrt{n}} H_3(x) + o\left(\frac{1}{\sqrt{n}}\right)\right].$$

It is tempting to conclude that the distribution of Y_n has an Edgeworth series expansion with error of order $o(1/\sqrt{n})$. This conclusion is valid provided that $Y_n = \hat{Y}_n + o_p(1/\sqrt{n})$ implies that

$$\Pr(Y_n \leq t) = \Pr(\hat{Y}_n \leq t) + o(1/\sqrt{n}).$$

Unfortunately, this result does not hold in general. The following example illustrates the type of problem that may be encountered.

***Example* 14.10.** Let X denote a standard normal random variable and, for each $n = 1$, $2, \ldots$, let Z_n denote a random variable such that

$$\Pr(Z_n = 1) = 1 - \Pr(Z_n = 0) = \delta_n$$

where $\delta_n, n = 1, 2, \ldots$, is a sequence in $[0, 1]$ such that $\lim_{n\to\infty} \delta_n = 0$. Note that, for any $\epsilon > 0$,

$$\Pr(\sqrt{n}|Z_n| > \epsilon) = \Pr(Z_n = 1) = \delta_n$$

so that $Z_n = o_p(1/\sqrt{n})$.

Assume that X and Z_n are independent for any $n = 1, 2, \ldots$ and let $Y_n = X + Z_n$, $n = 1, 2, \ldots$. Then $Y_n = X + o_p(1/\sqrt{n})$; however,

$$\begin{aligned}
\Pr(Y_n \leq t) &= \Pr(X + Z_n \leq t) \\
&= \Pr(X + Z_n \leq t | Z_n = 0)(1 - \delta_n) + \Pr(X + Z_n \leq t | Z_n = 1)\delta_n \\
&= \Phi(t)(1 - \delta_n) + \Phi(t - 1)\delta_n.
\end{aligned}$$

Hence,

$$\Pr(Y_n \leq t) = \Pr(X \leq t) + O(\delta_n)$$

and if $\delta_n \to 0$ slowly, $\Pr(X \leq t)$ is a poor approximation to $\Pr(Y_n \leq t)$.

The problem is that the condition that $Z_n = o_p(1/\sqrt{n})$ holds provided only that $\Pr(\sqrt{n}|Z_n| > \epsilon) \to 0$ for any ϵ, while an approximation to the distribution of Y_n depends on the rate at which this probability approaches 0. This issue does not arise in the first-order approximation since $Z_n = o_p(1)$ implies that $\Pr(|Z_n| > \epsilon) = o(1)$; however, $Z_n = o_p(1/\sqrt{n})$ does not imply that $\Pr(\sqrt{n}|Z_n| > \epsilon) = o(1/\sqrt{n})$.   □

In spite of this negative result, in many cases in which a random variable Y_n has an expansion of the form $Y_n = \hat{Y}_n + o_p(1/\sqrt{n})$ the distribution functions of Y_n and $\hat{Y}_n$ do agree to order $o(1/\sqrt{n})$. However, some additional structure for the $o_p(1/\sqrt{n})$ term is required for this to hold.

Here we consider the case in which X_n is a sample mean based on n independent, identically distributed random variables and $Y_n = f(X_n)$ where f is a smooth function. Then, using a Taylor's series approximation,

$$Y_n = f(\mu) + f'(\mu)(X_n - \mu) + \frac{1}{2} f''(\mu)(X_n - \mu)^2 + \cdots$$

where $\mu = E(X_n)$. In this case, the approximating random variable $\hat{Y}_n$ may be taken to be a polynomial in $X_n - \mu$. Hence, before considering the distribution of Y_n in this scenario, we give a preliminary result on the distribution of a quadratic function of a random variable whose distribution follows an Edgeworth series expansion.

Lemma 14.2. *Let $Z_1, Z_2, \ldots$ denote independent, identically distributed random variables, each with mean 0 and standard deviation 1. Assume that the distribution of Z_1 is absolutely continuous and satisfies the conditions of Theorem 14.2.*
Let $X_n = \sum_1^n Z_j / \sqrt{n}$ and, for a given constant c, let

$$Y_n = X_n + \frac{c}{\sqrt{n}} X_n^2.$$

Then F_n, the distribution function of Y_n, satisfies the following: for any sequence $y_1, y_2, \ldots$ such that $y_n = y + a/\sqrt{n} + o(1/\sqrt{n})$ for some $a, y \in \mathbf{R}$,

$$F_n(y_n) = \Phi(y + (a - c)/\sqrt{n}) - \frac{\kappa_3 + 6c}{6\sqrt{n}} H_2(y + (a - c)/\sqrt{n})$$

$$\times \ \phi(y + (a - c)/\sqrt{n}) + o\left(\frac{1}{\sqrt{n}}\right)$$

where κ_3 denotes the third cumulant of Z_1.

Proof. Let H_n denote the distribution function of X_n. Then, by Theorem 14.2,

$$H_n(x) = \Phi(x) - \frac{\kappa_3}{6\sqrt{n}} H_2(x)\phi(x) + o\left(\frac{1}{\sqrt{n}}\right)$$

uniformly in x.

Suppose $c > 0$. Then the event that $Y_n \le y_n$ is equivalent to the event that

$$\frac{c}{\sqrt{n}} X_n^2 + X_n - y_n \le 0,$$

which is equivalent to the event that $a_n \le X_n \le b_n$, where

$$a_n = \frac{-1 - [1 + 4cy_n/\sqrt{n}]^{\frac{1}{2}}}{2c/\sqrt{n}} \quad \text{and} \quad b_n = \frac{-1 + [1 + 4cy_n/\sqrt{n}]^{\frac{1}{2}}}{2c/\sqrt{n}};$$

note that, for sufficiently large n, a_n and b_n are both real.

Hence,

$$\Pr(Y_n \le y_n) = \Phi(b_n) - \frac{\kappa_3}{6\sqrt{n}} H_2(b_n)\phi(b_n) - \Phi(a_n) + \frac{\kappa_3}{6\sqrt{n}} H_2(a_n)\phi(a_n) + o\left(\frac{1}{\sqrt{n}}\right).$$

It is straightforward to show that

$$b_n = y + a/\sqrt{n} - cy^2/\sqrt{n} + O\left(\frac{1}{n}\right)$$

and

$$a_n = -\sqrt{n}/(2c) + O(1).$$

Hence, $\Phi(a_n)$ and $\phi(a_n)$ are both $o\left(1/\sqrt{n}\right)$ and

$$\Pr(Y_n \le y_n) = \Phi(y) + \frac{a}{\sqrt{n}}\phi(y) - \frac{cy^2}{\sqrt{n}}\phi(y) - \frac{\kappa_3}{6\sqrt{n}}H_2(y)\phi(y) + o\left(\frac{1}{\sqrt{n}}\right);$$

rearranging this expression yields

$$\Pr(Y_n \le y_n) = \Phi(y) - \frac{\kappa_3 + 6c}{6\sqrt{n}}H_2(y)\phi(y)\frac{a-c}{\sqrt{n}}\phi(y) + o\left(\frac{1}{\sqrt{n}}\right).$$

The result now follows by noting that

$$\Phi(y + (a-c)/\sqrt{n}) = \Phi(y)\frac{a-c}{\sqrt{n}}\phi(y) + o\left(\frac{1}{\sqrt{n}}\right),$$

$$\phi(y + (a-c)/\sqrt{n}) = \phi(y) + o\left(\frac{1}{\sqrt{n}}\right)$$

and

$$H_2(y + (a-c)/\sqrt{n}) = H_2(y) + o\left(\frac{1}{\sqrt{n}}\right). \qquad \blacksquare$$

The following theorem shows that if a function f has a Taylor's series expansion, and the distribution of $\bar{Z}_n$ has an Edgeworth series expansion, then an approximation to the distribution of $f(\bar{Z}_n)$ can be obtained by approximation of f by a quadratic and then applying the result in Lemma 14.2.

Theorem 14.5. *Let* $Z_1, Z_2, \ldots$ *denote independent, identically distributed random variables, each with mean* 0 *and standard deviation* 1. *Assume that the distribution of* Z_1 *is absolutely continuous and satisfies the conditions of Theorem 14.2; in addition, assume* $E(Z_1^4) < \infty$. *Let* $\bar{Z}_n = \sum_{j=1}^n Z_j/n$ *and let* $X_n = f(\bar{Z}_n)$, *where* f *is a three-times differentiable function satisfying the following conditions:*

(i) $|f'(0)| > 0$
(ii) There exists a $\delta > 0$ *such that*

$$M \equiv \sup_{|x|<\delta} |f'''(x)| < \infty.$$

Let

$$Y_n = \sqrt{n}(X_n - f(0))/f'(0)$$

and let F_n *denote the distribution function of* Y_n. *Then*

$$F_n(y) = \Phi(y - c/\sqrt{n}) - \frac{\kappa_3 + 6c}{6\sqrt{n}}H_2(y - c/\sqrt{n})\phi(y - c/\sqrt{n}) + o\left(\frac{1}{\sqrt{n}}\right)$$

where $c = f''(0)/[2f'(0)]$ *and* κ_3 *denotes the third cumulant of* Z_1.

Proof.   We begin by showing that, for any $\alpha < 1/8$, and any random sequence m_n such that $|m_n| \le |\bar{Z}_n|$,

$$\Pr\left(|\frac{1}{n}\frac{f'''(m_n)}{f'(0)}(\sqrt{n}\bar{Z}_n)^2| > \frac{1}{n^{\frac{1}{2}+\alpha}}\right) = o\left(\frac{1}{\sqrt{n}}\right) \quad \text{as } n \to \infty. \qquad (14.8)$$

Let δ denote the constant in condition (ii) of the theorem. Then

$$\Pr\left(\left|\frac{1}{n}\frac{f'''(m_n)}{f'(0)}(\sqrt{n}\bar{Z}_n)^2\right| > \frac{1}{n^{\frac{1}{2}+\alpha}}\right)$$

$$\leq \Pr\left(\left|\frac{1}{n}\frac{f'''(m_n)}{f'(0)}(\sqrt{n}\bar{Z}_n)^2\right| > \frac{1}{n^{\frac{1}{2}+\alpha}} \cap |\bar{Z}_n| < \delta\right) + \Pr(|\bar{Z}_n| \geq \delta).$$

By Chebychev's inequality,

$$\Pr(|\bar{Z}_n| \geq \delta) \leq \frac{1}{n\delta^2} = O\left(\frac{1}{n}\right).$$

Note that when $|\bar{Z}_n| < \delta$, then $|m_n| < \delta$, so that $|f'''(m_n)| < M$. Hence,

$$\Pr\left(\left|\frac{1}{n}\frac{f'''(m_n)}{f'(0)}(\sqrt{n}\bar{Z}_n)^2\right| > \frac{1}{n^{\frac{1}{2}+\alpha}} \cap |\bar{Z}_n| < \delta\right) \leq \Pr\left(\frac{\frac{M}{nf'(0)}|\sqrt{n}\bar{Z}_n|^3 > 1}{n^{\frac{1}{2}+\alpha}}\right)$$

$$\leq \frac{E[n^2\bar{Z}_n^4]}{(n^{\frac{1}{2}-\alpha})^{\frac{4}{3}}(f'(0)/M)^{\frac{4}{3}}}.$$

Using the facts that $E(\bar{Z}_n^4) = O(1/n^2)$ and $\alpha < 1/8$, it follows that

$$\frac{E[n^2\bar{Z}_n^4]}{\left(n^{\frac{1}{2}-\alpha}\right)^{\frac{4}{3}}(f'(0)/M)^{\frac{4}{3}}} = o\left(\frac{1}{\sqrt{n}}\right),$$

proving (14.8).

Now consider the proof of the theorem. Using a Taylor's series expansion, we can write

$$X_n = f(\bar{Z}_n) = f(0) + f'(0)\bar{Z}_n + \frac{1}{2}f''(0)\bar{Z}_n^2 + \frac{1}{6}f'''(m_n)\bar{Z}_n^3$$

where m_n lies on the line segment connecting X_n and 0; hence, $|m_n| \leq |\bar{Z}_n|$. Then

$$Y_n = \sqrt{\bar{Z}_n} + \frac{1}{2}\frac{f''(0)}{f'(0)}(\sqrt{n}\bar{Z}_n)^2\frac{1}{\sqrt{n}} + \frac{1}{6}\frac{f'''(m_n)}{f'(0)}(\sqrt{n}\bar{Z}_n)^3.$$

Let

$$\hat{Y}_n = \sqrt{\bar{Z}_n} + \frac{1}{2}\frac{f''(0)}{f'(0)}(\sqrt{n}\bar{Z}_n)^2\frac{1}{\sqrt{n}}$$

and

$$R_n = \frac{1}{6}\frac{f'''(m_n)}{f'(0)}(\sqrt{n}\bar{Z}_n)^3.$$

Then, for $0 < \alpha < 1/8$,

$$\Pr(Y_n \leq y) = \Pr\left(Y_n \leq y \cap |R_n| \leq \frac{1}{n^{\frac{1}{2}+\alpha}}\right) + \Pr\left(Y_n \leq y \cap |R_n| \geq \frac{1}{n^{\frac{1}{2}+\alpha}}\right)$$

$$= \Pr\left(\hat{Y}_n + R_n \leq y \cap |R_n| \leq \frac{1}{n^{\frac{1}{2}+\alpha}}\right) + o\left(\frac{1}{\sqrt{n}}\right)$$

$$\leq \Pr\left(\hat{Y}_n > y + \frac{1}{n^{\frac{1}{2}+\alpha}}\right) + o\left(\frac{1}{\sqrt{n}}\right).$$

Let $y_n = y - 1/n^{\frac{1}{2}+\alpha}$. Then

$$\Pr(\hat{Y}_n \le y_n) = \Pr\left(\hat{Y}_n \le y_n \cap |R_n| \le \frac{1}{n^{\frac{1}{2}+\alpha}}\right) + o\left(\frac{1}{\sqrt{n}}\right)$$

$$= \Pr\left(Y_n - R_n \le y_n \cap |R_n| \le \frac{1}{n^{\frac{1}{2}+\alpha}}\right) + o\left(\frac{1}{\sqrt{n}}\right)$$

$$\le \Pr\left(Y_n \le y_n + \frac{1}{n^{\frac{1}{2}+\alpha}}\right) + o\left(\frac{1}{\sqrt{n}}\right)$$

$$\le \Pr(Y_n \le y) + o\left(\frac{1}{\sqrt{n}}\right).$$

Hence,

$$\Pr\left(\hat{Y}_n \le y - \frac{1}{n^{\frac{1}{2}+\alpha}}\right) + o\left(\frac{1}{\sqrt{n}}\right) \le \Pr(Y_n \le y) \le \Pr\left(\hat{Y}_n \le y + \frac{1}{n^{\frac{1}{2}+\alpha}}\right) + o\left(\frac{1}{\sqrt{n}}\right).$$

The result now follows from Lemma 14.2.   ∎

The following corollary extends the result given in Theorem 14.5 to the case in which the underlying random variables do not have mean 0 and standard deviation 1. The proof is straightforward and, hence, is left as an exercise.

Corollary 14.1. *Let $W_1, W_2, \ldots$ denote independent identically distributed random variables, each with mean μ and standard deviation σ. Assume that the distribution of $(W_1 - \mu)/\sigma$ is absolutely continuous and satisfies the conditions of Theorem 14.2; in addition, assume $E(W_1^4) < \infty$. Let $\bar{W}_n = \sum_{j=1}^{n} W_j/n$ and let $X_n = f(\bar{W}_n)$, where f is a three-times differentiable function satisfying the following conditions:*

(i) $|f'(\mu)| > 0$

(ii) There exists a $\delta > 0$ such that

$$M \equiv \sup_{|x-\mu|<\delta} |f'''(x)| < \infty.$$

Let

$$Y_n = \frac{\sqrt{n}(X_n - f(\mu))}{\sigma f'(\mu)}$$

and let F_n denote the distribution function of Y_n. Then

$$F_n(y) = \Phi(y - c/\sqrt{n}) - \frac{\kappa_3/\sigma^3 + 6c}{6\sqrt{n}} H_2(y - c/\sqrt{n})\phi(y - c/\sqrt{n}) + o\left(\frac{1}{\sqrt{n}}\right)$$

where $c = \sigma f''(0)/[2f'(0)]$ and κ_3 denotes the third cumulant of W_1.

It is useful to note that the approximation given in Corollary 14.1 is identical to the one obtained by the following informal method. Using a Taylor's series expansion,

$$X_n - f(\mu) = f'(\mu)(\bar{W}_n - \mu) + \frac{1}{2}f''(\mu)(\bar{W}_n - \mu)^2 + \cdots.$$

Hence,

$$Y_n = \frac{\sqrt{n}(X_n - f(0))}{\sigma f'(0)} = \sqrt{n}(\bar{W}_n - \mu)/\sigma + \frac{\sigma}{2}\frac{f''(0)}{f'(0)}[\sqrt{n}(\bar{W}_n - \mu)/\sigma]^2 \frac{1}{\sqrt{n}} + \cdots.$$

Consider the truncated expansion

$$\sqrt{n}(\bar{W}_n - \mu)/\sigma + \frac{\sigma}{2}\frac{f''(0)}{f'(0)}[\sqrt{n}(\bar{W}_n - \mu)/\sigma]^2 \frac{1}{\sqrt{n}}.$$

Neglecting terms of order $O(n^{-1})$, the first three cumulants of this random variable are

$$\sigma\frac{f''(0)}{2f'(0)}\frac{1}{\sqrt{n}}, \quad 1, \quad \text{and} \quad \frac{\kappa_3/\sigma^3 + 3\sigma f''(0)/f'(0)}{\sqrt{n}},$$

respectively. Using these cumulants in an Edgeworth expansion leads directly to the result given in Corollary 14.1.

Hence, this approach yields the correct approximation, even though the cumulants of the truncated expansion are not necessarily approximations to the cumulants of the distribution of Y_n; in fact, the cumulants of the distribution of Y_n may not exist.

***Example* 14.11 (*Estimator of the rate parameter of an exponential distribution*).** Let $W_1, \ldots, W_n$ denote independent random variables each with a standard exponential distribution. This distribution has cumulant-generating function $-\log(1 - t)$ for $|t| < 1$; the first three cumulants are therefore 1, 1, and 2, respectively. Let $\bar{W}_n$ denote the sample mean of the W_j and let $X_n = 1/\bar{W}_n$; this statistic may be viewed as an estimator of the rate parameter of the underlying exponential distribution.

Here $X_n = f(\bar{W}_n)$, where $f(t) = 1/t$. Hence, $f'(\mu) = -1$

$$c = \sigma\frac{f''(\mu)}{2f'(\mu)} = -\frac{1}{\mu} = -1.$$

Since $\kappa_3 = 2$, it follows that the distribution function of $Y_n = \sqrt{n}(1 - 1/W_n)$ may be expanded

$$\Phi(y + 1/\sqrt{n}) + \frac{2}{3\sqrt{n}}H_2(y + 1/\sqrt{n})\phi(y + 1/\sqrt{n}) + o\left(\frac{1}{\sqrt{n}}\right). \qquad \square$$

14.5 Approximation of Moments

Stochastic asymptotic expansions can also be used to approximate the moments of a random variable. Suppose that $Y_1, Y_2, \ldots$ and $\hat{Y}_1, \hat{Y}_2, \ldots$ are two sequences of real-valued random variables such that

$$Y_n = \hat{Y}_n + R_n/n^\alpha, \quad n = 1, 2, \ldots$$

for some random variables $R_1, R_2, \ldots$. Then, under some conditions, we can approximate the expected value of Y_n by the expected value of $\hat{Y}_n$. The following lemma gives a basic result of this type that is often used in this context; the proof is left as an exercise.

***Lemma* 14.3.** *Suppose that* $Y_n = \hat{Y}_n + R_n/n^\alpha$, *where* $\hat{Y}_n$ *and* R_n *are* $O_p(1)$, *such that* $E(|R_n|) < \infty$, *and* $\alpha > 0$.

If either $E(|Y_n|) < \infty$ *or* $E(|\hat{Y}_n|) < \infty$, *then*

$$E(Y_n) = E(\hat{Y}_n) + E(R_n)/n^\alpha.$$

One commonly used application of this idea is in a variation of the δ-method. Suppose that $\bar{Y}_n$ is a sample mean based on n observations and f is a smooth function. Then, using a Taylor's series expansion,

$$f(\bar{Y}_n) = f(\mu) + f'(\mu)(\bar{Y}_n - \mu) + \frac{1}{2}f''(\mu)(\bar{Y}_n - \mu)^2 + \cdots,$$

where $\mu = E(\bar{Y}_n)$. Then $E[f(\bar{Y}_n)]$ can be approximated by the expected value of the leading terms in the expansion of $f(\bar{Y}_n)$.

The difficulty in applying this idea is in controlling the order of the remainder term in the expansion. Here we give two results of this type. The first puts strong conditions on the function f, specifically that the fourth derivative of f is bounded, but weak conditions on the distribution of the underlying random variables. The second puts weak conditions on the function, but strong conditions on the distribution of the random variables.

Theorem 14.6. *Let* $Y_1, Y_2, \ldots$ *denote independent, identically distributed, real-valued random variables, each with range* $\mathcal{Y}$. *Assume that the distribution of* Y_1 *has mean* μ, *variance* σ^2, *third cumulant* κ_3, *and fourth cumulant* κ_4. *Let* $\bar{Y}_n = \sum_{j=1}^n Y_j/n$ *and consider* $f(\bar{Y}_n)$, *where* f *is a real-valued four-times differentiable function on* $\mathcal{Y}_0$, *a convex subset of* $\mathbf{R}$ *containing* $\mathcal{Y}$ *such that*

$$\sup_{y \in \mathcal{Y}_0} |f^{(4)}(y)| < \infty.$$

Then

$$E[f(\bar{Y}_n)] = f(\mu) + \frac{1}{2}f''(\mu)\frac{\sigma^2}{n} + O\left(\frac{1}{n^2}\right) \quad as \ \ n \to \infty.$$

Proof.   By Taylor's theorem,

$$f(\bar{Y}_n) = f(\mu) + f'(\mu)(\bar{Y}_n - \mu) + \frac{1}{2}f''(\mu)(\bar{Y} - \mu)^2$$
$$+ \frac{1}{6}f'''(\mu)(\bar{Y} - \mu)^3 + \frac{1}{24}f^{(4)}(m_n)(\bar{Y} - \mu)^4$$

where $|m_n - \mu| \leq |\bar{Y}_n - \mu|, n = 1, 2, \ldots$. Note that $E(\bar{Y}_n - \mu) = 0$,

$$E[(\bar{Y}_n - \mu)^2] = \frac{\sigma^2}{n} \quad \text{and} \quad E[(\bar{Y}_n - \mu)^3] = \frac{\kappa_3}{n^2}.$$

Using the fact that $f^{(4)}$ is bounded,

$$E\{|f^{(4)}(m_n)|(\bar{Y} - \mu)^4\} \leq M \, E\{(\bar{Y} - \mu)^4\}$$

for some constant M, so that

$$E\{|f^{(4)}(m_n)|(\bar{Y} - \mu)^4\} = O\left(\frac{1}{n^2}\right) \quad as \ \ n \to \infty.$$

The result now follows from Lemma 14.3.   ∎

***Example* 14.12 (*Function of a Poisson mean*).** Let $X_1, X_2, \ldots$ denote independent, identically distributed random variables, each with a Poisson distribution with mean λ. Consider the function $\exp(-\bar{X}_n)$, where $\bar{X}_n = \sum_{j=1}^n X_j/n$; this function can be interpreted as an estimator of $\Pr(X_1 = 0)$.

Let $f(t) = \exp(-t)$; clearly, f is four-times differentiable and $f^{(4)}(t) = \exp(-t)$ is bounded for $t \geq 0$. It follows from Theorem 14.6 that

$$\mathrm{E}[\exp(-\bar{X}_n)] = \exp(-\lambda) + \frac{1}{2}\exp(-\lambda)\frac{\lambda}{n} + O\left(\frac{1}{n^2}\right) \quad \text{as } n \to \infty. \qquad \square$$

***Example* 14.13 (*Function of a normal mean*).** Let $Y_1, Y_2, \ldots$ denote independent, identically distributed random variables, each with a standard normal distribution. Consider approximation of $\mathrm{E}[\Phi(\bar{Y}_n)]$, where $\bar{Y}_n = \sum_{j=1}^n Y_j/n$ and $\Phi(\cdot)$ denotes the standard normal distribution function.

Let $f(t) = \Phi(t)$. Then $f''(t) = -t\phi(t)$ and $f^{(4)}(t) = (t^3 - 3t)\phi(t)$; here ϕ denotes the standard normal density function. It is straightforward to show that $|f^{(4)}(t)|$ is bounded for $t \in \mathbf{R}$. Hence, it follows from Theorem 14.6 that

$$\mathrm{E}[\Phi(\bar{Y}_n)] = \frac{1}{2} + O\left(\frac{1}{n^2}\right). \qquad \square$$

As noted above, the conditions required in Theorem 14.6 are very strong; for instance, they are much stronger than the conditions required by the higher-order version of the δ-method. The main reason for this is that approximating an expected value, which can be greatly influenced by the tails of the distribution, is a much more difficult problem than approximating a probability.

The conditions on the function f can be weakened, provided that stronger conditions are placed on the distribution of the underlying random variables. The following theorem gives one example of this type of result.

***Theorem* 14.7.** *Let $Y_1, Y_2, \ldots$ denote independent, identically distributed, real-valued random variables, each with range $\mathcal{Y}$. Assume that moment-generating function of the distribution of Y_1 exists.*

Let $\bar{Y}_n = \sum_1^n Y_j/n$ and consider $f(\bar{Y}_n)$, where f is a real-valued four-times differentiable function on $\mathcal{Y}_0$, a convex subset of $\mathbf{R}$ containing $\mathcal{Y}$ such that

$$|f^{(4)}(y)| < \alpha \exp(\beta|y|), \quad y \in \mathcal{Y}_0$$

for some constants α and β.

Then

$$\mathrm{E}[f(\bar{Y}_n)] = f(\mu) + \frac{1}{2}f''(\mu)\frac{\sigma^2}{n} + O\left(\frac{1}{n^2}\right) \quad \text{as } n \to \infty,$$

where $\mu = \mathrm{E}(Y_1)$ and $\sigma^2 = \mathrm{Var}(Y_1)$.

Proof. The proof follows the same general argument used in the proof of Theorem 14.6; hence, the result holds provided that

$$\mathrm{E}\{f^{(4)}(m_n)[\sqrt{n}(\bar{Y}_n - \mu)]^4\}$$

is $O(1)$ as $n \to \infty$, where m_n is a random point on the line segment connecting μ and $\bar{Y}_n$.

Note that, using the Cauchy-Schwarz inequality,

$$\left| E\{ f^{(4)}(m_n)[\sqrt{n}(\bar{Y}_n - \mu)]^4 \} \right| \le E\{ |f^{(4)}(m_n)|[\sqrt{n}(\bar{Y}_n - \mu)]^4 \}$$

$$\le E\{ |f^{(4)}(m_n)|^2 \}^{\frac{1}{2}} E\{ [\sqrt{n}(\bar{Y}_n - \mu)]^8 \}^{\frac{1}{2}}.$$

By the condition in the theorem,

$$|f^{(4)}(m_n)| \le \alpha \, \exp\{\beta |m_n|\}, \quad n = 1, 2, \ldots.$$

Note that $|m_n| \le |\mu| + |\bar{Y}_n|$. Hence,

$$E\big[|f^{(4)}(m_n)|^2 \big] \le \alpha^2 \, \exp\{2\beta |\mu|\} E[\exp\{2\beta |\bar{Y}_n|\}]$$

and, since

$$|\bar{Y}_n| \le \frac{1}{n} \sum_{j=1}^{n} |Y_j|,$$

$$E\big[|f^{(4)}(m_n)|^2 \big] \le \alpha^2 \, \exp\{2\beta |\mu|\} E[\exp\{2\beta |Y_1|/n\}]^n = \alpha^2 \, \exp\{2\beta |\mu|\} M_{|Y_1|}(2\beta/n)^n,$$

where $M_{|Y_1|}$ denotes the moment-generating function of $|Y_1|$. Since

$$n \log M_{|Y_1|}(2\beta/n) = n \left[E(|X_1|) \frac{2\beta}{n} + o\!\left(\frac{1}{n}\right) \right],$$

$$\lim_{n \to \infty} M_{|Y_1|}(2\beta/n)^n = \exp\{2\beta E(|X_1|)\}.$$

By Theorem 4.18,

$$E\{ [\sqrt{n}(\bar{Y}_n - \mu)]^8 \} = O\!\left(\frac{1}{n^4}\right) \quad \text{as} \ \ n \to \infty.$$

It follows that

$$\left| E\{ f^{(4)}(m_n)[\sqrt{n}(\bar{Y}_n - \mu)]^4 \} \right| = O\!\left(\frac{1}{n^2}\right)$$

proving the result. ∎

Example 14.14 (*Function of a geometric mean*). Let $X_1, X_2, \ldots$ denote independent, identically distributed random variables, each with the discrete distribution with frequency function

$$p(x) = \left(\frac{1}{2}\right)^x, \quad x = 1, 2, \ldots;$$

this is a geometric distribution. This distribution has mean 2 and variance 2. Consider the function $\exp(\sqrt{\bar{X}_n})$, where $\bar{X}_n = \sum_{j=1}^{n} X_j/n$.

Let $f(t) = \exp(\sqrt{t})$; f is four-times differentiable with

$$f''(t) = \frac{1}{4}(t^{-1} - t^{-\frac{1}{2}}) \exp(\sqrt{t});$$

the fourth derivative of f is of the form $g(t) \exp(\sqrt{t})$, where g is a polynomial in $1/\sqrt{t}$. Hence, the conditions of Theorem 14.7 are satisfied and

$$E[\exp(\sqrt{\bar{X}_n})] = \exp(\sqrt{2}) - \frac{1}{8}(\sqrt{2} - 1)\exp(\sqrt{2})\frac{1}{n} + O\left(\frac{1}{n^2}\right) \quad \text{as} \quad n \to \infty. \qquad \square$$

***Example* 14.15 (*Power of the mean of exponential random variables*).** Let $Y_1, Y_2, \ldots,$ Y_n denote independent, identically distributed, standard exponential random variables; it follows that $E(Y_1) = \mathrm{Var}(Y_1) = 1$. Consider $E[\bar{Y}_n^r]$ for some $r > 0$. To obtain an approximation to this quantity, we can apply Theorem 14.7 with $f(t) = t^r$. Note that $f''(t) = r(r-1)t^{r-2}$ and

$$f^{(4)}(t) = r(r-1)(r-2)(r-3)t^{r-4}.$$

Hence, in order for the conditions of Theorem 14.7 to be satisfied, we must assume that $r \geq 4$. It follows that, for $r \geq 4$,

$$E\left[\bar{Y}_n^r\right] = 1\frac{1}{2}r(r-1)\frac{1}{n} + O\left(\frac{1}{n^2}\right) \quad \text{as} \quad n \to \infty.$$

In this example, exact computation of $E(\bar{Y}_n^r)$ is possible. Recall that $\sum_{j=1}^n Y_j$ has a standard gamma distribution with index n; see Example 7.15. Hence,

$$E(\bar{Y}_n^r) = \frac{1}{n^r}E\left[\left(\sum_{j=1}^n Y_j\right)^r\right] = \frac{1}{n^r}\int_0^\infty t^r \frac{1}{\Gamma(n)}t^{n-1}\exp(-t)\,dt = \frac{1}{n^r}\frac{\Gamma(n+r)}{\Gamma(n)}.$$

Hence,

$$\frac{\Gamma(n)}{\Gamma(n+r)} = \frac{1}{n^r}\left[1 + \frac{1}{2}r(r-1)\frac{1}{n} + O\left(\frac{1}{n^2}\right)\right]^{-1}$$

$$= \frac{1}{n^r}\left[1 - \frac{1}{2}r(r-1)\frac{1}{n} + O\left(\frac{1}{n^2}\right)\right],$$

as $n \to \infty$, which is in agreement with the result in Example 9.5. $\quad\square$

Note that, by Example 9.5, the expansion for $E[\bar{Y}_n^r]$ given in Example 14.15 continues to hold for $0 < r < 4$. The reason that Theorem 14.7 does not apply in this case is that the theorem is designed to handle the case in which $|f^{(4)}(x)|$ is large for large $|x|$, while for $0 < r < 4$, $|f^{(4)}(x)|$ is large for x near 0. It would not be difficult to give a result analogous to that given in Theorem 14.7 that is designed to handle this situation.

Functions of a random vector

We now consider the case in which $Y_1, Y_2, \ldots$ are d-dimensional random vectors and f is a real-valued function defined on a subset of $\mathbf{R}^d$. In order to generalize the results presented for the case in which the Y_j are real-valued, we must consider derivatives of $f(y)$ with respect to y. For simplicity, we will denote these derivatives by $f'(y)$, $f''(y)$, and so on, as in the case in which y is a scalar. Hence, $f'(y)$ is a vector, $f''(y)$ is a matrix, $f'''(y)$ is a three-dimensional array, and so forth. We will use subscripts to denote the elements of these arrays, so that, for example, $f'''(y)$ has elements $f'''_{ijk}(y)$ for $i, j, k = 1, \ldots, d$. For a

q-dimensional array C define

$$C[t, \ldots, t] = \sum_{i_1, \ldots, i_q} C_{i_1 \cdots i_q} t_{i_1} \cdots t_{i_q}.$$

Here we present a generalization of Theorem 14.6, which provides an expansion for $E[f(\bar{Y}_n)]$ under the assumption that the fourth derivative of f is bounded.

Theorem 14.8. *Let $Y_1, Y_2, \ldots$ denote independent, identically distributed random vectors. Assume that the first four cumulants of the distribution of Y_1 exist and are finite; let $\mu = E(Y_1)$ and $\Sigma = \mathrm{Cov}(Y_1)$. Let $\bar{Y}_n = \sum_{j=1}^{n} Y_j / n$ and consider $f(\bar{Y}_n)$, where $f : \mathcal{Y}_0 \to \mathbf{R}$ is a four-times differentiable function, where $\mathcal{Y}_0$ is a convex subset of $\mathbf{R}^d$ for some $d = 1, 2 \ldots$ such that $Pr(Y_1 \in \mathcal{Y}_0) = 1$. Assume that*

$$\sup_{i,j,k,\ell=1,\ldots,d} \sup_{y \in \mathcal{Y}_0} |f_{ijk\ell}^{(4)}(y)| < \infty.$$

Then

$$E[f(\bar{Y}_n)] = f(\mu) + \frac{1}{2}\mathrm{tr}\{\Sigma f''(\mu)\}\frac{1}{n} + O\left(\frac{1}{n^2}\right) \quad as \ n \to \infty.$$

Proof. By Taylor's theorem,

$$f(\bar{Y}_n) = f(\mu) + f'(\mu)[\bar{Y}_n - \mu] + \frac{1}{2}f''(\mu)[\bar{Y}_n - \mu, \bar{Y}_n - \mu]$$

$$+ \frac{1}{6}f'''(\mu)[\bar{Y}_n - \mu, \bar{Y}_n - \mu, \bar{Y}_n - \mu]$$

$$+ \frac{1}{24}f^{(4)}(m_n)[\bar{Y}_n - \mu, \bar{Y}_n - \mu, \bar{Y}_n - \mu, \bar{Y}_n - \mu]$$

where, for each $n = 1, 2, \ldots$, there exists a $t \equiv t_n$ such that $m_n = t\mu + (1 - t)\bar{Y}_n$.

The proof now follows along the same lines as the proof of Theorem 14.6, using the facts that $E\{f'(\mu)[\bar{Y}_n - \mu]\} = 0$,

$$E\{f''(\mu)[\bar{Y}_n - \mu, \bar{Y}_n - \mu]\} = \frac{1}{n}\mathrm{tr}\{\Sigma f''(\mu)\}. \qquad \blacksquare$$

Example 14.16 (Product of squared means). Let $(X_1, W_1), (X_2, W_2), \ldots$ denote a sequence of independent, identically distributed two-dimensional random vectors such that X_1 has mean μ_X and variance σ_X^2, W_1 has mean μ_W and variance σ_W^2, and let ρ denote the correlation of X_1 and W_1. Assume that the first four cumulants of the distribution of (X_1, W_1) exist.

Consider $E(\bar{X}_n^2 \bar{W}_n^2)$, where $\bar{X}_n = \sum_{j=1}^{n} X_j / n$ and $\bar{W}_n = \sum_{j=1}^{n} W_j / n$. Let $Y_j = (X_j, W_j)$, $j = 1, 2, \ldots$; then $E(\bar{X}_n^2 \bar{W}_n^2)$ is of the form $E[f(\bar{Y}_n)]$, $\bar{Y}_n = \sum_{j=1}^{n} Y_j / n$, where $f(t) = t_1^2 t_2^2$, $t = (t_1, t_2) \in \mathbf{R}^2$.

Note that

$$f''(t) = \begin{pmatrix} 2t_2^2 & 4t_1 t_2 \\ 4t_1 t_2 & 2t_1^2 \end{pmatrix}.$$

It is straightforward to show that the fourth derivative of f is bounded. Hence, by Theorem 14.8,

$$
\begin{aligned}
\mathrm{E}(\bar{X}_n^2 \bar{W}_n^2) &= \mu_X^2 \mu_W^2 + \frac{1}{2} \mathrm{tr} \left[\begin{pmatrix} \sigma_X^2 & \rho\sigma_X\sigma_W \\ \rho\sigma_X\sigma_W & \sigma_W^2 \end{pmatrix} \begin{pmatrix} 2\mu_W^2 & 4\mu_X\mu_W \\ 4\mu_X\mu_W & 2\mu_X^2 \end{pmatrix} \right] \frac{1}{n} + O\left(\frac{1}{n^2}\right) \\
&= \mu_X^2 \mu_W^2 + \left(\mu_W^2 \sigma_X^2 + \mu_X^2 \sigma_W^2 + 4\rho\mu_X\mu_W\sigma_X\sigma_W \right)\frac{1}{n} + O\left(\frac{1}{n^2}\right).
\end{aligned}
$$

Since

$$
\mathrm{E}(\bar{X}_n^2) = \mu_X^2 + \frac{\sigma_X^2}{n}
$$

and

$$
\mathrm{E}(\bar{W}_n^2) = \mu_W^2 + \frac{\sigma_W^2}{n},
$$

it follows that

$$
\mathrm{Cov}(\bar{X}_n^2, \bar{W}_n^2) = 4\rho\mu_X\mu_W\sigma_X\sigma_W \frac{1}{n} + O\left(\frac{1}{n^2}\right). \qquad \square
$$

14.6 Exercises

14.1 Let $X_1, X_2, \ldots$ denote independent, identically distributed random variables, each with an absolutely continuous distribution with density function

$$
\frac{\sqrt{\theta_1}}{\sqrt{(2\pi)}} \exp[\sqrt{(\theta_1\theta_2)}]x^{-\frac{3}{2}} \exp\left\{ -\frac{1}{2}(\theta_1 x + \theta_2/x) \right\}, \quad x > 0
$$

where $\theta_1 > 0$ and $\theta_2 > 0$. This is an inverse Gaussian distribution with parameters θ_1 and θ_2. Find the Edgeworth series approximations with error $o(1/\sqrt{n})$ to the density function and distribution function of $\sum_{j=1}^n X_j/n$ for the case $\theta_1 = \theta_2 = 1$.

14.2 Let $X_1, X_2, \ldots$ denote independent, identically distributed random variables, each uniformly distributed on the interval $(0, 1)$. Find the Edgeworth series approximations with error $o(1/n)$ to the density function and distribution function of $\sum_{j=1}^n X_j/n$.

14.3 Let $X_1, X_2, \ldots$ denote independent, identically distributed random variables, each with an absolutely continuous distribution, and assume that $\mathrm{E}(|X_1|^3) < \infty$. Let $\hat{F}_n$ denote the Edgeworth series approximations with error $o(1/\sqrt{n})$ to the distribution function of $\bar{X}_n = \sum_{j=1}^n X_j/n$.

For each $n = 1, 2, \ldots$, let $Y_j = a + bX_j$, for some constants $a, b, b > 0$, and consider two approaches to approximating the distribution function of $\bar{Y}_n = \sum_{j=1}^n Y_j/n$. One approach is to use Edgeworth series approximations with error $o(1/\sqrt{n})$, as described in Theorem 14.2. Another approach is to approximate the distribution function of $\bar{X}_n$ by $\hat{F}_n$ and then use the relationship between the distribution function of $\bar{Y}_n$ and the distribution function of $\bar{X}_n$ resulting from the fact that $\bar{Y}_n = a + b\bar{X}_n$. Are these two approximations the same? If not, which approximation would you expect to be more accurate?

14.4 The Edgeworth expansion of the density of a random variable has the undesirable feature that, for some values of the argument, the approximation given by the Edgeworth expansion may be negative. To avoid this problem, it is sometimes useful to approximate a density function by expanding the log of the density and then exponentiating the result to obtain an approximation for the density itself. Determine an Edgeworth expansion for the log-density with error $o(1/\sqrt{n})$.

14.5 Consider a random variable with mean 0 and standard deviation 1. Suppose the density of this random variable is approximated by an Edgeworth expansion with error $o(1/\sqrt{n})$. Using this approximation as a genuine density function, find approximations for the median and mode of the distribution.

14.6 Let $X_1, X_2, \ldots$ denote independent, identically distriubuted random variables with mean 0 and standard deviation 1. Let F_n denote the distribution function of $\sum_{j=1}^{n} X_j/\sqrt{n}$ and assume that there is a Edgeworth series approximation to F_n. Show that

$$\Pr\left\{-a < \sum_{j=1}^{n} X_j/\sqrt{n} < a\right\} = F_n(a) - F_n(-a)$$

$$= \Phi(a) - \Phi(-a) + o\left(\frac{1}{\sqrt{n}}\right) \quad \text{as } n \to \infty.$$

That is, for approximating the probability that $\sum_{j=1}^{n} X_j/\sqrt{n}$ lies in an interval that is symmetric about 0, the normal approximation given by the central limit theorem has error $o(1/\sqrt{n})$.

14.7 Let Y_n denote a statistic with a density function that has an Edgeworth series approximation of the form

$$\phi(y)\left[1 + \frac{\kappa_3}{6} H_3(y)\frac{1}{\sqrt{n}} + \left(\frac{\kappa_4}{24} H_4(y) + \frac{\kappa_3^2}{72} H_6(y)\right)\frac{1}{n} + \cdots\right].$$

Let $X_n = Y_n^2$. Find an Edgeworth series approximation to the density function of X_n by using the approximation to the density function of Y_n, together with the change-of-variable formula given in Chapter 7. Relate the terms in the approximation to the chi-squared distribution.

14.8 Let $X_1, X_2, \ldots$ denote independent, identically distributed random variables, each with an absolutely continuous distribution with density function

$$\frac{1}{\sqrt{(2\pi)}} \exp(1)x^{-\frac{3}{2}} \exp\left\{-\frac{1}{2}(x + 1/x)\right\}, \quad x > 0;$$

see Exercise 14.1. Find approximations with error $o(1/\sqrt{n})$ to the quantiles of the distribution of $\sum_{j=1}^{n} X_j/n$.

14.9 In the proof of Theorem 14.1, it is stated that the Fourier transform of

$$\hat{p}_n(x) = \phi(x) + \frac{\kappa_3}{6\sqrt{n}} H_3(x)\phi(x)$$

is

$$\exp\left\{-\frac{1}{2}t^2\right\}\left[1 + \frac{\kappa_3}{6\sqrt{n}}(it)^3\right].$$

Prove this fact.

14.10 Let $X_1, X_2, \ldots$ denote independent, identically distributed random variables, each normally distributed with mean μ and standard deviation σ. Find the saddlepoint approximation to the density function of $\sum_{j=1}^{n} X_j/n$.

14.11 Let $X_1, X_2, \ldots$ denote independent, identically distributed random variables, each with an absolutely continuous distribution with density function

$$\frac{1}{\sqrt{(2\pi)}} \exp(1)x^{-\frac{3}{2}} \exp\left\{-\frac{1}{2}(x + 1/x)\right\}, \quad x > 0;$$

see Exercise 14.1.

Find the saddlepoint approximation to the density function of $\sum_{j=1}^{n} X_j/n$.

14.12 Let $Y_1, Y_2, \ldots$ denote independent, identically distributed random variables, each with an absolutely continuous distribution with density function

$$\frac{1}{2} \exp\{-|y|\}, \quad -\infty < y < \infty.$$

Find an approximation to $\Pr\{\sum_{j=1}^{5} Y_j \geq 4\}$ based on the saddlepoint approximation. Compare the result to the approximation based on the central limit theorem.

14.13 Let $X_1, X_2, \ldots$ denote independent, identically distributed random variables, each with a distribution satisfying the conditions of Theorem 14.2. Let $\bar{X}_n = \sum_{j=1}^{n} X_j/n$. For a given constant b let

$$Y_n = \bar{X}_n + b\bar{X}_n^2.$$

Find b so that the distribution of $\sqrt{n} Y_n$ is standard normal, with error $o(1/\sqrt{n})$.

14.14 Prove Corollary 14.1.

14.15 Let $Y_1, Y_2, \ldots$ denote independent, identically distributed random variables, each with a chi-squared distribution with 1 degree of freedom. Using Theorem 14.5, find an approximation to the distribution function of $[\sum_{j=1}^{n} Y_j/n]^{\frac{1}{3}}$, suitably normalized. Based on this result, give an approximation to $\Pr(\chi_n^2 \leq t)$, where χ_n^2 denotes a random variable with a chi-squared distribution with n degrees of freedom.

14.16 Prove Lemma 14.3.

14.17 Let $Y_1, Y_2, \ldots$ denote independent, identically distributed random variables, each with density

$$\alpha y^{\alpha-1}, \quad 0 < y < 1$$

where $\alpha > 0$. Find an approximation to $\mathrm{E}[(1 + \bar{Y}_n)^{-1}]$, where $\bar{Y}_n = \sum_{j=1}^{n} Y_j/n$.

14.18 Let $Y_1, Y_2, \ldots$ denote independent, identically distributed, real-valued random variables, each with range $\mathcal{Y}$. Assume that the distribution of Y_1 has mean μ, variance σ^2, third cumulant κ_3, and fourth cumulant κ_4. Let $\bar{Y}_n = \sum_{j=1}^{n} Y_j/n$. Find expansions for

$$\mathrm{E}\left\{ f\left(\bar{Y}_n - \mu \right) \right\}$$

and

$$\mathrm{E}\{ f(\frac{\bar{Y}_n - \mu}{\sigma}) \}$$

where f is a real-valued four-times differentiable function on $\mathcal{Y}_0$, a convex subset of $\mathbf{R}$ containing $\mathcal{Y}$, such that

$$\sup_{y \in \mathcal{Y}_0} |f^{(4)}(y)| < \infty.$$

14.19 Let $Y_1, Y_2, \ldots$ denote independent, identically distributed, real-valued random variables. Assume that the distribution of Y_1 has mean μ, variance σ^2, third cumulant κ_3. Let $\bar{Y}_n = \sum_{j=1}^{n} Y_j/n$ and let

$$f(y) = ay^3 + by^2 + cy + d$$

for some constants a, b, c, d. Find an exact expression for $\mathrm{E}\{f(\bar{Y}_n)\}$ and compare it to the approximation given in Theorem 14.6.

14.20 Let $Y_1, Y_2, \ldots$ denote independent, identically distributed, real-valued random variables, each with range $\mathcal{Y}$. Assume that moment-generating function of the distribution of Y_1 exists. Let $\bar{Y}_n = \sum_{1}^{n} Y_j/n$ and consider $f(\bar{Y}_n)$, where f is a real-valued four-times differentiable function on $\mathcal{Y}_0$, a convex subset of $\mathbf{R}$ containing $\mathcal{Y}$. Assume that

$$|f^{(4)}(y)| < \alpha \, \exp(\beta|y|), \quad y \in \mathcal{Y}_0$$

for some constants α and β. Find an expansion for $\mathrm{Var}\{f(\bar{Y}_n)\}$.

14.7   Suggestions for Further Reading

Wallace (1958) gives a concise overview of asymptotic approximations to probability distributions and Ghosh (1994) discusses the use of higher-order asymptotic theory in statistical inference.

Edgeworth series approximations are discussed in Feller (1971, Chapter XVI) and Kolassa (1997, Chapter 3); a comprehensive treatment is given by Bhattacharya and Rao (1976). Edgeworth series approximations also apply to vector-valued random variables. The derivation is essentially the same as in the univariate case. The cumulant-generating function of the normalized sample mean can be expanded up to a specified power of $n^{-\frac{1}{2}}$. The truncated expansion may then be inverted to yield an approximation to the density function of the standardized sample mean. The leading term in the approximation is the multivariate normal density and the later terms are based on higher-order cumulant arrays. See Kolassa (1997, Chapter 6) and McCullagh (1987, Chapter 5) for further details.

Saddlepoint approximations are considered in detail in Jensen (1995) and Kolassa (1997, Chapters 3 and 4); see also Barndorff-Nielsen and Cox (1989) and Daniels (1954). Reid (1988) surveys the use of saddlepoint methods in statistical inference and Field and Ronchetti (1990) consider the application of saddlepoint methods to the problem of approximating the distribution of statistics more general than sample means. The material in Section 14.4 on stochastic asymptotic expansions is based on Hall (1992, Chapter 2); see also Barndorff-Nielsen and Cox (1989, Chapter 3) and Skovgaard (1981).

Appendix 1

Integration with Respect to a
Distribution Function

A1.1 Introduction

In this appendix we consider integration with respect to a distribution function, a concept which plays a central role in distribution theory. Let F denote a distribution function on $\mathbf{R}^d$; the integral of a function g with respect to F is denoted by

$$\int_{\mathbf{R}^d} g(x)\, dF(x). \tag{A1.1}$$

Our goal is to define and describe the properties of integrals of the form (A1.1).

One approach is to use results from measure theory and the general theory of integration with respect to a measure. The distribution function F defines a measure μ on $\mathbf{R}^d$. The integral (A1.1) may then be written as

$$\int g(x)\mu(dx) \quad \text{or} \quad \int g\, d\mu$$

and the properties of these integrals follow from standard results in the theory of measure and integration. See, for example, Ash (1972) or Billingsley (1995).

The purpose of this appendix is to present a brief summary of this theory for those readers who have not studied measure theory.

Consider a distribution function $F : \mathbf{R}^d \to [0, 1]$ and let $g, g_1, g_2, \ldots$ denote real-valued functions on $\mathbf{R}^d$. Let X denote a random variable with distribution function F. There are several properties that any definition of integration with respect to a distribution function should satisfy:

I1. *Integration of indicator functions*

Let A denote a subset of $\mathbf{R}^d$ and let $\mathrm{I}_{\{x \in A\}}$ denote the indicator function of A. Then

$$\int_{\mathbf{R}^d} \mathrm{I}_{\{x \in A\}}\, dF(x) \equiv \int_A dF(x) = \Pr(X \in A).$$

I2. *Linearity*

Let a_1, a_2 denote constants. Then

$$\int_{\mathbf{R}^d} [a_1 g_1(x) + a_2 g_2(x)]\, dF(x) = a_1 \int_{\mathbf{R}^d} g_1(x)\, dF(x) + a_2 \int_{\mathbf{R}^d} g_2(x)\, dF(x).$$

I3. *Nonnegativity*

If $g \geq 0$, then

$$\int_{\mathbf{R}^d} g(x) \, dF(x) \geq 0.$$

I4. *Continuity*

Suppose that, for each $x \in \mathbf{R}^d$, the sequence $g_1(x), g_2(x), \ldots$ is an increasing sequence with limit $g(x)$. Then

$$\int_{\mathbf{R}^d} g(x) \, dF(x) = \lim_{n \to \infty} \int_{\mathbf{R}^d} g_n(x) \, dF(x),$$

if the limit of the integrals exists and

$$\int_{\mathbf{R}^d} g(x) \, dF(x) = \infty$$

otherwise.

These requirements may be used to construct a general definition of integration.

A1.2   A General Definition of Integration

Define a *simple function* to be a function of the form

$$g(x) = \sum_{i=1}^{m} \alpha(g)_i I_{\{x \in A_i\}}$$

where $\alpha(g)_1, \alpha(g)_2, \ldots, \alpha(g)_m$ are given real numbers, $A_1, A_2, \ldots, A_m$ are disjoint subsets of $\mathbf{R}^d$, and m is a given positive integer.

By (I1) and (I2), if g is a simple function then

$$\int_{\mathbf{R}^d} g(x) \, dF(x) = \sum_{i=1}^{m} \alpha(g)_i \Pr(X \in A_i).$$

Furthermore, if $g_1, g_2, \ldots$ is an increasing sequence of simple functions such that, for each x, $g_1(x), g_2(x), \ldots$ converges to $g(x)$, then, by (I4), the integral of g is given by

$$\int_{\mathbf{R}^d} g(x) \, dF(x) = \lim_{n \to \infty} \int_{\mathbf{R}^d} g_n(x) \, dF(x),$$

provided that the limit exists.

Hence, if a nonnegative function g may be written as a limit of simple functions in this manner then the value of the integral

$$\int_{\mathbf{R}^d} g(x) \, dF(x)$$

may be determined, although that value may be ∞. It may be shown that the value of such an integral is unique: if g may be written as the limit of two different sequences of simple functions, then those two sequences must lead to the same value for the integral. Call a nonnegative function *integrable* if it may be written as the limit of simple functions in this manner and the value of its integral is finite. Thus, the integral of any nonnegative integrable function is well-defined and finite.

If g is a general function, not necessarily nonnegative, we may write $g = g^+ - g^-$ where g^+ and g^- are nonnegative functions, called the positive and negative parts of g, respectively. By the linearity condition on the integral,

$$\int_{\mathbf{R}^d} g(x)\,dF(x) = \int_{\mathbf{R}^d} g^+(x)\,dF(x) - \int_{\mathbf{R}^d} g^-(x)\,dF(x).$$

We will say that g is integrable if its positive and negative parts are integrable. The integral of a integrable function g is given by

$$\int_{\mathbf{R}^d} g(x)\,dF(x) = \int_{\mathbf{R}^d} g^+(x)\,dF(x) - \int_{\mathbf{R}^d} g^-(x)\,dF(x).$$

It is important to note that if a function g is not integrable, it does not mean that its integral does not exist. The integral of g exists provided that *either* g^+ or g^- is integrable. The function g is integrable provided that *both* g^+ and g^- are integrable; since $|g| = g^+ + g^-$, g is integrable provided that $|g|$ is integrable.

Consider the class of functions whose positive and negative parts can be written as the limits of increasing sequences of simple functions; call such a function an *extended simple function*. The integral of an extended simple function, given that it exists, can be determined using the method described above. Clearly for this approach to be useful the class of extended simple functions must be sufficiently broad.

There is a close connection between extended simple functions and the measurable sets discussed in Section 1.2. In particular, a function $g : \mathbf{R}^d \to \mathbf{R}$ is an extended simple function provided that, for each measurable subset of $\mathbf{R}$, A, the set given by

$$\{x \in \mathbf{R}^d : g(x) \in A\}$$

is measurable; such a function is said to be *measurable*. As in the case of measurable sets, nearly every function of practical interest is measurable and we will proceed as if all functions are measurable.

The general integral described above has a number of useful properties. Several of these are given below without proof; for further details, see, for example, Ash (1972), Billingsley (1995), or Port (1994). In these properties, the term *almost everywhere* (F), written $a.e.\,(F)$, is used to denote a property that holds for all $x \in A$ where A is some set such that the probability of A, under the distribution with distribution function F, is 1.

Some basic properties

Let g_1, g_2 denote functions on $\mathbf{R}^d$.

(i) If $g_1 = 0$ $a.e.\,(F)$ then

$$\int_{\mathbf{R}^d} g(x)\,dF(x) = 0.$$

(ii) If $g_1 = g_2$ $a.e.\,(F)$ and

$$\int_{\mathbf{R}^d} g_1(x)\,dF(x)$$

exists, then

$$\int_{\mathbf{R}^d} g_2(x)\,dF(x)$$

exists and

$$\int_{\mathbf{R}^d} g_1(x)\,dF(x) = \int_{\mathbf{R}^d} g_2(x)\,dF(x).$$

(iii) If $g_1 \geq 0$ and

$$\int_{\mathbf{R}^d} g_1(x)\,dF(x) = 0$$

then $g_1 = 0\ a.e.\ (F)$.

Change-of-variable

Consider a random variable X taking values in $\mathcal{X} \subset \mathbf{R}^d$ and let F_X denote the distribution function of the distribution of X. Let g be a function defined in Let $g : \mathcal{X} \to \mathbf{R}^q$ and let $Y = g(X)$. Let F_Y denote the distribution function of the distribution of Y. Then

$$\int_{\mathbf{R}^d} y\,dF_Y(y) = \int_{\mathbf{R}^d} g(x)\,dF_X(x).$$

A1.3 Convergence Properties

The properties described in this section are all concerned with the following question. Suppose $g_1, g_2, \ldots$ is a sequence of functions defined on $\mathbf{R}^d$. How is the convergence of the sequence of integrals

$$\int_{\mathbf{R}^d} g_n(x)\,dF(x), \quad n = 1, 2, \ldots$$

related to the convergence of the sequence of functions g_n, $n = 1, 2, \ldots$?

Fatou's lemma

If there exists a function g such that $g_j \geq g$, $j = 1, 2, \ldots$, and

$$\int_{\mathbf{R}^d} g(x)\,dF(x) > -\infty,$$

then

$$\liminf_{n \to \infty} \int_{\mathbf{R}^d} g_n(x)\,dF(x) \geq \int_{\mathbf{R}^d} \liminf_{n \to \infty} g_n(x)\,dF(x).$$

If there exists a function g such that $g_n \leq g$, $n = 1, 2, \ldots$, and

$$\int_{\mathbf{R}^d} g(x)\,dF(x) < \infty,$$

then

$$\limsup_{n \to \infty} \int_{\mathbf{R}^d} g_n(x)\,dF(x) \leq \int_{\mathbf{R}^d} \limsup_{n \to \infty} g_n(x)\,dF(x).$$

Dominated convergence theorem

If there exists a function h such that $|g_n| \le h, n = 1, 2, \ldots,$

$$\int_{\mathbf{R}^d} h(x) \, dF(x) < \infty,$$

and

$$\lim_{n \to \infty} g_n(x) = g(x) \quad a.e. \, (F),$$

then

$$\lim_{n \to \infty} \int_{\mathbf{R}^d} g_n(x) \, dF(x) = \int_{\mathbf{R}^d} g(x) \, dF(x).$$

Beppo Levi's theorem

Suppose that $g_1, g_2, \ldots$ is an increasing sequence of functions with limit g such that

$$\sup_n \int_{\mathbf{R}^d} g_n(x) \, dF(x) < \infty.$$

Then

$$\int_{\mathbf{R}^d} g(x) \, dF(x) < \infty$$

and

$$\lim_{n \to \infty} \int_{\mathbf{R}^d} g_n(x) \, dF(x) = \int_{\mathbf{R}^d} g(x) \, dF(x).$$

A1.4 Multiple Integrals

Consider two random variables, X and Y, with ranges $\mathcal{X} \subset \mathbf{R}^d$ and $\mathcal{Y} \subset \mathbf{R}^p$, respectively. Let F denote the distribution function of the distribution function of (X, Y) and let g denote a function defined on $\mathcal{X} \times \mathcal{Y}$. Suppose we are interested in the integral

$$\int_{\mathbf{R}^{p+d}} g(x, y) \, dF(x, y).$$

The following results state that this integral may be computed by first integrating with respect to x and then with respect to y, leading to an iterated integral.

Fubini's theorem

Suppose $F = F_X F_Y$ where F_X is the distribution function of the distribution of X and F_Y is the distribution function of the distribution of Y. Then

$$\int_{\mathbf{R}^{p+d}} g(x, y) \, dF(x, y) = \int_{\mathbf{R}^p} \left[\int_{\mathbf{R}^d} g(x, y) \, dF_X(x) \right] dF_Y(y).$$

Note that the condition that $F = F_X F_Y$ implies that X and Y are independent. The following result is an extension of Fubini's Theorem to the case in which independence does not hold.

Suppose F may be written $F(x, y) = F_X(x; y)F_Y(y)$ for all x, y, where for each $y \in \mathcal{Y}$, $F_X(\cdot; y)$ defines a distribution function on $\mathbf{R}^p$ and F_Y denotes a distribution function on $\mathbf{R}^q$. Then

$$\int_{\mathbf{R}^{p+d}} g(x, y)\, dF(x, y) = \int_{\mathbf{R}^p} \left[\int_{\mathbf{R}^d} g(x, y)\, dF_X(x; y) \right] dF_Y(y).$$

A1.5 Calculation of the Integral

Although the properties of the integral with respect to a distribution function have been discussed in detail, we have not discussed the actual calculation of the integral

$$\int_{\mathbf{R}^d} g(x)\, dF(x)$$

for given choices of g and F. Although, in principle, the integral may be calculated based on the definition given above, that is generally a difficult approach. Fortunately, there are some basic results that make this calculation routine in many cases of practical interest. Here we consider only the case in which F is a distribution function on the real line; integrals with respect to distribution functions on $\mathbf{R}^d$, $d > 1$, can be calculated as iterated integrals.

Two types of distribution function F arise often in statistical applications. One is a step function; the other is an absolutely continuous function. The integral of an integrable function g with respect to either type of function is easy to calculate using results from basic calculus.

A distribution function F defined on $\mathbf{R}$ is a step function if there is a partition of $\mathbf{R}$,

$$x_0 < x_1 < x_2 < \cdots$$

such that F is constant on each interval $[x_{j-1}, x_j)$. Suppose F is a step function and define

$$\alpha_j = \begin{cases} \lim_{d \to 0^+} F(x_1 + d) - F(x_1) & \text{for } j = 1 \\ \lim_{d \to 0^+} F(x_j + d) - \lim_{d \to 0^+} F(x_j - d) & \text{for } j = 2, \ldots, n-1 \\ F(x_n) - \lim_{d \to 0^+} F(x_n - d) & \text{for } j = n \end{cases}.$$

Then α_j represents the size of the jump at x_j so that

$$F(x) - F(\tilde{x}) = \alpha_j \quad \text{for } x_j \geq x < x_{j+1} \text{ and } x_{j-1} \leq \tilde{x} < x_j.$$

If g is integrable with respect to F then

$$\int_{\mathbf{R}} g(x)\, dF(x) = \sum_{j=1}^{\infty} g(x_j)\alpha_j.$$

Now suppose that F is absolutely continuous. Recall that a real-valued function h defined on $[a, b]$ is said to be *absolutely continuous* if for every $\epsilon > 0$ there exists a $\delta > 0$ such that for all positive integers n and all disjoint subintervals of $[a, b]$, $(a_1, b_1), \ldots, (a_n, b_n)$ such that

$$\sum_{j=1}^{n} |b_j - a_j| \leq \delta$$

implies that

$$\sum_{j=1}^{n} |h(b_j) - h(a_j)| \leq \epsilon.$$

A function $h : \mathbf{R} \to \mathbf{R}$ is absolutely continuous if the restriction of h to $[a, b]$ is absolutely continuous for any $a < b$. A sufficient condition for h to be absolutely continuous is that there exists a constant M such that

$$|h(x_1) - h(x_2)| \leq M|x_1 - x_2|$$

for all x_1, x_2 in $[a, b]$. This is called a *Lipschitz condition*. Under this condition,

$$\sum_{j=1}^{n} |h(b_j) - h(a_j)| \leq M \sum_{j=1}^{n} |b_j - a_j| \leq M\delta$$

so that δ may be taken to be ϵ/M.

There is a close connection between absolutely continuous functions and functions defined by integrals. If $h : [a, b] \to \mathbf{R}$ is given by

$$h(x) - h(a) = \int_a^x g(t)\,dt$$

for some function g, then h is absolutely continuous. Conversely, if $h : [a, b] \to \mathbf{R}$ is an absolutely continuous function then there exists a function g such that

$$h(x) - h(a) = \int_a^x g(t)\,dt.$$

Hence, if F is absolutely continuous then there exists a nonnegative function f such that

$$F(t) = \int_{\mathbf{R}} I_{\{-\infty < x \leq t\}} f(x)\,dx, \quad -\infty < t < \infty.$$

In this case,

$$\int_{\mathbf{R}} g(x)\,dF(x) = \int_{\mathbf{R}} g(x) f(x)\,dx;$$

the integral

$$\int_{\mathbf{R}} g(x) f(x)\,dx$$

may be taken to be the usual Riemann integral studied in elementary calculus, provided that the Riemann integral of gf exists. For instance, the Riemann integral of a bounded function h over an interval $[a, b]$ exists if the set of discontinuities of h in $[a, b]$ is countable.

A1.6 Fundamental Theorem of Calculus

Let F denote a distribution function on the real line.

(i) Suppose there exists a function f such that

$$F(x) = \int_{-\infty}^x f(t)\,dt, \quad -\infty < x < \infty.$$

If f is continuous at x, then $F'(x)$ exists and is equal to $f(x)$.

(ii) Suppose $F'(x)$ exists for all $-\infty < x < \infty$ and

$$\int_{-\infty}^{\infty} F'(x) < \infty.$$

Then

$$F(x) = \int_{-\infty}^{x} F'(t)\,dt, \quad -\infty < x < \infty.$$

(iii) Suppose that

$$F(x) = \int_{-\infty}^{x} f(t)\,dt, \quad -\infty < x < \infty,$$

for some nonnegative, integrable function f. Then $F'(x) = f(x)$ for almost all x.

For proofs of these results, and much further discussion, see, for example, Billingsley (1995, Section 31).

A1.7   Interchanging Integration and Differentiation

Let A be an open subset of $\mathbf{R}$, and let f denote a continuous, real-valued function on $\mathbf{R}^d \times A$. Assume that

(i) for each $t \in A$,

$$\int_{\mathbf{R}^d} |f(x,t)|\,dF(x) < \infty$$

(ii) for each $t \in A$,

$$\frac{\partial}{\partial t} f(x,t)$$

exists and

$$\int_{\mathbf{R}^d} |\frac{\partial}{\partial t} f(x,t)|\,dF(x) < \infty$$

(iii) there exists a function h on $\mathbf{R}^d$ such that

$$|f(x,t)| \le h(x), \quad t \in A$$

and

$$\int_{\mathbf{R}^d} |h(x)|\,dF(x) < \infty.$$

Define

$$g(t) = \int_a^b f(x,t)\,dF(x), \quad t \in A.$$

Then g is differentiable on A and

$$g'(t) = \int_a^b \frac{\partial f}{\partial t}(x,t)\,dF(x), \quad t \in A.$$

See Lang (1983, Chapter 13) for a proof of this result.

Appendix 2

Basic Properties of Complex Numbers

This appendix contains a brief review of the properties of complex numbers; in distribution theory, these are used in connection with characteristic functions.

A2.1 Definition

A *complex number* x is an ordered pair (x_1, x_2) where x_1 and x_2 are real numbers. The first component, x_1, is called the *real part* of the complex number; the second component, x_2, is called the *imaginary part*. Complex numbers of the form $(x_1, 0)$ are said to be real and are written simply as x_1. Two complex numbers $x = (x_1, x_2)$ and $y = (y_1, y_2)$ are considered equal if $x_1 = y_1$ and $x_2 = y_2$.

Addition of complex numbers is defined by

$$x + y = (x_1 + y_1, x_2 + y_2)$$

and multiplication is defined by

$$xy = (x_1 y_1 - x_2 y_2, x_1 y_2 + x_2 y_1).$$

The absolute value or *modulus* of a complex number x is the nonnegative number, denoted by $|x|$, given by

$$|x|^2 = x_1^2 + x_2^2.$$

The complex number $(0, 1)$ is called the *imaginary unit* and is denoted by i. Hence, the complex number $x = (x_1, x_2)$ may be written $x = x_1 + i x_2$. Using the definition of multiplication,

$$i^2 \equiv ii = (0, 1)(0, 1) = (-1, 0) = -1.$$

A2.2 Complex Exponentials

For a complex number $x = x_1 + i x_2$, the exponential $\exp(x)$ is defined as

$$\exp(x) = \exp(x_1 + i x_2) = \exp(x_1)[\cos(x_2) + i \, \sin(x_2)].$$

It may be shown that, if x and y are complex numbers,

$$\exp(x) \exp(y) = \exp(x + y).$$

Also, if x is purely imaginary, that is, is of the form $(0, x_2)$, then

$$|\exp(x)| = 1.$$

Let z denote a real number. Then the cosine, sine, and exponential functions have the following power series representations:

$$\cos(z) = \sum_{j=0}^{\infty} (-1)^j \frac{z^{2j}}{(2j)!}, \quad \sin(z) = \sum_{j=0}^{\infty} (-1)^j \frac{z^{2j+1}}{(2j+1)!},$$

and

$$\exp(z) = \sum_{j=0}^{\infty} \frac{z^j}{j!}.$$

Since $i^{2j} = (-1)^j$, we may write

$$\cos(z) = \sum_{j=0}^{\infty} \frac{(iz)^{2j}}{(2j)!} \quad \text{and} \quad \sin(z) = -i \sum_{j=0}^{\infty} \frac{(iz)^{2j+1}}{(2j+1)!}.$$

It follows that

$$\exp(iz) = \cos(z) + i \, \sin(z) = \sum_{j=0}^{\infty} \frac{(iz)^{2j}}{(2j)!} - i^2 \sum_{j=0}^{\infty} \frac{(iz)^{2j+1}}{(2j+1)!}$$

$$= \sum_{j=0}^{\infty} \frac{(iz)^j}{j!}.$$

Hence, for any complex number x,

$$\exp(x) = \sum_{j=0}^{\infty} \frac{x^j}{j!}.$$

Lemma A2.1. *We may write*

$$\exp\{it\} = \sum_{j=0}^{n} \frac{(it)^j}{j!} + R_n(t)$$

where

$$|R_n(t)| \le \min\{|t|^{n+1}/(n+1)!, \ 2|t|^n/n!\}.$$

Proof. First note that, using integration-by-parts,

$$\int_0^t (t - s)^{n+1} \cos(s) \, ds = (n+1) \int_0^t (t - s)^n \sin(s) \, ds$$

and

$$\int_0^t (t - s)^{n+1} \sin(s) \, ds = -(n+1) \int_0^t (t - s)^n \cos(s) \, ds + t^{n+1}.$$

Hence,

$$\int_0^t (t-s)^n \exp\{is\}\, ds = \frac{t^{n+1}}{n+1} + \frac{i}{n+1} \int_0^t (t-s)^{n+1} \exp\{is\}\, ds. \qquad \text{(A2.1)}$$

It follows that

$$\int_0^t \exp\{is\}\, ds = t + i \int_0^t (t-s) \exp\{is\}\, ds$$

$$= t + i \left[\frac{t^2}{2} + \frac{i}{2} \int_0^t (t-s)^2 \exp\{is\}\, ds \right].$$

Continuing in this way shows that

$$i \int_0^t \exp\{is\}\, ds = \sum_{j=1}^n \frac{(it)^j}{j!} + i^{n+1} \int_0^t (t-s)^n \exp\{is\}\, ds$$

and, hence, that

$$\exp\{it\} = \sum_{j=0}^n \frac{(it)^j}{j!} + \frac{i^{n+1}}{n!} \int_0^t (t-s)^n \exp\{is\}\, ds.$$

Consider the remainder term

$$\frac{i^{n+1}}{n!} \int_0^t (t-s)^n \exp\{is\}\, ds.$$

Since, by (A2.1),

$$\frac{i}{n} \int_0^t (t-s)^n \exp\{is\}\, ds = \int_0^t (t-s)^{n-1} \exp\{is\}\, ds - \int_0^t (t-s)^{n-1}\, ds,$$

it follows that

$$\frac{i^{n+1}}{n!} \int_0^t (t-s)^n \exp\{is\}\, ds = \frac{i^n}{(n-1)!} \int_0^t (t-s)^{n-1} [\exp\{is\} - 1]\, ds$$

and, hence, that

$$\left| \frac{i^{n+1}}{n!} \int_0^t (t-s)^n \exp\{is\}\, ds \right| \le \frac{2}{n!} t^n.$$

Also,

$$\left| \frac{i^{n+1}}{n!} \int_0^t (t-s)^n \exp\{is\}\, ds \right| \le \frac{1}{n!} \operatorname{sgn}(t) \int_0^t |t-s|^n\, ds \le \frac{|t|^{n+1}}{(n+1)!}.$$

The result follows. ∎

Lemma A2.2. *Let u and v denote complex numbers. Then*

$$|\exp(u) - (1+v)| \le [|u-v| + |u|^2/2] \exp(|u|).$$

Proof.   The proof is based on the following result. Let t denote a complex number. Using the power series representation for $\exp(t)$,

$$\left| \exp(t) - (1+t) \right| = \left| \sum_{j=2}^{\infty} \frac{t^j}{j!} \right| \leq \frac{|t|^2}{2} \sum_{j=0}^{\infty} |t|^j \frac{2}{(j+2)!} \right|$$

$$\leq \frac{|t|^2}{2} \left| \sum_{j=0}^{\infty} \frac{|t|^j}{j!} \right| = \frac{|t|^2}{2} \exp(|t|). \qquad (A2.2)$$

Now consider the proof of the lemma. Note that

$$| \exp(u) - (1+v) | = | \exp(u) - (1+u) + (u-v) | \leq | \exp(u) - (1+u) | + | u - v |.$$

The result now follows from (A2.2) along with the fact that $\exp(|u|) \geq 1$.   ∎

A2.3   Logarithms of Complex Numbers

If x is a real number, then $\log(x)$ is defined by

$$\exp\{\log(x)\} = x.$$

First note that this same approach will not work for complex numbers. Let $x = x_1 + i x_2$ denote a complex number and suppose that $y = y_1 + i y_2$ denotes $\log(x)$. Then

$$\exp(y) = \exp(y_1)[\cos(y_2) + i \sin(y_2)]$$

so that we must have

$$\cos(y_2) \exp(y_1) = x_1 \quad \text{and} \quad \sin(y_2) \exp(y_1) = x_2;$$

that is,

$$y_1 = \log(|x|) \quad \text{and} \quad \tan(y_2) = x_2/x_1.$$

Since if y_2 satisfies $\tan(y_2) = x_2/x_1$ so does $y_2 + 2n\pi$ for any integer n, the requirement $\exp(\log(x)) = x$ does not uniquely define $\log(x)$.

The *principal logarithm* of a complex number x is taken to be

$$\log |x| + i \, \arg(x)$$

where $\arg(x)$ denotes the *principal argument* of x. The principal argument of $x = x_1 + i x_2$ is defined to be the real number θ, $-\pi < \theta \leq \pi$, satisfying

$$x_1 = |x| \cos(\theta), \quad x_2 = |x| \sin(\theta).$$

Whenever the notation $\log(x)$ is used for a complex number x, we mean the principal logarithm of x.

Let x denote a complex number. If $|x - 1| < 1$, then $\log(x)$ may also be defined by

$$\log(x) = \sum_{j=1}^{\infty} (-1)^{j+1} (x-1)^j / j.$$

Appendix 3

Some Useful Mathematical Facts

This appendix contains a review of some basic mathematical facts that are used throughout this book. For further details, the reader should consult a book on mathematical analysis, such as Apostal (1974), Rudin (1976), or Wade (2004).

A3.1 Sets

A3.1.1 Basic definitions. A *set* is a collection of objects that itself is viewed as a single entity. We write $x \in A$ to indicate that x is an element of a set A; we write $x \notin A$ to indicate that x is not an element of A. The set that contains no elements is known as the *empty set* and is denoted by $\emptyset$.

Let A and B denote sets. If every element of B is also an element of A we say that B is a *subset* of A; this is denoted by $B \subset A$. If there also exists an element of A that is not in B we say that B is a *proper* subset of A. If A and B have exactly the same elements we write $A = B$. The *difference* between A and B, written $A \setminus B$, is that set consisting of all elements of A that are not elements of B.

A3.1.2 Set algebra. Let S denote a fixed set such that all sets under consideration are subsets of S and let A and B denote subsets of S. The *union* of A and B is the set C whose elements are either elements of A or elements of B or are elements of both; we write $C = A \cup B$. The *intersection* of A and B is the set D whose elements are in both A and B; we write $D = A \cap B$. Sets A and B are said to be *disjoint* if $A \cap B = \emptyset$. The *complement* of A, denoted by A^c, consists of those elements of S that are not elements of A.

Let $\mathcal{F}$ denote an arbitrary collection of sets. The union of all sets in $\mathcal{F}$ is that set consisting of those elements that belong to at least one of the sets in $\mathcal{F}$; this set is written

$$\bigcup_{A \in \mathcal{F}} A.$$

The intersection of all sets in $\mathcal{F}$ is that set consisting of those elements that belong to every set in $\mathcal{F}$; this set is written

$$\bigcap_{A \in \mathcal{F}} A.$$

DeMorgan's Laws state:

$$\left(\bigcup_{A \in \mathcal{F}} A \right)^c = \bigcap_{A \in \mathcal{F}} A^c$$

and

$$\left(\bigcap_{A \in \mathcal{F}} A \right)^c = \bigcup_{A \in \mathcal{F}} A^c.$$

A3.1.3   Ordered sets. Let S denote a set and let $<$ denote a relation on S with the following properties:

  (O1) If $x, y \in S$ then exactly one of the following statements is true: $x < y$, $y < x$, $x = y$.
  (O2) If $x, y, z \in S$ and $x < y$, $y < z$, then $x < z$.

The relation $<$ is known as an *order* on S; a set S together with an order $<$ is known as an *ordered set*. We write $x \leq y$ to mean either $x < y$ or $x = y$.

Let S denote an ordered set and let $A \subset S$. We say that A is *bounded above* if there exists an element $b \in S$ such that $x \leq b$ for all $x \in A$; b is said to be an *upper bound* of A. We say that A is *bounded below* if there exists an element $a \in S$ such that $a \leq x$ for all $x \in A$; a is said to be a *lower bound* of A.

Let b be an upper bound of A with the following property: if c is another upper bound of A then $b < c$. Then b is called the *least upper bound* or *supremum* of A; *supremum* is generally denoted by *sup* and we write

$$b = \sup A.$$

Similarly, if a is lower bound of A with the property that any other lower bound c satisfies $c < a$, then a is known as the *greatest lower bound* or *infimum* of A and we write

$$a = \inf A.$$

A3.1.4   Open and closed sets. Let S denote a set such that, for any two elements $x, y \in S$, there is associated a real number $d(x, y)$ called the *distance* between x and y. The function d is said to be a *metric* on S provided that it satisfies the following properties:

  (M1) $d(x, x) = 0$ and $d(x, y) > 0$ if $x \neq y$
  (M2) $d(x, y) = d(y, x)$
  (M3) for any $z \in S$, $d(x, z) \leq d(x, y) + d(y, z)$.

The set S together with a metric d is known as a *metric space*. The most commonly used metric space is the set of real numbers **R**, together with the metric $d(x, y) = |x - y|$.

Let S denote a metric space with metric d. A *neighborhood* of $x \in S$ is the set consisting of all points $y \in S$ such that $d(x, y) < \epsilon$ for some $\epsilon > 0$; ϵ is known as the *radius* of the neighborhood. Let $A \subset S$; a point $x \in S$ is called a *limit point* of the set A if every neighborhood of x contains a point $y \in A$ such that $y \neq x$. The set A is said to be *closed* if every limit point of A is an element of A. The point x is said to be an *interior point* of A if there exists a neighborhood of x that is a subset of A. The set A is said to be *open* if every point of A is an interior point of A. Let $B \subset S$ and let C denote the set of all limit points of B; the *closure* of B is the set $B \cup C$.

A3.1.5   Cartesian products. Consider two sets, A and B. The *cartesian product* of A and B, denoted by $A \times B$, is the set of all ordered pairs (a, b) with $a \in A$ and $b \in B$. The cartesian product of sets $A_1, \ldots, A_k$, denoted by $A_1 \times \cdots \times A_k$, is the set of all ordered k-tuples $(a_1, \ldots, a_k)$ such that $a_1 \in A_1, a_2 \in A_2, \ldots, a_k \in A_k$.

For instance, for any positive integer k, k-*dimensional Euclidean space*, denoted by $\mathbf{R}^k$, is the cartesian product $\mathbf{R} \times \cdots \times \mathbf{R}$ of k copies of the set of real numbers $\mathbf{R}$. This is a metric space with metric

$$d(x, y) = \left[\sum_{j=1}^{k} (x_j - y_j)^2 \right]^{\frac{1}{2}},$$

where $x = (x_1, \ldots, x_k)$ and $y = (y_1, \ldots, y_k)$.

Let A and B denote sets. A subset C of $A \times B$ is called a *relation*. The set of all elements $x \in A$ that occur as first members of pairs in C is called the *domain* of C; the set of all elements $y \in B$ that occur as second members of pairs in C is called the *range* of C. A *function* is a relation C such that no two elements of C have the same first member. That is, if C is a function with domain A and range that is a subset of B, then associated with any element $x \in A$ is a unique element $y \in B$; this association is given by the ordered pair (x, y). A function C is said to be *one-to-one* if, for any two pairs $(x_1, y_1), (x_2, y_2) \in C$, $y_1 = y_2$ if and only if $x_1 = x_2$.

Two sets A and B are said to be *similar*, written $A \sim B$, if there exists a one-to-one function with domain A and range B. Similar sets have the same number of elements, in a certain sense.

A3.1.6   Finite, countable, and uncountable sets. A set S is said to be *finite* if it contains a finite number of elements. If S is finite then $S \sim \{1, 2, \ldots, n\}$ for some $n = 0, 1, 2, \ldots$; in this case, n is the number of elements in S. A set that is not finite is said to be an *infinite set*.

A set S is said to be *countably infinite* if it can be put into a one-to-one correspondence with the set of positive integers; that is, if

$$S \sim \{1, 2, \ldots\}.$$

A *countable* set is one that is either finite or countably infinite. A set which is not countable is said to be *uncountable*.

A3.1.7   Compact sets. Let S denote a metric space with metric d and let $A \subset S$. The set A is said to be *compact* if the following condition holds. Let $\mathcal{F}$ denote a collection of open subsets of S such that

$$A \subset \bigcup_{B \in \mathcal{F}} B;$$

such a collection is said to be an *open covering* of A. Then there exists a finite collection of sets $B_1, B_2, \ldots, B_m, B_j \in \mathcal{F}, j = 1, 2, \ldots, m$, such that

$$A \subset \bigcup_{j=1}^{m} B_j.$$

That is, A is compact if, for any open covering $\mathcal{F}$ of A, there exists a finite number of sets in $\mathcal{F}$ that also cover A.

Suppose that $S = \mathbf{R}^n$ for some $n = 1, 2, \ldots$. A set $B \subset \mathbf{R}^n$ is said to be *bounded* if there exists a real number c such that

$$B \subset \left\{ x \in \mathbf{R}^n : \sum_{j=1}^{n} x_j^2 \leq c^2 \right\}.$$

A subset A of $\mathbf{R}^n$ is compact if and only if it is closed and bounded.

A3.1.8 Sets of measure 0. Let S denote a subset of $\mathbf{R}$. We say that S *has measure 0* provided that the following condition holds. For any $\epsilon > 0$ there exist real numbers $a_1, b_1, a_2, b_2, \ldots, a_j \leq b_j, j = 1, 2, \ldots$, such that

$$S \subset \bigcup_{n=1}^{\infty}(a_n, b_n) \quad \text{and} \quad \sum_{n=1}^{N}(b_n - a_n) < \epsilon, \quad N = 1, 2, \ldots.$$

Note that these conditions are given for the case in which there is a countably infinite number of pairs a_n, b_n; if there is a finite number, the conditions can be modified in the obvious manner. For $a < b$, the set (a, b) consists of all $x, a < x < b$.

If S is a countable set, or is the union of a countable number of sets each of measure 0, then S has measure 0.

If a particular condition holds except for those x in a set S which has measure 0, we say that the condition holds *almost everywhere* or *for almost all x*.

A3.2 Sequences and Series

A3.2.1 Convergent sequences. Let S denote a metric space with metric d. A *sequence* in S is a function with domain $\{1, 2, \ldots\}$ and with the range of the function a subset of S. Such a sequence will be denoted by $x_1, x_2, \ldots$, where $x_j \in S, j = 1, 2, \ldots$.

A sequence in S is said to *converge* if there exists an element $x \in S$ such that for every $\epsilon > 0$ there exists a positive integer N with the property that $n \geq N$ implies that $d(x_n, x) < \epsilon$ for all $n \geq N$. In this case, we say that $x_1, x_2, \ldots$ *converges to x* or that x is the *limit* of x as $n \to \infty$; this is written as

$$x = \lim_{n \to \infty} x_n$$

or $x_n \to x$ as $n \to \infty$.

If $x_1, x_2, \ldots$ does not converge then it is said to *diverge*. It is important to note that there are two ways in which a given sequence may diverge: it may fail to approach a limiting value or it may approach a limiting value that is not an element of the set S. For instance, suppose that $S = (0, 1)$. The sequence $1/4, 1/2, 1/4, 1/2, \ldots$ does not approach a limiting value; the sequence $1/2, 1/3, 1/4, \ldots$ approaches the limiting value 0, but $0 \notin S$. Both sequences diverge.

A sequence $x_1, x_2, \ldots$ is called a *Cauchy sequence* if for every $\epsilon > 0$ there is an integer N such that $d(x_n, x_m) < \epsilon$ whenever $n \geq N$ and $m \geq N$. Every convergent sequence is a Cauchy sequence; however, not all Cauchy sequences are convergent. If every Cauchy sequence in a metric space S converges, S is said to be *complete*. The Euclidean spaces $\mathbf{R}^k$, $k = 1, 2, \ldots$, are complete.

Let $x_1, x_2, \ldots$ denote a sequence of real numbers. Suppose that, for every $M \in \mathbf{R}$, there exists a positive integer N such that

$$x_n \geq M \quad \text{for} \ n \geq N.$$

Then x_n grows indefinitely large as $n \to \infty$; in this case, we write $x_n \to \infty$ as $n \to \infty$. If for every $M \in \mathbf{R}$ there exists a positive integer N such that

$$x_n \leq M \quad \text{for} \ n \geq N,$$

we write $x_n \to -\infty$ as $n \to \infty$.

A3.2.2 Subsequences. Let $x_1, x_2, \ldots$ denote a sequence of real numbers and let $n_1, n_2, \ldots$ denote an increasing sequence of positive integers. Then the sequence $y_1, y_2, \ldots$, where $y_k = x_{n_k}$, is called a *subsequence* of $x_1, x_2, \ldots$. The sequence $x_1, x_2, \ldots$ converges to x if and only if every subsequence of $x_1, x_2, \ldots$ also converges to x. A sequence $x_1, x_2, \ldots$ is said to be *bounded* if there exists a real number M such that $|x_n| \leq M$ for all $n = 1, 2, \ldots$. Every bounded sequence of real numbers contains a convergent subsequence.

Let $x_1, x_2, \ldots$ denote a sequence of real numbers and let S denote the subset of $\mathbf{R} \cup \{-\infty, \infty\}$ such that for each $x \in S$ there exists a subsequence $x_{n_1}, x_{n_2}, \ldots$ such that

$$x_{n_j} \to x \quad \text{as} \ j \to \infty.$$

Then $\sup S$ is called the *upper limit* or lim sup of $x_1, x_2, \ldots$; this is written

$$\limsup_{n \to \infty} x_n.$$

Similarly, $\inf S$ is called the *lower limit* or lim inf of $x_1, x_2, \ldots$ and is written

$$\liminf_{n \to \infty} x_n.$$

Note that

$$\liminf_{n \to \infty} x_n \leq \limsup_{n \to \infty} x_n.$$

The sequence $x_1, x_2, \ldots$ converges to $x \in \mathbf{R}$ if and only if

$$\liminf_{n \to \infty} x_n = \limsup_{n \to \infty} x_n = x.$$

If $x_1, x_2, \ldots$ and $y_1, y_2, \ldots$ are two sequences such that $x_n \leq y_n$ for all $n \geq N$ for some positive integer N, then

$$\liminf_{n \to \infty} x_n \leq \liminf_{n \to \infty} y_n$$

and

$$\limsup_{n \to \infty} x_n \leq \limsup_{n \to \infty} y_n.$$

A3.2.3 *O* and *o* notation. Let $a_1, a_2, \ldots$ and $b_1, b_2, \ldots$ denote sequences of real numbers such that $b_n > 0$ for all $n = 1, 2, \ldots$. We write $a_n = O(b_n)$ as $n \to \infty$ if there exists a constant M such that $|a_n| \leq M b_n$ for all $n = 1, 2, \ldots$. We write $a_n = o(b_n)$ as $n \to \infty$ if

$$\lim_{n \to \infty} \frac{a_n}{b_n} = 0.$$

A3.2.4   Series. Let $a_1, a_2, \ldots$ denote a sequence of real numbers. Define a new sequence $s_1, s_2, \ldots$ by

$$s_n = \sum_{j=1}^{n} a_j, \quad n = 1, 2, \ldots.$$

The sequence $s_1, s_2, \ldots$ is called a *series* and s_n is called the nth *partial sum* of the series. The sum

$$\sum_{j=1}^{\infty} a_j$$

is called an *infinite series*.

The series converges or diverges depending on whether the sequence $s_1, s_2, \ldots$ converges or diverges. If

$$\lim_{n \to \infty} s_n = s$$

for some real number s, we write

$$s = \sum_{j=1}^{\infty} a_j;$$

however, it is important to keep in mind that s is actually defined as

$$s = \lim_{n \to \infty} \sum_{j=1}^{n} a_j.$$

We say that the series *converges absolutely* if the series

$$\sum_{j=1}^{\infty} |a_j|$$

converges. If the series converges absolutely and

$$\sum_{j=1}^{\infty} a_j = s$$

then any rearrangement of $a_1, a_2, \ldots$ also has sum s.

A3.2.5   Double series. Let $a_{jk}, j = 1, 2, \ldots, k = 1, 2, \ldots,$ denote real numbers and let

$$s_{mn} = \sum_{j=1}^{m} \sum_{k=1}^{n} a_{jk}, \quad m = 1, 2, \ldots, \quad n = 1, 2, \ldots.$$

The double sequence given by s_{mn} is called a *double series*. If

$$\lim_{m \to \infty, n \to \infty} s_{mn} = s$$

the double series is said to converge to s.

Let

$$\bar{s}_{mn} = \sum_{j=1}^{m} \sum_{k=1}^{n} |a_{jk}|, \quad m = 1, 2, \ldots, \quad n = 1, 2, \ldots.$$

If

$$\lim_{m\to\infty, n\to\infty} \bar{s}_{mn}$$

exists, then the double series converges absolutely and

$$\sum_{j=1}^{\infty}\sum_{k=1}^{\infty} a_{jk} = \sum_{k=1}^{\infty}\sum_{j=1}^{\infty} a_{jk}.$$

Suppose that $a_{jk} \geq 0$ for all j, k. If

$$\sum_{k=1}^{\infty} a_{jk}$$

converges for each $j = 1, 2, \ldots$ and

$$\sum_{j=1}^{\infty}\sum_{k=1}^{\infty} a_{jk}$$

converges, then

$$\sum_{j=1}^{\infty}\sum_{k=1}^{\infty} a_{jk} = \sum_{k=1}^{\infty}\sum_{j=1}^{\infty} a_{jk}.$$

A3.3   Functions

A3.3.1   Continuity. Let A and B be metric spaces. Recall that a function C is a subset of the cartesian product $A \times B$; elements of this subset are of the form $(x, f(x))$ where $x \in A$ and $f(x) \in B$. The domain of the function is the set of $x \in A$ for which f is defined; denote this set by D. The range of the function is the set of $f(x) \in B$ corresponding to $x \in D$. We may describe the function by giving f, together with the set D and, hence, we will refer to f as the function, writing $f : D \to B$ to indicate the domain of the function. A function $f : D \to B$ where B is a subset of the real line **R** is said to be a *real-valued* function.

Let d_A denote the metric on A and let d_B denote the metric on B. We say that $f : D \to B$ is *continuous* at a point $x \in D$ if for every $\epsilon > 0$ there exists a $\delta > 0$ such that

$$d_B(f(x), f(z)) < \epsilon$$

for all $z \in D$ such that $d_A(x, z) < \delta$. If f is continuous at x for all $x \in D$ we say that f is *continuous on D*.

The function f is said to be *uniformly continuous* on D if for every $\epsilon > 0$ there exists a $\delta > 0$ such that

$$d_B(f(x), f(z)) < \epsilon$$

for all x and z for which $d_A(x, z) < \delta$.

Suppose that x is a limit point of D and that f is continuous at x. Then, for any sequence $x_1, x_2, \ldots$ such that $x_n \to x$ as $n \to \infty$,

$$\lim_{n\to\infty} f(x_n) = f(x).$$

We write this property as

$$\lim_{y \to x} f(y) = f(x).$$

For the case in which A and B are subsets of $\mathbf{R}$, a function $f : D \to B$ is continuous on D if and only if for each $x \in D$ and $\epsilon > 0$ there exists a $\delta > 0$ such that

$$|f(x) - f(z)| < \epsilon$$

for all $z \in D$ such that $|x - z| < \delta$. The function f is uniformly continuous on D if for every $\epsilon > 0$ there exists a $\delta > 0$ such that

$$|f(x) - f(z)| < \epsilon$$

for all x and z in D for which $|x - z| < \delta$.

Let $D = (a, b)$ be a subset of $\mathbf{R}$ and consider a function $f : D \to \mathbf{R}$. Let x satisfy $a \le x < b$. We write

$$\lim_{y \to x+} f(y) = c \quad \text{or} \quad f(x+) = c$$

if for any sequence $h_1, h_2, \ldots$ such that $h_n > 0$ for all $n = 1, 2, \ldots$, $h_n \to 0$ as $n \to 0$,

$$\lim_{n \to \infty} f(x + h_n) = c.$$

If $f(x+) = f(x)$ we say that f is *right continuous* at x. Similarly, if $a < x \le b$, we write

$$\lim_{y \to x-} f(x) = d \quad \text{or} \quad f(x-) = d$$

if for any sequence $h_1, h_2, \ldots$ such that $h_n > 0$ for all $n = 1, 2, \ldots$ and $h_n \to 0$ as $n \to 0$,

$$\lim_{n \to \infty} f(x - h_n) = d.$$

If $f(x-) = f(x)$ we say that f is *left continuous* at x.

If $f(x+)$ and $f(x-)$ both exist and

$$f(x+) = f(x-) = f(x)$$

then f is continuous at x. Suppose that f is not continuous at x; that is, suppose that x is a discontinuity point of f. If $f(x+)$ and $f(x-)$ both exist, f is said to have a *simple discontinuity* at x. If $f(x+) = f(x-) \ne f(x)$ the discontinuity is said to be *removable* since the discontinuity disappears if we redefine the value of $f(x)$ to be the common value of $f(x+)$ and $f(x-)$.

A3.3.2 Continuous functions on a compact set. Let A and B be metric spaces and let $f : D \to B$ denote a continuous function, where D is a subset of A. If D is compact then the range of f is a compact subset of B and f is uniformly continuous on D.

Suppose that f is real-valued so that B may be taken to be $\mathbf{R}$. If D is compact, then there exists a constant c such that $|f(x)| \le c$ for all $x \in D$; in this case, we say that f is *bounded*. Let

$$M = \sup_{x \in D} f(x) \quad \text{and} \quad m = \inf_{x \in D} f(x).$$

Then there exist points $x_0, x_1 \in D$ such that

$$M = f(x_0) \quad \text{and} \quad m = f(x_1).$$

A3.3.3 Functions of bounded variation.

Let f denote a real-valued function defined on the real interval $[a, b]$ consisting of all x, $a \leq x \leq b$, and let

$$\{x_0, x_1, \ldots, x_n\}$$

denote elements of $[a, b]$ such that

$$a = x_0 < x_1 < \cdots < x_{n-1} < x_n = b.$$

If there exists a number M such that

$$\sum_{j=1}^{n} |f(x_j) - f(x_{j-1})| \leq M$$

for all such sets $\{x_0, x_1, \ldots, x_n\}$ for all $n = 1, 2, \ldots$, then f is said to be *of bounded variation*.

A function f is of bounded variation if and only if it can be written as the difference of two nondecreasing functions.

A3.3.4 Convex functions.

Let f denote a real-valued function defined on a real interval (a, b). The function f is said to be *convex* if for any $x, y \in (a, b)$ and any α, $0 \leq \alpha \leq 1$,

$$f(\alpha x + (1 - \alpha)y) \leq \alpha f(x) + (1 - \alpha)f(y).$$

Every convex function is continuous.

A3.3.5 Composition of functions.

Let A, B, C denote metric spaces. Let f denote a function from A_0 to B where $A_0 \subset A$; let $B_0 \subset B$ denote the range of f and let $g : B_0 \to C$ denote a function on B_0. The function $h : A_0 \to C$ defined by

$$h(x) = g(f(x)), \quad x \in A_0,$$

is known as the *composition* of f and g. If f and g are both continuous, then h is continuous.

A3.3.6 Inverses.

Let A and B be metric spaces and let $f : D \to B$ denote a function, where D is a subset of A. Recall that f is said to be a one-to-one function if, for all $x, y \in D$, $f(x) = f(y)$ if and only if $x = y$. If f is one-to-one we may define the *inverse of* f as the function $g : B_0 \to D$, where B_0 is the range of f, satisfying $g(f(x)) = x$, $x \in B_0$. The notation $g = f^{-1}$ is often used. If g is the inverse of f, then g is also one-to-one and f is the inverse of g.

Let D_0 denote a subset of D. The *restriction of f to D_0* is the function $f_0 : D_0 \to B$ such that $f_0(x) = f(x)$, $x \in D_0$. The restriction of f to D_0 may be invertible even if f itself is not.

A3.3.7 Convergence of functions.

Let A and B be metric spaces and let D be a subset of A. Consider a sequence of functions $f_1, f_2, \ldots$ where, for each $n = 1, 2, \ldots, f_n : D \to B$.

We say that the sequence $f_1, f_2, \ldots$ *converges pointwise* to a function f if for each $x \in D$,

$$\lim_{n \to \infty} f_n(x) = f(x).$$

This means that, for each $x \in D$ and each $\epsilon > 0$, there exists a positive integer N such that

$$d_B(f_n(x), f(x)) < \epsilon \quad \text{whenever } n \geq N;$$

here d_B denotes the metric on B. Note that N may depend on both x and ϵ.

We say the sequence $f_1, f_2, \ldots$ *converges uniformly* to f if N does not depend on the x under consideration. That is, $f_1, f_2, \ldots$ converges uniformly to f if, for each $\epsilon > 0$, there exists a positive integer N such that $n \geq N$ implies that

$$d_B(f_n(x), f(x)) < \epsilon \quad \text{for all } x \in D.$$

For each $n = 1, 2, \ldots$, let

$$M_n = \sup_{x \in D} d_B(f_n(x), f(x)).$$

Then $f_1, f_2, \ldots$ converges uniformly to f if and only if

$$\lim_{n \to \infty} M_n = 0,$$

that is, if and only if

$$\lim_{n \to \infty} \sup_{x \in D} d_B(f_n(x), f(x)) = 0.$$

In some cases, we may have uniform convergence only on a subset of D. Let D_0 be a subset of D such that

$$\lim_{n \to \infty} \sup_{x \in D_0} d_B(f_n(x), f(x)) = 0;$$

in this case, we say that $f_1, f_2, \ldots$ converges uniformly to f on D_0. Suppose that $f_1, f_2, \ldots$ converges uniformly to f on a set D_0 and let $x \in D_0$. If f_n is continuous at x for each $n = 1, 2, \ldots$, then f is continuous at x.

A3.3.8 Weierstrass approximation theorem. Let f denote a real-valued function defined on a subset $[a, b]$ of the real line, where $-\infty < a < b < \infty$. There exists a sequence of polynomials $p_0, p_1, \ldots$, where for each $n = 1, 2, \ldots$, p_n is a polynomial of degree n, such that $p_0, p_1, \ldots$ converges uniformly to f on $[a, b]$.

One such sequence of polynomials is given by the *Bernstein polynomials*. Suppose that $a = 0$ and $b = 1$; it is straightforward to modify the Bernstein polynomials to apply in the general case. For each $n = 0, 1, \ldots$, let

$$B_n(x) = \sum_{k=0}^{n} f\left(\frac{k}{n}\right) \binom{n}{k} x^k (1 - x)^{n-k}, \quad 0 \leq x \leq 1.$$

Note that B_n is a polynomial of degree n. Then

$$\lim_{n \to \infty} \sup_{0 \leq x \leq 1} |B_n(x) - f(x)| = 0.$$

A3.3.9 Power series. Consider a function f of the form

$$f(x) = \sum_{n=0}^{\infty} a_n(x - x_0)^n \qquad (A3.1)$$

where x is a real number and $a_0, a_1, \ldots$ and x_0 are constants. Series of this type are called *power series* and the series in (A3.1) is said to be a *power series expansion* of f; the function f is said to be an *analytic function*.

Let

$$\lambda = \limsup_{n \to \infty} |a_n|^{\frac{1}{n}}$$

and let $r = 1/\lambda$. Then the series

$$\sum_{n=0}^{\infty} a_n(x - x_0)^n$$

converges absolutely for all x such that $|x - x_0| < r$ and the series diverges for all x such that $|x - x_0| > r$. The quantity r is known as the *radius of convergence* of the series and the interval $(x_0 - r, x_0 + r)$ is known as the *interval of convergence* of the series.

Let

$$f(x) = \sum_{n=0}^{\infty} a_n(x - x_0)^n$$

with radius of convergence $r_f > 0$ and let

$$g(x) = \sum_{n=0}^{\infty} b_n(x - x_0)^n$$

with radius of convergence $r_g > 0$. Let $r = \min(r_f, r_g)$ and let $x_1, x_2, \ldots$ be a sequence such that $|x_n - x_0| < r$ for all $n = 1, 2, \ldots$ and $x_n \to x_0$ as $n \to \infty$. If $f(x_n) = g(x_n)$, $n = 1, 2, \ldots$, then $f(x) = g(x)$ for all x such that $|x - x_0| < r$.

Consider a function f with power series expansion

$$f(x) \sum_{n=0}^{\infty} a_n(x - x_0)^n$$

and let r denote the radius of convergence of the series, where $r > 0$. Then f is continuous at x for any $x \in (x_0 - r, x_0 + r)$.

A3.4   Differentiation and Integration

A3.4.1   Definition of a derivative. Let f denote a real-valued function defined on a real interval (a, b) and let x_0 satisfy $a < x_0 < b$. Then f is said to be *differentiable at x_0* whenever the limit

$$\lim_{x \to x_0} \frac{f(x) - f(x_0)}{x - x_0}$$

exists. The value of limit is called the derivative of f at x_0 and is denoted by $f'(x_0)$; thus,

$$f'(x_0) = \lim_{x \to x_0} \frac{f(x) - f(x_0)}{x - x_0},$$

provided that the limit exists. If f is differentiable at x_0 then f is continuous at x_0.

We say that f is *differentiable on* (a, b) if f is differentiable at x_0 for each $x_0 \in (a, b)$. Suppose that f is differentiable on (a, b). If $f'(x) \geq 0$ for all $x \in (a, b)$ then $f(x_1) \geq f(x_0)$ for all $x_1 \geq x_0$; if $f'(x) \leq 0$ for all $x \in (a, b)$ then $f(x_1) \leq f(x_0)$ for all $x_1 \geq x_0$; if $f'(x) = 0$ for all $x \in (a, b)$ then f is constant on (a, b).

A3.4.2 Chain rule. Let f denote a real-valued function defined on (a, b) and let g denote a function defined on the range of f. Consider the composite function $h(x) = g(f(x))$. Suppose that f is differentiable at $x_0 \in (a, b)$ and that g is differentiable at $f(x_0)$. Then h is differentiable at x_0 and

$$h'(x_0) = g'(f(x_0))f'(x_0).$$

A3.4.3 Mean-value theorem. Let f denote a real-valued function defined on an interval $[a, b]$. Suppose that f is continuous on $[a, b]$ and is differentiable on (a, b). Then there exists a point $x_0 \in (a, b)$ such that

$$f(b) - f(a) = f'(x_0)(b - a).$$

A3.4.4 L'Hospital's rule. Let f and g denote real-valued functions defined for all x, $a \leq x \leq b$, $-\infty \leq a < b \leq \infty$. Suppose that

$$\lim_{x \to b^-} f(x) = c \quad \text{and} \quad \lim_{x \to b^-} g(x) = d.$$

If $d \neq 0$, then

$$\lim_{x \to b^-} \frac{f(x)}{g(x)} = \frac{c}{d}.$$

If $d = 0$ and $c \neq 0$, then

$$\frac{f(x)}{g(x)} \to \infty \quad \text{as} \quad x \to b^-.$$

If $d = 0$ and $c = 0$, then

$$\frac{f(x)}{g(x)} \to \frac{0}{0} \quad \text{as} \quad x \to b^-.$$

The ratio $0/0$ is said to be an *indeterminate form* and there is no value associated with $0/0$.

Suppose that

$$\frac{f(x)}{g(x)} \to \frac{0}{0} \quad \text{as} \quad x \to b^-.$$

Assume that f and g are differentiable on (a, b), $g'(x) \neq 0$ for each $x \in (a, b)$, and

$$\lim_{x \to b^-} \frac{f'(x)}{g'(x)}$$

exists. Then

$$\lim_{x \to b^-} \frac{f(x)}{g(x)} = \lim_{x \to b^-} \frac{f'(x)}{g'(x)}.$$

This result is known as *L'Hospital's rule*. It also holds, with obvious modifications, for limits as $x \to a$ or for limits as x approaches an interior point of (a, b).

Another indeterminate form is ∞/∞. Suppose that

$$f(x) \to \infty \quad \text{and} \quad g(x) \to \infty \quad \text{as} \quad x \to b^-.$$

Then

$$\frac{f(x)}{g(x)} \to \frac{\infty}{\infty} \quad \text{as} \quad x \to b^-$$

and this ratio has no value. If f and g are differentiable on (a, b), $g'(x) \neq 0$ for each $x \in (a, b)$, and

$$\lim_{x \to b^-} \frac{f'(x)}{g'(x)}$$

exists, then

$$\lim_{x \to b^-} \frac{f(x)}{g(x)} = \lim_{x \to b^-} \frac{f'(x)}{g'(x)}.$$

A3.4.5 Derivative of a power series. Let

$$f(x) = \sum_{n=0}^{\infty} a_n (x - x_0)^n$$

with radius of convergence $r > 0$. Then, for any x in the interval $(x_0 - r, x_0 + r)$, the derivative $f'(x)$ exists and is given by

$$f'(x) = \sum_{n=1}^{\infty} n a_n (x - x_0)^{n-1};$$

the radius of convergence of this series is also r.

A3.4.6 Higher-order derivatives. Let f denote a real-valued function defined on an interval (a, b); suppose that f is differentiable on (a, b) and let f' denote the derivative. If f' is differentiable on (a, b) then the derivative of f', denoted by f'', is known as the *second derivative* of f. In the same manner, we can define the nth *derivative* of f, denoted by $f^{(n)}$, $n = 1, 2, \ldots$.

A3.4.7 Leibnitz's rule. Let f and g denote real-valued functions defined on a real interval (a, b) and assume that, for some $n = 1, 2, \ldots$ and some $x \in (a, b)$, $f^{(n)}(x)$ and $g^{(n)}(x)$ both exist. Let $h = fg$. Then $h^{(n)}(x)$ exists and

$$h^{(n)}(x) = \sum_{j=0}^{n} \binom{n}{j} f^{(j)}(x) g^{(n-j)}(x).$$

A3.4.8 Taylor's theorem. Let f denote a real-valued function defined on an interval (a, b). Suppose that the nth derivative of f, $f^{(n)}$, exists on (a, b) and that $f^{(n-1)}$ is continuous on $[a, b]$. Let $x_0 \in [a, b]$. Then, for every $x \in [a, b]$,

$$f(x) = f(x_0) + \sum_{j=1}^{n-1} \frac{f^{(j)}(x_0)}{j!}(x - x_0)^j + R_n(x)$$

where

$$R_n(x) = \frac{f^{(n)}(x_1)}{n!}(x - x_0)^n$$

and $x_1 = tx_0 + (1 - t)x$ for some $0 \le t \le 1$. It is often stated that x_1 "lies between" x_0 and x.

An alternative form of Taylor's Theorem expresses the remainder term $R_n(x)$ in terms of an integral. Suppose that $f^{(n)}$ is continuous almost everywhere on (a, b). Then

$$R_n(x) = \frac{1}{(n-1)!} \int_a^x (x - t)^{n-1} f^{(n)}(t) \, dt.$$

A3.4.9 Riemann integral. Let $-\infty < a < b < \infty$ and let f be a bounded, real-valued function on $[a, b]$. Consider a partition $P = \{x_0, x_1, \ldots, x_n\}$,

$$a = x_0 < x_1 < \cdots < x_{n-1} < x_n = b$$

of $[a, b]$ and let

$$\|P\| = \max_{i=1,\ldots,n} |x_i - x_{i-1}|.$$

The *Riemann integral* of f over $[a, b]$ is given by the number A satisfying the following condition: for every $\epsilon > 0$ there exists a $\delta > 0$ such that, for every partition P with $\|P\| < \delta$ and for every choice of points $t_1, \ldots, t_n$ such that $x_{i-1} \le t_i \le x_i$, $i = 1, \ldots, n$,

$$\left| \sum_{j=1}^n f(t_j)|x_j - x_{j-1}| - A \right| < \epsilon.$$

If such a number A does not exist, then f is not Riemann-integrable.

The Riemann integral of f over $[a, b]$ exists if either f is bounded and continuous almost everywhere on $[a, b]$ or if f is of bounded variation.

A3.4.10 Riemann–Lebesgue lemma. Consider a real-valued function f defined on the set (a, b), where $-\infty \le a < b \le \infty$ and assume that the integral

$$\int_a^b f(t) \, dt$$

exists. Then, for any $c \in \mathbf{R}$,

$$\lim_{\alpha \to \infty} \int_a^b f(t) \sin(\alpha t + c) \, dt = 0.$$

A3.4.11 A useful integral result. Let f denote a function of bounded variation such that

$$\int_{-\infty}^{\infty} f(t)\,dt$$

exists and is finite. If f is continuous at x_0 then

$$\lim_{\alpha \to \infty} \frac{1}{\pi} \int_{-\infty}^{\infty} f(x_0 + t)\frac{\sin(\alpha t)}{t}\,dt = f(x_0).$$

A3.4.12 Mean-value theorem for integrals. Let f and g denote continuous, real-valued functions on a real interval (a, b) and assume that g is nonnegative. Then there exists $x_0, a < x_0 < b$, such that

$$\int_a^b f(x)g(x)\,dx = f(x_0)\int_a^b g(x)\,dx.$$

A3.5 Vector Spaces

A3.5.1 Basic definitions. Let V denote a set and let $f : V \times V \to V$ denote a function. Thus, for each $x, y \in V$, $f(x, y)$ is an element of V; this element will be denoted by $x + y$. Assume that the following conditions hold for all $x, y, z \in V$: $x + y = y + x$ and $(x + y) + z = x + (y + z)$. Furthermore, assume that V contains a null element, which we will denote by 0, such that $x + 0 = x$ for all $x \in V$.

Let $g : \mathbf{R} \times V \to V$ denote a function. Thus, for each $\alpha \in \mathbf{R}$ and $x \in V$, $g(\alpha, x)$ is an element of V; this element will be denoted by αx. For all $\alpha, \beta \in \mathbf{R}$ and all $x, y \in V$ we assume that the following conditions hold: $\alpha(x + y) = \alpha x + \alpha y$, $(\alpha + \beta)x = \alpha x + \beta x$, and $\alpha(\beta x) = (\alpha \beta)x$. Furthermore, we assume that, for all $x \in V$, $1x = x$ and $0x = 0$; note that in the latter equation 0 is used as both a real number and the null element of V.

When these conditions are satisfied, V, together with the functions f and g, is called a *vector space* and the elements of V are called *vectors*. The operation given by $+$ is known as *vector addition*. The elements of $\mathbf{R}$ are known as *scalars* and the operation given by αx for $\alpha \in \mathbf{R}$ and $x \in V$ is known as *scalar multiplication*. Vector subtraction may be defined by $x - y = x + (-1)y$.

The most commonly used vector space is $\mathbf{R}^n$. Elements of $\mathbf{R}^n$ are n-tuples of the form $(x_1, \ldots, x_n)$ where $x_j \in \mathbf{R}$, $j = 1, 2, \ldots, n$. In this case, vector addition is simply component-wise addition of real numbers and scalar multiplication is the component-wise multiplication of real numbers. The null element of $\mathbf{R}^n$ is the vector in which each element is 0.

A3.5.2 Subspaces. Let V denote a vector space and let M denote a subset of V. We say that M is a *subspace* of V if for each $x, y \in M$ and all $\alpha, \beta \in \mathbf{R}$, $\alpha x + \beta y \in M$. If M_1 and M_2 are both subspaces of V then their intersection $M_1 \cap M_2$ is also a subspace of V.

The *sum* of two subspaces M_1 and M_2, denoted by $M_1 + M_2$, is the set of all vectors of the form $x + y$ where $x \in M_1$ and $y \in M_2$. The set $M = M_1 + M_2$ is also a subspace of V. If each element of M has a unique representation of the form $x + y$, where $x \in M_1$ and $y \in M_2$, then M is said to be the *direct sum* of M_1 and M_2 and we write $M = M_1 \oplus M_2$.

Let $x_1, x_2, \ldots, x_n$ denote elements of a vector space V. The set of all vectors of the form

$$\alpha_1 x_1 + \alpha_2 x_2 + \cdots + \alpha_n x_n,$$

where $\alpha_j \in \mathbf{R}$, $j = 1, 2, \ldots, n$, is a subspace of V, known as the subspace *generated by* $\{x_1, \ldots, x_n\}$ and the vectors $x_1, \ldots, x_n$ are said to *span* the subspace.

A3.5.3 Bases and dimension. A set of vectors $x_1, x_2, \ldots, x_n$ in a vector space V is said to be *linearly independent* if for any elements of $\mathbf{R}$, $\alpha_1, \ldots, \alpha_n$,

$$\sum_{j=1}^{n} \alpha_j x_j = 0$$

implies that $\alpha_1 = \cdots = \alpha_n = 0$.

Let $S = \{x_1, \ldots, x_n\}$ where $x_1, \ldots, x_n$ are linearly independent. If the subspace generated by S is equal to V, then $x_1, x_2, \ldots, x_n$ is said to be a *basis* for V. In this case, every element of $x \in V$ may be written

$$x = \alpha_1 x_1 + \alpha_2 x_2 + \cdots + \alpha_n x_n$$

for some scalars $\alpha_1, \alpha_2, \ldots, \alpha_n$, which are known as the *coordinates* of x with respect to $x_1, \ldots, x_n$. If such a basis exists, the vector space is said to be *finite dimensional*. A finite-dimensional vector space V has many different bases. However, each basis has the same number of elements; this number is known as the *dimension* of V.

For the vector space $\mathbf{R}^n$ the dimension is n. The canonical basis for $\mathbf{R}^n$ is the vectors $(1, 0, \ldots, 0), (0, 1, 0, \ldots, 0), \ldots, (0, 0, \ldots, 0, 1)$. Thus, the coordinates of a vector $x = (a_1, \ldots, a_n)$ with respect to the canonical basis are simply $a_1, a_2, \ldots, a_n$.

A3.5.4 Norms. Let V denote a vector space. A *norm* on V is a function, denoted by $|| \cdot ||$, with the following properties: $||x|| \geq 0$ for all $x \in V$ and $||x|| = 0$ if and only if $x = 0$; for $x, y \in V$,

$$||x + y|| \leq ||x|| + ||y||;$$

for each scalar α and each $x \in V$, $||\alpha x|| = |\alpha| \, ||x||$. A norm represents a measure of the "length" of a vector. A vector space together with a norm is known as a *normed vector space*.

For $x, y \in V$, let $d(x, y) = ||x - y||$. Then d is a metric on V and V together with this metric forms a metric space. If this metric space is complete, then V, together with the norm $|| \cdot ||$, is known as a *Banach space*.

For the vector space $\mathbf{R}^n$, the most commonly used norm is

$$||x|| = \left[\sum_{j=1}^{n} a_j^2 \right]^{\frac{1}{2}}, \quad x = (a_1, \ldots, a_n).$$

A3.5.5 Linear transformations. Let V and W denote vector spaces and let $T : V \to W$ denote a function; such a function is often called a *transformation*. Suppose that the domain of T is the entire space V and for any $x_1, x_2 \in V$ and any scalars α_1, α_2,

$$T(\alpha_1 x_1 + \alpha_2 x_2) = \alpha_1 T(x_1) + \alpha_2 T(x_2);$$

in this case, T is said to be a *linear transformation*. When T is linear, we often denote $T(x)$ by Tx. The transformation T is said to be *affine* if the transformation T_0 given by

$$T_0(x) = T(x) - T(0)$$

is linear.

Let $T : V \to W$ denote a linear transformation. The *range* of T is the set

$$\mathcal{R}(T) = \{y \in W : y = Tx \text{ for some } x \in V\};$$

$\mathcal{R}(T)$ is a subspace of W. The *nullspace* of T, denoted by $\mathcal{N}(T)$, is the set of $x \in V$ for which $Tx = 0$; $\mathcal{N}(T)$ is a subspace of V.

A linear transformation $T : V \to V$ is said to be *idempotent* if $T^2 x \equiv T(Tx) = Tx$ for all $x \in V$. The linear transformation such that $Tx = x$ for all $x \in V$ is known as the *identity transformation* and will be denoted by I.

Suppose that V is finite-dimensional. A linear transformation $T : V \to W$ is invertible if and only if either $\mathcal{N}(T)$ consists only of the null vector or $\mathcal{R}(T) = W$. The *rank* of T is the dimension of $\mathcal{R}(T)$.

Consider a linear transformation T from $\mathbf{R}^n$ to $\mathbf{R}^n$. Let $S = \{x_1, \dots, x_n\}$ denote a basis for $\mathbf{R}^n$. Then, since for any $j = 1, 2, \dots, n, Tx_j \in \mathbf{R}^n$,

$$Tx_j = \sum_{i=1}^{n} \beta_{ij} x_i, \quad j = 1, 2, \dots, n$$

for some constants $\beta_{ij}, i, j = 1, 2, \dots, n$. Thus, if a vector x has coordinates $\alpha_1, \dots, \alpha_n$ with repect to S, the coordinates of Tx with respect to S, $\gamma_1, \dots, \gamma_n$, may be obtained using matrix multiplication:

$$\begin{pmatrix} \gamma_1 \\ \gamma_2 \\ \vdots \\ \gamma_n \end{pmatrix} = \begin{pmatrix} \beta_{11} & \beta_{12} & \cdots & \beta_{1n} \\ \beta_{21} & \beta_{22} & \cdots & \beta_{2n} \\ \vdots & \vdots & & \vdots \\ \beta_{n1} & \beta_{n2} & \cdots & \beta_{nn} \end{pmatrix} \begin{pmatrix} \alpha_1 \\ \alpha_2 \\ \vdots \\ \alpha_n \end{pmatrix}.$$

It follows that, for a given basis, the linear transformation T may be represented by an $n \times n$ matrix $m(T)$. The same ideas may be applied to linear transformations from $\mathbf{R}^m$ to $\mathbf{R}^n$; in this case $m(T)$ is an $n \times m$ matrix. It is important to note that the matrix associated with a linear transformation depends not only on the transformation, but on the basis under consideration.

Let T_1 and T_2 denote linear transformations from $\mathbf{R}^n \to \mathbf{R}^n$ and let $T = T_2(T_1) = T_2 T_1$ denote the composition of T_1 and T_2. Then $m(T)$ may be obtained from $m(T_1)$ and $m(T_2)$ by matrix multiplication:

$$m(T) = m(T_2) m(T_1).$$

A3.5.6 Inner products. Let V denote a vector space. Let $\langle \cdot, \cdot \rangle$ denote a real-valued function defined on $V \times V$ with the following properties: for all $x, y, z \in V$, $\langle x, y \rangle = \langle y, x \rangle$,

$$\langle x + y, z \rangle = \langle x, z \rangle + \langle y, z \rangle,$$

for all scalars α, $\langle \alpha x, y \rangle = \alpha \langle x, y \rangle$, and $\langle x, x \rangle \geq 0$ with $\langle x, x \rangle = 0$ if and only if $x = 0$. Such a function is known as an *inner product* and the vector space V together with an inner product is known as an *inner product space*.

An inner product can be used to define a norm $|| \cdot ||$ by

$$||x||^2 = \langle x, x \rangle$$

and, hence, an inner product space is a metric space. If this metric space is complete, it is known as a *Hilbert space*.

Let V denote a Hilbert space with inner product $\langle \cdot, \cdot \rangle$. Vectors $x, y \in V$ are said to be *orthogonal* if

$$\langle x, y \rangle = 0;$$

this is often written $x \perp y$. Let M_1 and M_2 denote subspaces of V. A vector $x \in V$ is said to be orthogonal to M_1, written $x \perp M_1$, if x is orthogonal to every vector in M_1. Subspaces M_1 and M_2 are said to be orthogonal, written $M_1 \perp M_2$, if every vector in M_1 is orthogonal to every vector in M_2.

Let M be a given subspace of V. The set of all vectors in V that are orthogonal to M is called the *orthogonal complement* of M and is denoted by $M^\perp$. If M is finite-dimensional, then $V = M \oplus M^\perp$.

The *Cauchy-Schwarz* inequality states that, for $x, y \in V$,

$$|\langle x, y \rangle| \leq ||x|| \, ||y||$$

with equality if and only if either $x = \alpha y$ for some $\alpha \in \mathbf{R}$ or $y = 0$.

Let $x_1, x_2, \ldots, x_n$ denote a basis for V. This basis is said to be *orthonormal* if

$$\langle x_i, x_j \rangle = \begin{cases} 0 & \text{if } i \neq j \\ 1 & \text{if } i = j \end{cases}.$$

Let T denote a linear transformation from $\mathbf{R}^n$ to $\mathbf{R}^n$. The *adjoint* of T is the linear transformation $T' : \mathbf{R}^n \to \mathbf{R}^n$ such that, for all $x, y \in \mathbf{R}^n$,

$$\langle x, Ty \rangle = \langle T'x, y \rangle.$$

The matrix corresponding to T' is given by $m(T') = m(T)^T$, the matrix transpose of $m(T)$. Let T_1 and T_2 denote linear transformations from $\mathbf{R}^n$ to $\mathbf{R}^n$. Then $(T_1 + T_2)' = T_1' + T_2'$, $(T_1 T_2)' = T_2' T_1'$, and $(T_1')' = T_1$. Analogous results hold for $m(T_1)$ and $m(T_2)$.

A linear transformation T is said to be *self-adjoint* if $T' = T$; in this case, the matrix $m(T)$ is a symmetric matrix.

A3.5.7 Projections. Let V denote a finite-dimensional Hilbert space with inner product $\langle \cdot, \cdot \rangle$ and let M denote a subspace of V. Then any vector $x \in V$ may be uniquely written $x = m + e$ where $m \in M$ and $e \perp M$. The vector m is known as the *orthogonal projection of x onto M* or, simply, the *projection of x onto M*. Note that, since the projection of x depends on the definition of orthogonality, it depends on the inner product under consideration.

There exists a linear transformation $P : V \to M$ such that for each $x \in V$, Px is the orthogonal projection of x onto M. Clearly, for $x \in M$, $Px = x$ so that $P^2 = P$; that is, P is idempotent. Also, P is self-adjoint. Conversely, any linear transformation on V that is idempotent and self-adjoint represents orthogonal projection onto some subspace of V.

A3.5.8 Orthogonal transformations. Let V denote a finite-dimensional Hilbert space with inner product $\langle \cdot, \cdot \rangle$. A linear transformation $T : V \rightarrow V$ is said to be *orthogonal* if either $T'T = I$ or $TT' = I$; either of these two conditions implies the other. If T is orthogonal then, for all $x, y \in V$,

$$\langle Tx, Ty \rangle = \langle x, y \rangle$$

and $\|Tx\| = \|x\|$.

Let $x_1, x_2, \ldots, x_n$ and $y_1, y_2, \ldots, y_n$ denote two sets of orthonormal basis vectors. Suppose there exists a linear transformation T such that, for each $j = 1, 2, \ldots, n$, maps $Tx_j = y_j$. Then T is an orthogonal transformation.

Conversely, if $x_1, x_2, \ldots, x_n$ is an orthonormal basis for V and T is an orthogonal transformation, then $Tx_1, \ldots, Tx_n$ is also an orthonormal basis for V.

Clearly, if T is an orthonormal transformation then the matrix corresponding to T satisfies $m(T)m(T)^T = m(T)^T m(T) = m(I)$.

A3.5.9 Eigenvalues and eigenvectors. Let V denote a finite-dimensional vector space and let $T : V \rightarrow V$ denote a self-adjoint linear transformation. A scalar λ is called an *eigenvalue* of T if there exists $v \in V$ such that $Tv = \lambda v$; a nonzero vector $v \in V$ is called an *eigenvector* of T if there exists a scalar λ such that $Tv = \lambda v$. Note that if v is an eigenvector of T then αv is an eigenvector of T for any $\alpha \neq 0$. When V is a normed linear space, it is convenient to standardize the eigenvectors so that they have norm equal to 1.

Let v_1 and v_2 denote eigenvectors of T such that $Tv_j = \lambda_j v_j$, $j = 1, 2$. If $\lambda_1 \neq \lambda_2$, then v_1 and v_2 are orthogonal.

Let λ denote an eigenvalue of T and let $M(\lambda)$ denote the set of vectors $x \in V$ such that $Tx = \lambda x$; then $M(\lambda)$ is a subspace of V. The dimension of $M(\lambda)$ is known as the *multiplicity* of λ.

Suppose that V is a Hilbert space. Let T denote a self-adjoint linear transformation and let $\lambda_1, \ldots, \lambda_m$ denote the eigenvalues of T, including multiplicities; that is, if the multiplicity of a particular eigenvalue is r, then the sequence $\lambda_1, \ldots, \lambda_m$ contains r occurrences of that eigenvalue. Then there exist orthonormal vectors $e_1, e_2, \ldots, e_m$ such that, for any $x \in V$,

$$Tx = \lambda_1 \langle e_1, x \rangle e_1 + \cdots + \lambda_m \langle e_m, x \rangle e_m. \tag{A3.2}$$

If λ_j is unique, that is, if it has multiplicity 1, then e_j is a vector of norm 1 satisfying $Te_j = \lambda_j e_j$. If $\lambda_j = \lambda_{j+1} = \cdots = \lambda_{j+r}$, then $e_j, e_{j+1}, \ldots, e_{j+r}$ are orthonormal vectors spanning the subspace $M(\lambda_j)$.

A3.5.10 Quadratic forms. Let V denote a finite-dimensional vector space and let $T : V \rightarrow V$ denote a self-adjoint linear transformation. A *quadratic form* on V is a function $x \mapsto \langle x, Tx \rangle$. The transformation T is said to be *positive-definite* if the corresponding quadratic form is always positive for nonzero x: $\langle x, Tx \rangle > 0$ for all $x \in V$, $x \neq 0$. Similarly, T is said to be *nonnegative-definite* if $\langle x, Tx \rangle \geq 0$ for all $x \in V$, $x \neq 0$. The terms *negative-definite* and *nonpositive-definite* are defined in an analogous manner.

Using (A3.2), it follows that

$$\langle x, Tx \rangle = \lambda_1 \langle e_1, x \rangle^2 + \cdots + \lambda_m \langle e_m, x \rangle^2.$$

Suppose $\lambda_1 \geq \lambda_2 \geq \cdots \geq \lambda_m$. Then

$$\inf_{x \in V : ||x||=1} \langle x, Tx \rangle = \lambda_m$$

and

$$\sup_{x \in V : ||x||=1} \langle x, Tx \rangle = \lambda_1.$$

It follows that T is positive-definite if and only if all eigenvalues are positive and T is nonnegative-definite if and only if all eigenvalues are nonnegative.

A3.5.11 Determinants. Let V denote a finite-dimensional vector space and let $T : V \to V$ denote a linear transformation. The *determinant* of T, denoted by $\det(T)$, is a real number associated with T; that is, $\det(\cdot)$ is a function that associates a real number with each linear transformation from V to V.

Here we give the basic properties of the function $\det(\cdot)$. Let T, T_1, and T_2 denote linear transformations from V to V. Then $\det(T_1 T_2) = \det(T_1) \det(T_2)$. Let α be a scalar. Then $\det(\alpha T) = \alpha^n \det(T)$, where n is the dimension of V. The identity transformation I has determinant 1. The transformation T and its adjoint, T', have the same determinant: $\det(T') = \det(T)$. The transformation T is invertible if and only if $\det(T) \neq 0$; when $\det(T) = 0$ the transformation is said to be *singular*, otherwise it is *nonsingular*. The eigenvalues of T are those values of $\lambda \in \mathbf{R}$ for which $T - \lambda I$ is singular, that is, for which $\det(T - \lambda I) = 0$.

References

Abramowitz, M., and Stegun, I. (1964). *Handbook of Mathematical Functions*. Wiley, New York.

Anderson, T.W. (1975). *The Statistical Analysis of Time Series*. Wiley, New York.

Anderson, T.W. (1984). *An Introduction to Multivariate Analysis*, 2nd ed. Wiley, New York.

Andrews, G.E., Askey, R., and Roy, R. (1999). *Special Functions*. Cambridge University Press, Cambridge.

Apostol, T. (1974). *Mathematical Analysis*, 2nd ed. Addison-Wesley, Reading.

Arnold, B.C., Balakrishnan, N., and Nagaraja, H. N. (1992). *A First Course in Order Statistics*. Wiley, New York.

Ash, R.B. (1972). *Real Analysis and Probability*. Academic Press, New York.

Ash, R.B., and Gardner, M.F. (1975). *Topics in Stochastic Processes*. Academic Press, New York.

Barndorff-Nielsen, O.E. (1978). *Information and Exponential Families*. Wiley, Chichester.

Barndorff-Nielsen, O.E., and Cox, D.R. (1989). *Asymptotics Techniques for Use in Statistics*. Chapman and Hall, London.

Bhattacharya, R.N., and Rao, R.R. (1976). *Normal Approximation and Asymptotic Expansions*. Wiley, New York.

Bickel, P.J., and Doksum, K.A. (2001). *Mathematical Statistics*, Vol. 1, 2nd ed. Prentice-Hall, Upper Saddle River.

Billingsley, P. (1961). The Lindberg-Lévy theorem for martingales. *Proceedings of the American Mathematical Society* **12**, 788–792.

Billingsley, P. (1968). *Convergence of Probability Measures*. Wiley, New York.

Billingsley, P. (1995). *Probability and Measure*, 3rd ed. Wiley, New York.

Bleistein, N., and Handelsman, R. A. (1986). *Asymptotic Expansions of Integrals*. Dover, New York.

Breitung, K.W. (1994). *Asymptotic Approximations for Probability Integrals*. Springer-Verlag, Berlin.

Brillinger, D.R. (1969). The calculation of cumulants via conditioning. *Annals of the Institute of Statistical Mathematics* **21**, 215–218.

Brown, B.M. (1971). Martingale central limit theorems. *Annals of Mathematical Statistics* **42**, 59–66.

Brown, L.D. (1988). *Fundementals of Statistical Exponential Families. IMS Lecture Notes Monograph Series* **9**. IMS, Hayward.

Capinski, M., and Kopp, E. (2004). *Measure, Integral, and Probability*, 2nd ed. Springer-Verlag, New York.

Casella, G., and Berger, R.L. (2002). *Statistical Inference*, 2nd ed. Wadsworth, Pacific Grove.

Chernoff, H., and Teicher, H. (1958). A central limit theorem for sums of interchangeable random variables. *Annals of Mathematical Statistics* **29**, 118–130.

Cox, D.R., and Miller, H.D. (1965). *The Theory of Stochastic Processes*. Chapman and Hall, London.

Cramér, H. (1946). *Mathematical Methods of Statistics*. Princeton University Press, Princeton.

Cramér, H., and Leadbetter, M.R. (1967). *Stationary and Related Stochastic Processes*. Wiley, New York.

Daniels, H.E. (1954). Saddlepoint approximations in statistics. *Annals of Mathematical Statistics* **25**, 631–650.

David, H.A. (1981). *Order Statistics*, 2nd ed. Wiley, New York.

Davis, P.J., and Rabinowitz, P. (1984). *Methods of Numerical Integration*, 2nd ed. Academic Press, Orlando.

de Bruijn, N.G. (1956). *Asymptotic Methods in Analysis*. North-Holland, Amsterdam.

Doob, J.L. (1953). *Stochastic Processes*. Wiley, New York.

Eaton, M.L. (1988). *Group Invariance Applications in Statistics*. IMS, Hayward.

Erdélyi, A. (1953a). *Higher Transcendental Functions*, Vol. 1. A. Erdélyi, ed. McGraw-Hill, New York.

Erdélyi, A. (1953b). *Higher Transcendental Functions*, Vol. 2. A. Erdélyi, ed. McGraw-Hill, New York.

Erdélyi, A. (1956). *Asymptotic Expansions*. Dover, New York.

Evans, M., and Swartz, T. (2000). *Approximating Integrals via Monte Carlo and Deterministic Methods*. Oxford University Press, Oxford.

Feller, W. (1968). *An Introduction to Probability Theory and Its Applications*, Vol. I. Wiley, New York.

Feller, W. (1971). *An Introduction to Probability Theory and Its Applications*, Vol. II. Wiley, New York.

Ferguson, T. (1996). *A Course in Large Sample Theory*. Chapman and Hall, London.

Field, C., and Ronchetti, E. (1990). *Small Sample Asymptotics. IMS Lecture Notes – Monograph Series* **13**. IMS, Hayward.

Freedman, D. (1971). *Brownian Motion and Diffusion*. Holden-Day, San Francisco.

Freud, G. (1971). *Orthogonal Polynomials*. Pergamon Press, Oxford.

Fuller, W.A. (1976). *Introduction to Statistical Time Series*. Wiley, New York.

Galambos, J. (1978). *The Asymptotic Theory of Extreme Order Statistics*. Wiley, New York.

Ghosh, J.K. (1994). *Higher Order Asymptotics*. IMS, Hayward.

Gourieroux, C., and Monfort, A. (1989). *Statistics and Econometric Models*, Vol. 1. Cambridge University Press, Cambridge.

Hall, P. (1992). *The Bootstrap and Edgeworth Expansions*. Springer, New York.

Hammersley, J.M., and Handscomb, D.C. (1964). *Monte Carlo Methods*. Wiley, New York.

Hoeffding, H., and Robbins, H. (1948). The central limit theorem for dependent random variables. *Duke Mathematical Journal* **15**, 773–778.

Hoffmann-Jorgenson, J. (1994). *Probability with a View Towards Statistics*, Vol. 2. Chapman and Hall, New York.

Ibragimov, I.A. (1963). A central limit theorem for a class of dependent random variables. *Theory of Probability and its Applicaions* **8**, 83–89.

Jackson, D. (1941). *Fourier Series and Orthogonal Polynomials*. Mathematical Association of America, Oberlin.

Jensen, J.L. (1995). *Saddlepoint Approximations*. Oxford University Press, Oxford.

Johnson, R.A., and Wichern, D.W. (2002). *Applied Multivariate Statistical Analysis*, 5th ed. Prentice-Hall, Upper Saddle Point.

Karlin, S. (1975). *A First Course in Stochastic Processes*. Academic Press, New York.

Karlin, S., and Taylor, H.M. (1981). *A Second Course in Stochastic Processes*. Academic Press, New York.

Karr, A.F. (1993). *Probability*. Springer-Verlag, New York.

Kingman, J.F.C. (1993). *Poisson Processes*. Oxford University Press, New York.

Kolassa, J.E. (1997). *Series Approximation Methods*, 2nd ed. Springer, New York.

Lang, S. (1983). *Real Analysis*, 2nd ed. Addison-Wesley, Reading.

Lehmann, E.L. (1999). *Elements of Large-Sample Theory*. Springer, New York.

Lukacs, E. (1960). *Characteristic Functions*. Hafner, New York.

Manski, C.F. (2003). *Partial Identification of Probability Distributions*. Springer-Verlag, New York.

McCullagh, P. (1987). *Tensor Methods in Statistics*. Chapman and Hall, London.

McCullagh, P., and Nelder, J. (1989). *Generalized Linear Models*, 2nd ed. Chapman and Hall, London.

McLeish, D.L. (1974). Dependent central limit theorems and invariance principles. *Annals of Probability* **2**, 620–628.

Norris, J.R. (1997). *Markov Chains*. Cambridge University Press, Cambridge.

Pace, L., and Salvan, A. (1997). *Principles of Statistical Inference*. World Scientific, Singapore.

Parzen, E. (1962). *Stochastic Processes*. Holden-Day, San Francisco.

Port, S.C. (1994). *Theoretical Probability for Applications*. Wiley, New York.

Rao, C.R. (1973). *Linear Statistical Inference and Its Applications*, 2nd ed. Wiley, New York.

Rao, C.R., and Toutenburg, H. (1999). *Linear Models: Least-Squares and Alternatives*, 2nd ed. Springer-Verlag, New York.

Reid, N. (1988). Saddlepoint methods and statistical inference (with discussion). *Statistical Science* **3**, 213–238.

Ripley, B.D. (1987). *Stochastic Simulation*. Wiley, New York.

Robert, C.P., and Casella, G. (1999). *Monte Carlo Statistical Methods*. Springer-Verlag, New York.

Ross, S.M. (1985). *Introduction to Probability Models*, 3rd ed. Academic Press, Orlando.

Ross, S.M. (1995). *Stochastic Processes*, 2nd ed. Wiley, New York.

Rubinstein, R.Y. (1981). *Simulation and the Monte Carlo Method*. Wiley, New York.

Rudin, W. (1976). *Principles of Mathematical Analysis*, 3rd ed. McGraw-Hill, New York.

Schervish, M.J. (1995). *Theory of Statistics*. Springer-Verlag, New York.

Sen, P.K., and Singer, J.M. (1993). *Large Sample Methods in Statistics*. Chapman and Hall, London.

Serfling, R.J. (1980). *Approximation Theorems of Mathematical Statistics*. Wiley, New York.

Skovgaard, I.M. (1981). Transformation of an Edgeworth expansion by a sequence of smooth functions. *Scandinavian Journal of Statistical* **8**, 207–217.

Snell, J.L. (1988). *Introduction to Probability*. Random House, New York.

Stroud, A.H., and Secrest, D.H. (1966). *Gaussian Quadrature Formulas*. Prentice-Hall, Englewood Cliffs.

Stuart, A., and Ord, J.K. (1994). *Kendall's Advanced Theory of Statistics*, Vol. 1, 6th ed. Wiley, New York.

Szegö, G. (1975). *Orthogonal Polynomials*, 4th ed. American Mathematical Society, Providence.

Temme, N.M. (1982). The uniform asymptotic expansion of a class of integrals related to cumulative distribution functions. *SIAM Journal of Mathematical Analysis* **13**, 239–253.

Temme, N.M. (1996). *Special Functions*. Wiley, New York.

Thisted, R.A. (1988). *Elements of Statistical Computing*. Chapman and Hall, London.

van der Vaart, A.W. (1998). *Asymptotic Statistics*. Cambridge University Press, Cambridge.

Wade, W.R. (2004). *Introduction to Analysis*, 3rd ed. Prentice-Hall, Upper Saddle River.

Wallace, D.L. (1958). Asymptotic approximations to distributions. *Annals of Mathematical Statistics* **29**, 635–653.

Whittaker, E.T.T., and Watson, G.N. (1997). *A Course of Modern Analysis*, 4th ed. Cambridge University Press, Cambridge.

Widder, D.V. (1971). *An Introduction to Transform Theory*. Academic Press, New York.

Wong, R. (1989). *Asymptotic Approximations of Integrals*. Academic Press, San Diego.

Woodroofe, M. (1975). *Probability with Applications*. McGraw-Hill, New York.

Yaglom, A.M. (1973). *An Introduction to the Theory of Stationary Random Functions*. Dover, New York.

Name Index

Subject Index